World
Vegetables

Join Us on the Internet

WWW: http://www.thomson.com
EMAIL: findit@kiosk.thomson.com

thomson.com is the on-line portal for the products, services and resources available from International Thomson Publishing (ITP). This Internet kiosk gives users immediate access to more than 34 ITP publishers and over 20,000 products. Through *thomson.com* Internet users can search catalogs, examine subject-specific resource centers and subscribe to electronic discussion lists. You can purchase ITP products from your local bookseller, or directly through *thomson.com*.

Visit Chapman & Hall's Internet Resource Center for information on our new publications, links to useful sites on the World Wide Web and an opportunity to join our e-mail mailing list. Point your browser to: **http://www.chaphall.com/chaphall.html** or **http://www.chaphall.com/chaphall/lifesce.html** for Life Sciences

A service of

Vincent E. Rubatzky
Mas Yamaguchi

Department of Vegetable Crops
University of California, Davis

World
Vegetables

Principles, Production, and Nutritive Values

Second Edition

CHAPMAN & HALL

I (T)P® **International Thomson Publishing**
Thomson Science

New York • Albany • Bonn • Boston • Cincinnati • Detroit • London • Madrid • Melbourne
Mexico City • Pacific Grove • Paris • San Francisco • Singapore • Tokyo • Toronto • Washington

Cover design: Curtis Tow Graphics

Printed in the United States of America

Chapman & Hall
115 Fifth Avenue
New York, NY 10003

Chapman & Hall
2-6 Boundary Row
London SE1 8HN
England

Thomas Nelson Australia
102 Dodds Street
South Melbourne, 3205
Victoria, Australia

Chapman & Hall GmbH
Postfach 100 263
D-69442 Weinheim
Germany

International Thomson Editores
Campos Eliseos 385, Piso 7
Col. Polanco
11560 Mexico D.F
Mexico

International Thomson Publishing–Japan
Hirakawacho-cho Kyowa Building, 3F
1-2-1 Hirakawacho-cho
Chiyoda-ku, 102 Tokyo
Japan

International Thomson Publishing Asia
221 Henderson Road #05-10
Henderson Building
Singapore 0315

2 3 4 5 6 7 8 9 10 XXX 01 00 99 98 97

Library of Congress Cataloging-in-Publication Data

Rubatzky, Vincent E.
    World vegetables : principles, production, and nutritive values /
Vincent E. Rubatzky and Mas Yamaguchi. -- 2nd ed.
        p.   cm.
    Completely rev. and updated ed. of: World vegetables / Mas
Yamaguchi. 1983.
    Includes bibliographical references and index.
    ISBN 0-412-11221-3 (alk. paper)
    1. Vegetables.   I. Yamaguchi, Mas.   II. Yamaguchi, Mas. World
vegetables. III. Title.
SB320.9.R83   1996
635--dc20   96-23732
                                                            CIP

British Library Cataloguing in Publication Data available

To order this or any other Chapman & Hall book, please contact **International Thomson Publishing, 7625 Empire Drive, Florence, KY 41042.** Phone: (606) 525-6600 or 1-800-842-3636. Fax: (606) 525-7778. e-mail: order@chaphall.com.

For a complete listing of Chapman & Hall titles, send your request to **Chapman & Hall, Dept. BC, 115 Fifth Avenue, New York, NY 10003.**

Contents

v

Part A—Vegetables Consisting of Starchy Roots, Tubers, and Fruits

15   Sweet Corn

16   Plantain, Starchy Banana, Breadfruit, and Jackfruit

Part B—Vegetables Consisting of Succulent Roots, Bulbs, Leaves, and Fruits

Preface

The authors decided to revise and expand the first edition of *World Vegetables* because of the interest expressed by students and colleagues. The first edition, published in 1983, was well received nationally and internationally, and a Japanese translation was published. The book continues to be used in horticultural and food science courses domestically and abroad. We sense an increasing global awareness about vegetables and an interest for more information.

Students and colleagues have indicated that an expansion of the information in the first edition and the inclusion of more crops and illustrations would be welcomed. We have done that by expanding previous chapters and adding new chapters for aquatic vegetables, mushrooms, and herbs and spices. We also added many new illustrations.

This completely revised edition expands and updates information in the previous publication. The book presents many aspects about vegetables from a worldwide perspective using some of broad and long experiences of the authors' vegetable backgrounds, research, sabbatical studies, and travel. The authors' experiences in teaching courses at Davis was applied to explaining the different physiological complexities of many vegetables. We proudly share the benefits gained from our association with fellow faculty in the Department of Vegetable Crops, University of California and with other professionals in the United States and many other countries.

The book's first five chapters examines the current role of vegetables as world food crops, plant origins, their classification, their value in human nutrition, toxicants, and folklore. A chapter on toxic substances, folklore and medicinal use of vegetables is included because the subject appears to be of interest to many people. It is included not to emphasize the medicinal or toxic properties of plants used as vegetables, but to provide an appropriate perspective of their therapeutic or toxic involvement. Given as background material are two chapters about the basic principles for crop growth and production under adverse conditions.

The main body of the book describes the major and minor vegetables, their origin, taxonomy, botany, physiology, production, harvest and postharvest handling, composition, and use. Current world production statistics are provided for many crops, and important diseases, insects, and other pests are listed for many family groups.

Chapters 9 through 14 and 16 emphasize those crops which are important vegetable staples for much of the world's population. Remaining chapters consider other important vegetables, many of which are discussed in detail. Selected references are provided to give the reader an entry for additional information.

One appendix table lists over 350 vegetables according to family, genus, and species with the common English names and French, Spanish, Chinese, Japanese and local names where appropriate. Another table lists the approximate nutritive values for more than 260 vegetables, and a third table provides recommended storage conditions and periods for many vegetables.

The edibility of vegetables mentioned in this book, especially those containing toxins, varies greatly. Toxic effects, if any, of crop plants varies with species, cultivars, growth period and environment, cultural conditions, stage of crop development, plant portions, method of food preparation, the amount ingested, as well as age, size, sex, genetic makeup, and health of individuals consuming the vegetable. The information presented is for the enlightenment of the reader. Therefore, we disclaim any responsibility for discomfort, illness, or death caused by ingestion of toxin-containing vegetables mentioned in this publication.

Additionally, chemicals and procedures stated in the book are from observations of our research and that of others and may not apply to all conditions and crops. Therefore, *always* consult product labels for current recommended uses and registrations.

Acknowledgments

The preparation of this book gave the authors much pleasure and satisfaction for which we thank the many who assisted our effort. We gratefully acknowledge the advice and other contributions received from our colleagues in the Vegetable Crops Department of the University of California at Davis and other colleagues throughout the horticultural profession. These individuals generously gave their expertise and constructive criticism. Chapter reviews were graciously provided by people prominent in specific subject whenever requested. Our manuscript reviewers who, deservedly, we again thank and recognized were: Marikis N. Alvarez, Carlos Arbizu, Robert F. Becker, Brian L. Benson, Rupert Best, Vito V. Bianco, Paul W. Bosland, John C. Bouwkamp, James L. Brewbaker, Edward E. Butler, Marita I. Cantwell-De-Trejo, Edward E. Carey, Richard W. Chase, Richard Collins, Wanda W. Collins, Joe N. Corgan, Dermot P. Coyne, Lyle E. Craker, Stafford M.A. Crossman, Lesley Currah, Michael H. Dickson, David Douches, Elmer E. Ewing, DeLance Franklin, Walton C. Galinat, Paul L. Gepts, Irwin L. Goldman, Sang K. Hahn, Anthony E. Hall, Melvin R. Hall, Miguel Holle, George L. Hosfield, Richard A. Jones, John A. "Jack" Juvik, Stanley J. Kays, Bor S. Luh, Vitangelo Magnifico, Hector R. Marti, Franklin W. Martin, Donald N. Maynard, James D. McCreight, Laura C. Merrick, Nicholas D. Molenaar, Teddy E. Morelock, Henry M. Munger, Stephen K. O'Hair, Innocent C. Onwueme, Manuel C. Palada, Leonare M. Pike, John T.A. Proctor, Carlos F. Quiros, Christopher Ramcharan, Kenneth V.A. Richardson, Charles M. Rick, Laura B. Roberts-Nkrumah, Richard W. Robinson, Phillip Rowe, Edward J. Ryder, Cathy Sabota, Robert Scheuerman, John W. Scott, Jonathan R. Schultheis, Brian T. Scully, Joseph Sieczka, Philipp W. Simon, Paul G. Smith, Gene L. Spain, Kenji Takayanagi, Christopher Tankou, Anson E. Thompson, Edward C. Tigchelaar, Herman Timm, William F. Tracy, William Waycott, Paul H. Williams, Hector R. Valenzuela, Jill E. Wilson, Dirk Vuylsteke, Shang Fa Yang, and Frank W. Zink.

Shared sabbatical studies with Drs. Vito V. Bianco, Jose Laborde, and Vitangelo Magnifico were a stimulant for the preparation of this book and they are thanked for their encouragement and advice. We offer our appreciation to Janet Williams, an accomplished illustrator, who produced most of the line drawings and graphs. Our appreciation also extends to the many individuals who provided photographs.

Our wives, Verna Rubatzky and Ida Yamaguchi, cannot be excused from their mighty contributions of devoted patience and continuous encouragement. Nor can we excuse ourselves from full responsibility for errors or erroneous statements or interpretations contained in this volume.

Part I

Introduction

1

World Population, Land Area, and Food Situation

WORLD FOOD SITUATION

No discussion about food supply can be separated from its interaction with human populations. The earth's population was slightly less than six billion in 1996 and is projected to double in less than 50 years, providing the annual increase of about 1.5% annually continues. Clearly, food production must increase accordingly, in order to accommodate such population growth.

Throughout recent history, localized famines have periodically occurred because of droughts, floods, periods of temperature extremes, and diseases and pests that affected crop yields. Wars and other conflicts have also been responsible for periods and regions of food inadequacy. However, the threat of extensive worldwide famine, as theorized by Thomas Malthus has not materialized. Population increases have been accommodated by increased food production which was assisted by technological advances that increased productivity per land area and also by arable land expansion, the latter considered a limited option.

Until recently, the Food and Agriculture Organization (FAO) of the United Nations for statistical purposes had categorized countries into two groups: economically developed and developing countries. That distinction is no longer used after 1992. However, it was useful in distinguishing the disparity in food production and availability in regions of differing economic status. Significant differences in levels of food productivity and availability exist which demonstrates that economically developed countries are clearly more productive than developing countries. Table 1.1 presents comparisons of population of these two groupings. Another general distinction identifies developing countries as agriculturally based economies, and the economies of developed countries as industrially based.

TABLE 1.1. WORLD POPULATION, ANNUAL GROWTH RATE, AND PERCENT OF WORKING POPULATION IN AGRICULTURE WITHIN DEVELOPED AND DEVELOPING COUNTRIES DURING 1980 AND 1992, AND IN REGIONS OF THE WORLD IN 1994

Year	Region	Population (10^6)	Annual growth rate (%)	No. of years for population to double	Percentage of population in agriculture (%)
1980[a]	World	4447	1.8	39	51
	Developed	1170	0.8	88	13
	Developing	3277	2.2	32	66
1992[a]	World	5480	1.7	41	46
	Developed	1270	0.7	100	8
	Developing	4210	2.1	33	59
1994[b]	World	5630	1.5	46	45
	Africa	708	2.8	25	61
	North and Central America	449	1.3	53	10
	South America	314	1.7	41	21
	Asia	3333	1.6	44	57
	Europe	506	0.26	269	8
	Oceania	28	1.5	46	15

Note: Information calculated from data obtained from:
[a]1992 FAO Production Yearbook, Vol. 46. FAO, Rome, 1993.
[b]1994 FAO Production Yearbook, Vol. 48. FAO, Rome, 1995.

A country's total land area alone is not a good indication of food production capabilities, as there are vast regions where extreme temperature, water deficiency, and topography make the land unsuitable for crop production. Generally, arable land per capita is a good indicator of the food-producing capability and economic development of a country.

The proportion of a nation's population active in agricultural production has a direct relationship to the efficiency of food production. In developed countries, the population involved in food production in 1992 averaged less than 8%, whereas in developing countries the percentage was near 60%. Additionally, the agriculturally involved populations may be producing nonfood crops or food crops for export. Nevertheless, overall improved efficiencies in food production has resulted in the involvement of fewer workers. That trend continues in developed and developing countries.

Table 1.2 presents a range of comparisons that can be made regarding populations total and arable land areas between developed and developing countries during 1992.

In 1992, the population of the developed countries was about 23% of the world population, but these countries have about three times the amount of arable land per capita. There is no consistent relationships between the economic development of a nation with its population size or arable land. It is evident that the population involved in agriculture is

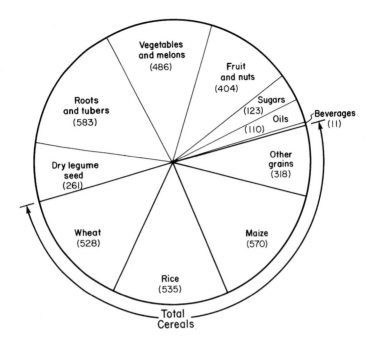

FIG. 1.1. World production of principal food crops; figures are in thousands of metric tons, 1994.

able sources of essential amino acids. Leafy and other vegetables are also good suppliers of vitamins, minerals, and dietary fiber. Vegetables can play an even more important role in the nutritional quality of diets. This can be accomplished through better dissemination about their nutritional value and through changes in eating habits that will benefit people, especially those on marginal diets. Nevertheless, vegetable production in developing countries, unfortunately, often takes a secondary role to high-calorie grain crops.

Table 1.3 shows that food from plant sources provide the major proportion of caloric as well as protein intake. It is also evident that the reliance on plant sources is of major importance for the developing countries. Vegetables of all types make up a significant contribution of foods derived from plants.

Table 1.4 compares world food production and consumption of some important vegetables between developed and developing countries. Not all vegetables are represented by these FAO data; some of those not included are asparagus, table beet, celery, chicory, sweet corn, leek,

TABLE 1.2. WORLD AND NATIONAL POPULATIONS, TOTAL AND ARAB AND LAND AREAS PER CAPITA, 1992

	Population (×10⁶)	Total land area (×10⁶ ha)		l (ha
		Total	Arable	Tota
World	5,480	13,042	1,347	2.3
Developed countries total	1,270	5,455	644	4.3(
Australia	18	764	47	42.4₄
Canada	27	922	46	34.1₈
France	57	55	18	0.9₆
Germany	80	35	12	0.44
Italy	58	29	9	0.50
Japan	124	38	4	0.31
Netherlands	15	3.4	0.9	0.23
Switzerland	7	4	0.4	0.57
United Kingdom	58	24	6.6	0.41
United States	255	917	186	3.60
Developing countries total	4,210	7,587	703	1.80
Brazil	154	846	52	5.49
Chile	14	75	4	5.36
China	1,188	933	93	0.79
Egypt	55	100	2	1.82
India	880	297	166	0.34
Indonesia	191	181	16	0.95
Iran	62	164	14	2.65
Mexico	88	191	23	2.17
Nigeria	116	91	30	0.78
Philippines	65	30	5	0.46

Source: 1992 FAO Production Yearbook, Vol. 46. FAO, Rome, 1993.

very much higher in developing countries than in those of econom developed countries (Table 1.1).

The Netherlands is an example where intensive and high-v agricultural production supports a large population relative to th ble land, and in fact enables that country to be an important exporting country. In contrast, Japan, also with a large populatioı a small amount of arable land per capita, imports food. The Japa economy is able to do so because of exports of manufactured goo(

Figure 1.1 illustrates the proportion of world production of the m food crop groupings. The volume of vegetable commodities produce shown in the figure comprises 27% of the total. This figure does include many important vegetables that are not considered in FAO tistics.

However, Fig. 1.1 is somewhat oversimplified because it does fully account for the essential nutrient content of the foods produc Many vegetable commodities meet human caloric demands because the carbohydrates they contain, and legume crops are especially va

TABLE 1.3. WORLD AND REGIONAL DAILY PER CAPITA CALORIC AND PROTEIN CONSUMPTION AND THE PERCENTAGES OF EACH FROM PLANT SOURCES

Year	Region	Energy		Protein	
		Calories total	% from plants	Grams total	% from plants
1961–63[a]	World	2287	84	62.6	68
	Developed	3031	73	90.4	51
	Developing	1940	93	49.7	82
1979–81[a]	World	2579	84	67.5	66
	Developed	3287	71	98.6	44
	Developing	2473	90	60.6	77
1992[b]	World	2718	84	70.8	65
	Africa	2282	93	56.0	79
	North and Central America	3383	72	96.9	42
	South America	2689	81	67.4	54
	Asia	2585	89	64.3	76
	Europe	3410	68	100.6	42
	Oceania	3129	67	92.8	36

Sources:
[a] 1992 FAO Production Yearbook, Vol. 46, FAO, Rome, 1993.
[b] 1994 FAO Production Yearbook, Vol. 48, FAO, Rome, 1995.

TABLE 1.4. ESTIMATED SELECTED VEGETABLE PRODUCTION AND DAILY PER CAPITA UTILIZATION DEVELOPED AND DEVELOPING COUNTRIES, 1992

Vegetable commodity	Developed countries		Developing countries	
	Production (10^6 t)	Utilization (g/day)	Production (10^6 t)	Utilization (g/day)
White potato	184.8	399.0	83.7	54.5
Sweet potato	2.0	4.3	126.0	82.1
Cassava	0	0	152.2	99.1
Yam	0.2	0.5	27.6	18.0
Taro and tannia	0.4	0.8	5.2	3.5
Cabbage	20.0	43.0	18.2	11.8
Artichoke	1.1	2.3	0.2	0.1
Tomato	33.7	72.7	36.8	23.9
Cauliflower	3.1	6.6	2.4	1.5
Pumpkin and squash	1.6	3.4	5.9	3.8
Cucumber and gherkin	5.6	12.1	9.0	5.8
Eggplant	1.1	2.3	4.7	3.0
Peppers	2.7	5.6	7.0	4.5
Onion, dry	11.0	23.7	17.3	11.2
Garlic	0.7	1.5	2.7	1.7
Beans, green	1.4	3.1	1.6	1.1
Peas, green	3.9	8.3	1.1	0.7
Carrot	9.3	20.0	4.8	3.1
Watermelon	9.2	19.8	19.2	12.5
Other melons	3.6	7.8	8.8	5.7
Plantain	0	0	26.8	17.4

Source: 1992 FAO Production Yearbook, Vol. 46. FAO, Rome 1993.

lettuce, mustards, okra, radish, spinach, and turnip. These and still others not mentioned collectively make significant food contributions.

White potato is by far the most consumed vegetable in the developed countries, followed by tomatoes and cabbage (FAO statistics for cabbage also include Chinese cabbage, kale, brussels sprouts, and sprouting broccoli). In developing countries, starchy root and tuber vegetables are the commodities most consumed. These include cassava, sweet potato, and white potato. The consumption of plantain, onion, watermelon, and tomato is also important.

INCREASING WORLD FOOD SUPPLY AND AVAILABILITY

Presently, the annual increase in world food supply of about 2% per year appear sufficient to support a similar increase in population. Approximately 20% of this increase is achieved from crop land expansion into regions of less favorable temperatures, such as arid areas and previously forested regions of the tropics. This contribution is unlikely to continue. The "slash and burn" practices in tropical rain forests have regional and international ecologists and environmentalists very concerned regarding soil erosion, loss of genetic diversity, disturbances of the CO_2 balance in the earth's atmosphere, and numerous economic and societal concerns. The rain forest's ability to utilize CO_2 in photosynthesis helps to stabilize the CO_2 concentration of the atmosphere. With the decline in forested and other vegetative areas, and in addition to increased industrial emissions, the CO_2 levels have risen. Continued increase in atmospheric CO_2 levels are believed to be a prelude to global warming and its consequences.

Relinquishment of prime arable lands and/or optimum plant growth climates to accommodate population growth and urbanization remains a major concern for every country. As the availability of arable production areas decline, marginal land, by necessity, will increasingly be used for food production. However, this alternative requires higher inputs of all types in order to achieve satisfactory levels of productivity.

The other 80% of the increased food supply has resulted primarily because of technological advancements. The broadened use of improved cultivars, fertilization, as well as better management of soil, water, weed, diseases, pests, and improved cultural practices have increased production efficiencies and yields. High-density plantings, growth regulators, and various kinds of machines are additional factors that have contributed to greater productivity. Further yield and efficiency advances will be achieved by greater exploitation of germplasm resources through conventional breeding and genetic engineering. Future gains in food production will depend on the application of new and improved technologies rather than from increases in cultivated land.

Reducing the cost and effort involved in food production while increasing productivity can benefit all populations. The technology that permits improvement needs to be available to all nations. This is especially important for developing countries where the proportion of disposable income spent for food is much higher than in developed countries. Inadequate food distribution from areas of surplus production to areas of deficit intensifies the problem. Trade barriers and food habits are partly responsible for some of the disparities of available food. If these impediments were removed, further benefits could be achieved in worldwide food availability.

INCREASING NUTRIENT PRODUCTION EFFICIENCY

All aspects of the food chain from production to consumption are important in considering the effectiveness of nutrient production. Production of plants for animal feed utilizes a considerable land area, and the recovery of some of the expended crop nutrients in the form of animal food requires some time. The nutrients provided to the animal are always considerable greater than the nutrients the animal food provides; that ratio varies greatly between animal species. This chain of events may be depicted as plant → animal → human. The feeding of crops to animals to obtain animal food products directly detracts from the supply available for direct human consumption. If the animal were removed from this chain, the efficiency of the land to produce food for humans would be increased. Figure 1.2 shows that less time is required when vegetative portions of crop plants are harvested directly for human consumption. Annual crops reach harvest stage in a relatively short time, and some can be produced more than once in a year, thereby effectively improving land productivity compared to biennial or perennial crops. Many vegetable species meet this criterion. Crop selection to produce the highest amount of essential food nutrients per unit area of land in the shortest period of time would the most efficient way to increase nutrient production. In addition to yield in terms of time and area, the quality of a crop's nutrient composition is important in determining its total food value.

Achieving high nutrient production efficiencies faces many chal-

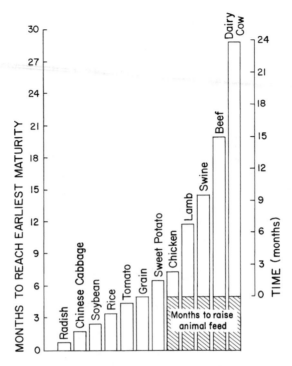

FIG. 1.2.   Time required for production of various vegetable and animal foods.

lenges beyond those that are physical, namely those posed by the food eating habits and preferences of different populations.

MAXIMIZING ARABLE LAND USE

For continued productivity, arable land already in food production must be protected against high salinity, poor drainage, soil erosion, and contamination, which can limit yields. Cultural and other managerial practices that favor the sustainable use of land for food production need to be expanded. Some cropping procedures that can maximize arable land usage for food production are accomplished by the following: (1) Sequential cropping—growing two or more crops in sequence within the growth period. (2) Intercropping—growing two or more crops simultaneously on the same land and depending on crop interaction, either as mixed, row, or strip plantings. (3) Relay intercropping—simultaneous growth of two or more crops during part of the growth period of each; usually one being transplanted or seeded after suitable development of the first crop but before the first is harvested. (4) Ratoon cropping— whereby regrowth of the same crop after harvest occurs from suckers or adventitious shoots.

PREVENTION OF PRODUCTION AND POSTHARVEST LOSSES

In many countries much of the crop production is lost because of unfavorable weather caused by temperature extremes, droughts, winds, or floods. These situations are difficult to avoid, although an expansion of irrigation use can circumvent some drought conditions. Effective water utilization is often a key factor in avoiding production losses. Wind breaks, flood control and temperature modification also can lessen losses.

Other production losses are due to weed competition, disease, and insect infestations. Improvement in plant resistance and effective pest management with pesticides and biological controls can limited some of these losses. Birds, rodents, and other animals through consumption or contamination account for substantial losses. Inappropriate harvest and postharvest handling practices further contribute to physical and quality losses of products. Postharvest crop perishability most commonly results from moisture loss, decay, physical damage, respiratory loss, as well as other factors such as the development of off-flavors or discoloration. Deterioration losses also occur because of inadequate or poor storage and/or inability to accommodate surplus production or to process some of the production.

NEW FOOD SOURCES

The development of new foods from other than most of the conventional sources can increase the overall food supply. The oceans and seas offer huge resources of potential food. In many Asian countries algae and other lower forms of plant life are harvested to supplement the food supply. Additionally some of the presently discarded culls and/ or residues from existing crops can be salvaged and/or reformulated to make edible products. Development of new vegetable species from germplasm resources not currently utilized can make significant contributions. Biotechnological applications offer a huge potential for the development of new cultivars as well as the improvement of many presently cultivated crops.

DEVELOPMENT OF FAVORABLE AND EQUITABLE TRADE PROGRAMS

Unlike grain crops, most vegetables are bulky, succulent, and more perishable than cereals. Accordingly, they must be consumed or processed within a relatively short period before substantial quality loss occurs. Similarly, their bulk and perishability makes long-distance transport complicated and costly. Therefore, it is not surprising that in most countries, vegetables are usually produced near consumption centers and often are not widely distributed because of inadequate and/or expensive transportation.

However, when quality and prices can favorably compete with those produced in other areas, some vegetable commodities are transported greater distances either domestically or internationally. Export opportunities exist when surplus and/or off-season production can find additional markets, and countries capable of surplus food production actively seek these markets. For other situations, certain crops are grown expressly for export. Such opportunities are important for a nation's economic development.

Rapid-transport equipment, improved packaging, refrigerated containers, and controlled-atmosphere handling and storage are some of the reasons for improved export possibilities. Some high-value commodities can justify the use of air transportation. Unfortunately, these resources and appropriate technologies may not be available to some nations.

National trade polices often are an additional limitation. Some governments, in the name of national food security, have established policies and support programs intended to enhance domestic food production. Also, in order to protect national agricultural interests, domestic labor, or for other reasons, many countries also establish policies and practices that present barriers to imports of food products.

National agricultural polices in some countries must undergo major changes in order to increase productivity. Increased support from governmental, private, and/or public/private sources will be required for research and to adapt technology to meet agricultural production needs and to compete in international markets.

To better accommodate access to and distribution of world food supplies, many present international trade policies need to be changed to encourage such trade. For example, the European Economic Community (EEC) has overcome many trade barriers among countries of that community, with generally shared benefits to the participants. The North American Free Trade Agreement (NAFTA) among Canada, Mexico, and the United States is another example of the intent to broaden international trade. Although the General Agreement on Tariffs and Trade (GATT) has global implications, many new regional and individual trade agreements are certain to be established among various countries with the aim to improve food availability and to enhance economic development. The trend toward lowering trade barriers will likely continue and allow for improved distribution of food supplies. Lowered trade barriers can increase demand for food products from developing countries and through greater export opportunities, advance the economies and living standards of the exporting nations. When trade barriers are lowered, the incentive for increased food production and export is improved with the net result of a greater supply. Many consumers, especially those in economically developed countries, have exhibited an increased demand for a greater variety of vegetables. For these clientele, value-added and semiprepared vegetable commodities will find a ready market.

INTERNATIONAL AND REGIONAL RESEARCH CENTERS

To take advantage of production technologies achieved in the developed nations for the betterment of less developed nations, several international agricultural research centers were established. The Rockefeller Foundation and the Ford Foundation in cooperation with the governments of Mexico and the Philippines created centers which were precursors to a broadened global program of additional agricultural research centers. The Consultative Group on International Agricultural Research (CGIAR) comprises a network of about 13 organizations dedicated to agricultural research. These centers are located throughout the world. Their objectives are the improvement of food production and thereby to enhance the economic growth of less developed regions and countries. A first and major research emphasis was the improvement of staple food production. The "green revolution" is a term that is testimony to the research accomplishments that resulted in enormous yield increases in crop staples such as rice, wheat, and maize. India is one example of a nation moving from a situation of variable and inadequate food supplies to a position of self-sufficiency and, in some situations, a food-exporting country.

Most of the international centers are orientated to plant crop research. Whereas the International Maize and Wheat Improvement Center (CIMMYT) and International Rice Research Institute (IRRI) deal with cereals, others that have vegetable crop involvement are as follows:

CIAT   International Center for Tropical Agriculture—Cali, Colombia

ICRISAT   International Crop Research Institute for the Semi-Arid Tropics—Patancheru, India

ICARDA   International Center for Agricultural Research in the Dry Areas—Aleppo, Syria

IITA   International Institute of Tropical Agriculture—Ibadan, Nigeria

CIP   International Potato Center—Lima, Peru

AVRDC   Asian Vegetable Research and Development Center—Shanhua, Tainan, Taiwan

Additional international centers are contemplated. However, these are not intended as substitutes for the ongoing research of other agricultural research agencies within individual countries.

The participation and contributions of the Food and Agricultural Organization (FAO), its various agencies, and the World Bank are also recognized for their international efforts that have improved food production and availability throughout the world. The United States Agency for International Development (USAID) and Japan International Cooperation Agency (JICA) and those of other nations provide valuable assistance for agricultural research to other countries.

SUMMARY

An obvious present solution to resolving world food inadequacy is stabilizing population growth and increasing consumer income. This is not an easy task. Amelioration of world food concerns requires a diverse portfolio of factors that need to be addressed by concerted efforts to make improvements. Because these factors are greatly interrelated, their solution is difficult.

Some demographers maintain that the food production carrying capacity of the earth cannot accommodate the predicted expansion of world population. This view is not shared by other demographers who believe that the earth's resources are capable of supporting human populations well in excess of predicted growth. Their view is that available resources, properly utilized and managed, are capable of greater productivity than that currently obtained. How much more is uncertain. Time will reveal which prophecy was correct. However, in addition to excellent husbandry of the earth's resources, economic and societal changes are also required if expanded population growth is to be accommodated with adequate and equitable food supplies.

Expanding populations and urbanization will continue to increase the competition for land, water, and energy resources. Therefore, it is paramount that production capabilities and efficiencies be increased to meet food needs. Operating within shrinking land, water, and energy resources will be challenging to all food producers.

Some resolution of world food concerns would involve the following:

Better stewardship through sustainable agricultural practices to assure that land and other necessary resources are used effectively. Productive land must be protected from erosion, from pollution, and, when appropriate, from nonagricultural purposes.

Better utilization of produced food. Postharvest losses can be limited by the application of present knowledge and technology, and alternative uses found for surplus crops. Correction of marketing practices and facilities that are presently inadequate for efficient food distribution.

Change in food-consumption patterns, traditions, and customs which interfere with efficiencies in food distribution and use. Broader education for all people can enable them to make more informed decisions concerning food production and availability.

Implementation of governmental policies that favor economic growth. National polices should also consider their global impact. Policies favorable to expanded global trade and distribution of foods can benefit many populations.

Increasing the purchasing power, especially of the populations in developing countries would improve the equitable distribution of foods.

The worldwide transfer of crop production and food use information through rapid and inexpensive electronic communication.

Research for a better understanding of biology and related interactions that will contribute to greater food productivity. Better access to the diversity of global genetic resources is an important goal. Application of new technologies to provide improvements in crop productivity.

These are some of the considerations which must be addressed if future world food needs are to be satisfied. International involvement is necessary if progress is to be achieved.

SELECTED REFERENCES

Eicher, C.K. and Staatz, J.M., eds. 1990. Agricultural Development in the Third World, 2nd ed. Johns Hopkins University Press, Baltimore, MD.

Garbus, L., Pritchard, A., and Knudsen, O., eds. 1991. Agricultural Issues in the 1990s. Proc. 11th Agric. Sector Symposium. The World Bank, Washington, DC.

Kula, E. 1994, Economics of Natural Resources, the Environment and Policies, 2nd ed. Chapman & Hall, London.

MacFie, H.J.H., and Thomson, D.M.H., eds. 1994. Measurement of Food Preferences. Blackie Academic & Professional/Chapman & Hall, Glasgow. Solomon, A.M., and Shugart, H.H., eds. 1993. Vegetation Dynamics and Global Change. Chapman & Hall, New York.

United Nations World Economic Survey. 1993. Current Trends and Policies in the World Economy, 1993. United Nations, New York.

Unklesbay, N. 1992. World Food and You. Food Products Press and Haworth Press, New York.

Van Driesche, R.G., and Bellows, T.S., Jr. 1996. Biological Control. Chapman & Hall, New York.

2

Origin, Evolution, Domestication, and Improvement of Vegetables

As recently as the mid-1800s relatively little was known about the origin of our important economic crops. For many scientists at that time, their origin was thought to be an "impenetrable secret." Alphonse de Candolle's classic work, *Origin of Cultivated Plants,* published in 1886 and more recent studies of the Russian botanist/geneticist Nikolai Vavilov and the many efforts of other scientists contributed to identifying likely centers of origin, diversity, and/or domestication of most of the world's important crops. Also deserving recognition are the early plant explorers and observers such as von Humboldt and Captain James Cook, and more recent scientists Carl Sauer, Jack Harlan, and Jack Hawkes for their theories and studies concerning crop evolution and domestication of many useful plants. Although full agreement about the theories proposed is lacking, the contributions of these people were essential for our knowledge about many plant species. Continuing studies using new technologies should reveal further information and possibly close some existing gaps about the origin and development of certain crop plants.

HUMANS AND AGRICULTURE

To understand the origin and evolution of crop plants, it is useful to also understand the processes that are believed to have occurred during the development of agriculture. Humans have existed on earth for over two million years, and as early inhabitants they were food gathers and hunters. Early humans followed a nomadic or seminomadic existence in their search for food. It is reasonable to assume that where and when the food supply was adequate, a somewhat sedentary life form could be followed, with various plants supplementing food obtained from hunting and fishing. This premise is supported by discoveries found in many archaeological sites.

PLANT CULTIVATION AND DOMESTICATION

It is generally believed that *in situ* cultivation of plants began about 8–10 thousand years ago—only a fraction of the human existence period. Why did it take so long? Probably because there were no examples or few experiences to follow and upon which to build. Because food searching was a major survival activity, to which much time and effort was devoted, there probably was little time to think about crop cultivation, even if aware of cultivation practices. Furthermore, experience and knowledge about plant cultivation could not be widely disseminated because travel between settlements and communication was minimal. However, within communities, plant cultural knowledge could be passed to subsequent generations.

Conditions Considered Necessary for Early Plant Domestication

Areas with an abundant supply of game and a marked diversity of edible food plants were where early agriculture started. These conditions allowed for some degree of permanence of settlements. With less effort required to roam and search for food, more effort and thought could be directed to the domestication and management of plants and animals.

Geographer botanist Carl Sauer theorized that certain conditions had to be present in order for plant domestication to have occurred. He proposed that the first cultivated lands would have been those that required little preparation. Not all land required clearing, and relatively flat well-drained land was more appropriate for early cultivation than situations that might require drainage or other modifications.

Another required condition would be the control of fire, necessary for cooking, warmth, and protection from predators, but also useful for land clearing. Even today in some regions, "slash and burn" practices continue to be widely practiced to prepare land for cultivation. Continuing conversion of rain forests and other land for arable cropping or animal grazing that use slash and burn procedures are causing serious regional and international environmental and sociological concerns. Early experiences would have shown that river valleys that faced periodic flooding or arid areas were less desirable. Archaeological evidence shows that where feasible, supplemental irrigation was employed relatively early in agricultural history.

Centers of Origin of Vegetables

Others in addition to Sauer suggested that the cradle of civilization was in the broad region stretching from the eastern Mediterranean and eastward to Indonesia. These regions have rainfall and temperatures favorable for plant growth and are likely responsible for the large diversity of food plants found in that region. Other regions of the earth with favorable conditions for plant growth also showed diversity. That diversity enabled the domestication of certain native plants for human use and agricultural development. Vavilov identified centers of origin for many plant species, in addition to vegetables based largely on the diversity he found in these locations (Fig. 2.1). Major centers of origin for vegetables Vavilov identified were as follows:

A   Chinese center—central and western China
      Soybean, *Glycine max*
      Chinese yam, *Dioscorea opposita*
      Radish, *Raphanus sativus*
      Chinese cabbage, *Brassica rapa*
      Rakkyo, *Allium chinense*
      Japanese bunching onion, *Allium fistulosum*
      Cucumber, *Cucumis sativus*
B   Indian Malaysian centers

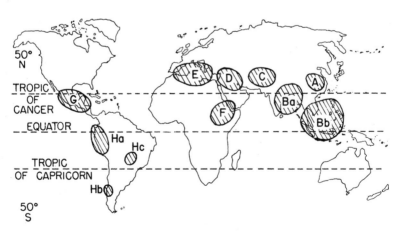

FIG. 2.1.   Main centers of origin of cultivated plants according to Vavilov.

Ba   Northeast India and Myanmar (Burma) center
    Mung bean, *Vigna radiata*
    Cowpea, *Vigna unguiculata*
    Eggplant, *Solanum melongena*
    Taro, *Colocasia esculenta*
    Cucumber, *Cucumis sativus*
    Yam, *Dioscorea alata*

Bb   Indochina and Malay Archipelago center
    Banana, *Musa acuminata, M. balbisiana*
    Breadfruit, *Artocarpus altilis*

C   Central Asia center—India, Afghanistan
    Pea, *Pisum sativum*
    Horse bean, *Vicia faba*
    Mung bean, *Vigna radiata*
    Mustard, *Brassica juncea*
    Onion, *Allium cepa*
    Garlic, *Allium sativum*
    Spinach, *Spinacia oleracea*
    Carrot, *Daucus carota*

D   Near-Eastern center—Asia Minor (Iran, Turkey)
    Lentil, *Lens culinaris*
    Lupine, *Lupinus albus*

E   Mediterranean center
    Pea, *Pisum sativum*
    Garden beet, *Beta vulgaris*
    Cabbage, *Brassica oleracea*
    Turnip, *Brassica rapa*
    Lettuce, *Lactuca sativa*
    Celery, *Apium graveolens*
    Chicory, *Cichorium intybus*
    Asparagus, *Asparagus officinalis*
    Parsnip, *Pastinaca sativa*
    Rhubarb, *Rheum officinale*

F   Ethiopian center
    Cowpea, *Vigna unguiculata*
    Garden cress, *Lepidium sativum*
    Okra, *Hibiscus esculentus*

G   South Mexico and Central America center
    Maize, *Zea mays*
    Common bean, *Phaseolus vulgaris*
    Lima bean, *Phaseolus lunatus*
    Malabar gourd, *Cucurbita ficifolia*

      Winter pumpkin, *Cucurbita moschata*
      Chayote, *Sechium edule*
      Sweet potato, *Ipomoea batatas*
      Arrowroot, *Maranta arundinacea*
      Pepper, *Capsicum annuum*
H   South America centers
Ha Ecuador, Peru, and Bolivia centers
      White potato, *Solanum tuberosum*
      Andean potato, *Solanum tuberosum* subsp. *andigena*
      Starchy maize, *Zea mays*
      Lima bean, *Phaseolus lunatus* (secondary center)
      Common bean, *Phaseolus vulgaris* (secondary center)
      Edible canna, *Canna edulis*
      Pepino, *Solanum muricatum*
      Tomato, *Lycopersicon lycopersicum*
      Ground cherry, *Physalis pubescens*
      Pumpkin, *Cucurbita maxima*
      Pepper, *Capsicum annuum*
Hb Chiloe center
      White potato, *Solanum tuberosum* subsp. *tuberosum*
Hc Brazilian–Paraguayan center
      Cassava, *Manihot esculenta*

Other scientists generally agreed with most of Valilov's classification, but several have suggested modifications that are not totally based on diversity because some species were discovered to have been domesticated in locations other than those of their origin. Thus, some locations once thought to be the origin of a species are instead centers of diversity. Following the reports of Vavilov, many additional crop plants have been identified with regard to their centers of origin and/or diversity. Further study should expand our knowledge about the regions of origin for many more plant species.

Early Cultivated Plants

Obviously, the first cultivated plants were obtained from the wild, some of which would have been utilized *in situ* and others managed near settlements. For cultivation to be practiced, there had to be a benefit. Early humans must have found that cultivation gave higher and more predictable returns for their efforts than that applied to food searching and gathering in the wild. Plants taken from their original environment may be more successful in a new habitat because of lessened competition and/or injurious pests and diseases.

Vegetative Culture

It is logical that the first domesticated plants were those easiest to propagate and would usually have included those that were vegetatively self-propagated rather than those seed propagated. Wild species adaptation, especially in humid and tropical regions, would benefit more from vegetative rather than seed propagation as a reproductive method.

For example, during the harvest of wild starchy roots or tubers, all plant portions usually are not removed and some remaining serve as propagules for subsequent growth. In warm and high-rainfall regions, such propagules, often without dormancy features, could grow and produce new roots or tubers, as well as, probably, spreading and expanding the growth area. This method of propagation is prevalent in the humid tropics and semitropical lowlands of the Americas, southeast Asia, and Africa.

A probable scheme advocated by some anthropologists and plant geographers is that some roots or tubers brought to settlements might inadvertently be dropped or even intentionally discarded in nearby clearings or refuse dumps. There, they could become established and grow rapidly because of less competition, and perhaps also benefit from the relatively high fertility contributed from discarded food or vegetative matter, as well as animal or human wastes developed in the settlement. Plants recognized as useful would be allowed to grow and would be protected from foraging animals. Much less effort would be required to obtain food from this source than to go into the wild to search and gather.

Seed Culture

In other regions where the environment was favorable for plants to flower and permit seed maturation, propagation could also rely on seed use. With favorable conditions, large quantities of seed can be produced. Archeobotanical evidence indicates that the origin of cereals in the Old and New World occurred in mountainous regions of the subtropics having marked wet and dry seasons. Like vegetative propagules, seed could fall into the soil near human settlements. With the onset of rain, the seed can germinate and grow. Similarly, it was recognized that other seed-bearing plants would reproduce from naturally dispersed seed; therefore, these seed were also collected and used in more systematic cultivation practices. After thousands of years of such random harvests, the practice of deliberately placing seed into soil evolved.

In order to be successful, seed had to be harvested and stored until used. Thus, for the early plant cultivators, seed propagation would have been a more demanding and restrictive practice than vegetative propagation. However, a notable advantage of seed propagation is that seed reserves can be stored and thus provide some food security in the event of crop failures.

CHANGES IN WILD PLANTS DURING DOMESTICATION

Gigantism

Some level of gigantism is usually exhibited by domesticated plants compared to their wild relatives which tend to be smaller. Human selections during domestication would be directed to maximizing edible portions. Therefore, cultivated plants typically are larger, with broader and thicker leaves, less stem branching, and larger flowers and fruit. A frequent cause of gigantism is polyploidy. For some domesticated species, such as the white and sweet potatoes, this is a common and preferred state.

Seed

Through human involvement, the seed of many species has increased in size. Large seeds have greater food reserves and favor rapid and vigorous seedling growth, and if the seeds are eaten, a large size is preferred. For the wild species, nonuniformity of germination enhances survival ability. Usually accompanying seed size increase is a decrease in the numbers of seed produced. In contrast, wild species tend to have many seeds.

Another significant result of domestication was a reduction in seed dispersal. Although highly desirable for the wild state, seed scattering leads to seed harvest difficulties and a loss of seed yield. Another change of wild species characteristics was the reduction or loss of dormancy and hard seed testa. Again, these changes conflict with survival abilities in the wild. In order to facilitate and maximize seed harvest, selection for uniformity of seed maturity was an important factor still another feature detrimental for the wild species.

Response to Temperature and Photoperiod

Because of their adaptation to the environment in which they grow, wild species are able to continue their existence. Plants that cannot adapt to the existing climatic and day length conditions do not survive.

Domestication of crop species requires either continued environmental adaptation or modifications either to the plant or environment that will allow survival and continuance of the species.

In addition to the obvious direct influence on growth rate, temperature also influences the plant's inherited growth habit. Thus, with biennial plants, flowering which might interfere with vegetative growth is usually deferred until temperatures appropriate for floral induction occur. Annuals are usually favored by long, warm, and dry growth conditions that can provide a growth period long enough to produce mature seed. Mature seed can be stored and remain viable through adverse growing conditions and continue the survival of the species when favorable conditions recur. Perennials offer the advantage of continued growth by one or more means of self-propagation.

Photoperiodism is a survival mechanism in which flowering and seed production usually occurs during favorable periods that allow for the full development and, thus, its survival. Asexual reproduction, bud break, and resprouting in some species are also influenced by day length.

Morphological Changes

In the process of domestication, various plant portions are enhanced in shape and/or size. During domestication, there is a tendency to select for features that improved the plant's productivity and usefulness. Even within species, considerable morphological differences can appear. An excellent example are the different cole crops groups of *Brassica oleracea*. Through natural interchange and from human involvement, kale, kohlrabi, cabbage, cauliflower, broccoli, and brussels sprouts have evolved, each morphologically very different, yet still of one species. Another example of one species with different forms is *Beta vulgaris* which includes table beet, Swiss chard, and sugar beet, each different in appearance and uses.

Other changes in shape have been made to accommodate specific usage and preferences. Thus, there are globe and flat onions, round and oblong melons, thin, round, and flat snap bean pods, and many other examples.

Reduction or Loss of Survival Ability

Reduction in the numbers of seed produced, nonshattering, loss of seed dormancy, and seed coat impermeability all contribute to reduction in the survival of species. Selection against characteristics such as thorns or barbs and the reduction or absence of bitter and/or toxic substances, which are deterrents to foraging animals as well as man, are factors that reduce plant survival when grown in the wild. The domestication of maize, *Zea mays,* can illustrate how the present-day forms of this species could not survive in the wild. The maize husk completely encloses the seed and prevents shattering as well as the opportunity for the seed to grow because it may not contact the soil. Thus, without human intervention to harvest, store, and plant seed, maize self-propagation is prevented.

Mechanisms of Change from Wild to Cultivated Forms

In addition to changes brought about by selection, mutation is one mechanism by which plants can change. In nature, gene mutations occur infrequently in some species and rarely in others. Mutations most often are disadvantageous, and if the altered plant fails to adapt and survive, the mutation is eliminated. Some favorable and unfavorable mutations may be lost, but occasionally others are retained and can be passed on to succeeding generations.

Through mutations and natural crossings, domesticated plants can gradually accumulate more of the valuable characters and thus increase their value to humans. These procedures are slow and require considerable time, and even then may only provide a minor part of the total variability of a species.

The systematic improvement of plants by controlled plant breeding involving the transfer of genetic characteristics from one plant and their incorporation into another is, in historical time, a relatively new procedure. Plant breeders have made tremendous progress in improving the usefulness of plants for humans. Developing technologies are allowing even greater and more rapid improvements in altering and developing new forms of plants.

Cultivated Plant Characteristics

The following characteristics are those generally associated with cultivated plants:

A reduced ability to compete with other plants, and usually with a preference for growth in open habitats.

A wide range of morphological variability

A wide range of physiological adaptation to specific sites

A suppression of the natural mechanism for seed distribution

A suppression of protective features, such as thorns and bitter and/ or toxic compounds

A tendency for reduced sexual fertility of vegetatively propagated crops

Changes in growth habit such as indeterminate to determinate flowering, fewer stems and less branching, and changes from biennial and perennial plants to annual-like characteristics

A tendency for seed to be more uniform in germinating

A tendency for inbreeding to replace outbreeding

PLANT IMPROVEMENT

To continue the evolution, domestication, and improvement of useful plants it is necessary that regional as well as global genetic resources be appropriately managed. The potential usefulness of existing biological diversity and that which may be created hopefully will be properly directed for the benefit of all peoples.

Wild species continue to be found, although expectedly at a lower rate, as natural native habitats are altered by human encroachment. Endangered and other species should be conserved for their potential contributions, even if these have yet to be determined. The usefulness of these genetic resources continues to be discovered and identified. Sometimes, their value is readily apparent, and in other circumstances, the value is determined by structured plant breeding or other genetic transfer technologies.

Wild species and other germplasm sources can be conserved by *in situ* or *ex situ* maintenance. Seed of some species can be conserved for long periods when using appropriate storage conditions. Other maintenance practices are used for species with recalcitrant seed or species that are reproduced by vegetative propagation.

Certain facilities, often called "gene banks," safely stored large collections of various species containing wild, domesticated, and other germplasm. Some tend to specialize in relatively few species, others such as the United States Department of Agriculture in Fort Collins, Colorado and the N.I. Vavilov All-Russian Institute of Plant Industry in Saint Petersburg are very extensive in their collections. Excellent collections and repositories exist in many other countries, and some level of plant germplasm collecting occurs in nearly every nation.

Some of these facilities conduct periodic surveys and collection trips with trained plant explorers in order to obtain additional samples of plant diversity. Samples are also obtained by submissions from various sources. The many samples collected and stored represent presently used materials and heirloom and landrace cultivars, in addition to wild species and related germplasm. These materials are cataloged, and relevant information is entered into a database for use by investigators, plant breeders, and others. Exchange programs are often arranged, and duplicate collections are encouraged in order to provide greater security of collected materials. When seed viability declines, arrangements are made to renew the stored material, usually by the production of new seed or plant material. Measures are taken to assure the purity and trueness of the samples.

In the past, exploration and collection trips often provided little compensation to the country of origin for the subsequent use of the genetic material acquired. That practice has changed; investigators now share plant material and relevant information with the source country and, in some situations, provide more direct compensation via licenses or other arrangements.

The question of priority is under continued debate, whether it deals with wild species or products of plant breeding. Plant breeders seek to protect and prolong their proprietary interests via plant patents or other protection rights.

Assisting the discovery, collection, conservation, and application of world genetic resources are various international and domestic institutions. The International Plant Genetic Resources Institute (IPGRI, previously IBPGR) the Food and Agricultural Organization of the United Nations (FAO), the various centers of the Consultative Group on International Agricultural Research (CGIAR), the World Bank, and many other organizations are important supporters of these activities. In many countries, ministries of agriculture, various public and private research organizations, and the seed trade are also active in the collection, maintenance, and utilization of germplasm.

The continuing emergence of biotechnical information, its communication, and improved computer-assisted cataloging procedures will greatly enhance the ability to create useful changes in plant materials used to produce food crops. Improved quality, pest resistance, and greater productivity are a few examples of desired objectives.

SELECTED REFERENCES

Bliss, F.A., Strosnider, R.E., Pike, L.M., Innes, N.L., and Ryder, E.J. 1984. The role of public and private plant breeders in horticultural crop improvement. HortScience 19, 797–811.

Blumler, M.A., and Byrne, A.R. 1991. The ecological genetics of domestication and the origins of agriculture. Curr. Anthropol. 32, 23–54.

Cohen, M.N. 1977. The Food Crisis in Prehistory; Overpopulation and the Origins of Agriculture. Yale University Press, New Haven, CT.

Holden, J.H.W., and Williams, J.T., eds. 1984. Crop Genetic Resource: Conservation and Evaluation. George Allen and Unwin, London.

Frankel, O.H., ed. 1973. Survey of Crop Genetic Resources in their Centers of Diversity: First Report. FAO/IBP, Rome.

Frankel, O.H., and Hawkes, J.G., eds. 1975. Crop Genetic Resources for Today and Tomorrow. Cambridge University Press, Cambridge, Vol. 2.

Frankel, O.H., and Soule, M.E. 1981. Conservation and Evolution. Cambridge University Press, Cambridge.

Harlan, J.R. 1992. Crops and Man, 2nd ed. American Society of Agronomy, and Crop Science Society of America, Madison, WI.

Hawkes, J.G. 1983. The Diversity of Crop Plants. Harvard University Press, Cambridge, MA.

Heiser, C.B. 1973. Seed to Civilization: The Story of Man's Food. W.H. Freeman, San Francisco.

Rindos, D. 1984. The Origins of Agriculture: An Evolutionary Perspective. Academic Press, Orlando, FL.

Sauer, C.O. 1969. Agricultural Origins and Dispersals. The MIT Press, Cambridge, MA.

Sauer, J.D. 1993. Historical Geography of Crop Plants—A Select Roster. CRC Press, Boca Raton, FL.

Simmonds, N.W. 1976. Evolution of Crop Plants. Longman, London.

Smith, C.E., Jr. 1968. The new world centers of origin of cultivated plants and archeological evidence. Econ. Bot. 22, 253–266.

Smith, C.E., Jr. 1969. From Vavilov to the present—A review. Econ. Bot. 23, 2–19.

Vavilov, N.I. 1951. The origin, variation, immunity and breeding of cultivated plants. Chron. Bot. 13, 1–366 (translated from Russian by K. Starr Chester).

Wann, E.V., Sink, K.C., Park, W.D., Bliss, F.A., and Lower, R.L. 1984. Genetic engineering. HortScience 19, 32–51.

Zohary, D., and Hopf, M. 1988. Domestication of Plants in the Old World. Clarendon Press, Oxford.

3

Vegetable Classification

IMPORTANCE AND BASIS FOR CLASSIFICATION

A systematic method of grouping different plants is essential in identifying and cataloging the voluminous information gathered about the many different plants known. When properly classified, orderly and efficient use of this information can result.

Relatively few plants are used as vegetables; among the hundreds of thousands of plants known, only several hundred are used as vegetables. However, in order to manage information about these plants, some system of classification, preferably having universal applicability, is necessary. Many methods of classification can be developed, but the value of any system depends on its usefulness. The system needs to be easy to use, accessible to all, and stable.

CLASSIFICATION ACCORDING TO CLIMATE (TEMPERATURE)

Climatic classification may have been one of the earliest attempts to logically group plants. Through experience, plant response to growing temperatures were recognized, and different plants could then be grouped for optimum growing temperatures and placed into either a cool or warm season category. Cool season vegetables show a preference during most of their growth for mean growing temperatures between 10°C and 18°C. Some have frost and even freezing tolerance, and for most, the edible products are leaf, stem, or root tissues. Familiar examples are cabbage, lettuce, spinach, potato, and carrot. Warm season crops are those that exhibit a preference for mean temperatures within a range of 18–30°C during most of their growth and development. Warm season vegetables are intolerant of frost, and botanically, the edible portions usually are the fruit or fruit products. Examples are tomato, melons, and beans.

Classification with regard to growing temperature or seasonal periods has some useful generalities, but there are overlaps and some exceptions. Temperature classification has more usefulness in temperate-zone conditions, because in the tropics, the distinction between warm and cool season vegetables is less clear.

OTHER CLASSIFICATIONS

Vegetable crops can also be loosely grouped according to their post-harvest and storage temperature characteristics (Appendix, Table D). Other observations would suggest separation into classes that emphasis different responses to soil acidity (pH), salinity, nutritional requirements, and drainage. Plant response to any of these conditions exhibit a range of variability. Plant rooting depth, seed germination, and day length responsiveness are other characteristics that are used to separate plants, but such classifications have limited usefulness.

Additional classifications that identify differences in edible plant parts (botanical organs) are also limiting. Likewise, crop use categorizes are also ambiguous because of the duplicity in how vegetables are prepared and used. Classification according to plant life span, such as annual, biennial, or perennial habit, or by preferred habitat, whether aquatic, xerophytic, or mesophytic, is insufficient to precisely identify different plants.

BOTANICAL CLASSIFICATION

It becomes evident that all of these classifications mentioned result in huge overlaps, and although they may have some general usefulness, they are inadequate for precise identification. The system that is most precise and useful is that of botanical classification.

Botanical classification is largely based on the variability among plants with reference to flower type, morphology, and sexual compatibility. The basic groupings most useful are family, genus, species, and cultivars. This classification system, best known as the Latin binomial, was published as *Species Plantarum* in 1753 by Linnaeus and tends to be the most exact, and it is most widely accepted internationally.

The Latin binomial botanical classification begins with the plant kingdom to which all plants belong. The classification continues as follows:

Division
  a. Algae and fungi (Thallophyta)
  b. Moses and liverworts (Bryophyta)
  c. Ferns (Pteridophyta)
  d. Seed plants (Spermatophyta)
    Classes of seed plants
      a. Cone-bearing (Gymnosperm)
      b. Flowering (Angiosperm)
        Subclass of flowering plants
          a. Monocotyledon
          b. Dicotyledon
            Order
              Family
                Genus
                  Species
                    Variety or Group (botanical)
                      Cultivar (horticultural variety)
                        Strain (horticultural)

An example of the above classification as applied to the cabbage cultivar Golden Acre YR (YR = yellows resistant) is

Division: Spermatophyta
  Class: Angiospermae
    Subclass: Dicotyledonae
      Order: Rhoeodales
        Family: Brassicaceae (Cruciferae)
          Genus: *Brassica*
            Species: *oleracea* L.
              Group: Capitata
                Cultivar: Golden Acre
                  Strain: Golden Acre YR

The complete Latin binomial name actually has a third element, that being the name, often abbreviated, of the individual who first described the species. The "L." following the species name indicates that C. Linnaeus (considered the father of the Latin binomial classification system) was the authority. For reasons of brevity, attached authorship for species mentioned in this publication are omitted except where the crop is discussed in detail.

Latin was chosen as the appropriate language, as it was widely used in scholarly circles. Additionally, Latin was a language unlikely to change, and therefore the identification and classification of the various plants would be stable. An International Code of Botanical Nomenclature helps to assure stability and resolves disagreement concerning plant classification.

DEFINITIONS USED IN BOTANICAL CLASSIFICATION

Family is an assemblage of genera that closely or uniformly resemble one another in general appearance and technical characters.

Genus identifies a more or less closely related and definable group of plants that may include one or more species. The species in the genus are usually structurally or phylogenetically related.

Species a group of similar organisms capable of interbreeding and are, more or less, distinctly different in morphological or other characteristics, usually reproductive parts, from other species in the same genus.

Variety is a subdivision of a species consisting of a population with morphological characteristics distinct from other species forms and is given a Latin name according to the rules of the International Code of Botanical Nomenclature. Variety was and continues to be used erroneously when the correct term should be cultivar. **Forma** is a subdivision of a species, ranking after variety. It is the lowest rank and usually designates a trivial variation.

Group is a category of cultivated plants at the subspecies level that have the same botanical binomial but have one or more characteristics sufficiently different to merit a name that distinguishes them from another category. The term is used for horticultural convenience and has no botanical recognition. Thus, it can be seen that botanical variety and group are similar and, therefore, often interchanged.

Cultivar, sometimes known as "horticultural variety," is a term that denotes certain cultivated plants that are alike in most important aspects of growth but are clearly distinguishable from others by one or more definite characteristics. When reproduced, they retain their distinguishing characteristics.

Cultigen refers to a plant or grouping of plants known only in cultivation, without a determined nativity, presumably having originated in the presently known form under domestication. Cultigen is not synonymous with cultivar.

Other terms having similar meaning as cultivar are as follows: **Clone** identifies material derived from a single individual and maintained by vegetative propagation. All members of the population are genetically identical and can be maintained essentially uniform with relatively little selection. **Line** refers to a uniform sexually reproduced population, usually self-pollinated, that is seed propagated and maintained to the desired standard of uniformity by selection. **Strain** is a term used to identify plants of a given cultivar that possess similar characteristics but differ in some minor feature or quality.

Appendix Table B lists the botanical name of many of the better known world vegetables.

USEFULNESS OF BOTANICAL CLASSIFICATION

For biologists, the binomial botanical classification permits the establishment of plant relationships and their origins and serves as a positive identification of plants, regardless of language. For horticulturists, it allows the identification and/or recognition of some general associations with regard to plant adaptation, cultivation, pest control, handling, storage, and usage.

SELECTED REFERENCES

Bailey, L.H., and Bailey, E.Z. 1976. Hortus Third (revised by staff of Liberty Hyde Bailey Hortorium). Macmillan Co., New York.

Brickell, C.D., Voss, E.G., Kelly, A.F., Schneider, F., and Richens, R.H., eds. 1980. International Code of Nomenclature for Cultivated Plants—1980. Regnum Vegetabile *104*.

Jones, S.B., Jr., and Luchsinger, A.E. 1986. Plant Systematics, 2nd ed. McGraw-Hill Book Co., New York.

Lanjouw, J., ed. 1966. International Code of Botanical Nomenclature. International Bureau for Plant Taxonomy and Nomenclature of the International Association for Plant Taxonomy, Utrecht.

van der Maesen, L.J.G., ed. 1986. First International Symposium on Taxonomy of Cultivated Plants. Wageningen, Netherlands: Acta Horticulturae No. 182.

4

Importance of Vegetables in Human Nutrition

VEGETABLES IN THE DIET

Humans require and always have been interested in food. Eating is a natural behavior in which nutrients are ingested, digested, absorbed, and utilized for sustenance of life. By definition, nutrition is a science and a process dealing with the utilization of nutrients through various biochemical pathways for growth, development, and maintenance. Food provides energy and nourishment, and all foods come directly or indirectly from plants, of which considerable amounts are from plants classified as vegetables. About two-thirds of the world population relies on a largely vegetarian diet. Table 4.1 presents world food sources that contribute to the average diet.

Dietary patterns are difficult to alter, as strong preferences exist and food use customs have cultural and religious aspects which greatly influence nutritional status and life-styles. These views are deeply ingrained and difficult to change. Availability and cost factors largely determine the choice of the foods in diets of individuals and families.

TABLE 4.1. WORLD FOOD SOURCES AND AVERAGE ANNUAL PER CAPITA FOOD CONSUMPTION, AND ITS DAILY CALORIE, PROTEIN AND FAT CONTRIBUTION, 1986–1988

Food source	Annual volume (kg)	(cal/day)	Protein (g/day)	Fat (g/day)
Cereals	188	1371	33.4	5.7
Starchy roots and tubers	62	141	2.0	0.3
Vegetables	69	46	2.5	0 .4
Sugar	25	237	0.1	0.0
Pulses	6	58	3.6	0.4
Nuts and oils	7	50	2.4	3.6
Fruits	53	64	0.8	0.4
Meats and nonplant sources	188	710	25.6	55.5

Source: U.N. World Economic Survey, 1993.

In some regions and countries, food choices are often limited, and the tendency is to meet caloric demands first, regardless of other nutritional needs. In such situations, people may be undernourished or malnourished.

The term "undernourished" is used to designate inadequate caloric intake. The term "malnourished" designates the lack of a minimum daily balance of nutrients that may include proteins and essential amino acids, fatty acids, vitamins, and minerals. Undernourished people are likely to be malnourished. Improper diet can lead to starvation, obesity, poisoning, disease, or death. Deficiency of any essential nutrient leads to specific physiological responses. For example, marasmus is a nutritional disease caused by partial starvation from chronic caloric deficiency. Kwashiorkor is a disorder caused by protein–calorie malnutrition. Increased susceptibility to infection and disease, and stunting of physical and mental growth are common results of poor or inadequate nutrition, especially evident in children. Improper nutrition results in reduced physical activities and intellectual capacity that limit the productive potential of humans. This is a profound effect, often receiving less attention than some other human malfunctions.

Proper nutrition depends on the consumption of a variety of foods of sufficient quantity, nutritional quality, and balance that permits growth and maintains proper body functions. The body's nutrient requirement varies with size, age, sex, health, activity, and genetic and biochemical characteristics of the individual. Pregnancy and lactation can modify nutritional requirements of females. The Recommended Daily Dietary Allowances established by the National Academy of Science, National Research Council (U.S.A.) are considered the adequate and safe levels which reflect the current state of nutritional knowledge.

Proper diets contain an adequate energy source, nutrients, and other dietary factors such as vitamins and minerals. Carbohydrates, fats, essential fatty acids, and proteins are considered macronutrients. Micronutrients include the essential vitamins and mineral elements. Water is a vital part of every diet, and fiber, and although not a nutrient, is beneficial for good health.

To support growth, development, and maintenance, diets must contain sufficient energy. The body obtains energy from food by the relatively efficient metabolism of carbohydrates, fats, and proteins; see Table 4.2.

Energy intake and expenditure are commonly measured in kilocalories (kcal) or kilojoules (kJ); 1 kcal = 4.184 kJ. Minimum caloric requirements vary as much as individual do. From 2400 to 2600 calories daily is generally regarded as adequate for most populations. However, for

TABLE 4.2.   METABOLIC EFFICIENCY OF DIFFERENT FOOD CLASSES

	Heat of combustion (kcal/g)	Metabolic energy (kcal/g)
Carbohydrate	4.1	4.0
Fat	9.4	9.0
Protein	5.6	4.0

some populations, daily consumption is chronically less than 2000 calories, whereas for other populations, daily intake exceeds 3000 calories.

NUTRIENTS IN FOODS

Carbohydrates and Other Substances

Sugars and Starches

Carbohydrates are the principal energy source, mostly derived from cereals and starchy vegetables. Carbohydrates consist of carbon, hydrogen, and oxygen, and structurally they are either simple or complex. Monosaccharides and disaccharides are examples of simple carbohydrates; polysaccharides, such as starches and fructosans, are complex carbohydrates.

Cellulose

Cellulose is a polymer of glucose that is not utilized by humans for energy. Certain animals, the ruminants, can utilize cellulose because of the presence of microbial flora in their digestive tracts.

Fiber

Fibers are composed of celluloses, hemicelluloses, pentosans, pectins, and mixtures of polysaccharides and lignins which affect the texture of plant foods. They are largely complex compounds that the body cannot convert into energy. Two basic fiber types, soluble and insoluble, occur. These are not directly absorbed but provide bulk and do affect digestion and nutrient absorption, and they act as "roughage" in bowel regulation. Vegetables are a major sources of both forms of fiber.

Lipids: Fats and Oils

Lipids are organic compounds, insoluble in water, and provide a concentrated energy source. They are structural components of cell membranes, act in the absorption of vitamins, the regulation of blood pressure, and smooth-muscle control, and interact with enzymes and hormones. Essential fatty acids, those which cannot be synthesized by the body, are linoleic acid and gamma linolenic acid.

Fats are esters of long-chain organic acids and alcohols such as glycerol. Most fats occur as triglycerides consisting of three fatty acid chains attached to a glycerol molecule. The number of double bonds determines the degree of saturation. Saturated fats are those without a double bond; those with one double bond are monounsaturated. Polyunsaturated fats are those with two or more double bonds. Plant fatty acids are more often monounsaturated or polyunsaturated than those from animal sources. The latter tend to be solid at room temperatures, whereas fats from plant sources are likely to be liquids.

Protein and Amino Acids

Proteins consist of carbon, oxygen, hydrogen, nitrogen, and sulfur. These are combined to form amino acids, comprised of a carboxyl (–COOH) and amino (–NH$_2$) groups, which when linked in various combinations, produce a specific protein. Next to water, proteins are the most abundant substance in the human body, comprising about half of body dry weight. The presence and proportion of various amino acids is a characteristic of each protein. There are some 36 naturally occurring amino acids, and 22 of these are incorporated into various food proteins. Proteins function as the building blocks for cells and are important component of enzymes. The body is able to synthesize all but 8 of about 20 amino acids. These eight are called "essential amino acids" and must be obtained from various foods. Table 4.3 lists the essential amino acids.

Protein biological values vary with the balance of the amino acids contained in the food. Proteins from animal sources generally have a higher "protein quality" than those from plants. Nevertheless, the contribution from plant sources is very important because large numbers of the world's population have limited access to animal proteins.

TABLE 4.3. ESSENTIAL AND NONESSENTIAL AMINO ACIDS

Amino acids required by the human infant and adult		Amino acids synthesized in vivo (nonessential amino acids)	
Arginine[a]	Phenyalanine	Alanine	Glutamine
Histidine[a]	Threonine	Asparagine	Glycine
Isoleucine	Tryptophane	Aspartic acid	Hydroxyproline
Leucine	Valine	Cysteine	Proline
Lysine		Cystine	Serine
Methionine		Glutamic acid	Tyrosine

[a]May not be required by human adult.

Vitamins

Vitamins are essential components of enzymes and enzyme systems involved in cellular energy production and the transfer and biosynthesis of many compounds. Although required in small amounts, they are essential for proper growth, development, and health. Almost all are obtained from the diet or are supplemented, as most are not synthesized by the body except for biotin, folic acid, and vitamin K, which are synthesized by intestinal tract bacteria. Vitamins are classified according to their fat or water solubility. The body stores fat-soluble vitamins, but water-soluble vitamins are not stored in significant amounts.

The fat-soluble vitamins are A, D, E, and K. Water-soluble vitamins include vitamin C (ascorbic acid) and those of the B-complex group: biotin, choline, folate, niacin, panthothenic acid, riboflavin, thiamin, vitamin B_6, and vitamin B_{12}. Both fat- and water-soluble vitamins are found in many leafy and leguminous vegetables, in cereals, and from animal sources.

Minerals

In addition to being components of body tissues, minerals are inorganic elements that also function as components of enzyme systems. Minerals interact with vitamins and hormones, and in other physiological functions. Their presence is essential, even though required in small quantities. They are classified as macrominerals or microminerals. The macrominerals include calcium, phosphorus, magnesium, sodium, potassium, and chlorine. Microminerals or trace minerals include iron, zinc, iodine, copper, manganese, fluoride, chromium, selenium, molybdenum, and cobalt. These are present in many foods; vegetables are a very important source.

Water

Water is the principal component in diets as well as in the human body. Its importance for maintaining cellular osmotic balance, for metabolic processes, and in the conveyance and elimination of bodily wastes is often overlooked. Sources are from direct intake of water, other liquids, food consumed, and water derived from metabolic processes. For example, 1 g each of starch, fat, or protein when completely metabolized produces about 0.6, 1.1, and 0.4 g of water, respectively.

FACTORS AFFECTING NUTRIENTS FROM PLANTS

The genetic potential of the plant is the primary control that determines what nutrients and how much of each is produced. That potential is often not realized because of other factors affecting plant development, yield, and quality. Through plant breeding and gene manipulations, the genetic potential of crop plants has been and will further be improved. As examples, breeders have almost doubled the beta-carotene content of some carrot cultivars, making them a better source of pro vitamin A. Amino acid quality and content and seed oil yields of many species have similarly been improved.

The growing environment is another important factor influencing plant nutrients. Temperature, moisture, light, and crop nutrition individually and collectively affect plant growth; when any component is limiting, growth is adversely compromised. Atmospheric pollution, cultural practices, plant-to-plant competition, and crop maturation also can exert an influence. Diseases and pests reduce yield and therefore, the available nutrient content.

The crop's nutrient content is highest at harvest; thereafter, that amount cannot be increased but will decrease. The rate of decrease is affected by time and holding conditions. Losses can occur during harvest and handling procedures because of crop damage. Time, temperature, and light can reduce the nutrient level and quality of harvested products. It is important that appropriate harvest and handling practices are used that minimize such losses. Losses also occur in storage and during processing. Proper temperature and relative humidity are essential to minimize product and nutrient losses. Losses also occur in preparing food for consumption. Washing, cutting, peeling, blanching, and cooking cause leaching and oxidative losses of some nutrients.

VEGETABLES AS A NUTRIENT SOURCE

In addition to the contribution of valuable nutrients, vitamins, and minerals, vegetables add variety, taste, color, and texture to diets. With certain exceptions, vegetables are generally low in proteins and fats, and many have a high moisture and low dry matter content. Often vegetables alone are not enough to satisfy the daily nutrient requirement, and large amounts would need to be ingested in order to supply the necessary nutrition. As an example, carrots are an excellent source of pro-vitamin A, and only about 45 g will provide the minimum daily requirement for vitamin A. However, one would have to consume about 3 kg of carrots to supply the minimum daily protein requirement. This amount provides 66 times the minimum daily vitamin A requirement, which would become toxic if eaten daily. Additionally, that volume of carrots would leave little room for other foods.

Totally vegetarian diets can provide adequate and proper nutrition. However, judgment in the selection of various plant sources is required to ensure that a balance of the required nutritional components is provided. In order to meet minimum caloric needs, a more concentrated source of carbohydrates, proteins, and fat may be needed. For many diets this need is met by a balance of starchy vegetables, cereals, fruits, nuts, and other nonanimal food sources. Unfortunately, an appropriate balance is not always achieved, sometimes because of availability but also because of an unawareness of what vegetables are proper sources of the necessary nutrients.

In many areas of the world, noncultivated edible vegetation is not utilized because of lack of information, some taboos or personal preferences. Sometimes, only when faced with starvation do some populations resort to consuming vegetative crops or plants. The food potential of noncultivated vegetation can be enormous and is available simply for the gathering. For various reasons, cultivation of vegetables is not as large as it could be because many populations prefer to obtain their nutrition from cereal and leguminous grain crops, and more affluent populations consume foods of animal sources. Plants are the least expensive food source, whereas proteins from animal sources are usually the most expensive. Conversion of cereal grains and other feedstuffs into animal food is a luxury for much of the world's population (Fig. 1.2) A greater recognition of the nutritional benefits and diversity provided by vegetable consumption would have a large impact on improving world health for all populations.

Although not an inclusive listing, some vegetables can be grouped according to the primary type of nutrient they provide.

Carbohydrate

white potato	sweet potato
dry beans	cassava
yam	taro (aroids)
plantain	

Fat

mature seed of some legumes and cucurbits

Protein

beans	peas
sweet corn	leafy crucifers

Pro-vitamin A

carrot	orange/yellow flesh sweet potato

squash
pepper
green peas

green leafy vegetables
green beans

Vitamin C
crucifers
tomato
pepper
melons

immature bean seed
bean sprouts
white potato
many leafy vegetables

Minerals
crucifers

most other leafy vegetables

SELECTED REFERENCES

MacFie, H.J.H., and Thomson, D.M.H., eds. 1994. Measurement of Food Preferences. Blackie Academic and Professional/Chapman Hall, Glasgow.

Pietrzik, K., ed. 1991. Modern Lifestyles, Lower Energy Intake and Micronutrient Status. Springer-Verlag, London.

Spallholz, J.E. 1992. Nutrition Chemistry & Biology. Prentice-Hall, Englewood Cliffs, NJ.

5

Toxic Substances and Some Folk and Medicinal Uses of Vegetables

Overall, vegetables are generally recognized as safe when appropriately used. They provide many nutritional benefits that contribute to health and its maintenance. Briefly described are some medicinal benefits of several vegetables, the types and functions of toxicants found in plants, how they affect humans, and those vegetables that contain toxic compounds.

MEDICINAL PLANTS AND USES

Humans have utilized plants as the main source of medicine. Natural drugs were the mainstay of medicine, and it is only recently that extensive use of synthetic drugs has occurred. Many compounds used in present-day or "Western"-type medicines are either derived directly from plants or as a synthesized form. Actually, many of the synthetic products are based on natural products. Throughout history, humans have learned much about the avoidance and curing of illness through the consumption and application of plants and their products. History reveals that some products have accidentally harmed humans and animals; confusion about common name usage has probably lead to some of these occurrences. Also, some plant products have been used intentionally to harm humans.

Considerable documentation exists about ancient customs and folk remedies passed from generation to generation about the uses of various plants by the many societies and peoples of the world. Some practices and customs, thousands of years old, are still used. The ancient Chinese, Indian, and Egyptian civilizations utilized herbal medicines, as did the Greeks and Romans. From the beginning of the Christian era to the present, other societies practiced herbal medicine. Herbalism remains active, more so in some countries than others.

Abundant literature exists about folk and medicinal usage of plants. Many putative benefits are claimed; some medical benefits are verified, but many cannot be substantiated. Literature reviews about herbal medicine give the impression that almost every disorder or malady is somehow treatable by a plant product or products. On the other hand, some plants are also well known for their undesirable and/or toxic effects.

Witch doctors, high priests, medicine men, and the early physicians of Greece and Rome all acquired some knowledge about the beneficial and also the undesirable effects of various plant compounds to humans and animals. Such knowledge was initially acquired by trial, and the observed results to the medicated subjects was passed on from generation to generation. Safety and effectiveness was learned from repeated use over time.

Many ancient writings, including those of the Bible, mentioned various plants used for a multitude of medicinal purposes. Even in ancient times, prevention of many types of cancer or its cure was a frequent benefit claimed; many vegetable species were reported to have this property. Other frequent benefits proposed and dealt with aphrodisiac properties, and, not infrequently, the opposite effect might be claimed for the same plant. The range of treatable disorders is questionable and might include poor appetite, bad breath, fever, headache, and freckles. There seems to be an appropriate plant to resolved almost any imaginable situation. It is not practical, and no attempt is made, to list the numerous maladies treatable by various plant compounds or the many plants reported to be involved.

Presently, herbal medicine is practiced to some extent in all societies, and most widely and prominently in China and India. Certain plant compounds in appropriate amounts can have beneficial pharmacological uses and, as such, constitute the basis for herbal medicinal practices. Plants contain thousands of compounds; a few are beneficial and most have functions that are still unknown. The compounds can function independently or in concert with other compounds to cause some effect. For some situations, the effect may be physiological; in others, the medical cure or relief may be psychological. Benefits are acknowledged, but when improperly practiced, herbal medicines have the potential to be dangerous. Therefore, it is best prescribed or used by those with specific knowledge, training, and experience. Self-administration without adequate and accurate information is unwise and strongly discouraged. In time and through research, some claims will be verified, others will be discounted, and new ones will be proposed. The importance is sufficiently great that medical research will continue to seek more and better information about the role of plants for human health.

Generally, there is no concern, aside from specific allergies, about the consumption of most vegetables, and as used, very few vegetables have significant medicinal or toxic properties. Major health benefits in the consumption of vegetables are for the nutrition they provide.

PLANT FAMILIES AND SPECIES HAVING MEDICINAL PROPERTIES

The list of medicinal plants is almost endless, and vegetables as we know them are a small part. Some examples of usage and medicinal properties of various vegetable families follow.

Fungi

Some fungi consumed as vegetables have medicinal qualities, others are known for their hallucinogenic attributes. Their use as an internal or topical medicine is ancient and lengthy. Several mushrooms are known for their blood anticoagulant properties; the woodear mushroom, *Auricularia auricula* being an example, whereas, corn smut, *Ustilago maydis,* has been used to control postchildbirth and other bleeding.

Alliaceae (Amaryllidaceae)

The Codex Ebers and Egyptian medical papyrus dating to about 1550 B.C. mentioned onion and garlic being used to treat headaches, parasites, and heart and circulatory disorders. The Romans used garlic for the expulsion of intestinal parasites. In India, garlic served as an antiseptic for wounds and skin ulcers. In China, onion tea treated headache, fever, cholera, and dysentery. Albert Schweitzer used garlic in Africa for treatment of amoebic dysentery. Many examples of folklore mention the "blood purification" properties of onion, garlic, leek, and other vegetable alliums, as well as their aphrodisiac properties. The antithrombotic properties of onion and garlic are recognized, and compounds contained in these species have antibacterial and antifungal activity and can also function as an insect and animal repellent. The antibiotic effects of alliums are due to the sulfur compounds, the bactericide being allicin. A drink of macerated garlic and wine was thought to provide immunity to diseases. Other claims suggest garlic can be used to treat high blood pressure, emphysema, tumors, cancer, as well as urinary infections and rheumatic disorders. Garlic was used as a confluent for smallpox, accomplished by applying cut pieces of cloves to the feet following the onset of the disease.

Apiaceae (Umbelliferae)

Many claims have been made for the cancer prevention properties of carrots. The carotenes in carrot roots have many nutritional benefits related to the avoidance of bone, skin, and sight disorders. Dill was used to improve appetite, remove bad breath, and also for some cancerous conditions. Celery seed and seed of other Apiaceae species are used for their stimulant effects. Asiatic pennywort is claimed to enhance longevity and memory and also cure leprosy and some skin wounds.

Araliaceae

The ginseng root has been used as a drug in the Orient for thousands of years for ailments ranging from heart disease to insomnia, as well as a love potion. Physiological experiments indicate that extracts from this plant have a mild stimulating effect.

Asteraceae (Compositae)

Greek and Roman physicians used extracts of wild lettuce leaves to combat fevers, headaches, and jaundice. Extracts were also used for cancers of the uterus, liver, and other tumors. Boiled dandelion leaves were similarly used and also have use as a diuretic and for urinary, kidney, and liver problems. Chicory was used for wart removal. Extracts from globe artichokes are used for the treatment of gastric disorders and for other therapeutic purposes. Folklore is extensive about burdock for cancer treatment and other curative properties.

Brassicaceae (Cruciferae)

Many claims have also been made for the cancer preventive properties of various plants of the Brassicaceae. Plants of the genus *Brassica* contain S-methylcysteine sulfoxide which has been shown to lower blood cholesterol levels. The Greeks recommended cabbage juice as an antidote for poisonous mushrooms. Mustard plasters are used to produce contact heating to reduce the pain of muscle strain or bruises. A compound called sulforaphane isolated from broccoli and some other brassicas, blocks the growth of tumors in mice which had been treated with a cancer-causing toxin. This may have future application in human cancer treatment.

Chenopodiaceae

Spinach has a long history of various medicinal uses. In England, spinach extracts were used to cleanse wounds and cure warts.

Convolvulaceae

In New Zealand, the Maori population used an infusion from sweet potato to lower fever, and crushed leaf tissues are used to treat skin diseases.

Cucurbitaceae

In parts of the Orient, dried fruit of bitter melon is a medicine for hemorrhoids, gout, rheumatism, and parasites. It is also used as antitumor compounds and for skin disorders and burns. Watermelon seed have been used to treat urinary tract infections, poor blood circulation, and a variety of disorders. Seed of *Cucumis melo* were used as a treatment for stomach cancer. Consumption of melons during very high-temperature conditions was believed to prevent sunstroke. *Momordica cochinchinensis* has been used to treat many types of wounds.

Cycadaceae

Natives of several southwest Pacific islands use cycads for curing skin ulcers, and in India cycad seed are used as a laxative.

Dioscoreaceae

An important heart medicine, digitalis is extracted from yams, as are some steroid compounds used for manufacturing certain drugs.

Fabraceae (Leguminosae)

Roman physicians used peas boiled in seawater to cure erysipelas, a bacterial infection of subcutaneous tissue. A mixture of chick peas and rosemary was prescribed for jaundice and the treatment of edema.

Ginkgoaceae

Consumption of raw seed were used to destroy cancer and to treat urinary disorders.

Liliaceae

Asparagus shoots were used to cure jaundice and to "cleanse the bowels" and for other disorders.

Poaceae (Graminae)

Various tonics prepared from bamboo sap were used to treatment rheumatism and to increase the flow of mother's milk.

Polygonaceae

Rhubarb was recommended for liver diseases and stomach disorders and to eradicate ringworm.

Solanaceae

Capsaicin, the compound largely responsible for pepper pungency, is known to ease neuronal mechanisms of pain and reduce symptoms of psoriasis, arthritis, and contact allergy.

Some of the medicinal effects of the various plant families mentioned are considered valid, but as it is with folklore, many proposed attributes require more evidence to establish effectiveness. Other plant species in families different from those preceding also have medicinal usage. An equally large or larger number of culinary herbs are also associated with various real and putative medicinal properties, but for brevity, these are not mentioned.

Of the many plants having medicinal properties, those used as vegetables are less involved than other species. Moreover, it is recognized that the major medical benefit of vegetable plants for humans is providing the energy and essential nutrients to sustain life and good health. Through proper nutrition, many illness can be avoided or minimized, whereas improper or poor nutrition is a contributor to poor health.

PLANT TOXICANTS

Even beneficial substances when consumed or used in excess can have harmful effects, whereas some toxic compounds can be tolerated only at low levels. The adage that dosage makes the poison is quite appropriate. What is important is whether and where the toxic compound accumulates and how and where it is metabolized or acts in the body.

One reason plants contain toxic substances is believed to be part of their evolutionary development. In the adaptation to their environment, various repellent structures such as spines, odors, bitter flavors, or toxins are produced as protective mechanisms against possible predators. A species could become extinct from persistent animal foraging, and the structures or compounds the plant produces or contains would tend to discourage predation.

Through domestication, most present-day food plants have been improved by selection, which enhanced their desirable characteristics and reduced or eliminated undesirable features. However, many domesticated vegetables still retain some chemical and physical similarities of the wild ancestors. In some situations, these vegetables may require specific handling or detoxification to render them safe for consumption.

TYPES OF TOXICANTS CONTAINED IN VEGETABLE PLANTS

Alkaloids are probably one of the most frequently found toxic compounds in vegetables. The chemical structure varies, although most have a ringlike configuration and contain nitrogen. These toxicants usually affect the nervous system.

Glycosides usually consist of a simple sugar and a nonsugar. The nonsugar moiety becomes toxic when the glycoside is metabolized. Some glycosides are cyanogenic, where the nonsugar portion is converted to hydrogen cyanide. Other glycosides have as a nonsugar component a steroid, saponin, coumarin, or sulfur-containing compound. Glycosides are frequently involved as respiratory inhibitors.

Proteinaceous compounds are inhibitory to many metabolic processes and are also allergens. These include specific proteins, polypeptides, and amines; certain amines are poisonous.

Alcohols. Some alcohols are neurovascular poisons.

Non-amino organic acids are carbon-based acids not containing nitrogen, but usually associated with a soluble salts, sodium oxalate for example. Their toxic effects result from ionic imbalances and kidney damage.

Resinoids, tannins, phenols, and **terpenoids** are compounds of diverse structure. They are often a cause of skin irritations. Tannins are known to decrease protein digestibility.

Mineral toxins play many roles, often interfering with vitamin functioning and with the absorption of certain nutrients that can affect ionic balance. Nitrate accumulation can interfere with respiratory function, and high accumulations of selenium, mercury, or cadmium are poisonous.

PHYSIOLOGICAL FUNCTIONS AFFECTED BY TOXICANTS

Many toxic effects are the result of the interactions of various compounds and, therefore, the mode of action is not always understood. The toxic reactions that follow are only examples of those associated with vegetables and do not represent the much wider spectrum of reactions that other plant toxins may cause.

Allergies

Allergens are usually induced by proteinaceous substances and they are widespread throughout the plant kingdom. Ingested proteins or even pollen on mucous membranes can cause physiological reactions, often with the production of histamines or histaminelike substances. In many cases, cooking or heating denatures the protein and lessens the degree of allergic reaction.

Enzyme Inhibitors

Because proteins are so often involved with allergenic or other toxic effects, it is not surprising that proteins affect and interact with enzymes. Inhibitors of protease affect its proteolytic activity which catalyzes the hydrolysis of proteins to amino acids. Some inhibitors may be heat stable, whereas others are heat liable. Several known inhibitors act on the following enzymes: amylase, trypsin, chymotrypsin, cholinesterase, carboxylpeptidase, and invertase. Protease inhibitors are found in jack fruit, beet root, turnip, chick pea, taro, in species of *Vigna* and *Phaseolus,* as well as winged bean. They are also found in sweet and white potatoes, lettuce, and maize.

Respiratory Inhibitors

Cyanogens are formed enzymatically from parent glycosides. Cyanide is formed from the cyanogens and inhibits the respiratory enzyme, cytochrome oxidase, and other enzymes.

Nervous Systems

Toxins affecting the nervous system often are alkaloids, and most often glycoalkaloids.

Nutrient/Mineral Absorption

Consumption of certain plants of Chenopodiaceae, Polygonaceae, and Araceae contain oxalates which interfere with calcium absorption. Selenium ingestion and accumulation is responsible for gastroenteritis, fatigue, dizziness, and dermatitis.

Hormonal Interference

Various goitrogens and steroids interfere with hormonal functions.

Antimetabolites

Hemagglutinins and other proteinaceous compounds are examples of antimetabolites. These substances can disrupt normal biochemical reactions, causing adverse physiological changes such as poor food absorption and low nitrogen retention.

Antivitamins

These compounds interfere with the utilization of different vitamins. Some plants, peas and beans for example, have compounds that interfere with vitamin E utilization. Other vitamins are also interfered with or inactivated by some plant compounds.

Carcinogens, Tumorigens, Teratogens

Several glycoalkaloids are incriminated as cancer-causing compounds. Cycasin in cycads is an identified carcinogen.

Physiological Disorganization and Irritants

Some mushrooms species possess specific respiratory and neurological toxins. A number of plant compounds can cause different forms of poisoning. Seed of grass or chick pea, *Lathyrus sativus,* contain lathrogens which can cause "lathyrism," resulting in lameness, paralysis, and deformity. Parsley and several other Apiaceae species produce phytotoxic compounds, such as psoralens. Terpenoids and some saponins cause contact irritation. The allyl isothiocyanates in mustard can cause dermatitis.

Birth Defects
Solanaceous alkaloids are believed to be responsible for causing certain birth defects, and some may have teratogenic effects, in addition to other toxicities.

Mechanical Injury
Contact with aroid raphids and cacti glochids can cause tissue injury and irritation.

VEGETABLE PLANT FAMILIES CONTAINING TOXICANTS

It is important to recognize that most of the plant products discussed become toxic when consumed in excess. Common sense is required in the consumption of most foods, including vegetables. Doing so usually eliminates possible toxicities.

Although not grouped according to botanical families, fungi are important vegetables. Obviously, cultivated mushrooms are free of toxic compounds or are present in minute quantities. Some species are known for their hallucinatory effects. Those of the genus *Amanita* contain phallin, a cell-destroying glucoside which can be inactivated by cooking. Other compounds are not destroyed by heating and can affect the liver, kidney, and heart.

Mushrooms toxicants are neurotoxins resulting in cell breakdown and also work on the gastrointestinal system. Some toxins are peptides or alkaloids, poisons that affect the nervous systems. Other proteinaceous substances have additional toxicities.

When gathering noncultivated mushrooms, each mushroom should be examined to verify that it is not a poisonous species, because it is not uncommon for edible and poisonous mushroom species to grow in the same area.

Apiaceae (Umbelliferae)

Anethole, present in dill and fennel, can cause dermatitis. Parsley and celery contain psoralens and terpenoids responsible for dermatitis. Some species also contain alkaloids and a cholinesterase inhibitor. Carrots contain a polyacetylenic alcohol called carotatoxin which can cause neurotoxic symptoms.

Araceae

Taro and other edible aroids contain oxalates which are associated with acridity and are toxic when present in high concentrations. They also contain raphids, which can cause tissue irritation, and a amylase inhibitor, which inhibits starch hydrolysis.

Araliaceae

Ginseng contains oxalic acid.

Asteraceae (Compositae)

Lettuce is known to be a nitrate nitrogen accumulator. Some members of this family contain alkaloids.

Brassicaceae (Cruciferae)

Species of *Brassica* contain a number of different goitrogens that can cause goiter, as well as cholinesterase inhibitors. Glucosinolates present in mustards can cause digestive disorders.

Chenopodiaceae

Beets, Swiss chard, and spinach contain oxalic acid. Spinach contains saponins and tends to readily accumulate high levels of nitrates. When spinach is ingested, nitrate ions oxidize blood hemoglobin, forming methemoglobin which interferes with blood oxygen transport. Nitrates also form nitrosamines which are carcinogenic compounds.

Convolvulaceae

The sweet potato contains a bitter substance called ipomeamarone.

Cucurbitaceae

Many cucurbit species contain various compounds, usually glycosides, that impart a bitter taste, and at high levels are very toxic. Pumpkins and squashes contain inhibitors of cholinesterase. Watermelon contains serotonin, which can cause elevated blood pressure.

Cycadaceae

Cycads contain glycosides, which when eaten in large quantities can sometimes be lethal. These cause symptoms of paralysis. The incidence of amyotropic lateral sclerosis, a neurological disease, was found to be significantly greater among individuals eating cycads. Cycads are suspected of having carcinogenic activity. The starch must be detoxified by heating or leaching, and leaves by cooking.

Dioscoreaceae

Some yam species contain diosgenin which is poisonous; most edible species contain little or none. Yams also contain saponins and estrogen-like steroids.

Euphorbiaceae

Bitter cultivars of cassava contain cyanogenic glycoside, which requires detoxification before consumption.

Fabaceae (Leguminosae)

Many members of this family contain compounds that are allergens. Some species contain cyanogenic glycosides, hemagglutinins, and trypsin, amylase, and/or glucose 6-phosphate dehydrogenase (G6PD) inhibitors, in addition to compounds having antivitamin properties for vitamins A, E, and D. An antivitamin E compound is contained in raw kidney beans and peas; a factor that interferes with zinc availability. Some species contain saponins. Grass or chick peas contain lathrogen, a glycoside that can cause lathyrism. Favism, an acute hemolytic anemia, is caused when a glucoside interferes with the functioning of the G6PD enzyme. It is estimated about 2% of the world's population may be susceptible to some degree.

A social disease known as flatulence is highly correlated with ingestion of seed of various legumes.

Ginkgoaceae

Ginkgo biloba seed contain skin- and gastric-irritating compounds.

Liliaceae

Asparagus contains saponins and a cholinesterase inhibitor.

Musaceae

Plantain and cooking bananas contain the presser amines, serotonin and norepinephrine, that constrict blood vessels and cause elevation of blood pressure.

Poaceae (Graminae)

Bamboo shoots contain a cyanogenic glucoside, which, when ingested, is potentially poisonous because of the cyanide released. Exposure of the emerging shoot to light enhances the bitterness. Parboiling and leaching are necessary to render the shoot tissue nontoxic.

Polygonaceae

Rhubarb and sorrel contain high levels of oxalic acid, which reacts with soluble calcium to form insoluble calcium oxalate, and thus prevents the absorption of this essential mineral. Rhubarb leaf blades contain a very toxic substance, which when ingested in large quantities can result in death. Rhubarb also contains a highly irritant glucoside.

Polypodiaceae

The bracken fern, *Pteridium aquilinum* is known to cause bone marrow damage and intestinal mucosa disorders, and also contains thiaminase, which hydrolyzes thiamine, making it inactive as a vitamin. This fern is also thought to contain carcinogenic compounds.

Portulacaceae

Some species contain oxalic acid.

Solanaceae

Many species contain bitter-tasting alkaloids: solanines in potatoes and tomatine in tomatoes. Some contain several protease inhibitors. Potatoes contain an invertase inhibitor which prevents the hydrolysis of sucrose. Pungent peppers contain the amide or resinoid, capsaicin, which is sometimes responsible for causing skin irritation or gastric discomfort.

Tetragoniaceae

New Zealand spinach contains oxalic acid.

Since written history, various plant substances have received considerable attention in numerous publications, a trend that has not diminished. Documented and undocumented information is immense, and discussion about plant medicinal and toxic substances continues to expand and appears to be boundless. To properly address this subject is beyond the scope of this book. Therefore, we limit our effort to a minimum discussion of medicinal and toxic properties of plants having vegetable usage, which, when compared to the number of species in the plant kingdom, is very small.

SELECTED REFERENCES

Blackwell, W.H. 1990. Poisonous and Medicinal Plants. Prentice-Hall, Englewood Cliffs, NJ.

Block, E. 1985. The chemistry of garlic and onions. Scientific American 252 (3), 114–119.

D'Mello, J.P.F., Duffus, C.M., and Duffus, J.H., eds. 1991. Toxic Substances in Crop Plants. Royal Society of Chemistry, Cambridge.

Duke, J.A. 1983. Medicinal Plants of the Bible. Trado-Medic Books, London.

Duke, J.A., and Ayansu, E.S. 1985. Medicinal Plants of China. Reference Publications, Inc., Algonac, MI.

Fenwick, G.R., Heaney, R.K., and Mawson, R. 1989. Glucosinolates. In Toxicants of Plant Origin. P.R. Cheeke, ed. CRC Press, Boca Raton, FL, Vol. II.

Ivie, G., Wayne, D.L., Holt, L., and Ivey, C.C. 1981. Natural toxicants in human foods: Psoralens in raw and cooked parsnip root *(Pastinaca sativa)*. Science *213,* 909–910.

Kingman, A.D., ed. 1977. Toxic Plants. Columbia University Press, New York.

Li, S.-C. 1973. Chinese Medicinal Herbs. Georgetown Press, San Francisco.

Liener, I.E. 1969. Toxic Constituents of Plant Foodstuffs. Academic Press, New York.

MacGregor, J.T. 1986. Naturally occurring toxicants in horticultural food crops. Acta Hort. *207,* 9–19.

Muenscher, W.C. 1981. Poisonous Plants of the United States. Macmillan, New York.

National Academy of Sciences. 1973. Toxicants Occurring Naturally in Foods. National Academy of Sciences, Washington, D.C.

Payne-Jackson, A., and Lee, J. 1993. Folk Wisdom and Mother Wit. Greenwood Press, Westport, CT.

Wertheim, A.H. 1974. The Natural Poisons in Natural Foods. Lyle Stuart, Inc., Secaucus, NJ.

Part II

Vegetable-Growing Principles

6

Environmental Factors Influencing the Growth of Vegetables

CLIMATIC FACTORS

Growth and development of plants depend on climatic as well as soil factors. Weather is the state of the atmosphere with respect to temperature, moisture, solar radiation, air movement, and other meteorological phenomena over a short period of time. Climate is the average course of weather at a specific location over a period of many years and is the integrated effect of weather. In addition, climate also influences weathering and the development and condition of the soil.

The kinds or plant species that will or can be grown in a given region is predicated by the climate, and the growth of an individual plant is directly dependent on the weather during its life cycle.

TEMPERATURE

Temperature has profound effects on all living organisms, favoring or limiting growth and thereby influencing the distribution of both plants and animals. The source of this energy is the sun; each minute about 2.0 g cal/cm^2 (2.0 Langley) is transmitted to the earth.

Global Temperature Distribution

Although the world is a rotating sphere and each latitude receives its portion of the insolation, the temperature of any particular point on the earth is governed by its proximity to other land masses and to air and ocean currents. Thus, a world map of temperature is not uniform around the earth. The equatorial zone is consistently warmer, and polar regions consistently colder than the mid-latitudes. Due to the 66½° inclination of the earth to the plane of its orbit around the sun, portions of the earth receive different amounts of insolation, varying with the time of year and latitude. At high latitudes, the difference in insolation is very large, resulting in four seasons: winter, spring, summer, and autumn. If it were not for the tilt of the earth, there would not be distinctive seasons because temperature and climate would vary little and the day length would be uniform.

Seasonal temperatures can vary as little as 3°C year round in the tropics and more than 45°C in polar regions. The range of air-temperature variation is much larger over the continents than over or near the oceans or seas because of moderation provided by the large heat capacity of water as compared to soil.

Diurnal fluctuations occur due to the rotation of the earth, as solar radiation is not received at night, and from radiation losses which occur from the surface at night. Thus, each day has a temperature maximum which occurs shortly after noon and a minimum which occurs at sunrise. Temperatures also decrease with increase in altitude, usually a decrease of about 6°C occurs with each 1000 m in elevation.

Temperature Effects on Plants

Cardinal Temperatures

The cardinal temperatures are minimum, maximum, and optimum. Minimum or maximum temperature is where growth ceases, and optimum is where growth is most favorable and/or rapid. The cardinal temperatures are not the same for all plants and vary with different families, genera, and species. Plant breeders select and develop plants for tolerance to low temperature in cool regions and for high-temperature tolerance in regions where that is appropriate.

Length of Growing Season

One of the oldest and useful measures of evaluating and predicting when crops can be grown is the number of frost-free days. This is the average period between the last killing frost in the spring and the first killing frost in autumn.

Van't Hoff's Law

This physical law states that for every 10°C rise in temperature, the rate of dry matter production or growth doubles. This response is commonly called the Q_{10} factor. However, this is usually only applicable in the range of about 5–35°C and can vary depending on the organism.

Heat Units or Degree-Days

The heat unit or degree-day concept is based on the theory that plant development is dependent on the amount of heat experienced during growth and is calculated by subtracting the minimum threshold temperature from the average temperature for a given day. The minimum threshold temperature varies with different crops, but usually ranges from 5°C to 15°C. The daily average is obtained by adding the minimum and maximum temperature and dividing by two. Heat units or degree-days are equivalent. For each day, every degree of difference above the threshold has a value of 1. The units are accumulated for a desired period, usually from planting to harvest. For example, if the average temperature for a day is equal to or less than the threshold temperature, the degree-day value is 0. If the average were 5°C above the threshold value, the degree-day value would be 5.

This concept is useful for crop scheduling and harvest prediction and is generally more accurate than reliance on previous cropping histories, such as days from planting to harvest. For some vegetables such as peas and sweet corn, the degree-day concept functions very well. For others, clearly established thresholds with regard to temperature response are unclear or are yet to be identified.

Diurnal Change (Thermoperiodicity)

Generally, a large diurnal range is favorable for net photosynthesis. Night temperatures play an important role for some crops. High night temperatures increase respiratory rates, which result in an increased utilization of carbohydrates produced during the day. In such situations, the plant makes little, if any, progress in growth or development.

Vernalization

The exposure of certain plants to low temperatures induces or accelerates bolting and/or flowering. This inductive stimulus is called vernalization. The required length of low-temperature exposure to achieve this effect varies with species and stage of plant development. With some species, devernalization, a reversal of vernalization, can occur if the plant is immediately exposed to high temperatures, usually above 30°C, following low-temperature exposure. In some situations, reversal is partial or not observed. Figure 6.1 shows the growth phases representative of a biennial plant in a temperate region.

For some species, seedling and young plants are insensitive to low-temperature conditions that promote flowering in older plants. Such plants at that growth stage are considered juvenile or nonresponsive.

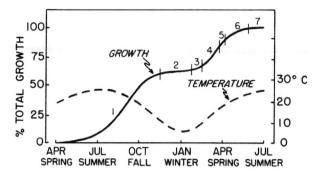

FIG. 6.1.   Growth phases in the life cycle of a biennial plant in a temperate region: 1-vegetative growth; 2-vernalization; 3-floral initiation; 4-seed stalk emergence and elongation; 5-flowering and fertilization; 6-seed maturation; 7-mature seed.

With other species, even imbibed or newly germinated seed can be vernalized. Some tubers, corms, and bulbs require chilling temperatures before dormant buds become active, and others require low temperatures before resumption of growth.

Rest and Dormancy in Seeds and Organs

Most mesophytic plants go through a phase of little or no growth at some stage in the life cycle. Seeds, buds, tuberous roots, tubers, rhizomes, bulbs, and corms often show this phenomenon. Under natural conditions, this period usually coincides with unfavorable environmental conditions such as low or high temperature or a lack or excess of moisture; sometimes, photoperiod is involved.

However, even when the environmental conditions are favorable, seeds and vegetatively reproductive organs are said to be at *rest* when the organ shows no signs of growth resumption. This rest period is also called internal or innate dormancy. Seeds and organs that have the potential to germinate or resume growth but do not because of unfavorable environmental conditions are termed *dormant*. Other terms describing this state are external, imposed, and quiescent dormancy.

Actually, the phases of rest and dormancy are not abrupt, but occur gradually as changes in inhibitor and/or hormonal concentrations occur over time. The rate of change is strongly influenced by temperature. Some plant physiologists describe several progressive stages such as early rest (predormancy) → rest (mid-dormancy) → after rest (postdormancy). Normally, when the organ is at rest, it may be extremely difficult to induce a resumption of growth. However, sometimes, high-temperature treatment at early rest may result in resumed growth. Following a period of rest, there is a transition period when it becomes progressively easier to induce growth. Temperature at which the organ is maintained can influence the rate of these changes; at high temperatures, each stage is shorter in duration than at low temperatures.

Freezing and Chilling Injury

Many cold-climate plants may be frozen by low temperatures without incurring injury. However, most vegetable plants are injured by temperatures at or slightly below freezing, unless they can be acclimated to tolerate these levels. Many tropical and subtropical plants can be damaged at nonfreezing temperatures below 10°C. This type of damage is known as chilling injury. Vegetables of tropical origin are sensitive to such injury when exposed to prolong periods of less than 10°C. A few hours at less than 10°C but above freezing usually is not harmful; generally, no harm occurs if warm temperatures soon follow the low-temperature exposure. These kinds of injury have a time and temperature interaction.

Susceptibility to cold and chilling injury varies with species and there can be large differences among cultivars of the same species. Susceptibility also varies with the stage of plant development; flowering and fruit development are highly susceptible periods.

Hardening

Plants can be somewhat modified (acclimated) and made relatively tolerant to cold temperature by subjecting the plant each day to lower and lower temperatures. This process, called hardening, causes an adaptation of cell protoplasm to low temperatures. Such a process takes place naturally in the autumn of the year. Hardening is not protection against freezing. Hardening also occurs when plants are subjected to gradual water stress or nutrient deprivation. Both conditions are useful to acclimate plants for high- or low-temperature stress.

High-Temperature Injury

In many arid and semiarid regions, high temperature may be a limiting factor in the economic production of some vegetables. Under high insolation and high humidity, leaf temperatures 8°C above that of ambient air temperature have been recorded. When the temperature rise is too great, heat destruction of protoplasm results in cell death. This often occurs in the range 45–50°C. For example, in some field conditions, with air temperatures above 38°C, tomato fruit temperatures of 49–52°C have been observed. Sun-exposed fruit in such conditions become damaged because of sunburn or sun scald.

As with cold resistance, plant's tissues, to a certain extent, can gradually become heat acclimated by slowly raising temperatures and increasing the exposure period to such conditions each day. Transpiration from leaf stomata helps cool leaves. It has been calculated that transpiration can reduce heating by about 15–25%.

Thermoclassification of Vegetables

Vegetables can be loosely grouped according to climatic preference, such as cool season and warm season crops. This classifies vegetables according to their optimum-temperature requirements and tolerance for certain minimum and maximum temperatures, and although not precise, it does have usefulness.

The cool season crops can be further divided into a grouping that includes crops such as spinach, beet, and many brassica vegetables that have a growth optimum between 16°C and 18°C and some tolerance to freezing. Another grouping has a similar temperature optima, but the crops are damaged by freezing temperatures. White potato, celery, lettuce, cauliflower, and Chinese cabbage are some of the crops typical of this group. Another cool season grouping involves crops such as onion and asparagus that have a higher optimum-temperature range (18–30°C) and some frost tolerance.

The warm season crops likewise can be separated into a group having temperature optima between 18°C and 30°C and intolerance to frost. Most vegetable legumes, tomato, pepper, and many cucurbit crops are representative of this group. Other crops such as sweet potato, yams, cassava, plantain, and okra have an optimum growth range between 21°C and to 35°C but will not grow well at temperatures less than 21°C. All can suffer chilling injury by exposure to extended periods of temperature less than 10°C.

MOISTURE

Hydrologic Cycle

Moisture is as important in climate as is temperature. Water is essential for life processes. Most of the earth's water is found circulating within the atmosphere from the surface to a height of about 15 km, and to a similar distance below the surfaces of land and ocean. The earth's moisture exists in liquid, gaseous (vapor), and solid (ice and snow) phases. The conversion from one phase to another is temperature dependent. The circulation and dispersion is called the hydrologic cycle; the cycle is sustained by solar energy. Evaporation from bodies of water, land, and plant surfaces (transpiration) constitute the vapor phase. When this source is transformed into a liquid or solid phase, it falls to the earth's surface as rain, snow, or hail. Condensation on the surface is dew.

World Precipitation and Distribution

The annual precipitation averaged over the earth is about 1000 mm and varies from a high of more than 12,000 mm in certain areas in India and Hawaii to a low of 0.5 mm at Arica in northern Chile.

The amount of rainfall is closely related to air pressure and winds. It is directly associated with moisture sources such as oceans or sources lacking moisture such as deserts. Heavy rain occurs along the equatorial low-pressure belt which is due to the abundant moisture from inflowing ocean air. The heaviest rainfall is usually near coasts or mountainous regions where moisture-laden air is cooled by assent.

Heavy rains commonly occur in the tropics between 20° latitudes north and south of the equator. The high-pressure areas between the 20° and 40° latitudes in each hemisphere are known as the horse latitudes and are generally regions of low rainfall. The Sahara, Arabian, Turkestan, and Gobi deserts, the Sonora desert of the southwest United States and northern Mexico, and others in Australia and portions of Chile are examples of arid regions in these latitudes. Low-rainfall regions such as Baja California, Peru, northern Chile, and areas of Senegal and Angola are caused by cold ocean currents, which allow little evaporation from the surface. Summer monsoons bring warm moist ocean air into the Indian subcontinent and southeastern Asian countries. This is due to periods of low pressure in the interior of Asia.

Between the 40° and 60° latitudes, precipitation is fairly heavy. Near the poles, high pressures again create conditions for low precipitation.

It is not only the amount of rainfall but also the distribution of rainfall that is important. About 500 mm of rain falls in south central Canada. This is sufficient for a good crop without irrigation because a large portion of the moisture occurs during the summer growing season. In contrast, about the same amount falls in central California, but nearly all of that is in the winter. Thus, irrigation is necessary to grow summer crops.

Humidity

Moisture in the atmosphere is measured as absolute humidity or as relative humidity. Absolute humidity is the amount of water present in a unit volume of air. Relative humidity (RH) is the amount of water present in air as a percentage of what could be held at saturation at the same temperature and air pressure. Dew point is the air temperature at which water vapor is at the saturation point. The dew point varies with the amount of water vapor in the air and the temperature. A temporary modification of local humidity can occur because of irrigation practices such as sprinkler irrigation.

Relative humidity is a important factor in plant growth and development because it has a strong influence on transpiration. Low humidity tends to increase transpiration; high humidity has the opposite effect. High humidity has other effects, such as favoring an increase in the incidence of many diseases and insect populations.

Classification of Regions According to Annual Precipitation

Arid	<250 mm
Semiarid	250–500 mm
Subhumid	500–1000 mm
Humid	1000–1500 mm
Wet	>1500 mm

Precipitation from Fog and Dew

Minute hygroscopic particles are effective in the condensation of water vapor into water droplets called fog. Fine salt particles or smoke are important in fog formation as nuclei for the condensation of water vapor. When high-moisture-containing air masses close to the earth are cooled, fog can form. Fog also forms when cold air comes into contact with warm waters. Moisture from fog can collect in appreciable amounts on plant leaves. This is especially important along coastal regions where seasonal rains are few, and thus plants often obtain enough moisture from fog to survive. Foggy conditions reduce moisture loss from transpiration.

Dew can result when heat is lost, usually by radiation at night from surfaces such as the ground or leaves. The air adjacent to the particular surface is cooled to the point of saturation, and water vapor condenses on the cooled surface. The temperature at which this occurs is called the dew point, and if above 0°C, the condensed water vapor is liquid (dew). However, if the surface temperature is below 0°C, frost or ice is formed by sublimation of the vapor.

Dew is very important in many arid regions of the world. In some areas, as much as 25 mm is deposited annually and is significant because the water can be directly absorbed by leaves. A single occurrence of "dewfall" can be as much as 0.3 mm. Under dry conditions, dew on the leaves increases cell turgor, which is important for growth. Excess dew can drip from plants and be absorbed by the soil. At or near freezing temperatures, the condensation of water as dew releases heat, which helps to prevent freezing, because when water vapor condenses to liquid, 540 cal of energy are released from each gram or milliliter of water.

The occurrence of dew can also be a disadvantage because it creates a high-humidity environment favorable for infection by pathogens.

Plant Ecological Classification According to Water Relationships

Hydrophytes

Hydrophytes are water-loving aquatic plants that normally grow in water or swamps. Some vegetable examples are taro, water convolvulus, water chestnuts, lotus root, and watercress; generally, most are tropical or semitropical plants.

Mesophytes

These are the most common of the terrestrial vegetables and prefer to grow in well-drained soils. Such plants wilt easily if stressed from low moisture.

Xerophytes

Xerophytes are the opposite of hydrophytes in preferring a drier habitat and are able to endure relatively long periods of low moisture without exhibiting wilt or damage. Some xerophytic species can lose from 50% to 75% of their moisture content without wilting; cacti are an example.

Physiology of Water in Plants

Water is the major constituent of plant tissues; it is the medium in which cell metabolic processes occur and the medium for transport between cells in plant tissues and organs. For these functions, relatively little water is actually required, but its availability is essential. It is estimated that less than 1% of the water that passes through the plant is utilized in the photosynthetic process. Nevertheless, in water-stressed plants, photosynthesis and growth are very much reduced. In fact, the photosynthetic rate is reduced well before the available soil moisture is depleted.

Most of the water absorbed by the plant is used in transpiration through minute leaf openings called stomata. As a result of transpiration, soil moisture and minerals are absorbed by roots and transported to other portions of the plant. Transpiration has a cooling effect on leaves, which is especially important in climates where air temperatures are very high.

Evapotranspiration is a term used to describe the combined evaporation from the soil surface, transpiration, and cuticular loss of water from plants. This is expressed as the rate of water loss from a given area and is useful in measuring and estimating the water requirement of growing plants. Factors influencing evapotranspiration include temperature, relative humidity, wind, plant leaf area, and the soil surface.

For most mesophytic vegetable species, little or no moisture stress during the entire growth period usually results in high yield and quality. Except for hydrophytes, flooding should be avoided, as this condition restricts oxygen to the roots and adversely affects growth.

LIGHT

Light is required for photosynthesis and, therefore, is another essential environmental component for plant life. Day length (actually the length of dark period) is important for some plants in inducing other physiological processes.

Light Intensity

Lux is a unit by which light intensity is measured. Maximum solar radiation on a clear cloudless day at noon varies from about 1.75 g cal/cm^2/min on high mountain tops to 1.50 g cal/cm^2/min at sea level, approximately 130,000 and 108,000 lux, respectively. Atmospheric conditions of smoke, dust, various gases, and clouds are commonly present and can reduce the amount of energy actually reaching the earth's surface.

Photosynthesis can occur at very low light intensities; although at about 5 lux, photosynthesis is negligible. The light compensation point for many plants is about 1000 lux. The compensation point is the light intensity at which the rate of photosynthesis equals the rate of respiration. The plant is light saturated when a further increase in light intensity does not increase photosynthesis. The light intensity at which saturation occurs can increase as the CO_2 concentration available to the plant increases. However, a point is reached where neither an increase in light intensity nor CO_2 concentration results in additional photosynthesis.

An individual leaf may be light saturated at an intensity of 30,000 lux, but the entire plant may not be light saturated at higher intensities because lower leaves may be below the light compensation point. In general, 10,000 lux is regarded as low light intensity, 50,000 or more as high intensity.

Light Quality

White light from the sun has a spectrum composed of many colors, from violet to deep red, with a wavelength range from 400 to 750 nm; this is the sensitive range of the human eye. Plants are responsive to a slightly wider range, from about 350 to 780 nm, which includes ultraviolet to far-red light.

Plant Physiological Responses to Certain Light Wavelengths

Response	Wavelength (nm)
Stem elongation	720–1000
Germination inhibition of certain seed	
Stimulation of onion bulbing	
Suppression of onion bulbing	650–690
Red pigment (lycopene) synthesis in tomato	
Flower stimulation of long-day plants	
Flower inhibition of short-day plants	
Promotion of germination of certain seeds	
Promotion of anthocyanins	
Photosynthesis	440–655
Chlorophyll formation	445–660
Phototropism	350–500

Duration of Light

Due to the earth's rotation on its axis, day and night occur, and because of the earth's tilt as it orbits the sun, the length of the light period at different times of the year varies from 0 to 24 h at the polar regions. At and near the equator, the length of the light period is relatively constant throughout the year. Figure 6.2 illustrates how day length changes with seasons at different latitudes.

The flowering response of plants to the relative length of the light period or its absence is called photoperiodism. Plants that develop and reproduce normally only when the photoperiod is less than a critical maximum are called short-day plants. Those that flower only when the photoperiod is greater than the critical minimum are long-day plants. For short-day plants, the duration of the dark period is the critical condition, rather than the duration of the light period. However, by tradition, the response is ascribed to the light period.

Phytochromes are plant pigments which absorb red and far-red light and are responsible for stimulating photoperiodic responses. For day-neutral plants, flowering appears not to be affected by photoperiod. Examples of vegetables having a long-day flowering response are spinach, some radishes, and Chinese cabbage. Some short-day vegetables are chayote, winged bean, and amaranth. Day-neutral vegetables include tomatoes, squash, beans, and many others.

In addition to flowering, there are other plant responses to photoperiod. For example, long days influence onion bulbing, and short days are responsible for tuber initiation of white potato and Jerusalem artichoke and the storage root enlargement of yams, cassava, and sweet potatoes.

Leaf Area Index

A knowledge of the photosynthetic efficiency of the leaf canopy of a plant is an important consideration in the evaluation of the dry matter production of a crop. In measuring efficiency, leaf canopy or leaf density is expressed as leaf area index (LAI). LAI is calculated as the total leaf area (leaf blades) subtended per unit area of land.

Most crops have an LAI ranging from 2 to 6; some monocotyledonous plants may have a leaf area index of 9, and some as high as 12. Crops with a vertical foliage orientation generally have a high LAI. The optimum LAI is not necessarily the maximum LAI. The lowest leaves usually are those with the least exposure to light and some could be below the compensation point. If so, these leaves will have to be partially maintained by the upper leaves, resulting in an overall decreased efficiency in total dry matter accumulation by the plant. Optimum LAI will vary with the intensity of solar radiation; it is larger for higher light intensities.

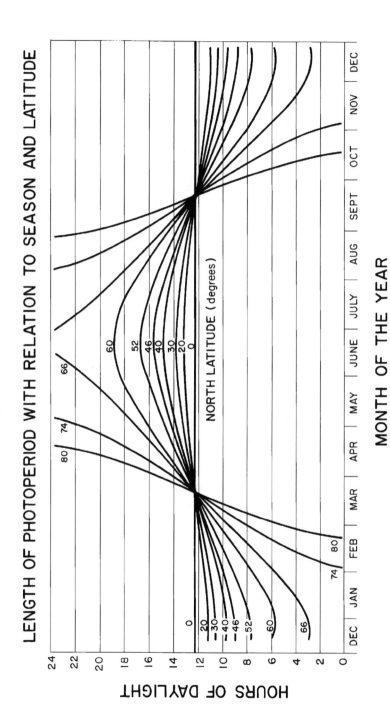

FIG. 6.2. Length of photoperiod relative to season and latitude.

TABLE 6.1.   WIND SPEED AND THE EFFECT UPON PLANTS

	Wind speed (km/h)	Effect
Calm	0–2	Smoke rises vertically
Light air	2–6	Smoke drifts
Light breeze	7–12	Leaf movement
Gentle breeze	13–19	Small twigs in motion
Moderate breeze	20–29	Small branches move
Fresh breeze	30–39	Small trees with leaves begin to sway
Strong breeze	40–50	Large branches move
Gales	51–100	At lower speeds, leaves are blown off; whole trees in motion. At moderate speeds twigs and branches break, and at high speeds, trees can be uprooted.
Storm	101–120	Severe damage; trees can be uprooted
Hurricane, typhoon	>120	Extensive damage; trees can be uprooted

WINDS

Winds are caused by differences in air pressure. Temperature differences produce pressure gradients which give rise to air movements. Air masses flowing from high- to low-pressure areas are winds. Wind speeds and possible physical effects on plants are shown in Table 6.1. In addition to physical movement, wind also affects plants in other ways, such as pollination and seed dispersal.

Winds affect atmospheric humidity by introducing humid or dry air masses, removal of humid air adjacent to leaf surfaces, increasing transpiration rate, and decreasing leaf temperature. Transpiration helps in the transport of mineral nutrients absorbed by the roots. When photosynthesis is rapid under a full leaf canopy, CO_2 can become limiting. Winds help to replenish the CO_2 supply to leaves, especially those deep within the canopy. However, strong winds can injure, break, and destroy above-ground portions of plants.

Winds causing the movement of warm or cold air masses can exert a considerable temperature change to the plant's growth environment. Wind is also important for pollen transport for certain species, functions in seed dispersion, and, in some situations, can carry vegetative portions to other locations where they serve as propagation material.

CLIMATE CLASSIFICATIONS USEFUL IN CROP ECOLOGY

Climate is the average course of weather at a location over a period of years. It is not just the average weather, as the variations from the mean are as important as the mean value itself. Weather components such as temperature, precipitation, light, and wind are variable from day to day and from season to season. Controls limiting the extent of these variations are latitude, the distribution of land and water, the semiper-manent low- and high-pressure locations responsible for wind patterns, storms, ocean currents, as well as altitude and mountain barriers.

The earliest and simplest classification is tropic, temperate, and frigid zones. However, this is not sufficient to study crop ecology. We need more details to learn about the climate of a particular region. Many geographers and climatologists have provided various classifications based on temperature, precipitation, seasonal characteristics, and natural vegetation. Many climatic classification systems have been proposed and applied. Other classifications using precipitation and evapo-transpiration data have been used in identifying suitable crop production locations. Also, the use of climatic analogs in which the comparison of climate similarities of one region are used to predict which crops of similar climatic characteristics might be successful at other world locations. Various technologies such as infrared aerial and satellite photography are employed to provide geographic and vegetation information useful to analyze and identify locations for crop production.

PHYSIOGRAPHIC AND EDAPHIC FACTORS

Physiographic factors include the topography (elevation and slope) of the land. Topography produces a marked effect on local climates. Mountain summits, slopes, and valleys each have different climates, which can differ within a relatively short distance. High-altitude locations have low air and soil temperatures and have greater wind velocities and exposure than valleys. High summits and slopes may get more or less rainfall than valleys, which can affect silting and erosion. Slopes have better drainage than level areas, and sun-facing slopes are warmer than those with less exposure to the sun.

Edaphic factors include the entire soil environment, its atmosphere, physical and chemical properties, and organisms. Plants are dependent on the soil for anchorage, moisture, and nutrients. Under natural conditions, soil characteristics may be of greater importance to plants than when under cultivation. Soil edaphic factors are altered by cultivation, fertilization, irrigation, drainage, and cropping.

Soil Composition

Soil is composed mainly of material derived from parent rock and developed largely through the interaction of the substratum with climate and living organisms. Typically, soils are composed of finely divided particles of modified parent material mixed with varying amounts of organic matter, ranging from 0% to 100%. A soil can be classified according to its texture. The texture is determined by the proportion of sand, silt, and clay. The particle sizes of these materials are 2.0–0.02, 0.02–0.002, and <0.002 mm in diameter, respectively. Figure 6.3 shows the various proportions of these three mineral fractions for different soil types.

Sand has very low moisture-holding capacity and is low in plant nutrients, whereas clay has very high moisture-holding capacity and usually a high mineral availability. Sand is important because it increases soil pore space which improves soil aeration. Generally, sandy loams are more easily managed than soils with a high clay content. Because of the tendency for clay soils to drain slowly, soils may be water saturated, a situation called waterlogging.

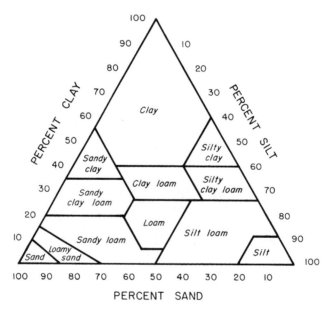

FIG. 6.3. Soil texture classified according to sand, silt, and clay content.

Soil Structure

Soil structure is the arrangement of an aggregation of soil particles. Soil aeration depends on soil structure. Good soil structure allows the adequate exchanges of CO_2 and O_2 needed for root activity. Moderately coarse textured soils have relatively large interstitial spaces (Fig. 6.4).

Compacted soils, often resulting from poor soil management, have reduced pore space and, accordingly, reduced rates of water penetration

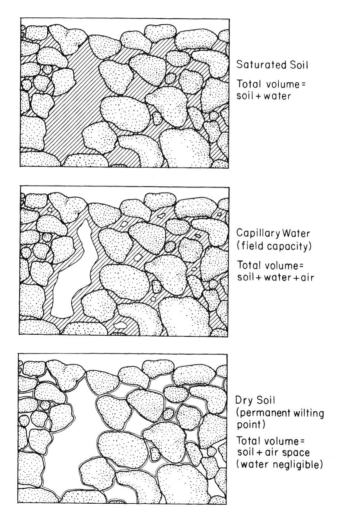

Saturated Soil

Total volume = soil + water

Capillary Water (field capacity)

Total volume = soil + water + air

Dry Soil (permanent wilting point)

Total volume = soil + air space (water negligible)

FIG. 6.4. Soil water content at different levels of moisture tension.

or drainage. Even with adequate water, plants grow poorly because roots are not well developed due to mechanical impedance, poor aeration, and excessive moisture.

Mineral Nutrients

Besides water, the soil is the main source of plant nutrients. Soil macronutrients include nitrogen, phosphorus, potassium, calcium, magnesium, and sulfur. Iron, copper, manganese, zinc, boron, cobalt, chlorine, and molybdenum are micronutients. Excessive amounts can cause plant toxicity, and insufficient amounts cause poor and/or abnormal growth. A proper balance of these nutrients is required for optimum growth and development.

Soil pH

An important factor is soil acidity or alkalinity measured as pH. Optimum pH for the growth of most vegetables is between 5.8 and 7.5. Some plants are "acid loving" and very few tolerate alkaline conditions. Soils in high-rainfall locations tend to be acid, whereas in arid conditions, soils frequently are alkaline. Depending on soil pH, certain mineral elements become more or less available, which can result in deficiencies or toxicities (Fig. 6.5). For example, iron and manganese become less available at high-pH levels, but in acid soils, these same elements can be available in excessive amounts.

Soil Temperature

Soil temperature is dependent on air temperature, and lags behind and is generally lower than ambient temperature in warm periods, whereas the opposite occurs during cold periods. However, it is possible for the temperature of dry soil surfaces to occasionally exceed air temperature.

Soil temperatures are critical for seed germination, root growth, and the development of underground storage organs. For example, when soil temperatures are above 30°C, tuber formation is suppressed for the white potato.

Soil Moisture

Soil moisture exists in various forms that include free, capillary, vapor, hygroscopic, and crystalline water. Plant roots can absorb free and capillary water, and even water vapor, but cannot utilize hygroscopic or crystalline water because of the strong adhesion to soil parti-

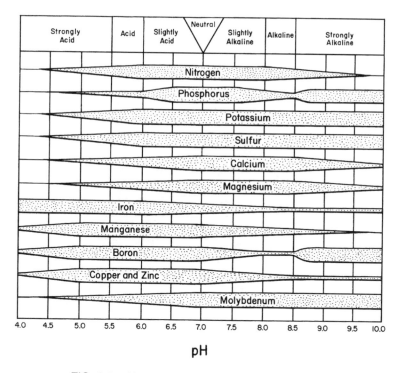

FIG. 6.5. Nutrient availability over soil pH range.

cles. In dry soils, some hygroscopic water can volatilize and become water vapor, but the quantity is too small to be significant for plant use. Crystalline water is held too strongly by chemical bonds to be available.

The soil is saturated when all the pores are filled with water. Free water readily percolates through the soil as the force of gravity acts upon it. After free water has drained, capillary water remains, and this condition is known as field capacity (FC), at about ⅓ atm diffusion pressure deficit (DPD). When most of the capillary water is depleted, the soil is at the permanent wilting point (PWP). At this level, plants cannot extract the remaining water from the soil, which is measured at about 15 atm DPD or moisture tension. These soil moisture phases are illustrated in Fig. 6.6.

Sandy soils hold very little water, but most of what is contained is available for plant use. Clays soils have a large water-holding capability. However, depending on the size of the clay colloids, a fair percentage of this moisture is hygroscopic water and is unavailable. This amount

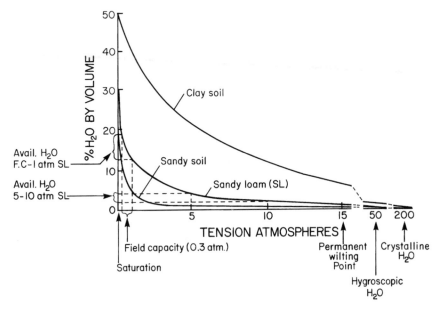

FIG. 6.6. Moisture content and energy required for water extraction from three soil types.

varies from 20% to 40% of total soil water. The different portions of soil water and their soil adhesive forces are depicted in Fig. 6.6.

In addition to the contribution to sustained crop growth, soil moisture has an important influence on seed germination. When moisture is abundant, most vegetable seed imbibe water and germination proceeds relatively rapidly. It is important to know the range of soil moisture in which seed can germinate; especially in low moisture conditions. For example, celery seed requires soil moisture to be at or near field capacity for germination. Other crops such as lima beans, peas, lettuce, and beets require relatively moist soils at the upper range of available soil moisture. Seeds of crops such as cucumber, carrot, onion, and snap bean will not germinate when soil moisture is near the PWP. Another group includes the seeds of many brassicas, cucurbit crops, and also tomato, pepper, and sweet corn, which can germinate over a broad range of available soil moisture from field capacity to near the PWP. These and other crop examples are not inclusive for any particular soil moisture grouping. It should also be apparent that at low soil moisture availability, the time for germination to occur is less rapid than when adequate moisture levels exists.

Excessive soil moisture for prolonged periods and compacted soils cause plant damage because insufficient oxygen interferes with water and mineral absorption by roots. In such conditions, it is possible for plants to wilt in spite of available free moisture. Furthermore, compacted soils are likely to have higher concentrations of ethylene, a plant growth regulator, that can affect plant functions.

Plant growth and photosynthesis are very much reduced before the PWP is reached. Lack of rain or other moisture deficit conditions may require irrigation to permit crop growth. Irrigation is supplied to crops by various application practices that include flooding, furrow, subbing, or overhead sprinkler methods. Drip irrigation is an application method that has increased from a relatively small beginning to become a major procedure. This method is able to effectively provide water uniformly with precision. Drip systems can reduce water consumption and may limit weed growth and some diseases. Additionally, the system can deliver some nutrients and pesticides during irrigation.

ROLE OF HUMANS IN AFFECTING THE ENVIRONMENT

Humans change the environment in whatever they do and this is inevitable as long as they inhabit the earth. Ideally, the changes made should not greatly disturb the ecological balance. As a result of human activities, the earth's atmosphere, soil, and water resources have changed and some changes have affected regional and global climates.

Fossil fuel usage has increased the CO_2 content of the atmosphere. Considerable speculation exists as to the future influence of the possible long-term increase of CO_2 as it might affect global temperatures and plant growth. The use of agricultural land for urban development has markedly changed local environments. It is not unusual for the temperature within urban areas to be several degrees higher than those in surrounding rural areas. Air pollution is injurious not only to human health but also to plants; some have a high sensitivity to pollution. Pollutants and particles in air can reduce the energy reaching the earth's surface, thus affecting photosynthesis and air temperatures.

Irrigation and drainage of land can cause profound changes in local climates. Intensively irrigated cultivation of some arid lands has resulted in measurable summer temperature decreases and humidity increases. Leaching of arid land has caused problems of excessively soluble salts to users of groundwater and downstream water. Additionally, fertilizers and pesticides, subjected to leaching, can increase pollution, especially of concern is nitrate nitrogen leaching into ground and surface waters.

Unless properly managed, clearing of new land for agriculture can create soil erosion problems. Heavy farm equipment increases soil compaction and adversely affects crop production.

Tropical rain forests play an important global role in maintaining the water, temperature, and CO_2 and O_2 balance that influences the earth's climate. In tropical rain forests where "slash and burn" agriculture is practiced, the soils become excessively leached and depleted of native nutrients. Such soils easily puddle because of impaired drainage and become difficult to manage. Land is abandoned after 4 or 5 years in favor of new land being cleared for cropping. Native vegetation may reestablish in the abandoned area, and after 15, 20, or more years, the soil might return to a manageable condition, but perhaps not. Soil erosion frequently occurs during this cycle. Proper management of the remaining natural land resources is especially important for maintaining the soil, water, heat, and gaseous balance that affects the overall climate of the earth.

SELECTED REFERENCES

Brady, N.C. 1984. The Nature and Properties of Soils, 9th ed. Macmillan, New York.

Craig, R.F. 1992, Soil Mechanics, 5th ed. Chapman & Hall, London.

Donahue, R.L., Miller, R.W., and Shickluna, J.C. 1983. Soils—an Introduction to Soils and Plant Growth, 5th ed. Prentice-Hall, Englewood Cliffs, NJ.

Ellis, S., and Mellor, A. 1995. Soils and Environment. Routledge, London.

Foth, H.D. 1984. Fundamentals of Soil Science, 7th ed. John Wiley & Sons, New York.

Frenkel, H., and Meiri, A., eds. 1985. Soil Salinity. Van Nostrand Reinhold, New York.

Hartmann, H., Kofranek, A., Rubatzky, V., and Flocker, W. 1986. Plant Science Growth, Development, and Utilization of Cultivated Plants, 2nd ed. Prentice-Hall, Englewood Cliffs, NJ.

Lockwood, J.G. 1985. World Climatic Systems. Edward Arnold Publ., London.

Okken, P.A., Swart, R.J., and Zwerver, S., eds. 1989. Climate and Energy. Kluwer Academic Publ., Dordrecht.

Sacher, J.A. 1973. Senescence and post-harvest physiology. Ann. Rev. Plant Physiol. 24, 197–224.

Salisbury, F.B. 1982. Photoperiodism. Hort. Rev. 4, 66–105.

Singer, M.J., and Munns, D.N. 1987. Soils—an Introduction. Macmillan, New York.

Solomon, A.M., and Shugart, H.H., eds. 1993. Vegetation Dynamics and Global Change. Chapman Hall, New York.

Wilsie, C.P. 1961. Crop Adaptation and Distribution. W.H. Freeman & Co., San Francisco.

Wyman, R.L., ed. 1991. Global Climate Change and Life on Earth. New York: Chapman and Hall.

Yamaguchi, M., Flocker, W.J., and Howard, F.D. 1967. Soil atmosphere as influenced by temperature and moisture. Soil Sci. Soc. Am. Proc. 31, 164–167.

7

Controlling Climate for Vegetable Production in Adverse Climates and During Off-Seasons

The production of crops at times other than during normal seasonal periods has only been practiced relatively recently. In the tropics and subtropics, many kinds of vegetables can be and are grown in abundance practically all year round, and usually little if any climate modifications is necessary. However, in temperate climates, certain crops are almost impossible to grow from late autumn through winter due to cold temperatures, and others are difficult to grow during high summer temperatures. In these regions some "off-season" vegetable production is possible when farmers are capable of modifying the temperature portion of a crop's growth environment.

For such off-season and certain specialized vegetable production, glass-covered or plastic-film-covered structures are used to provide protection and favorable climatic modifications. Plastic tunnels and various kinds of mulches are similarly used. Thousands of hectares of land are devoted to the culture of vegetables in these structures during the winter season, most notably in northern Europe, North America, and parts of Asia.

Three critical components in crop production are the plant which produces the desired product(s), the environment which supplies the energy and nutrition, and the human which manages the interrelationship. Factors affecting plant growth include light, temperature, moisture, carbon dioxide, nutrients, wind, pollutants, salinity, diseases, pests, weeds, and other competition. Humans exert management and control of these factors by supplementation, or minimizing their influence for the benefit of productive growth.

Reasons for modifying the plant microclimate is obvious to farmers, who must grow crops under adverse conditions. By providing a more favorable environment, the farmer can increase yield, quality, and economic return. Modification may allow a crop to be produced satisfactorily when it otherwise could not. For some situations, seasonal production can be extended by providing protection from adverse conditions and/or enhance growing conditions.

TEMPERATURE CONTROL AND ITS INFLUENCE

Topographical Effects

Although not feasible for large areas, some level of temperature control can be achieved by the management and utilization of natural local features, conditions, and microclimates.

Sun-facing hillside slopes in temperate zones receive more radiation during the day than level surfaces; conversely, those surfaces not facing the sun are cooler. Thus, the warmer surfaces are used for growing crops during cool weather and the cooler surfaces during warm or hot conditions. At night, cold air, being more dense than warm air, will drain down slopes, resulting in the cold air accumulating at the base of the sloping land while further up the slope, the air may be somewhat warmer. At times of threatening frost, a degree or two of temperature increase may be a significant difference.

Shape of Plant Beds

Although on a smaller scale, the same kind of effect as planting on hillside slope is achieved by planting on the side rather than the level or flat portion of a bed. The determination of bed or row orientation is another important factor. Row direction, whether east and west or north and south, depends on the position of the sun during the growth period. A diagrammatic representation of how the radiation is intercepted on the slope and top of the bed in the Northern Hemisphere is shown in Fig. 7.1. An example of soil temperature variation at different areas of a raised bed and how temperatures change during the day are shown in Fig. 7.2.

Wind Barriers and Shelter Belts

Winds can injure plants by physically whipping or blasting them with sand and dust. Wind barriers and shelter belts help to alleviate such damage. Shelter belts are barriers, often of living plants (trees and shrubs), of sufficient height to present an obstruction to winds

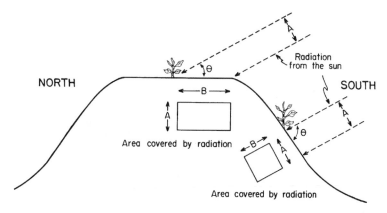

FIG. 7.1. Amount of radiation at top and side of a raised bed running east–west in the Northern Hemisphere. The illustration shows that the surface perpendicular to the sun receives a greater amount of radiation per unit of surface area.

(Fig. 7.3). Even a single row or corn or barley can serve to reduce wind velocity. Materials other than living plants can be used for the same purpose.

Decisions concerning bed or row orientation should consider the direction of prevailing winds. A good shelter belt can reduce the wind velocity 50% at a distance of 10 times the height of the barrier. Temperatures behind these barriers can be enhanced by 1.5–4°C depending on the

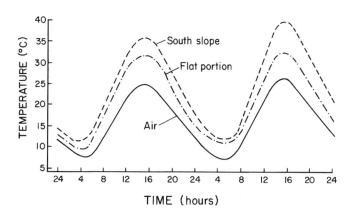

FIG. 7.2. Typical spring soil temperatures at 12 mm depth in raised beds running east–west in the Northern Hemisphere.

FIG. 7.3. Windbreak constructed with natural materials.

type of barrier and soil and moisture conditions. At night, however, more heat can be lost by radiation, particularly if the barrier is dense and causes the air to be stilled. Nevertheless, some increase in yields and early maturation of crops can be obtained by use of various wind barriers.

Factors Influencing Soil Temperature

Soil Color
Light-colored soils tend to be cooler than dark-colored soils.

Texture
Coarse soils tend to warm sooner than fine textured soils, and soils high in organic matter do not warm as fast as mineral soils.

Moisture
Moist soils warm up more slowly than dry soils because water has a higher heat capacity than soil. Conversely, moist soils cool off more slowly than dry soils. Evaporation of water from the soil surface has a cooling effect.

Mulches

Various kinds of mulches are used to modify soil temperature and moisture relationships. With a mulch of pulverized soil, the air between soil particles acts as insulation, and porous soils conduct less heat than compacted soils. Straw and similar relatively dry vegetative matter have been used as mulching materials for centuries. Such materials usually have a greater insulating effect than pulverized soil. On hot days, soil temperatures under a straw mulch can be as much as 17°C lower than another without mulch, and on some occasions during cold days it can maintain soil temperature above that of ambient air. Various materials are used to block or limit heat radiation loss during the night.

Various kinds of plastic films are used as mulches; clear, transparent plastic films generally produce more soil warming than black or opaque films (Fig. 7.4). Black plastic films are preferred for their effectiveness in suppressing weed growth. Other kinds of mulches are water emulsions of asphalt, vermiculite, and aluminum foil; the first two provide soil anticrusting properties, and the latter also has effectiveness as an aphid repellent to reduce insect damage and vector-transmitted viral diseases. Many different plant and nonplant materials are also used as mulches.

Frost Protection

Vegetable growers are occasionally confronted with frost and the potential of freeze damage. Various practices are used to avoid and minimize low-temperature injury, and, within limits, protection against this occurrence is possible.

Because water has a high heat capacity (1 cal/g), it can be effective in preventing frost and some freezing. Generally, the temperature of water is higher than air temperature on frosty nights and heat is released when water freezes. When 1 g of water freezes, 80 cal of heat are released to the immediate surrounding environment, preventing the temperature from dropping lower. Fogging or sprinkling plants with water can be effective in preventing freezing. This is especially useful when the temperature drop is rapid. The same principle applies with flooding or furrow irrigation although this use is less effective. Application should be continuous when air temperature is below freezing. Ice can form during fogging or sprinkling, and a heavy layer can cause crop damage from the weight of ice alone.

Another method is to add heat directly into the environment. Open fires or various kinds of heaters and/or smudge pots can warm the air. The smoke from combustion acts like clouds in hindering heat loss by

FIG. 7.4.   (a) Black plastic film used to raise soil temperature, retain soil moisture, and control weeds. (b) Clear plastic film covering a shallow soil trench in which early seedling plants are grown. The clear film is more effective than black film for increasing soil temperature, but can also increase weed growth. The film appears opaque due to moisture condensation.

radiation from plants and the ground to the sky. Because the volume of air that can feasibly be heated will be relatively small, heaters are most effective when there is a temperature inversion layer about 9–15 m above the surface. The inversion acts to trap the heated air beneath. Temperature inversion is a condition in which warm air overlies a cold air mass; this is contrary to normal conditions where the air gets colder with an increase in altitude.

Another way to transfer heat into the environment is by the movement and/or mixing of warm with cold air. For this purpose, air is moved with various kinds of large fans (wind machines) and in some situations with helicopters. Mixing with the inversion layer would result in some warming of the air near the plants. When heaters and wind machines are used together, more protection is achieved. Even under a strong temperature inversion, the gain in ground temperature is small, usually less than 3°C. This may be enough to avoid freezing. At temperatures of less than −3°C, it is unlikely that wind machines will provide adequate protection.

Plant Protective Materials and Growth-Favoring Structures

Mulches were probably among the first materials used when humans first became aware that plant growth could be improved by protection against an adverse environment. Some of the earliest protective forms were probably barriers against wind and used as heat traps, as well as to reduce radiation heat loss. Brushing or barriers made of plant materials such as reeds, leaves, or other plants are used to prevent winds from cooling soil and the immediate plant environment as well as serving as a heat trap (Fig. 7.5). Row covers, usually comprised of a porous nonwoven fabric, are placed directly over the crop during early growth. Their effect is to raise the temperature under the cover and to provide some protection from insects. The covers are removed when

FIG. 7.5.   Brushing acts much like a windbreak, but when angled toward the sun it will trap heat and improve heat retention by retarding night radiation.

ambient temperatures are satisfactory and/or the covers interfere with crop development. When properly installed, these devices are effective.

Evaporative cooling from wet soils and surfaces can be used to keep temperatures near plants lower than ambient air temperatures. This is possible because about 540 cal of heat is necessary to vaporize 1 g of liquid water to gaseous water. The rate of cooling is affected by relative humidity; at high humidity, less cooling can occur.

Hot caps and cloches provide protective enclosures for individual or groups of plants. Hot caps are tent-shaped covers made of glassine paper or plastic placed over one or several seedlings or small plants (Fig. 7.6). These have the effect of a miniature glasshouse, and being translucent, they allow light to enter. They also act as heat traps and reduce radiation. Cloches are similar and usually consist of small glass panels that are wood or wire braced and therefore are rigid and can withstand strong winds. These can be disassembled for reuse, whereas hot caps are generally disposable.

Shading devices protect plants from heat and undesirable light intensities. Baskets, shade covers, lath, and screen houses are used to reduce light intensity and accompanying temperature; screen houses also serve to exclude insect pests. Whitewash applications have been used to protect tomato and melon fruit from damage caused by intense sunlight; disagreement exists as to the benefits of such applications.

FIG. 7.6.   Hot caps; note oil-fed burner used for frost protection.

FIG. 7.7.   Low plastic tunnel constructed with two sheets of clear film over wire hoops which allows for ventilation by easily lowering one side.

Plastic tunnels have come into wide usage because they are easily constructed and relatively inexpensive. These are made of one or more sheets of plastic film supported on bamboo, wire, or wood to cover or enclose plants (Fig. 7.7). Films may be perforated or opened when necessary for ventilation. Generally less than 1 m in height, such tunnels may be of varying lengths.

Cold frames are semipermanent structures frequently made of wood or concrete and topped with a clear glass or plastic cover (Fig. 7.8). The sun's heat is trapped within the structure, making the air several degrees higher than the outside air as well as warming the soil. Additionally, straw or cloth covers can be used to reduce heat loss at night. The structure is ventilated by raising the cover slightly.

Warm frames (hot beds) are similar to cold frames but have provisions for heating. Heat sources may be decomposing manure or compost, or subsurface heating pipes or electric cables.

Glasshouses

Glasshouses, often called greenhouses, are not always constructed with glass. With the advent of large and wide fiberglass panels and plastic films, many plant growth structures were and continue to be

FIG. 7.8.   Cold frames used to accelerate early seedling growth.

constructed using these materials. Although less permanent than glass-houses, these structures are relatively comparable in function and their construction is less complicated and less expensive.

Often these structures are large enough for workers to easily maneuver and some are large enough for the operation of cultural equipment. Many provide heating, cooling, and other environmental management, often with sophisticated and automated instrumentation.

Heat can be applied as with warm frames, but circulated heated air or radiant heating is more common. To reduce night radiation heat loss, heat blankets or curtains are used. The interior space can be cooled with evaporative water coolers, and although not common, refrigerated air conditioning is used. Additional cooling may be provided with whitewash and other shading materials. Automatic controls for temperature, CO_2, relative humidity, ventilation, irrigation, and fumigation are also provided in some facilities.

TRANSPLANTING AND HARDENING

Plants can be started in various kinds of plant growth structures early in the season and transplanted when outdoor conditions become favorable. Transplanting of bare-rooted plants can be a severe shock to rapidly growing seedlings. To reduce transplant shock and better accommodate establishment, plants are hardened. Hardening before transplanting can be accomplished by withholding water and/or by exposure to lower temperatures. Hardened plants can withstand unfavorable conditions better than rapidly growing succulent transplants. For example, nonhardened cabbage seedling can be injured at $-2°C$, but hardened ones are not injured at $-6°C$.

ADDITIONAL REGULATION OF PLANT GROWTH

Control of light is an important factor for plant regulation. When light intensity is low, supplemental lighting is often necessary. Light intensities of at least 8,500–11,000 lux are required to grow most plants to maturity. Maximum efficiency for photosynthesis varies with temperature and light intensity.

Photoperiod

Alteration of photoperiod (day length) is another method of plant regulation (Fig. 6.2). Crops will mature earlier under long days because of longer time for photosynthesis than when grown under short-day conditions.

Day length also affects physiological processes such as flowering. With regard to flowering, some plants respond to light intensities of less than 550 lux. A long day of 18 h or more can be achieved by a 3-h light period in the middle of the dark period or by a flashing light for several seconds every minute during the dark period. Day lengths can be reduced by covering the plant with a dark cloth or moving it into a dark chamber.

Photosynthetic efficiency can also be increased by raising the CO_2 concentration. This can best be provided in an enclosed structure. Field applications, although beneficial, usually are less effective and generally not feasible. Nevertheless, where possible, arrange for air movement to help replenish the CO_2 supply in the plant vicinity.

Growth-Regulating Chemicals

The application of growth regulating chemicals can also alter plant growth. Such products are used to regulate plant height, flowering, and fruit set. During cool temperatures when pollinating conditions are poor, materials such as p-chlorophenoxyacetic acid (PCPA) and naphthaleneacetic acid (NAA) are applied to tomato flowers to improve fruit set. Ethephon, 2-chloroethyphosphonic acid, an ethylene-releasing compound, can increase the formation of female flowering in cucumbers and melons and hasten fruit ripening of climacteric vegetables such as tomatoes and melons. These vegetables are recognized by the characteristic rise in respiration rate of the fruit during ripening. Materials such as gibberellic acid and silver nitrate can preferentially induce male flowering of some cucurbit species.

Use of Insects for Pollination

Colonies of bees are used for pollination of cucurbits and other vegetables. Flies are useful for pollination in the seed production of Apiaceae and *Allium* species. In some situations, other means to effect pollination can be used. Hand-held vibrators and various kinds of shaker and/or blower devices are used in glasshouses as well as outdoors to perform pollination that would normally be accomplished by insects or wind.

Anticrusting Materials

To facilitate seedling emergence through soil crusts, anticrusting materials help improve soil aeration and lessen the incidence of root diseases. Gypsum is a common material applied to seed beds to minimize crusting. Other products such as vermiculite can be used alone or in combination with other products. These are typically applied with or to cover the seed (Fig. 7.9).

Enhancement of Seed Germination

Practices that improve seed germination are useful in modifying the earliness and success of planted crops. Seed priming is useful in this regard. In this procedure, seed are placed in a medium, which allows hydration and the initiation of the germination process. However, the osmotic level of the medium is sufficiently high so that germination cannot be completed. The seed are dried and can be used immediately or at a later time. This process improves earliness and uniformity of germination. Additionally, conditioning can improve performance during harsh germination conditions of high or low temperatures.

Climate Management Affecting Plants

Climatic conditions strongly influence the incidence of plant pests such as diseases, insects, and weed growth, usually to the detriment of crop production. High relative humidity, excessive soil wetness, and waterlogging are conducive to disease occurrences. Avoiding such condi-

FIG. 7.9.   Asphalt mulch serving as a soil anticrusting agent and it increases soil heat absorption, which enhances seed emergence and seedling growth.

tions enables producers to better manage disease control. High temperatures and humidity conditions similarly favor rapid insect development, and the ability to manage environmental conditions can enhance yield and/or limit crop losses. It is clear that climatic management of the environment can lessen various pest problems, and although not discussed, their importance should not be discounted.

SELECTED REFERENCES

Bezdicek, D.F., and Power, J.F. 1984. Organic Farming: Current Technology and its Role in Sustainable Agriculture. American Society of Agronomy, Madison, WI.

Brady, N.C. 1984. The Nature and Properties of Soils. Macmillan, New York.

Charles-Edwards, D.A., and Rimmington, G.M. 1986. Modeling Plant Growth and Development. Academic Press, New York.

Edens, T.C., Fridgen, C., and Battenfield, S. 1985. Sustainable Agriculture and Integrated Farming Systems. Michigan State University Press, East Lansing.

Hartmann, H.T., Kofranek, A.M., Rubatzky, V.E., and Flocker, W.J. 1988. Plant Science—Growth, Development, and Utilization of Cultivated Plants. Prentice-Hall, Englewood Cliffs, NJ.

Levitt, J. 1980. Responses of Plants to Environmental Stresses. Vol. 1, Chilling, Freezing, and High Temperature Stresses, 2nd ed. Academic Press, New York.

Levitt, J. 1980. Responses of Plants to Environmental Stresses. Vol. 2, Water, Radiation, Salt and Other Stresses, 2nd ed. Academic Press, New York.

Lockwood, J.G. 1985. World Climatic Systems. Edward Arnold Publ., London.

Loomis, R.S., and Connor, D.J. 1992. Crop Ecology: Productivity and Management in Agricultural Systems. Cambridge University Press, Cambridge.

Marachner, H. 1986. Mineral Nutrition in Higher Plants. Academic Press, New York.

Nakamura, S., Teranishi, T., and Aoki, M. 1982. Promoting effect of polyethylene glycol on germination of celery and spinach seed. J. Japan Soc. Hort. Sci. *50,* 461–467.

Neiburger, M., Edinger, J.G., and Bonner, W.D. 1982. Understanding Our Atmospheric Environment, 2nd ed. W.H. Freeman & Co., San Francisco.

Okken, P.A., Swart, R.J., and Zwerver, S., eds. 1989. Climate and Energy. Kluwer Academic Publ., Dordrecht.

Rosenberg, N.J., and Blad, B.L. 1983. Microclimate, The Biological Environment, 2nd ed. John Wiley & Sons, New York.

Solomon, A.M., and Shugart, H.H., eds. 1993. Vegetation Dynamics and Global Change. Chapman & Hall, New York.

Teare, I.D., and Peet, M.M. 1982. Crop–Water Relations. John Wiley & Sons, New York.

Wyman, R.L., ed. 1991. Global Climate Change and Life on Earth. Chapman & Hall, New York.

Part III

World Vegetables

8

Global View of Vegetable Usage

Food usage is influenced by its availability (supply and cost) and also by social and religious customs. Throughout history, access to food and its supply was a major human concern, and remains as a high priority. In many nations and regions of the world, food-consumption patterns are relatively fixed, and changes are relatively few and slow to occur. Food use generally closely follows those of preceding generations, whereby people continue to consume the same kinds of foods that were eaten during childhood.

In early human history, the food-producing and food-consumption patterns of various peoples throughout the world was fairly well established. Exceptions occurred because of travel, relocation, and sometimes as a result of invasion and occupation by enemies. One result of European exploration of the Americas was the introduction of many plant species from New World regions into the Old World and vice versa. Other plant interchange also occurred within and between Asian, African, and European regions prior to and following the pre-Columbian period. With extensive exploration and colonization by Europeans and others, crops from different parts of the world were introduced into regions other than their origin.

Mixing of different cultures, through travel and intermarriage, brought about additional changes in local and regional food patterns and the establishment of a greater mixture of crops, whether endemic or introduced. An example of such a changing pattern is Brazil, where the endemic vegetables of cassava, corn, and beans consumed by the native population are now complemented with vegetables, such as onions, garlic, carrot, and brassicas introduced by Portuguese traders and immigrants. Additional vegetables such as okra and watermelon were introduced along with the importation of slaves from Africa. In more recent times, vegetables common to the Orient have been introduced by Japanese immigrants to Brazil. Recent Asian and Hispanic immigrants have brought about dramatic changes in the diversity of vegetables consumed in the United States, especially in the localities where immigrant populations are concentrated. Many other countries can parallel changes similar to those experienced in Brazil and the United States.

Food use patterns in many parts of the world are rapidly experiencing changes due to the continuously increasing mobility of people. Broadening of trade and improvements in rapid transport and effective postharvest handling and processing of various vegetables will further contribute to changes in world food demand and supply.

SOURCES OF VEGETABLES USED FOR FOOD

The five variations of vegetable food sources suggested by the horticulturist G.J.H. Grubben are as follows:

1. Wild plants collected from spontaneous vegetation of which an estimated 1500 species are used. These are important in primitive areas of developing countries, and in a few parts of the world, they are a primary vegetable source.
2. Wild vegetables growing as spontaneous "weeds" in food crops or sometimes in protected compounds. These number about 500 species and are natural selections of what might be considered primitive cultivars. A small percentage of these comprise some of the vegetables used in developing countries.
3. About 200 species, of which many are primitive cultivars, are grown in home gardens and as mixed crops in fields. These comprise about 40% of the vegetables in developing countries and about 15% of those in developed countries.
4. Other vegetable species grown on a small scale and for labor-intensive market production are usually grown in monoculture, but where land is scare, they may be grown in multiple-cropping systems. In developing countries, these plant types comprise about 40% of the vegetable volume produced. In developed countries, they account for less than 10% of the total vegetables produced.
5. Relatively few vegetable species, perhaps as few as 35, are cultivated in highly intensive production systems. A small percentage of the total production of these vegetables is produced in developing countries, whereas they represent over 75% of the vegetable production in developed countries.

Listing the many wild and weedy vegetable plants referred to in sources 1 and 2 suggested by Grubben is beyond the intent of this book.

Several publications that address some of these plants are included at the end of this chapter. Many vegetables representative of those Grubben likely would include in his suggested sources 3 to 5 are discussed in the chapters that follow and are listed in Appendix Table B.

STARCHY VEGETABLES

Starchy vegetables are very important in supplying the energy food needs of many populations in tropical and semitropical regions and to a large extent also those in temperate zones. The principal root crops are sweet potatoes and cassava. The major tuberous crops are white potatoes and yams. Also important are the corm crops, of which taro and tannia predominate. Significant rhizome crops include arrowroot and canna. Important starchy fruits include plantain, starch banana, breadfruit, jackfruit, and sweet corn.

Table 8.1 shows the world production of major starchy vegetables and their relationship to major cereal crop production. Only in Africa are starchy vegetables produced in greater volume than cereals. The contribution of starchy vegetables, although overall smaller than cereals, is substantial and significant. Also important is that in many situations these starchy vegetables are less costly to produce. The efficiency of food calorie production for some starchy vegetable crops as measured by land, labor, and other production costs is superior to the production efficiency of several major cereals (Fig. 9.8). However, continuing to favor cereal production are many alternative food uses for animal feed as well as for various processed industrial products. Similar alternatives for many starchy vegetables is less frequent. However, white potato and cassava are increasingly being used for fuel alcohol production and other industrial purposes.

Among the starchy vegetables, the white potato is by far the most

TABLE 8.1. COMPARISON OF ENERGY, PROTEIN, AND CARBOHYDRATE VALUES OF STAPLE STARCHY VEGETABLES

Crop	Values per 100 g of edible portion			
	Water %	Energy (kcal)	Protein (g)	Carbohydrate (g)
White potato	78	82	2.1	18
Sweet potato	70	118	1.6	27
Cassava	60	148	1.2	35
Yams	70	108	1.9	25
Taro	73	99	2.0	23
Plantain	65	127	1.1	31

important; world production in 1994 was 265 million tons. Cassava followed with 152 million tons, and sweet potatoes with 124 million tons. Each of these food staples is important in different parts of the world—the white potato in temperate regions of North America and practically all of Europe. The sweet potato predominates in central and southern Asia, and cassava in central Africa, many areas of South America, and some southeast Asian countries. Yams and aroids are similarly largely confined to specific regions, their world production being about 30 million and 5.8 million tons, respectively. These crops are especially important in supplying the caloric and protein needs in certain areas of the tropics (Table 8.1)

The world production of sweet corn is difficult to ascertain. Sweet corn is an important commodity in the United States, where about 2.5 million tons was processed in 1993, and about 860 thousand tons was produced for the fresh market. World maize production in 1994 was about 570 million tons, but it is estimated that a small percentage of that production was probably harvested and consumed at an immature stage as a fresh vegetable. Considering the volume of maize produced, even a small percentage is a significant amount.

Banana and plantain production in 1994 was approximately 52 million and 29 million tons, respectively. A portion of immature dessert-type bananas are also used in many countries as a starchy cooked vegetable. Plantains and cooking bananas are predominately consumed in tropical regions and are a significant component in satisfying caloric requirements. The nutritive values of starchy and other vegetables are provided in Appendix Table C.

SUCCULENT VEGETABLES

Numerically, most of the worlds vegetables belong in the succulent category being relatively high in water content and low to intermediate in caloric content. However, a decided advantage is that many are high in human nutritional components such as pro-vitamin A, vitamin C, other vitamins, various minerals, and fiber. Legumes and many green leafy vegetables are good protein sources.

Many kinds of succulent vegetables are used by people throughout the world. Those that have the broadest usage depend on their availability, cost, and acceptance by consumers. A large proportion of succulent vegetables are tropical or subtropical in origin. A classical example of crop expansion is the tomato. Although of tropical origin, tomatoes are now widely grown throughout much of the world. Many other crops

TABLE 8.2. WORLD PRODUCTION OF ROOT AND TUBER VEGETABLES AND CEREALS

	Tuber and root vegetables			Cereals		
	Area (10³ ha)	Yield (t/ha)	Production (10³ t)	Area (10³ ha)	Yield (t/ha)	Production (10³ t)
World	48,419	12.03	582,687	689,146	2.83	1,950,599
Africa	16,018	7.65	122,519	84,676	1.24	104,737
North and Central America	1,274	22.68	28,889	95,629	4.57	437,420
South America	3,569	12.34	44,038	35,235	2.48	87,330
Asia	16,993	14.17	240,732	302,908	2.97	898,644
Europe	3,939	19.84	78,125	62,605	4.17	261,214
Oceania	279	11.42	3,184	11,743	1.30	15,278

Source: 1994 FAO Production Yearbook, Vol. 48. FAO, Rome, 1995.

parallel this example of being cultivated more extensively in areas other than those of their origin.

To give some prospective of the importance of vegetables, Table 8.2 and 8.3 present the world production and average per capita consumption of vegetables according to FAO crop reporting statistics. These do not include a large amount of unreported production and consumption.

In many situations, a greater consumption of vegetables would improve the nutrition of some populations of the developing more than those of the developed regions. Table 8.3 indicates that annual vegetable consumption per capita is considerably less in much of Africa and many regions of Asia, South America, and Oceania compared to Europe. Even though the cereal crops are the major source of concentrated food energy, starchy vegetables also contribute a significant supply. In temperate regions, the white potato is the principal source, whereas in warmer regions, other tuber and root crops dominate.

TABLE 8.3. WORLD PRODUCTION AND CONSUMPTION OF VEGETABLES AND MELONS

	1978		1994	
	Production (10³t)	Consumption (kg/capita/yr)	Production (10³t)	Consumption (kg/capita/yr)
World	340,342	79	485,550	86
Africa	22,217	49	34,076	82
North and Central America	33,182	91	46,861	104
South America	11,405	48	16,018	51
Asia	179,076	71	291,110	87
Europe	63,739	132	67,322	133
Oceania	1,633	73	2,650	94

Source: 1978 FAO Production Yearbook, Vol. 32. FAO, Rome; 1979 and 1994 FAO Production Yearbook, Vol. 48. FAO, Rome, 1995.

Overall, consumption in developed regions is about twice that of developing regions. Vegetable utilization is increasing because of better education and communication about the health benefits that many vegetables provide. This awareness appears to be more evident in the developed regions.

SELECTED REFERENCES

Altman, D.W., and Watanabe, K.N., eds. 1995. Plant Biotechnology Transfer to Developing Countries. R.G. Landers Co., Austin, TX.

Eicher, C.K., and Staatz, J.M., eds. 1990. Agricultural Development in the Third World, 2nd ed. John Hopkins University Press, Baltimore, MD.

Grubben, G.J.H. 1976 and 1994. Personal communications. Centre for Plant Breeding and Reproduction Research (CPRO-DLO), Wageningen, Netherlands.

Kula, E. 1994. Economics of Natural Resources, the Environment and Policies, 2nd rev. Chapman & Hall, London.

Loomis, R.S., and Connor, D.J. 1992. Crop Ecology: Productivity and Management in Agricultural Systems. Cambridge University Press. Cambridge.

Martin, F.W., and Ruberte, R.M. 1975. Edible Leaves of the Tropics. Antilliam College Press, Mayaguez, Puerto Rico.

Solomon, A.M., and Shugart, H.H., eds. 1993. Vegetation Dynamics and Global Change. Chapman & Hall, New York.

Unklesbay, N. 1992. World Food and You. Food Products Press and Haworth Press, New York.

Part A

Vegetables Consisting of Starchy Roots, Tubers, and Fruits

9

White or Irish Potato

Solanum tuberosum L. subsp. *tuberosum*
Family: Solanaceae

Origin and Development

The potato originated in the Andean regions of Peru and Bolivia. The Incas utilized potatoes at least 2000 years before the arrival of Spanish explorers. Carbon[14] dating of starch grains found in archaeological excavations indicated potatoes were used at least 8000 years ago.

Potato plants, wild and cultivated, have good *in situ* survival because tubers have a high moisture content, and starch and other nutrient reserves that allow regeneration. Unharvested tubers could remain dormant in the soil and sprout when growing conditions became favorable; thus, continued survival was assured. During early development and in primitive circumstances, the ability to store and preserve gathered tubers enhanced their usefulness as a food crop. As an example, *chuno,* a dehydrated storable potato product, was made by foot tramping and natural drying of tubers by repeated freezing and thawing in some high-elevation Andean regions.

The name "potato" is believed to be derived from the Inca name "papa"; the sound-alike, batata, a Caribbean Indian name for sweet potato is also suggested. The association with Ireland is thought to be responsible for the name "Irish potato," which is retained even though potatoes are grown in many countries. White potato is the most common name. Although some cultivars are white fleshed and have white skins (periderm), that name does not account for the internal and external color variations that occur. Nevertheless, although neither white nor Irish is accurate, that association persists.

The potato's introduction into Spain about 1570 from South America lead to an unimaginable growth and distribution of a new food crop with profound economic and historical results. From Spain, the potato was taken into neighboring European countries and in less than 100 years was being grown fairly extensively in many regions of Europe. Distribution beyond Europe soon occurred with the introduction into India about 1610, China in 1700, and Japan about 1766. Scotch-Irish immigrants introduced the potato into North America in the early 1700s.

However, before wide adoption and acceptance of the potato in Europe, there was considerable skepticism concerning its edibility. When first introduced into Europe, the potato was regarded as poisonous because of its foliar resemblance to nightshades (*Solanum* species). The tuber was considered unfit to eat, or fit only for the very poor and as animal feed. Acceptance was also poor because of low productivity. Andean introductions (*Solanum tuberosum* subsp. *andigena*) obtained from low-latitude regions performed poorly because they were not adaptable to European temperate latitudes, although in southern European regions, productivity was better. Herbarium specimens and drawings indicate that Andean and not Chilean (*S. tuberosum* subsp. *tuberosum*) were first introduced and that Chilean sources were not present until the 19th century.

During its early introduction, rural populations and tenant farmers in some European countries were encouraged, and sometimes even compelled, by landowners to produce potatoes. At the beginning of the industrial age, the crop became a subsistence staple for the peasant population. Its value as a human food soon was recognized, along with the potential to produce more calories at a lower cost than grain crops. Therefore, potatoes were increasingly grown to meet the food needs of the expanding European population.

The increased dependency on this food source resulted in an extension of production areas, a development that contributed to the severity of the potato crop failure and resulting Irish famine during 1845 and 1846. Many years of extensive cultivation, especially in Ireland, with limited crop rotation and increased land area in potato production made the potato crop highly vulnerable to diseases. The late blight fungus, *Phytophthora infestans,* became established, and with favorable conditions, fungal populations increased to epidemic proportions, partly because the most susceptible potato stocks then being grown were of Andean ancestry. About one million people died of starvation in Ireland during this period. Because of the famine, massive migrations of the population occurred, as well as considerable economic disruption in Ireland and other European countries. An effect of the crop failure was the introduction during the 19th century of better-adapted Chilean potato types replacing the initial Andean sources. This formed the genetic base which is now referred to as *S. tuberosum* subsp. *tuberosum.*

Taxonomy

Solanum is a large and diverse genus of over 1500 species, about 90 of which are tuber producers and very few are cultivated. The range of wild tuber-bearing *Solanum* species extends from southwestern United States, into highland areas of Mexico, Central America, and throughout the Andes with extensions into Chile and Argentina. Tubers of wild species are generally small and bitter tasting because of the alkaloids they contain. For the potato, most genetic diversity is found in the Andean regions of Peru and Bolivia. Wild tuber-bearing and the primitive cultivated species provide a large diverse gene pool to broaden adaptation for disease resistance and for other heritable characteristics to improve the cultivated species, particularly *S. tuberosum*.

The wild diploid *S. leptophyes* found in north central Bolivia may have been the primary ancestor of the cultivated species. The diploid *S. stenotomum* probably is a cultigen of *S. leptophyes* that, through natural introgression (hybridization) with other species and possibly some human involvement during cultivation, resulted in other cultigens, one being *S. tuberosum*, which is not found in the wild state.

S. tuberosum subsp. *tuberosum,* the potato of world commerce, is a tetraploid ($2n = 2X = 48$). In addition to *S. tuberosum* subsp. *andigena,* other recognized cultivated species are *S. anjanhuiri* (2X), *S. chaucha* (3X), *S. curtilobum* (5X), *S. juzepczukii* (3X), *S. phureja* (2X), *S. stenotomum* (2X), and *S. goniocalyx* (2X). Within the genus *Solanum,* about 70% of wild forms are diploids ($2n = 24$), which are mostly self-incompatible and about 15% are tetraploids, of which most are self-fertile.

S. tuberosum originated in the southern Peruvian and Bolivian highlands. It has been proposed that under cultivation by prehistoric Indians, it evolved as a tropical high-elevation subspecies classified as *S. tuberosum* subsp. *andigena* and a temperate-like lowland subspecies identified as *S. tuberosum* subsp. *tuberosum.* The latter, having a long-day tuberization response, would have better adaptation to European temperate climates. However, it is known that through selection, *S. tuberosum* subsp. *andigena* is capable of long-day adaptation.

Botany

Potatoes are dicotyledonous short-lived perennials that are typically cultivated as annuals for their edible enlarged underground tubers. Plants that are asexually produced from tubers develop finely branched, relatively shallow, and fibrous spreading adventitious roots; whereas plants grown from true seed form a slender taproot with many laterals. Above-ground stems are erect and initially smooth and become angular

and branched with continued growth. Plant growth ranges from compact to sprawling. Compound pinnate leaves with petioled leaflets vary
in size, shape, and texture. Underground stems (rhizomes), commonly
called stolons, accumulate and store the product of photosynthesis in
tuberous swelling near the terminus of the stolons (Fig. 9.1). Carbohydrates are translocated as sucrose to the stolons, where cell division
and enlargement result in tuber growth; the transported sucrose is
converted and stored in starch grains.

Flowers, clustered in a primary cymose inflorescence, have a five-
lobed fused corolla with colors ranging from white to pink to bluish
purple. Being nectarless, they are mostly cross-pollinated, usually by
wind, but insects can also pollinate. With some cultivars, floral abortion
frequently occurs, so fruit set is sparse. Fertilized flowers produce small
spherical green or purplish berries which are toxic because of the
glycoalkaloids they contain. Small flat seed, few to several hundred

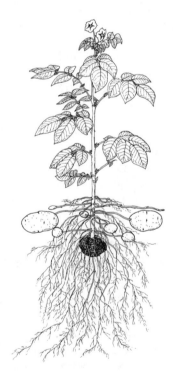

FIG. 9.1.   Representative illustration of potato plant.

embedded in the pulp, are oval or kidney shaped and yellow or yellowish brown (Fig. 9.2).

Morphologically, the tuber is a shortened, thickened, fleshy stem with leaves reduced to scales or scars subtending axillary buds known as "eyes." They form in a compressed spiral arrangement on the tuber surface, with a greater number closely spaced at the apical end. The "eyes" are located in leaf axil scars and remain dormant during tuber enlargement. Actually, each eye is a multiple bud cluster, and each bud is capable of producing a stem. Cultivars differ with regard to bud numbers. Apical buds are the first to sprout and tend to inhibit sprouting of other buds. Such apical dominance diminishes with distance or separation from the apical portion of the tuber; low temperatures and aging of the tuber also reduce apical dominance.

The anatomy of a typical tuber is shown in Fig. 9.3. The principal tissues of the tuber are the periderm, cortex, vascular cylinder of phloem and xylem, and regions of inner and outer medulla or pith tissues. The cambium produces little secondary tissue. Tuber surfaces can be smooth or rough because of netting or russeting with periderm color ranging from brown to light tan, red, or dark purple (Fig. 9.4). Flesh color usually is either light yellow or white; there are cultivars with either deep yellow, orange, red, or purple colored flesh. Tuber shapes vary; they may be elongated, blocky, round, or flattened.

FIG. 9.2. Photograph of true potato seed. Source: N. Pallais, Centro International de la Papa, Peru.

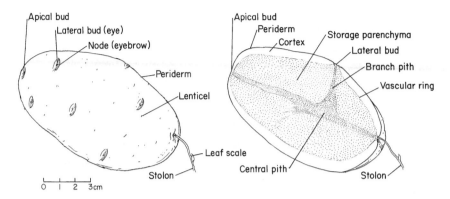

FIG. 9.3.   Anatomy of a potato tuber.

FIG. 9.4.   Variation of tuber periderm (skin) surfaces between russet and smooth-skin cultivars. Note, the skin feathering of lower tubers which is usually a result of immaturity and failure of proper skin set.

Culture

Soils

Well-drained, deep, medium to medium coarse textured friable soils that are slightly acid (pH 5.5–6.5) are preferred; good aeration is also important. Soil pH less than 5.4 helps to control common potato scab *(Streptomyces scabies)*. Soil texture and compaction have a strong influence on tuber shape as well as yield and quality. Crop rotations are important for disease and pest management, and consecutive plantings in the same field are to be avoided.

Moisture

Potato roots are highly branched and are relatively shallow. About 90% are found within 50 cm of the surface, which tends to increase susceptibility to soil moisture stress. High-moisture demand occurs following tuber initiation and during tuber enlargement. Uniformity of moisture supply is very important, especially for tuber shape. Knobbiness, which is usually caused by moisture stress, can be especially severe for some cultivars. Excessive soil moisture results in enlarged tuber lenticels, and excessive rain or prolonged leaf wetness and high humidity is conducive to foliar diseases. Crop moisture requirement varies greatly, ranging from 250 to more than 500 mm of water. Soil moisture should not fall below 60% of field capacity.

Nutrition

Potato plants are capable of producing high tuber yields and, accordingly, are heavy consumers of nutrients. In commercial production, fertilizers are routinely used to supplement native fertility. Fertilizers can be broadcasted before planting, by side dressing or in combination with irrigation. Nutrient availability is important for early plant establishment; the highest demand occurs during tuber enlargement. Sometimes, all the supplemental fertilizer is applied at planting, although incremental applications at certain plant growth stages are more effective. Delayed tuberization, tuber maturity, and low total tuber solids may result from excessive fertilizer applications, especially nitrogen.

Preplant soil analysis is useful to estimate fertilizer needs. Tissue analysis also is useful to identify the plant's nutrient status. Suggested potato leaf petiole tissue levels sufficient for good growth vary with the stage of plant growth. Nitrogen levels for dry petiole tissue at early season growth should be about 12,000 ppm, 2000 ppm for phosphorus, 11% for potassium. During tuber enlargement the respective levels are 5000 ppm N, 1000 ppm P, and 6% K.

Propagation

In potato grower terminology, seed refers to the seed tuber and not botanically true seed. Being highly heterozygous, true seed often produce plants and tubers unlike those of the parent plant and their use has been mostly limited to breeding purposes. Recently, as a result of plant breeding research, the use of true seed for propagation has increased. Vegetative propagation, using whole or tuber portions, remains the principal method; stem cuttings are used infrequently.

Although vegetative propagation has the advantage of trueness to type, major disadvantages are the bulk and the possibilities of disease transmission. Dependency on vegetative propagation places considerable importance on the availability of adequate supply and disease-free and pest-free materials.

For most of the world's production, growers use tubers from a preceding crop. Chronic occurrences of diseased propagating materials lead to the need to produce disease-free stocks. For this purpose, seed tuber production is commonly performed in cool regions unfavorable for disease occurrence and where virus vectoring insects are few or absent. Also in these environments, some foliar virus disease symptoms are usually more readily expressed, and because seed tuber production fields are inspected, diseased plants can be removed (rouged). Other diseases, such as bacterial wilt, and early blight are better expressed under high-temperature conditions. In addition to field testing, laboratory tests are also available for identification of viruses and other diseases. Tissue culture procedures using heat therapy and/or meristem micropropagation are used to obtain initially disease-free plant stocks. The succeeding regeneration is usually used for seed tuber production. In some countries, governments or other organizations certify the disease-free status of seed tubers.

Conditions experienced by seed tubers during their production, handling, and storage can have an important effect on subsequent performance. Time of planting, harvest, and storage temperatures influence the "physiological age" of the seed tuber, (Table 9.1). Late planting

TABLE 9.1.   INFLUENCE OF SEED TUBER PHYSIOLOGICAL AGE ON CROP DEVELOPMENT

	Young seed tuber	Old seed tuber
Emergence	Delayed	Early
Apical dominance	Strong	Weak
Main stem formation	Few	Multiple
Foliage production	Early	Delayed
Foliage biomass	Less	More
Foliage senescence	Delayed	Early
Root system	Large	Small
Tuber initiation	Late	Early
Tuber enlargement	Slow	Rapid
Tuber number	Low	High
Tuber size	Large	Small
Crop maturity	Late	Early
Crop yield potential	High	Low

and/or early harvested potatoes, having grown for a shorter time, are physiologically young; late harvests advances tuber age. Physiologically old seed tubers tend to produce plants having many stems and a large number of small tubers. Conversely, physiologically young seed tubers tend to produce plants having fewer stems and fewer but larger tubers.

High temperatures during plant growth can advance the physiological age of tubers. Such tubers when used for seed can affect subsequent plant growth and yield. For example, plants grown from seed tubers produced under cool (13–14°C) conditions usually have higher yields than plants produced from tubers grown under warm (26°C) temperatures; however, the response varies with cultivars. Accelerated aging can result when tubers are subjected to stress during growth and/or physical injury during postharvest handling and storage.

Seed Tuber Storage

An important facet of propagation is seed tuber storage. Because freshly harvested potatoes will not sprout, tubers intended for propagation should be stored after harvest in order to pass through the *rest** period. Sprout emergence and subsequent growth are influenced by storage period and temperature; there is variability among cultivars. Tubers are stored at low temperatures, 3–4°C, and high, 90%, RH with proper ventilation to maximize storage life and minimize premature sprout development. A storage period of 6 weeks is usually adequate for most cultivars to satisfy the rest requirement, although some may require a longer time. Temperatures lower than 2°C can cause injury to subsequent sprouting. Tubers stored at 12–22°C have shorter rest and dormancy periods and also exhibit stronger apical dominance compared to those stored at low temperatures. Exposure to light during low-temperature storage results in plants that produce many small tubers, whereas dark storage at high temperatures results in plants that produce fewer but larger tubers.

In warm climates, the use of natural diffused light is a low-cost alternative to temperature-controlled storage. Tuber exposure to diffused light has a similar effect as low temperature in restricting sprout growth and reducing apical dominance. Under these conditions, sprout

---

*The condition called *rest* (internal dormancy) can last from 5 to 20 weeks, depending on cultivar, holding temperature, and tuber maturity. Freshly harvested tubers usually do not sprout even when provided temperature and moisture conditions favorable for sprouting. The rest period can be shortened by high temperatures (21–27°C), high-humidity storage, low oxygen concentration, and wounding or cutting of tubers. In most situations, rest is dissipated with time. Rest differs from *dormancy,* where the lack of growth is due to unfavorable environmental conditions.

elongation is arrested. In some European countries, this practice is called "chitting" or green sprouting. Greening and solanine formation is not of concern because tubers are not consumed.

True Potato Seed in Commercial Production

Research, primarily at the International Potato Center in Peru, has lead to a greater reliability regarding true potato seed (TPS) for propagation (Fig. 9.2). TPS usage offers freedom from propagation transmission of most viruses and circumvents the disadvantages associated with the handling, storage, and transport of bulky seed tubers. Therefore, TPS offers great promise, especially for developing countries. One hundred grams of TPS is the propagation equivalent of 2 or 3 tons of seed tubers. However, TPS cultivars often lack uniformity of tuber size, shape, color, and quality, and yield poorly. Parental lines are being developed that have better uniformity. While TPS quality and use will continue to improve, these obstacles need to be overcome.

Because TPS exhibit dormancy, they are first used to produce transplants rather than being directly field-sown. The seedlings are planted after 4 or 5 weeks when they are better able to survive possible adverse growing conditions. This delay causes the growing period to be extended several weeks. TPS are also sown at high densities in nurseries to produce seed tubers, which, because of their small size, are usually planted whole.

In developing countries, seed tuber costs are a significant part of production inputs. The hand pollination and other costs for producing TPS are high but are expected to be lower than that for conventional seed tubers. TPS has the potential to expand potato growing to areas where seed tuber as propagating material is limited.

Planting

Prior to planting, seed tubers no longer in rest should be removed from low-temperature storage and placed at 10–13°C for several days. Transfer from low to high temperatures results in better sprout emergence following planting. Small tubers can be used whole; larger ones are cut into uniform pieces which should contain at least one eye, preferably two or three. Whole tubers are preferred because they produce more stems compared to an equivalent weight of a cut seed tuber piece. Whole seed tubers are easier to handle and reduce the spread of diseases associated with cutting.

Cut seed tuber pieces should be blocky and weigh between 40 and

60 g. Seed tuber size influences plant stem size and vigor. Large seed pieces produce large stems that grow faster and produce more leaf area and high yield, although with pieces larger than 60 g, the advantage diminishes. Roots develop from the base of the stems, not from seed tuber tissues (Fig. 9.1).

Depending on scale of operation, seed tuber cutting is done by hand or with automated equipment. Disease-free tubers and sanitary practices are important to avoid disease spread during cutting. After cutting, seed pieces can be planted directly, but usually are placed at a warm temperature (15–21°C) and high humidity for several days for cut surfaces to heal. The healing process consists of suberization, which is rapid at warm temperatures. Fungicides can also be applied to protect pieces from decay.

Presprouted whole or cut seed can hasten field emergence after planting, a useful practice where growing seasons are short. Tubers and/or seed pieces are spread out in light at warm temperatures, 20–25°C, and allowed to sprout. Exposed to light, sprouts tend to be short, thick, and green, and generally are not injured by or interfere with planting.

The quantity of seed tubers planted depends in part on production objectives, cultivar characteristics, seed availability, and cost. Size of tubers, whether whole or cut, determines plant spacings and the amount required. Close spacings tend to result in more, but smaller potatoes, whereas wide spacing result in fewer but larger potatoes. Cultivars with a heavy tuber set should be spaced wider than those with a light tuber set.

The effects of seed maturity on subsequent stem and tuber growth can, to some extent, be compensated by plant spacings and length of the growth period. Using old seed, spacings can be increased to compensate for the large number of tubers expected, and young seed are spaced more closely to adjust for the fewer number. When properly managed, equivalent yields can be obtained with seed of different physiological ages.

Plantings can be made as soon as soil temperatures permit, preferably above 15°C. Shoot emergence at 12°C may require 30–35 days. Between 22°C and 30°C, shoots emerge from seed pieces within 10 days. Planting methods range from hand placement to the use of highly automated equipment. Minimizing damage during planting is important to avoid decay. Whole tubers and/or seed pieces are uniformly spaced at depths between 5 and 15 cm. Planting depth varies with soil type, temperature, and cultivar. It is important to maintain adequate soil cover over the seed tuber and to avoid desiccation, greening, sunburn, and other injuries to developing tubers.

Growth Physiology, Photoperiod, Temperature, and Interactions

Beyond its origin, the potato plant has become broadly adapted and can be found growing throughout most temperate regions as far as 60° latitude. Lack of high-temperature adaptation limits production in the lowland tropics. The interactions of photoperiod and temperature are the most important factors affecting plant and tuber development. Photoperiod has a direct influence on tuber initiation. The influence can be modified by temperature, although cultivar response to day length and/or temperature can also vary.

Growth Stages

Foliage growth rapidly increases after sprouting and stem emergence, and fresh weight accumulation is rapid and often linear for as long as 90–100 days, and sometimes longer for late-maturing cultivars (Fig. 9.5). Thereafter, the rate of growth declines and stops with the onset of foliage senescence or if terminated by frost or harvest.

Foliage growth is more responsive to temperature effects than to photoperiod, but regardless of day or night temperatures, long days have a much greater effect than short days. That influence is maximized

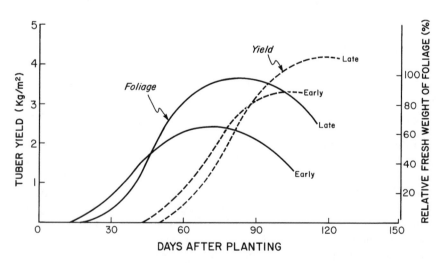

FIG. 9.5.   Relative growth pattern of early- and late-maturing cultivars.

when day temperatures are moderate to high (20–30°C) and night temperatures low (10–17°C).

Long days increase the period for photosynthesis, resulting in increased plant size and large tuber yield. Even in the extreme northern countries of Finland, Norway, and Sweden where days are very long but with a short growing season, yields of 20–30 t/ha are obtained. Conversely, short days restrict the length of the photosynthetic period, and, therefore, plants are small and less productive. Long periods of high temperature (greater than 35°C) can cause injury.

Although potatoes are considered short-day plants with respect to tuberization, interactions with temperature cannot be disregarded. Long days delay the start of tuberization and limit tuber number; warm temperatures (25–30°C) enhance tuber initiation, and temperatures above 30°C usually prevent tuber initiation. However, under long days and temperatures less than 20°C, tuber initiation is increased; about 12°C is optimal.

Cultivars vary in photoperiod/temperature response. An example of this is at mid-latitudes, where days are relatively long and temperatures are mild in the summer, which provide a climate ideal for potato production. Tubers are usually initiated about 45 days after planting. Following tuberization, tuber enlargement is ideal at mean temperatures of 17°C or slightly less.

High day and night temperatures reduce net assimilation rates; the amount of dry matter partitioned to the tubers is reduced because of high rates of respiration. Although mean day temperatures between 20°C and 30°C are most favorable for foliage growth, the optimum air temperature for maximizing yield (starch accumulation) is between 16°C and 18°C (Fig. 9.6). High night temperatures are more often responsible for low tuber yield than high day temperatures. A primary reason for the poor adaptation of potatoes to subtropical and tropical environments is due to prevailing high night temperatures.

In summary, temperatures between 20°C and 30°C enhance stem and foliage growth but are less favorable for tuber formation and development. Temperatures less than 20°C are favorable for tuber initiation and enlargement. Low night temperatures (10–17°C) can offset some of the effect of high day temperatures (25–30°C) with regard to tuberization and tuber development.

Neither photoperiod nor temperature are totally independent in their influence, and each affect above- and below-ground growth differently. Cultivar differences often play an important role in varied responses to photoperiod and temperature.

FIG. 9.6.   Effect of night-day soil temperatures on size, shape and development of 'Russet Burbank' tubers. Fahrenheit temperature conversion to Centigrade is as follows:

   50–55°F = 10.0–12.8°C
   60–65°F = 15.6–18.3°C
   70–75°F = 21.1–23.9°C
   80–85°F = 26.7–29.4°C

Source: Yamaguchi, Timm and Spurr, 1964.

Tuberization

Tuberization is initiated by the arresting of stolon elongation and by starch accumulation, which results in an increase of volume and weight. Tuberization is influenced collectively by day length, temperature, photosynthetic reserves, and cultivar. High light intensity and low plant nitrogen levels tend to enhance tuber formation.

The three phases of tuberization are as follows: initiation, occurring as differentiation of a bud on the stolon into a tuber primordium; enlargement, recognized by rapid cell division, accompanied by starch accumulation; and maturation, which occurs when the tuber enters the dormant stage. Tuber enlargement takes precedence over vegetative growth and new tuber initiation. Concurrent with tuber maturation is foliar senescence.

Many factors contribute to tuber yield; a major influence is the length of the growth period before leaf senescence occurs. Late-maturing cultivars typically yield more than early-maturing cultivars because of the ability of the foliage to produce high levels of photosynthate (Fig. 9.5).

Harvesting

Crops are harvested from 90 to 160 days after planting and this may vary with cultivars, production area, and marketing conditions. High yields are usually obtained with late-maturing cultivars and from long growing periods. Occasionally, harvesting becomes necessary before foliage senescence or frost-kill occurs and tubers are not fully developed. Existing foliage can interfere with harvest, especially when machinery is used. To reduce plant interference with harvest equipment, the tops are destroyed a week or two before harvest by mechanical shredding or with a desiccant. Foliage destruction tends to firm (set) the periderm tissue of immature tubers, thus improving resistance to possible injury during harvest.

Harvest practices vary from simple hand digging and placement into small containers, to the use of highly automated equipment that separates potatoes from soil and rapidly transfers large volumes into bulk containers or wagons. Mechanization greatly reduces labor and is responsible for the large scale of production in many countries. Proper soil moisture during harvest and soil temperatures above 20°C help reduce the incidence of bruising compared to harvesting during low temperatures and dry soil conditions.

Curing and Storage

Following harvest, potatoes, especially those intended for storage should be cured by holding at 15–20°C and at high RH for 10 or more days to enhance periderm formation and heal harvest wounds. Wound healing, the formation of a corklike layer of cells beneath damage tissues, occurs rapidly at 20°C and helps to restrict disease infection and moisture loss. After curing, the temperature is lowered; the amount lowered depends on the expected length of storage and intended use.

The potato is at its best culinary and processing quality at the time of harvest. Storage extends the availability and thereby assists with orderly marketing, distribution, and utilization. Whereas storage can extend the usefulness of harvested potato crops, quality does diminish proportionally to the length of storage. However, in well-constructed and well-managed storages, tubers of some cultivars can be stored in marketable condition for more than 10 months.

Storages are designed to prevent moisture loss, decay, and early sprouting while removing respiratory heat. Accurate temperature and ventilation management are the most important features. Other factors being equal, tuber quality is extended at temperatures of 2–4°C and high RH (90–95%). High temperature decreases storage life because of increased respiration. However, RH is also important, as about 90% of the weight loss is due to moisture loss and about 10% is because of respiration. Light is excluded to avoid chlorophyll development that results in tuber greening and the associated formation of toxic and bitter-tasting glycoalkaloids.

Many types of storages are used; those providing precise temperature and humidity control are ideal. Some are highly automated and may also provide controlled-atmosphere (CA) management. Others are very simple, such as *in situ* field holding, field clamps, placement in damp sand and various kinds of pits, cellars, and above- or below-ground covered structures that rely on ambient-temperature management (Fig. 9.7). Even simple storage facilities, if well designed and insulated, can provide satisfactory storage conditions. Storage facilities should be clean and, if necessary, disinfected to minimize diseases. Diseased and damaged potatoes should be excluded, and direct contact with moisture avoided, so as to limit the spread of decay. If tubers are to be washed, it is usually deferred until removal from storage.

In many modern bulk storages, potatoes are placed in large piles or compartments. If overly large, such piles can interfere with ventilation and cause crushing of tubers at the bottom. Wooden slatted floors or air ducts are commonly used to improve ventilation and to drain moisture that may accumulate. Some storages use large pallet boxes which improve ventilation, lessen damage, and greatly facilitate handling into and out of storage using forklifts.

Storage conditions vary depending on intended usage. Those stored for table use are typically maintained at a high relative humidity and at about 4°C. For processing, they are also held at a high relative humidity but at somewhat higher temperatures (10–16°C), because starch is converted to reducing sugars at low temperatures. The presence of reducing sugars increases the tendency for tissue darkening when potatoes are processed by frying or dehydration. At warm temperatures, reducing sugars are converted to starch. With a high ratio of starch to reducing sugars, tissue discoloration is minimized or avoided. In order to extend storage time and thereby provide a continuous supply

FIG. 9.7. In-field storage of potatoes which can be removed when convenient or desirable.

of potatoes, it is common for most producers to provide low-temperature storage. To remedy the starch to reducing sugar conversion, potatoes can be "reconditioned" before processing. Reconditioning involves removal from low-temperature storage and placement for several days or more at 18–21°C and 85–90% RH to accelerate the conversion. Most often, potatoes taken from low-temperature storage are not immediately consumed and thus are reconditioned by warm temperatures experienced during the marketing period. Potatoes having a high reducing sugar content expectedly have a somewhat sweet taste. There is a marked degree of difference between cultivars in their ability to accumulate sugars. Therefore, cultivar selection is important in producing acceptable quality processed potato products.

Depending on the length of the storage period, temperature management is not always totally adequate in controlling tuber sprouting. Therefore, to further minimize sprouting, chemical treatments can be applied. Maleic hydrazide is sprayed onto the foliage 2–3 weeks after full bloom or when most tubers have reached a size of 3–4 cm. An application of 1000–6000 ppm is effective in inhibiting sprouting. Chloroisopropyl-N-tetrachlorocarbamate (CIPC) can be applied as a dip or aerosol treatment to tubers after harvest and after injuries are healed. Inhibitors should not be applied to tubers intended for seed use.

Storage Disorders

Excluding early sprouting, most storage disorders are due to rough physical handling beginning at harvest and detrimental conditions existing within the storage. Disease incidence is usually traceable to preexisting tuber infection prior to storage, although inadequate disinfection of the storage can also be responsible.

Diseases and Pests

The frequency and magnitude of potato diseases and pests present significant management concerns for producers. Disease-free propagating materials, crop rotation, moisture management, the prudent use of pesticides in concert with integrated pest management practices, and storage sanitation are used to control diseases and pests. Following are some of the more important potato diseases and pests.

Important potato diseases and pests.

Bacteria
Black leg   *Erwinia carotovora* var. *atroseptica*

Brown rot   *Pseudomonas solanacearum*
Ring rot   *Corynebacterium sepedonicum*
Soft rot   *Erwinia carotovora* var. *carotovora*

Bacteria-like
Common scab   *Streptomyces scabies*

Fungi
Black leg   *Erwinia phytophthora*
Black scurf   *Rhizoctonia solani*
Early blight   *Alternaria solani*
Fusarium dry rot   *Fusarium solani*
Fusarium wilt   *Fusarium* spp.
Gangrene   *Phoma exigua*
Late blight   *Phytophthora infestans*
Phoma leaf spot   *Phoma andina*

Pink rot	*Phytophthora erythroseptica*
Powdery mildew	*Erysiphe cichoracearum*
Powdery scab	*Spongospora subterranea*
Pythium rot (leak)	*Pythium ultimum*
Septoria leaf spot	*Septoria lycopersici*
Silver scurf	*Helminthosporium solani*
Stem rot	*Sclerotium rolfsii*
Verticillium wilt	*Verticillium albo-atrum*
Wart	*Synchytrium endobioticum*
White mold	*Sclerotinia sclerotiorum*

Viruses, viroid, and phytoplasma (formally mycoplasma)
Aster yellows phytoplasma (AY)
Potato leaf roll virus (PLRV)
Potato spindle tuber viroid (PSTV)
Potato yellow dwarf virus (PYDV)
Potato virus X (PVX)
Potato virus Y (PVY)
Sugar beet curly top virus (BCTV)
Tobacco mosaic virus (TMV)
Tobacco rattle virus (TRV)
Tomato spotted wilt virus (TRSV)

Insect pests

Aster leafhopper	*Macrosteles quadrilineatus*
Beet leafhopper	*Circulifer tenellus*
Colorado potato beetle	*Leptinotarsa decemlineata*
European corn borer	*Ostrinia nubilalis*
Flea beetles	*Epitrix* spp.
Green peach aphid	*Myzus persicae*
Leaf hopper	*Empoasca* spp.
Potato aphid	*Macrosiphum euphorbiae*
Potato tuberworm	*Phthorimaea operculella*
Two-spotted spider mite	*Tetranychus urticae*
Wireworms	*Limonius* spp.

In addition to the tuberworm, a number of other Lepidoptera, and some wireworm species are also common potato pests.

Nematodes

Cyst	*Globodera* spp.
Golden	*Globodera rostochiensis*
Root knot	*Meloidogyne* spp.

Root lesion   *Pratylenchus* spp.
Stubby root   *Trichodorus* and *Paratricho-*
                            dorus spp.

Weeds compete with the crop by reducing plant growth and are possible hosts for insects and diseases; they also can reduce tuber quality and harvest efficiency. Crop rotation and hand and mechanical cultivation are primary weed-control methods in most of the world, although selective herbicides are widely used in developed countries.

Physiological Disorders

In addition to those mentioned in storage disorders, several other physiological disorders also affect potato production and quality. Low-temperature conditions during development cause injury to tubers; excessively high temperatures cause tissue damage, called *internal heat necrosis.* Other tuber defects are as follows: (1) *hollow heart,* due to cool temperatures at tuber initiation and/or excessively rapid enlargement often incited by high temperature during development; (2) *blackheart,* due to insufficient oxygen resulting from waterlogging or inadequate ventilation in storages; (3) *blackspot,* a tissue darkening resulting from enzymic oxidation of phenolic compounds initiated or resulting from impact injury or pressure bruising of susceptible tissue occurring usually at the stem end; (4) *irregular growth* (knobbiness), resulting from moisture stress; (5) *enlarged lenticels,* caused by excessive soil moisture; and (6) *greening,* from exposure to light.

Quality Characteristics and Uses

Potatoes are a useful crop because they are grown in many areas and produce, within a relatively short time, a high-quantity and high-quality food per unit of land. Other attributes include storage ability, ease of preparation, palatability with a high satiety value, and wide acceptability.

Important tuber quality factors include external appearance, size, shape, skin texture and pigmentation, flesh color, eye depth and number, defects and, importantly, dry matter.

The texture of the cooked potato is greatly influenced by its dry matter content (Table 9.2) and also by tuber cell size and the ratio of amylose to amylopectin starches. Culinary and processing uses are influenced by these features. In general, tubers with a high dry matter, high amylose to amylopectin ratio, small cell size, and low sugar content are preferred for most processing uses and for preparation by baking

TABLE 9.2. RELATIONSHIP OF SPECIFIC GRAVITY AND TOTAL SOLIDS TO POTATO TEXTURE AND USE

Specific gravity	Percentage total solids	Texture	Optimum usage
<1.06	<16	Very soggy	Pan frying, salads, canned processing
1.06–1.07	16–18	Soggy	Pan frying, salads, boiling, and canning
1.07–1.08	18–20	Waxy	Boiling, mashing, fair to good for chips or canned processing
1.08–1.09	20–22	Mealy, dry	Good for baking, chips, and french fries; some cultivars tend to slough when boiled
>1.09	>22	Very mealy, or dry	Good for baking, chips, and french fries; greater tendency for brittle chips and sloughing when boiled

Source: Modified from Chase et al. (1990).

or frying. Such potatoes, when boiled, tend to slough and have a mealy texture. Potatoes with a low dry matter are best used boiled because they tend to remain intact. The starch composition of many low dry matter potatoes tends to have a low amylose to amylopectin ratio. Such potatoes, when baked, tend to have a moist texture.

Besides water (about 78%), other components are carbohydrates and protein, usually about 18% and 2%, respectively. Potatoes also contain substantial quantities of minerals and vitamin C. Yellow-fleshed cultivars contain some carotene. When comparing food values, it is important to note that the high water content of potatoes must be considered in making a comparison with cereal grains or other high dry-weight foods. Because of their relatively high consumption, potatoes provide significant amounts of protein and calories despite their low absolute content. The proximate nutrient composition for potato is shown in Appendix Table C.

Toxic Components

Potato plants and tubers contain the toxic glycoalkaloids, alpha-solanine and alpha-chaconine, which act as cholinesterase inhibitors. When tubers are exposed to light, chlorophyll along with the glycoalkaloids are synthesized. The amount of glycoalkaloids formed depends on exposure length, intensity and light quality (mostly ultraviolet), and temperature; little is synthesized at temperatures below 5°C. These compounds taste bitter, and ingestion can cause illness and death in extreme cases; toxicity depends on the amount ingested. Mechanical injury also induces the formation of these substances.

Normally, the highest amounts of glycoalkaloids are found in tissues with high metabolic activity such as sprouts and flowers. The content in foliage and stems is higher than in tubers. The tuber skin (periderm) has the highest glycoalkaloid concentration; peeling removes most but not all of it. Mature tubers contain 2–6 mg/100 g fresh weight. The content is high early in tuber development; small immature tubers have the highest (14–28 mg/100 g) levels. Heat does not destroy these substances, although some can be leached during boiling. Glycoalkaloid content varies with cultivars. The introduction of some newly developed cultivars, promising in many beneficial characteristics, has been prevented because of high glycoalkaloid content. The acceptable concentration is less than 20 mg/100 g; at that level the bitter flavor is very apparent.

Breeding

During the early development of the potato, improvement was largely a matter of selection, much of it done by lay persons rather than by the structured activity by public or private breeders. For example, 'Russet Burbank' was selected from botanical seed progenies before 1890 and was rapidly and widely adopted, and continues to be a major cultivar in North America. Similarly, 'Bintje,' introduced about 1910 also continues to be widely produced in Europe. Early advancements relied on somatic mutations, most of which were not useful. Accordingly, relatively few major cultivars emerged. Conventional breeding is based on crossing two cultivars with complementary traits and selecting within the segregating population. The probability of finding improved cultivars has been very low. This is probably due to the complex tetrasomic inheritance patterns. Sexual breeding of potatoes is a relatively recent development that offers the opportunity to expand the development and introduction of improved cultivars. An interesting example is the cultivar Greta, having late blight resistant, that was developed by interspecific hybridization to S. demissum, a Mexican wild species.

Use of true seed has many potential benefits, 2n gametes and cytoplasmic male sterility provides a potential procedure for producing hybrid seed. The modern cultivated S. tuberosum spp. tuberosum potato has a narrow genetic base but is amenable to many biotechnological manipulations, such as embryo rescue, protoplast fusion, transformations, and anther culture, which should lead to improved germplasm for breeding purposes. Further exploitation of wild species should expand the genetic base. Major germplasm collections are maintained at the International Potato Research Center in Peru, also in Russia, the United States, Germany, and several other countries.

Production

In volume, potatoes are the world's fourth major food crop, following wheat, maize, and rice. Potato production of protein and carbohydrate per land area per day exceeds that of any single grain crop (Fig. 9.8). Furthermore, the high nutritive value of potatoes has encouraged expansion of production into many areas, even those where they are less productive.

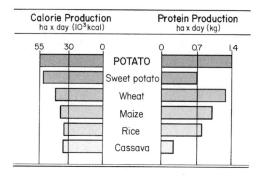

FIG. 9.8.   Calorie and protein production of potatoes compared with other major food sources. Source: N. Pallais, Centro International de la Papa, Peru.

TABLE 9.3.   WORLD POTATO PRODUCTION, 1994

	Area (ha × 10³)	Yield (tons/ha)	Production (tons × 10³)
World	18,191	14.6	265,436
Africa	745	10.7	7,981
North and Central America	795	32.7	26,003
South America	935	12.7	11,873
Asia	5,755	13.7	78,663
Europe	9,913	14.1	139,505
Oceania	49	28.8	1,411
Leading countries			
China	3,202[a]	12.5	40,039[a]
Russian Federation	3,400[a]	9.9	33,780
Poland	1,697	13.6	23,058
United States	557	37.4	20,835
Ukraine	1,527	10.5	16,102
India	1,000[a]	15.0	15,000[a]
Germany	293	31.6	9,257
Belarus	750[a]	11.0	8,241
Netherlands	172[a]	45.1	7,748[a]
United Kingdom	171[a]	41.4	7,065[a]

[a]Estimated
Source: 1994 FAO Production Yearbook, Vol. 48. FAO, Rome, 1995.

When listed as a single entity, the potato production statistics of the former Soviet Union were notable for its production of more than 25% of world production for many years. Even now, 3 of the former republics, Russian Federation, Ukraine, and Belarus, are among the 10 leading potato producing countries, collectively accounting for more than 22% of world volume. China, Poland, and the United States produce more than 15%, 8%, and 7%, respectively. A greater application of production technology, cultural, and disease and pest management is responsible for the considerably higher yields in western European countries and the United States compared to those of the former Soviet Union republics and China.

World total production has shown a declining trend for several decades, the decline is more noticeable in the industrialized countries. The overall decline is most apparent in reduced tableware consumption, whereas use as processed potato products has increased. FAO statistics estimate world utilization of current production as 45% for human food, 30% for animals, 15% for seed, 2% for starch, and about 8% as waste.

Presently, temperate-region countries are the largest producers and consumers of potatoes, whereas countries in the tropics and subtropics, which include many of the less economically developed countries, produce and consume less. Impediments to production in these areas are high day and night temperatures, high disease and pest susceptibility, and poor soils. Additionally, difficulties in obtaining clean adapted seed stocks, adequate storage, and other problems associated with the introduction of a new crop are obstacles that require resolution before production can be expanded. Plant breeders are continuing research efforts to provide greater adaptation in other regions such as the lowland tropics.

For many years, the International Potato Research Center (CIP) in Peru has developed and implemented valuable research findings. The efforts of CIP and those of other institutions will ensure that the potato crop continues its importance as a food staple.

SELECTED REFERENCES

Allen, E.J., and Scott, R.K. 1980. Analysis of growth of the potato crop. J. Agric. Sci. *94*, 583–606.

Chase, R.W., Silva, G., Douches, D., and Hammerschmidt, R. 1990. Selecting potato varieties for Michigan. Michigan State University Cooperative Extension Service, E-2222, East Lansing, MI.

CIP annual reports. Centro International de la Papa/International Potato Center, Lima, Peru.

Concilio, L., and Peloquin, S.J. 1987. Tuber yield of true potato seed families from different breeding schemes. Am. Potato J. *64,* 81–85.

Douches, D.S., and Jastrzebski, K. 1993. Potato, *Solanum tuberosum* L. In Genetic Improvement of Vegetable Crops. G. Kalloo and B.O. Bergh, eds. Pergamon Press, Oxford, pp. 605–644.

Gregory, L.E. 1965. Physiology of tuberization. In Encyclopedia of Plant Physiology. Springer-Verlag, Berlin, Vol. XV, Part 1, pp. 1328–1354.

Grun, P. 1990. The evolution of cultivated potatoes. Econ. Bot. *44* (Suppl 3), 39–55.

Harris, P. 1993. The Potato Crop. Van Nostrand Reinhold, New York.

Hawkes, J.G. 1992. The Potato, Evolution, Biodiversity and Genetic Resources. Behlaven Press, London.

Hooker, W.J., ed. 1981. Compendium of Potato Diseases. American Phytopathological Society Press, St. Paul, MN.

Horton, D.E. 1987. Potatoes: Production, Marketing and Programs for Developing Countries. Westview Press, Bolder, C. IT Publications, London.

Kozukue, N., Kozukue, E., and Mizuno, S. 1987. Glycoalkaloids in potato plants and tubers. HortScience *22,* 294–296.

Levy, D. 1984. Cultivated *Solanum tuberosum* L. as a source for the selection of cultivars adapted to hot climates. Trop. Agric. (Trinidad) *61,* 167–170.

Lisinska, G., and Leszczynski, W. 1989. Potato Science and Technology. Elsevier Science, New York.

Pallais, N. 1991. True potato seed: Changing potato propagation from vegetative to sexual. HortScience *26,* 239–241.

Sinden, S.I., Sanford, L.L., and Webb, R.E. 1984. Genetic and environmental control of potato glycoalkaloids. Am. Potato J. *61,* 146–156.

Sieczka, J.B., and Thornton, R.E., eds. 1993. Commercial Potato Production in North America. Potato Association of America, Orono, ME.

Talburt, W.F., and Smith, O. 1987. Potato Processing, 4th ed. Van Nostrand Reinhold, New York.

Yamaguchi, M., Timm, H., and Spurr, A.R. 1964. Effects of soil temperature on growth and nutrition of potato plants and tuberization, composition, and periderm structure of tubers. Proc. ASHS 84:412–423.

10

Sweet Potato

Ipomoea batatas (L.) Lam.
Family: Convolvulaceae

Origin

Most systematic botanists concur that the origin of sweet potato, *Ipomoea batatas,* was somewhere in tropical America in regions that include Panama, northern parts of South America, and the West Indies. The closest relative, *I. trifida,* is found wild in Mexico. Another related species, *I. tiliacea,* can be found in the West Indies. *Ipomoea batatas* is a hexaploid, but most of the approximate 400 *Ipomoea* species are diploids. The evolutionary steps in sweet potato domestication remain unclear.

Sweet potatoes were grown by the Mayans in Central America, Incas in Peru, and Maoris in New Zealand well before the 16th century. Sweet potato remains, perhaps of wild species, found in Peruvian caves were estimated to be more than 8000 years old. The plant was introduced into Spain by European explorers about A.D. 1600 to western Africa by Portuguese traders and later introductions into India, the East Indies, China, and Japan.

The early presence of sweet potatoes in the Philippines and Polynesia suggests those locations as possible centers of origin. However, other evidence suggest sweet potato clones were transferred from Central America and/or Mexico to the Philippines and from Peru to South Pacific islands. Although uncertain, transfer may have occurred by migrating people and traders or by natural dispersal of seed via ocean currents. Seed have extremely hard testa that could have provided protection to survive long periods of ocean exposure. Distinct differences existing between sweet potatoes in the West Indies compared with those in the Philippines and Polynesia support the view of dispersion from different Central American locations.

Botany

Sweet potato is a perennial dicotyledon with long trailing vines and smooth, flat, cordate to lobed leaves, borne on erect petioles. The central stem from which lateral branches form is usually prostrate and, depending on internode length, can appear more or less bushlike. Cultivar types, whether bush, intermediate, or vining are determined by internode rather than vine length; vine branching varies with cultivars. Edible portions are the enlarged storage roots, shoot tips, and young leaves. *Ipomoea aquatica* is grown specifically for its edible foliage.

Fibrous roots can develop adventitiously from opposite sides of each node on portions of the stem in contact with soil. The edible storage organs erroneously called tubers are tuberous roots of thickened secondary roots. Typically, about 15% of the roots formed will thicken and form the storage organs which develop fairly close, within 25 cm of the surface. Most of the storage root growth is normally achieved in about 2 months after planting. The diameter continues to increase as long as the foliage remains active. The major portion of the storage root consists of parenchyma tissues. A diagrammatic cross section of a sweet potato storage root is shown in Fig. 10.1.

Plants usually develop 4–10 storage roots (Fig. 10.2). Most commercially marketed roots range in weight between 100 and 400 g. Root enlargement, incorrectly referred to as tuberization, is a result of rapid cell division of the central parenchyma tissues, followed by cell enlargement and starch deposition. Root periderm color varies with cultivars, ranging from buff, yellow, orange, copper, or red to purple. Similarly, flesh colors may be white to light yellow, dark orange, red, or purple. Shape varies from elongated to almost round. Surface smoothness of storage roots varies with cultivars; some tend to develop surface ridges.

Flowers, 3–4 cm in diameter, are pale pink with a reddish, lavender, or purple trumpet-shaped throat resembling those of morning glory. They open in the early morning and close and wilt within hours; pollination is by insects. One to four seed form in a round capsule. Mature seed are hard, black, and flattened and often require scarification to facilitate germination.

Sweet potatoes are generally short-day plants and, depending on cultivar, have a photoperiod requirement of 11 h or less to flower. Flowering, although common in tropical areas, is infrequently observed in temperate locations. Flowering and seed production are important only for breeding. Most cultivars are self-incompatible. In order to improve flowering characteristics for sexual crossings, sweet potatoes can be grafted onto rootstock of other *Ipomoea* species.

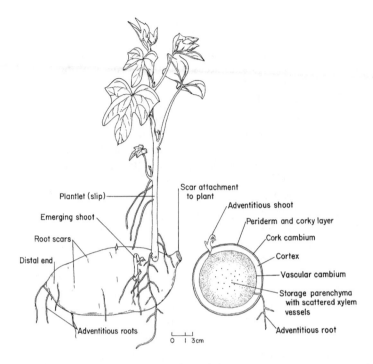

FIG. 10.1.   Intact and transverse section of the sweet potato storage root. The plantlets (slips) are removed for propagation.

Culture

Being a dependable and productive low-input crop with a relatively short growth period, sweet potatoes are an important commodity and are grown in many tropical, subtropical, and warm temperate regions during frost-free periods. The crop is grown in regions as far as 40° latitude. Growth is best in areas having warm days and nights.

Generally, high light and long days favor vine growth, preferential to storage root enlargement. Increased vine growth in response to long days is correlated with low yields. Growth periods range from 90 to 150 days in temperate areas. In the tropics, growth is continuous and harvests can be made when suitable root size is attained.

Well-drained, friable sandy loams with adequate aeration are preferred. Soil compaction adversely affects storage root shape and size. Sweet potatoes have a moderate tolerance to low pH and are adaptable to a range of 4.5–7.5; about 6.0–7.5 is optimum. Below pH 5.2, plants

FIG. 10.2.   Sweet potato storage roots of a single plant.

are less susceptible to pox and scurf diseases. Plants also have moderate tolerance to salinity.

A uniform water supply, about 25–30 mm each week, is optimum for active growth. Sweet potatoes tolerate some drought because of deep rooting. About half of the total root system is contained within the surface 30 cm; some are found as deep as 2 m. Plants are intolerant of flooding; waterlogged soils reduce yields. However, irregular heavy rains are tolerated if drainage is good. Moisture deficit can reduce yields when it occurs during root enlargement.

Sweet potatoes possess the ability for reasonable production even in infertile soils. The occasional association with mycorrhize (Glomus fasciculatum) enhances yields in low-phosphorus soils. The crop is often grown as a second or relay crop in order to utilize residual fertilizers from preceding crops. Table 10.1 shows the nutrient removal of sweet potatoes. Petiole tissue analyses can be used to determine crop nutritional status during growth.

When sweet potatoes are grown in multiple cropping systems, it is important to select shade-tolerant cultivars, as excessive shading reduces yield. The leaf area index is relatively low because growth is mostly horizontal, limiting light penetration below the canopy. However, some shade-tolerant cultivars actually perform better in slight shade than in full sunlight.

TABLE 10.1.   NUTRIENT REMOVAL BY SWEET POTATOES AVERAGE OF FOUR CULTIVARS

Nutrient	Root (kg/ha)	Vine (kg/ha)	Total
N	47	52	99
P	19	8	27
K	179	101	280
Ca	11	46	57
Mg	9	9	18

Source: Adapted from Scott and Bouwkamp (1974).

Propagation

Storage roots have no natural dormancy and have the ability to initiate adventitious sprouts which develop from the vascular cambium when temperatures and moisture are favorable. The sprouts, called shoots or slips, develop adventitious roots. Such plantlets are used as seedling transplants (Fig. 10.1).

In temperate regions, storage roots are used to produce sprouts for transplanting. Such usage requires substantial amounts of storage roots that might otherwise be consumed. Small storage roots, not suitable for market, are sometimes used for direct field plantings.

Because of proximal dominance, sprouts are produced earlier and in greater numbers at the storage root's "stem end." Dominance is strongest in freshly harvested storage roots but is weakened by aging or transverse division (cutting) of storage roots. Cured cut storage root pieces can be used because more total sprouts are produced than from intact storage roots. Storage root pieces placed at a density of 100/m^2 usually provide about 800–900 sprouts. About 500–700 kg of roots produce a sufficient number of sprouts to plant 1 ha.

If growing conditions permit, propagating material can be produced in the open field. However, early sprout production in temperate areas can be achieved from heated nursery beds that accelerate sprouting and growth. Soil temperatures of 28–30°C are best for producing sprouts, which usually emerge in 1–2 weeks. They are ready for pulling and transplanting after 4–6 weeks. Cold temperatures result in slow emergence and growth. Temperatures greater than 30°C result in thin leggy sprouts that establish poorly after transplanting.

Uniform intact sweet potatoes or cut pieces are laid as a single layer into the nursery bed. They are closely placed in order to maximize the number and uniformity of shoots produced (Fig. 10.3). After placement, these are covered with a 5–10-cm layer of soil. Nursery beds often have a sandy soil base to provide good drainage, easy placement of root

FIG. 10.3. Sweet potato plantlet production using low plastic film tunnels. The soil is heated by buried decomposing organic matter. Bags act as ballast to anchor the edge of the film which is extended over the wire supporting hoops to cover the bed.

pieces, and easy sprout removal. Nursery beds can also be covered with plastic for better temperature control. Often used are low plastic tunnels which can increase growing temperature as well as permit ventilation when necessary. Before pulling, sprouts are usually hardened for a day or two by uncovering and/or withholding water for a day or two; the length of time depends on temperature and soil type. In situations where transplanting is delayed, pulled plants can be temporarily stored at 16°C and high humidity; lower or higher temperatures are less favorable. Transplanted sprouts usually establish rapidly, and do best at soil temperatures of 21°C. At temperatures greater than 30°C, rooting is inhibited; the minimum critical temperature is 15°C. Transplants are usually placed in the soil at a depth of 5–10 cm.

In tropical areas, instead of storage roots, vine shoot tips and stem cuttings are commonly used for propagation. These materials have the advantage of reducing propagation costs by not using otherwise consumable storage roots. Shoot tips are also less likely to transmit disease or introduce soil pests such as nematodes, although weevils may be transferred. Cuttings can be removed from existing crops while the donor plants continue growth, or can be obtained from plants at harvest. Cuttings commonly used are 30–45 cm long with about eight or nine nodes. These are usually planted horizontally, or at a slight incline, 5–10 cm deep, with at least three or four nodes covered with soil. Rooting and establishment is usually rapid. The polarity of sweet potato cuttings or the presence or absence of the shoot apex do not influence yield. This is because the plant is prostrate and new shoots, arising from leaf nodes, maintain the polarity of the plant.

Hand transplanting of stem cuttings or rooted shoots is the most common method of field establishment. In the United States, transplanting is mechanized. The intended usage of the crop often determines planting densities. Wide spacings produce large roots, which are preferred for food processing and industrial uses, whereas high densities result in small roots, which are more suitable for home consumption. Plant populations commonly range between 20 and 30 thousand transplants per hectare.

Plantings are often established on raised beds or ridges to improve drainage and facilitate soil cover of developing storage roots. It is important that they are adequately covered with soil, as exposure to light prevents the initiation and enlargement of storage roots. In partial harvest practices, roots are recovered with soil for further enlargement of storage roots already initiated.

Meristem and tissue culture are used to obtain disease-free material for multiplication of clean propagating stocks. Programs have been established in the United States and elsewhere to produce certified disease-free materials. Although it is possible to use true seed for propagation, variable storage root qualities and yields may result. The primary use of true seed is for breeding purposes. The use of somatic embryo's as "synthetic seed" for propagation now under investigation may become feasible.

Growth and Development

In tropical regions, growth is perennial and vines continue to grow. Storage roots can be harvested when they attain a suitable size; other roots are allowed to develop and enlarge. Not having a distinct maturity, yields tend to increase with longer growing periods.

In temperate zones, sweet potatoes are grown as annuals; growth is stopped by cold weather or frost. The growth cycle has three major phases: (1) active fibrous root growth with moderate vine growth; (2) extensive vine growth with the formation of a large leaf area and the initiation of storage root development; and (3) storage root enlargement associated with a slowed rate of leaf and fibrous root growth.

Foliage development is generally rapid, and about 60 days after transplanting, or about early midgrowth, a full leaf canopy is usually achieved. The leaf area index at this time is about 5. Thereafter, the leaf area index gradually decreases (Fig. 10.4). Plants readily adapt to high-temperature extremes, although temperatures much above 30°C tend to inhibit growth and yield, especially when high night temperatures persist. The best growing temperatures are 29°C days and 21°C nights; a mean of about 24°C is optimum.

Initiation of storage root enlargement can begin as soon as 30–35 days after planting, and continues as the principal sink until harvest or termination of growth. High temperatures inhibit root growth more than vine growth. Optimum storage root enlargement occurs during 25°C day and 20°C night temperatures. Soil temperatures greater than 30°C tend to decrease yields.

Plants are intolerant of frost, and at less than 15°C, active growth can cease. At temperatures less than 10°C, chilling injury occurs and plants can be killed. Also, as the storage root gets larger, an anaerobic condition can occur because of the large oxygen gradient between interior and exterior tissues, causing storage tissue breakdown. The incidence of storage root rot can often increase following heavy rains.

Changes in the ratio of starch to sugar occur during growth. Starch

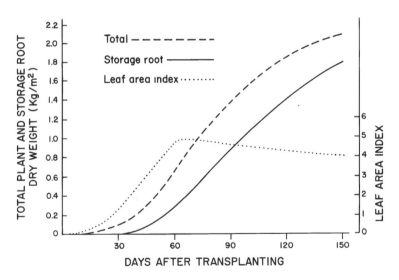

FIG. 10.4. Sweet potato plant growth following transplanting. Source: Adapted from Agata, 1982.

is low in very young roots and remains low during maximum growth, probably because carbohydrates are used for new tissue formation. Total sugars also decrease during the rapid growth period. With enlargement, further decreases in sugars occur while starch increases.

Harvest

A major proportion of the world's sweet potatoes is a subsistence crop that is harvested as needed, usually by hand digging. Roots can be harvested early without terminating plant growth by carefully cutting away some of the storage roots with minimum disturbance of the soil. This is a common practice in Papua New Guinea and several tropical countries. Unharvested mature storage roots lose palatability because of continuous fiber development.

In large-scale production for many temperate and subtropical areas, harvest usually occurs as a single event, about 90–150 days after transplanting. Sometimes, crops are harvested as early as 80 days, but at other times, harvests can be extended beyond 150 days. Time of commercial harvest is often influenced by market needs.

Foliage removal before digging facilitates harvesting of the storage root and it is believed to thicken (set) the tender skin (periderm). Avoiding injury throughout harvest is important because sweet potatoes are easily damaged. Adequate soil moisture and warm temperatures during harvest reduces susceptibility to physical damage. After digging, sunburn and desiccation should be prevented. Harvest mechanization is used in some countries, and the crop is often handled in pallet bulk containers to minimize injury. To avoid increased decay, sweet potatoes are not washed before curing or storage.

Curing

Curing of roots following harvest minimizes further damage by healing harvest wounds. Curing involves holding storage roots at temperatures between 27°C and 30°C and at high RH (85–95%) during a 4–7-day period in order to form a corky periderm layer below the damaged areas. Periderm synthesis is optimum at these conditions. After periderm formation, microbial invasion and water loss are limited. For indoor curing, good air exchange is important in preventing moisture condensation and elevated CO_2 levels. Excessive curing time or high temperatures results in shrinkage, reduced storage life, and a tendency for early sprouting. During curing, there is a loss in dry matter and an increase in sugars.

Field curing offers less management control but is widely used in many countries. Large piles of harvested roots are covered with vines thick enough to serve as insulation and also to raise the humidity. Respiration of the roots increases the temperature of the pile. Curing usually is completed in about 7–10 days; weight loss during curing can be as much as 5%.

Storage

Because of year-round supply in many tropical regions, there is little incentive to store roots. Nevertheless, for short-term holding, ground pit storage which provides good drainage and insulation are often used. Above-ground clamps are also used for short-term storage. For long-term storage, in temperate production regions, temperatures are reduced to 13–16°C following curing, and the RH is maintained at 85–90% to reduce respiration and moisture loss and to prevent sprouting. Long-term storage below 13°C results in chilling injury and increases interior tissue pithiness. Temperatures should not be less than 10°C, and storages ventilated to keep CO_2 low. Weight losses that average about 2% each month are possible; good storages are able to minimize such losses. Roots can lose as much as 10% of their fresh weight without the appearance of shriveling. Under optimal conditions, sweet potatoes can be stored for 6 months or longer.

Diseases and Other Pests

Sweet potato roots are affected by a number of diseases and pests in the field and following harvest. Although root damage is often more obvious, foliar injury can be serious. Vegetative propagation can result in transmission of diseases and pests. Prevention relies on healthy propagating materials, hygienic production practices, adequate crop rotation, and careful storage of properly cured roots. Plant resistance to some pests exists.

Important bacterial diseases are soft rot *Erwinia chrysanthemi,* bacterial wilt *Pseudomonas solanacearum,* and Pox (Scab), *Streptomyces ipomoea.*

Fungal pathogens attacking sweet potatoes include the following:

Alternaria leaf spot	*Alternaria* spp.
Black rot	*Ceratocystis fimbriata*
Cercospora leaf spot	*Cercospora* spp.
Charcoal rot	*Macrophomina phaseolina*
Foot rot	*Plenodomus destruens*
Fusarium surface rot	*Fusarium oxysporium*

Fusarium wilt	*Fusarium oxysporium,* f. sp. *batatas*
Fusarium storage root rot	*Fusarium solani*
Java black rot	*Diplodia tubericola*
Leaf and stem scab	*Sphaceloma batatas*
Red rust	*Coleosporium ipomoeae*
Rhizopus soft root rot	*Rhizopus stolonifer*
Scurf	*Monilochaetes infuscans*
Septoria leaf spot	*Septoria bataticola*
Southern blight	*Sclerotium rolfsii*
Stem canker	*Rhizoctonia solani*
Stem and Foliage scab	*Elsinoe batatas*
Violet root rot	*Helicobasidium mompa*
White rust	*Albugo ipomoeae-panduratae*

Aphid-vectored viruses responsible for reduced productivity include chlorotic leaf spot virus, internal cork virus, Russet crack virus, sweet potato feathery mottle virus (SPFMV), and sweet potato vein mosaic virus (SPVMV). Whitefly-vectored viruses are sweet potato mild mottle virus (SPMMV) and sweet potato yellow dwarf virus (SPYDV). A leaf-hopper-vectored mycoplasm-like disease is little leaf, also known as witches-broom.

Important nematode pests include the following:

Lesion, *Pratylenchus* spp.
Reniform, *Rotylenchulus reniformis*
Root knot, *Meloidogyne* spp.
Stubby root, *Paratrichodorus* spp.
Spiral, *Helicotylenchus* spp.
Stem and bulb, *Ditylenchus* spp.

Partial resistance is available for the root knot nematode, but resistance alone should not be relied on, but used in conjunction with fumigation and/or rotation to limit the resurgence of damaging populations.

Physiological disorders affecting sweet potatoes are souring, caused by oxygen deficiency because of flooding. Other crop and quality losses are due to growth cracks, chilling injury, and pithiness caused by prolonged warm and dry storage. Hardcore is a condition where interior storage root tissues remain hard after cooking. This is induced by root exposure to chilling followed by storage at nonchilling temperatures. Susceptibility varies with cultivar and severity and is a function of prior curing treatment and storage.

Important insect pests are as follows:

Cotton aphid	*Aphis gossypii*
Armyworm	*Spodoptera litura*
Cucumber beetles	*Diabrotica* spp.
Green peach aphid	*Myzus persicae*
Flea beetles	*Systena* spp.
Flower beetle	*Notoxus calcaratus*
Scarabee	*Euscepes postfasciatus*
Striped sweet potato weevil	*Alcidodes dentipes*
Sweet potato flea beetle	*Chaetocnema confinis*
Sweet potato hornworm	*Agrius convolvuli*
Sweet potato leaf beetle	*Typophorus nigritus viridicyaneus*
Sweet potato moth	*Herse convolvuli*
Sweet potato vine borer	*Omphia anastamosalis*
Sweet potato weevil	*Cylas formicarius elegantulus* *Cylas puncticollis*
Sweet potato whitefly	*Bemisia tabaci*
Tortoise beetles	*Metriona* spp.
West Indian sweet potato weevil	*Euscepes postfasciatus*
White grub	*Plectris aliena* and *Phyllopaga ephilda*
Wireworms	*Conoderus* spp.

Worldwide, the sweet potato weevil is the single most important insect pest both in the field and storage.

Sweet potato plants become competitive with weeds after 5–6 weeks when a full canopy is usually established; high plant densities are also helpful in reducing early weed competition. Effective herbicides are available, although for most of world production, hand weeding remains the primary practice. Cultivation, mechanical or by hand, must be shallow to avoid injury of the fibrous root system near the surface.

Uses and Quality Characteristics

Many uses are made of the high-energy value of sweet potatoes. They are a staple food for many millions, and in some countries, consumption may account for 70% of daily caloric intake. Large variations in taste and texture preferences exist between peoples in different regions. For most consumers, preference is given to high-starch, low-sugar, dry textured sweet potatoes. However, in some production areas, namely the United States, dark-skin, high-sugar cultivars with yellow or orange flesh are preferred (Fig. 10.5). In other countries, the preference is for light-skin, white-fleshed starchy cultivars.

There are three major sweet potato types which are characterized after cooking as (1) flesh that is firm, dry, and mealy, (2) flesh that is soft, moist, and gelatinous, and (3) flesh that is coarse and fibrous. The majority of production is of the first type with most cultivars being white fleshed. In addition to human use, a large amount of this type is also used for animal feeding and industrial products. The second type is mainly for human consumption, but in terms of volume, is of less importance. The third type is generally utilized for animal feeding and for industrial uses such as starch and alcohol production.

In U.S. markets, cultivars having a deep orange flesh color and soft, moist, sugary textures after cooking are erroneously called "yams." The true yam is a very different plant in the genus *Dioscorea* (Chapter 12). In the United States, yam is a marketing term that distinguishes dark red or purple skin, orange-fleshed sweet potatoes from those of light skin, yellow-fleshed roots that are less sweet and have a dry firm texture after cooking. The latter are often called 'Jersey' types.

Besides freshly cooked uses, sweet potatoes are processed by canning and used to make chips, noodles, flour, and candy. In Japan and Taiwan, much of the crop is fed to animals and some is used for starch and

FIG. 10.5.   Packaged 'Beauregard' sweet potatoes, a dark-skin, orange-flesh culti-var popular in the United States. Source: Courtesy of Jonathan R. Schulltheis.

alcohol manufacture; a substantial portion of the crop is also fed to animals in China.

Sweet potatoes provide a high-calorie contribution to diets: The average dry matter content is 30%, of which from 75–90% is carbohydrate; fat content averages about 0.4% (Appendix Table C). Sweet potato starch is composed of about one-third amylose and two-thirds amylopectin. During cooking, much of the starch is converted to maltose, which is responsible for the sweet taste.

The protein content of most cultivars is between 1.5% and 2.5%, which is low to average when compared to other vegetables. Although protein quality is fairly balanced, it is low in sulfur-containing amino acids. Sweet potatoes are a good source of vitamin C and supply a fair level of the B vitamins. Orange-fleshed sweet potatoes are an excellent sources of beta-carotene, containing more than yellow-flesh types, whereas white-flesh types have little or none. Raw sweet potatoes contain a trypsin inhibitor which lowers protein digestibility. However, because sweet potatoes are consumed after cooking the inhibitor is destroyed (see Chapter 5).

Tender leaves, petioles, and stem tips of sweet potatoes are also used as vegetable greens in Southeast Asia. These have a high pro-vitamin vitamin A and vitamin C content and are much higher in protein than storage roots, about 30% on a dry weight basis, and are comparable in food value to water convolvulus, *I. aquatica*.

Production

The sweet potato is very important in developing countries where about 98% of world production occurs. Sweet potatoes occupy about 19% of the total land area used for root and tuber vegetables and account for about 22% of world tonnage of these vegetables.

Despite its origin, sweet potatoes basically are now an Asian crop, where 92% of total world production occurs. China, with over 6 million hectares, produces more than 84% of the world's volume. (See Table 10-2.)

Sweet potatoes are being displaced proportionately by what are perceived as more prestigious crops such as maize, cereals, legumes, and leafy vegetables in many countries. A review of past production statistics show that the large reduction of land area used to produce the crop has been offset by increased yields, the result of improved cultivars and production technology. Consumption in less developed countries continues to remain high. However, a substantial volume for animal usage is likely to persist. Globally, sweet potatoes are ranked ninth

TABLE 10.2. WORLD SWEET POTATO PRODUCTION, 1994

	Area (ha × 10³)	Yield (t/ha)	Production (t × 10³)
World	9,380	13.3	124,339
Africa	1,384	5.0	6,944
North and Central America	166	6.9	1,140
South America	116	10.7	1,248
Asia	7,587	15.1	114,347
Europe	5	12.0	60
Oceania	121	5.0	600
Leading countries			
China	6,511[a]	16.2	105,180[a]
Vietnam	393[a]	6.5	2,541[a]
Uganda	478	4.5	2,151
Indonesia	197	9.4	1,854
Japan	51	24.6	1,264
India	138[a]	8.3	1,150[a]
Rawanda	160[a]	6.3	1,000[a]
Philippines	148[a]	4.7	700[a]
Kenya	66[a]	9.9	650[a]
Brazil	61[a]	10.3	630[a]

[a]Estimated.
Source: 1994 FAO Production Yearbook, Vol. 48, FAO, Rome, 1995.

among the most important food crops. They are mainly a subsistence crop and traditionally are consumed locally, and because of their low value, bulk, and perishability when not cured, sweet potatoes are not widely traded, except locally, and are not frequently found in international trade.

Crop Improvement

Because of its subsistence and food staple importance, considerable research for crop improvement are conducted by several international organizations. The International Potato Center (CIP) in Peru has an improvement program and is a germplasm repository. Germplasm repositories are also located at Fort Collins, Colorado and Griffin, Georgia in the United States.

Although the world average yield is over 13 t/ha, in many parts of the world, especially Africa, yields are much less, mainly because of subsistence agriculture. Researchers have developed cultivars with a potential for early production of high yields with limited or no supplemental fertilization. However, lack of weevil resistance in these cultivars has limited their acceptance.

Sweet potatoes rank high in available food value returned for the input of labor and materials. The crop is likely to continue as an important food in spite of its low status as a "poor man's food." Advancements will rely on improved breeding and crop production research. Intraspecific hybridization, because of crossing barriers and ploidy differences, is difficult, but substantial germplasm variability exists in the *Ipomoea* species. Therefore, the reliance of somatic mutation and selection for improvement can be surpassed with the use of new technologies for cultivar improvements, especially with regard to enhanced nutritional value. A wide range in crude protein content, as much as 10%, occurs in some germplasm sources, and suggests there is potential for significant quality improvement. Similarly, yield potential has not been fully exploited. The major breeding efforts are directed to improvement of flavor, weevil resistance, yield and yield stability, and virus resistance. Excellent breeding programs exists in the United States, Japan, Indonesia, and some other countries.

SELECTED REFERENCES

Agata, W. 1982. The characteristics of dry matter and yield production in sweet potato under field conditions. In Sweet Potato—Proceedings of the First International Symposium, 1981. R.E. Villareal and T.D. Griggs, eds. AVRDC, Taiwan, pp. 119–127.

Bouwkamp, J.C., ed. 1985. Sweet Potato Products: A Natural Resource for the tropics. CRC Press, Boca Raton, FL.

Clark, C.A., and Moyer, J.W. 1988. Compendium of Sweet Potato Diseases. American Phytopathological Society Press, St. Paul, MN.

Collins, W.W. 1988. Sweet potato production in the developing world. In Improvement of Sweet Potato *(Ipomoea batatas)* in East Africa: With Some References to Other Tuber and Root Crops. 1988 Workshop Report ILRAD, International Potato Center, Lima Peru.

Hahn, S.K., and Hozyyo, Y. 1984. Sweet potato. In The Physiology of Tropical Field Crops. P.R. Goldsworty and N.M. Fisher, eds., John Wiley & Sons, New York, pp. 551–567.

Hall, M.R., and Phatak, S.C. 1993. Sweet potato. In Genetic Improvement of Vegetable Crops. G. Kalloo and B.O. Bergh, eds. Pergamon Press, Oxford, pp. 693–708.

Hall, M.R. 1994. Yield of sweet potato cuttings is not influenced by shoot apex or polarity. HortScience *29,* 41.

Kays, S.J. 1985. The physiology of yield in the sweet potato. In Sweet Potato Products: A Natural Resource for the Tropics. J.C. Bouwkamp, ed. CRC Press, Boca Raton, FL, pp. 80–132.

Martin, F.W. 1985. Differences among sweet potatoes in response to shading. Trop. Agric. (Trinidad) *62,* 161–165.

O'Brien, P.J. 1972. The sweet potato: its origin and dispersal. Am. Anthropol. *74,* 342–365.

Scott, L.E., and Bouwkamp, J.C. 1974. Seasonal mineral accumulation by the sweet potato. HortScience *9,* 233–235.

Sherf, A.F., and Macnab, A.A. 1986. Vegetable Diseases and Their Control, 2nd ed. Wiley-Interscience, New York.

Villareal, R.L., Lin, S.K., Chang, L.S., and Lai, S.H. 1979. Use of sweet potato *(Ipomoea batatas)* leaf tips as vegetables. I. Evaluation of morphological traits. Expl. Agric. *15,* 113–116.
Woolfe, J.A. 1992. Sweet Potato: An Untapped Food Resource. Cambridge University Press, Cambridge.

11

Cassava

Manihot esculenta Crantz *(M. utilissima)*
Family: Euphorbiaceae
Other names: manioc, yuca, mandioca, tapioca

Origin

The genus *Manihot* is endemic only in the Western Hemisphere and is found mostly in the dry frost-free tropics. *M. esculenta,* a possible allotetraploid, $(2n = 36)$ is not found in the wild state. A center of cassava origin is believed to be somewhere in the northern Amazon region of Brazil, with dispersal thousands of years ago to adjoining areas. Southwest Mexico is also considered a region of origin or diversity. Considerable diversity is found between cassava forms in southern Mexico and Central America as compared to those in northeastern Brazil. In the period after Columbus, cassava was introduced into Africa in the late 16th century and to India during the early 19th century. In these locations as well as Southeast Asia, cassava rapidly became an important food staple.

Botany

Cassava is a monoecious dicot grown primarily for the highly digestible starch in storage roots, erroneously called tubers. A perennial bushy shrub, cassava grows 1–4 m tall with large deeply palmate leaves with lamina having five to nine lobes (Figs. 11.1, 11.2). Leaves, borne on long petioles are deciduous, lasting only a few months or less.

The stems exhibit distinctive branching patterns which vary with cultivars. Preference is for initial erect growth before branching occurs because of the ease of cultivation; excessive and especially basal branching is undesirable. Old stem portions exhibit prominent leaf scars; wide internodes denote rapid growth rate.

Seed propagated plants produce a prominent tap root. For vegetatively propagated plants, fibrous roots are initiated from the base of

FIG. 11.1.   Cassava plants growing in Nigeria. Source: Courtesy of I.C. Onwueme.

stem cuttings. Storage roots develop from the secondary thickening of previously fibrous adventitious roots. Enlargement occurs first at the proximal end, then advances toward the distal end. Storage roots vary in shape, and although most are cylindrical and tapering (Fig. 11.2); some exhibit branching. A short constricted woody textured portion connects the storage root to the stem. Storage roots range from 15 to 100 cm in length and from 3 to 15 cm in width. Depending on age, root weight ranges from several hundred grams up to 15 kg. Storage root enlargement is influenced by day length and can be initiated as soon as 8 weeks following planting. Plants commonly produce about 5–10 storage roots. The outer corklike periderm layer is often reddish in color, with variation from dark brown to off-white. The periderm layer encloses the thin cortex layer and thick fleshy parenchyma tissues and often shows cracking because of continued expansion. Interior flesh is usually white, although yellow and red-tinged colors are observed. A

FIG. 11.2.   Cassava foliage and storage roots.

prominent strand of tough vascular tissues is located in the center of the storage root (Fig. 11.3). Latex is present in the root as well as all plant parts.

Cultivars are often identified by the degree of bitterness, which is associated with the poisonous cyanogenic glucosides contained in the roots. "Sweet" types contain little of the glucosides and do not require detoxification before consumption. The amount of glucosides produced in roots is affected by the growing environment, and even sweet types have some bitterness.

Flowering is common for most cultivars and can occur as soon as 6 weeks after planting, yet some cultivars rarely flower. Flowers when formed are about 1 cm in diameter and occur in loose clusters on terminal panicles near the end of the branches. Flower color varies from greenish purple to light greenish yellow. Pistillate flowers are receptive well before male flowers open, thereby limiting self-pollination. Being out-crossers, *M. esculenta* and wild *Manihot* species exhibit

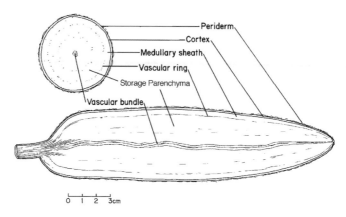

FIG. 11.3.   Cassava root anatomy, cross and median transverse sections.

inbreeding depression when selfed. Flowers produce nectar and insects perform cross-pollinations. Seed capsules are small and highly angled with narrow wings; they dehisce explosively and usually contain three flat hard seed that require scarification to improve germination.

Cultural Requirements

Best cassava growth occurs in regions between 15° latitude above and below the equator where mean temperatures are between 25°C and 27°C. However, adequate growth occurs within a temperature range of 16–30°C and at latitudes up to 30°C. Growth stops when temperatures are below 10°C and growth is reduced above 35°C; the plant is killed by frost. Low elevations are preferred, but temperature permitting, cassava can be grown up to elevations of 2000 m. The plant is very efficient at converting solar energy into carbohydrates, and high light intensity is important for high productivity. Cassava has the potential to accumulate and store more carbohydrates on an area–time basis than any of the major grains and most root or tuber crops.

Moisture needs are typically supplied by rainfall, and because of its relatively low cash value, supplemental irrigations are infrequent. Plantings are made early in the rainy season because adequate soil moisture is critical for crop establishment. Cassava has a broad adaptation to soil moisture and grows in areas differing in rainfall from 500 to 5000 mm, but most major production regions generally average between 1000 and 2000 mm. Plants tolerate long drought periods and cultural neglect by restricting growth through leaf shedding and assuming a dormantlike stage. Following rain, plants quickly develop new leaves.

Sandy or sandy loam soils are preferred. Deep loose soils allow better penetration for the developing storage roots; shallow and compacted soils inhibit root size and shape. Poorly aerated or waterlogged soils suppress growth and are conducive to root rot.

Although a slightly acid soil medium is optimum, cassava is tolerant to soil pH levels from 4 to 8; high salinity limits growth. Plants are also tolerant of low soil calcium and high available aluminum and manganese, conditions common to high-rainfall, acidic tropical soils, which most vegetables do not tolerate.

Cassava has a reputation of being soil nutrient depleting, which is not accurate because its nutrient requirements are similar to many other crops. That impression probably developed because in slash and burn agriculture, cassava is often the last crop grown before the land is abandoned. This reputation is given because of its ability to produce relatively well in nutrient-depleted soils.

Vegetative growth is promoted by high soil nitrogen and irrigation, and for high yields, several applications of nitrogen fertilizer, not to exceed 100 kg/ha during growth, are often suggested. However, excessive vegetative growth delays storage root development and, therefore, should be avoided.

Propagation

Cassava plants can produce viable seed, but these are not used for propagation because of variability. Instead, propagation is almost exclusively with stem cuttings, called stakes (Fig. 11.4). This method gives homogeneous plant populations.

Stakes, 20–30 cm long taken from a previous crop, are planted either horizontally, inclined, or vertically. If horizontal, the stake is covered with 5–10 cm of soil. Otherwise, one end of the stake is placed about 10–15 cm deep into the soil. Stake orientation does affect storage root formation; horizontal placement produces a greater number of storage roots closer to the surface than either upright or inclined stake orientations. From 10,000 to 20,000 stem pieces are planted per hectare.

Spacings typical for traditional ridge or raised-bed plantings are 50 cm between plants in ridges 1 m apart; flat plantings more often are equidistant at about 80–100 cm. A larger number of stakes are planted in soils of low fertility, and although total yield is not improved, the number of storage roots are increased. Cassava is planted in pure stands as well as interplanted with maize, groundnuts, or beans, in which case, plant spacings are widened.

FIG. 11.4.   Cassava stems (stakes) used for propagation.

Stakes should be selected from disease-free plants that exhibit high yields. Old, thick, and long stakes are recognized as being more productive because of greater carbohydrate reserves. However, old stakes are more likely to be virus infected. As a compromise, stakes are often taken from the midpoint of mature woody stems which have a lower probability of virus infection. Harvested stakes can be used immediately or, if necessary, stored before use. Where stake gathering is poorly organized, stake quality and health are frequently substandard. Nursery techniques for rapid production of virus-free stem cuttings are used by progressive growers. An advantage of using stem cuttings is that the crop's edible products are not used for propagation as with other underground starchy vegetable organs.

In traditional cultivation, stakes are planted in ridges or mounds that allow storage roots to develop and remain below the soil surface. Minimal tillage is common, although soil where the stake is planted may be loosened. While providing better drainage, ridge or mound plantings make hand harvesting easier. Large-scale producers able to mechanize their cultural and harvest practices frequently use flat planting procedures. Equipment is used to make shallow furrows into which stakes are placed horizontally and then covered with soil. Light textured soils are an advantage in mechanization.

Growth and Development

As the buds on the stem cuttings sprout and grow, callus tissue forms at the base of the cut surface of the stake. Roots soon emerge from this tissue as well as from nodes of new shoot portions in the soil. Depending on temperatures, sprouting usually occurs in 1–4 weeks, with an average of two to three sprouts developing from each stake. It is important that adequate soil moisture is available at this time. At first, leaves are small, but a full canopy and maximum leaf area index is generally achieved in 4–6 months after planting. A representative growth of a cassava plant during a usual production period demonstrating the change in leaf area index and total accumulated plant dry weight and of the storage root is shown in Fig. 11.5. First-year harvest generally occur at the conclusion of the rainy season, although harvest may be delayed until the second season of growth. At the end of the rainy season, leaf shedding occurs with a decline in leaf area index. Little or no growth occurs during the dry season, but with the resumption of rain, leaf growth resumes, and the leaf area index and storage root dry weight increase.

Cassava root production is largely affected by leaf development and its maintenance. Studies with different cultivar types indicate that plants which achieve a high leaf area index (LAI) produce high yields.

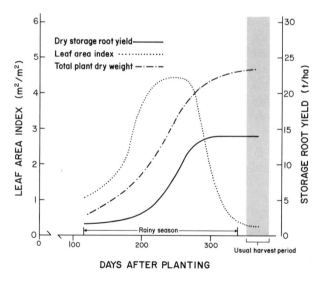

FIG. 11.5. Representative first year growth of the cassava plant. Source: (Drawn from data of Oka et al, (1989).

TABLE 11.1. GROWTH AND YIELD CHARACTERISTICS OF NONBRANCHING AND BRANCHING CULTIVAR TYPES OF CASSAVA

	Plant			Storage root	
Cultivar type	Height (m)	Stem apices	No.	dry wt. yield (t/ha)	% Starch content
Nonbranching	3.4	2.0	7.2	10.6	20.5
Branching	2.2	12.1	7.6	14.4	24.1

Source: Modified from data from Oka et al. (1989).

Branching habit influences plant height and the efficiency of foliage to produce photosynthate and thereby yield (Fig. 11.1). A comparison of an erect nonbranching cultivar and a branching cultivar shows the advantage of branching with regard to final storage dry root yield (Table 11.1).

Erect nonbranching plant types, although taller, produce fewer stems and leaves during early growth, resulting in an LAI less than that of plants with a branching habit. Stem apical dominance tends to suppress branching and axillary bud growth, a characteristic which varies with cultivars. Apical dominance is also weakened when flowers form at the apex. Cultivars with extensive branching and foliage development may produce more dry matter in leaves and stems to the disadvantage of storage root dry weight accumulation. The extent of the leaf canopy coverage is more important than canopy thickness, which may be greater in the tall plant but may not be beneficial if shading reduces the photosynthetic efficiency of lower-level foliage. Ideally, high-yielding cultivars should have a high LAI and high harvest index.

Root enlargement occurs under short days (less than 12 h). Roots developing from the central stem are long, cylindrical, and tapered. Once enlargement is initiated, the storage root decreases or ceases to function as an absorbing organ. Roots accumulate starch and show enlargement as early as 6–7 weeks after planting.

First-year storage roots range from 20 to 50 cm in length and with diameters between 5 and 10 cm. Size increases with continued growth; the longer the growth period and the greater the leaf area, the greater is the yield. However, extended growth periods can decrease edible quality because of increased fiber development. Nevertheless, as a hedge against famine and crop failures, growth may be allowed to continue for several years.

Harvesting

Because storage roots continually enlarge, they do not have a discrete maturity period. As a result, the harvest period can be variable, which partly explains the differences reported for crop yields that can range from less than 2 t/ha to more than 20 t/ha; about 10 t/ha is average. Harvests are usually made between 12 and 15 months after planting but can occur as early as 6 months, or even after 2 or 3 years. In general, sweet-type cultivars mature in 6–9 months compared to bitter cultivars that are grown for 12–18 months to obtain high yields. The latter are mainly used for processed foods, animal feed, and industrial products. Under ideal experimental conditions, yields of 80 t/ha have been achieved.

Harvest is achieved by cutting off stems and leaves, and leaving enough stem length to aid in uprooting and pulling the roots from the soil. In most of the world, hand harvesting is the usual practice, although for large-scale production, specialized equipment cuts stems and removes roots from the soil. In traditional cultivation, complete field harvest is not practiced because of the rapid perishability of storage roots after harvest. Cassava leaves are often used as potherbs and excessive defoliation and topping can also diminish yields.

Postharvest and Storage

Normally, cassava has a short postharvest life, with quality deterioration beginning soon after harvest. At ambient temperatures, quality loss is rapid; sweet-type storage roots deteriorate faster than bitter types. Internal discoloration (streaking) of vascular tissues can occur within as little as 2 or 3 days, and occasionally is followed by decay. The most common storage method is to leave roots unharvested until needed.

A storage practice in the tropics, during cool periods, is to place layers of harvested roots between layers of straw, leaves, or similar dry material. This is covered with a layer of soil about 15 cm thick. During warm periods, a thicker, 30–40-cm, layer of soil insulation is used. Vents are placed in the pile to provide air movement and cooling. The roots are removed as needed, but the storage period is relatively short. Alternatively, storage roots layered in moist sawdust and placed in plastic bags are stored in boxes or bins. If refrigeration is available, cassava can be held reasonably well for 1 or 2 weeks at 0–5°C with 85–95% RH.

Careful handling is necessary for any extended storage. If the easily bruised roots are damaged, they soon exhibit a grayish discoloration of the internal tissues and result in increased susceptibility to rots. Coating roots with wax or overwrapping with plastic film help reduce losses from desiccation and retard deterioration.

Diseases and Other Pests

Phytosanitary practices avoid and substantially reduce cassava diseases. Production and maintenance of disease-free propagating materials reduce crop losses. Although chemical pesticides are an important resource for pest and disease management, they are seldom used by subsistence farmers. Some integrated pest management programs utilize native predators and parasites in the control of insects such as mealybugs and green spider mites.

Important bacterial and fungal diseases affecting cassava include the following:

Bacteria

Cassava bacterial blight	*Xanthomonas campestris* pv. *manihotis*
Bacterial angular leaf spot	*Xanthomonas campestris* pv. *cassavae*
Bacterial stem rot	*Erwinia carotovora* pv. *carotovora*

Fungus

Anthracnose	*Colletotrichum manihotis, C. gloeosporioides*
Brown leaf spot	*Cercospora henningsii*
Cassava rust	*Uromyces manihotis*
Diplodia dry root and stem rot	*Diplodia manihotis*
Fusarium root and stem rot	*Fusarium* spp.
Phoma leaf spot or concentric-ring leaf spot	*Phoma* spp., also *Phyllosticta* spp.
Phytophthora soft root rot	*Phytophthora* spp.
Super-elongation	*Sphaceloma manihoticola,* also *Elsinoe brasiliensis*
White root or thread rot	*Fomes lignosus*

Cassava bacterial blight is the most widespread disease and is responsible for reduced growth and considerable loss of production. Super-elongation is an interesting disease that produces an exaggerated elongation of young stem internodes. This is the result of pathogen growth in intercellular spaces of the epidermis and cortex. The young shoots and petioles are distorted, and leaves are underdeveloped and die. Resistant cultivars have been developed and introduced.

Viruses are a prevalent problem for cassava production. African Cassava Mosaic Virus is the most important. Others also of importance are Cassava Common Mosaic Virus, Cassava Frog Skin Virus, Brown Streak Virus, and Cassava Vein Mosaic Virus.

Probably the most important insect pests are green spider mites *(Mononychellus tanajoa),* mealybug *(Phenacoccus* spp.), and white fly *(Bemisia tabaci).* Other pests include burrowing bug *(Cyrtomenus bergi),* red spider mite *(Tetranychus urticae),* hornworm *(Erinnysis ello),* cassava scale *(Aonidomytilus albus),* and variegated grasshopper *(Zonocerus variegatus),* as well as thrips, grubs and stem borers. Rodents, usually rats, and other vertebrate pests are also known to cause significant crop losses.

In the early 1970s, the mealybug inadvertently entered Africa production areas when a new cassava cultivar from South America was introduced for evaluation. Not having natural enemies in Africa, the mealybug flourished and in several years, the pest reportedly destroyed as much as 80% of the crop in some regions of Africa. To correct this problem, a tiny wasp from South America, a natural enemy of the mealybug, was brought to Africa, and within a few years, the problem was brought under control.

Nematodes are very damaging, especially when cassava is repeatedly grown in the same field. Important nematodes are the root knot *(Meloidogyne incognita* and *M. javanica)* and species of *Pratylenchus* and *Rotylenchulus.*

Early in its initial growth, cassava is not competitive with weeds and hoeing is generally practiced. Mulching is often used for weed and erosion control. The leaf canopy of cassava develops relatively rapidly, and once established, it is fairly effective in suppressing weed growth. Herbicides are available but are not widely used because the crop's value often does not justify the expense.

Use and Composition

Approximately 65% of production is used for human consumption in either fresh or processed form. Short postharvest life necessitates rapid utilization whether for direct consumption, marketing, or processing.

In Africa, roots are prepared for human consumption in many ways: dried pieces are ground into flour, grated, fermented, and roasted to produce gari and a sticky dough or porridge called fufu. Farinha de mandioca, a roasted flour, is a traditional and popular food in Brazil. Additional processed products are tapioca, fried chips, sun-dried flakes, and beer. Considerable amounts of cassava are used for animal feeding and industrial starch. The bitter cultivars are most often used to manufacture starch and in the production of fuel alcohol.

The major value of cassava is its high-caloric contribution; fresh roots contain about 35–40% dry matter, of which 90% is carbohydrate. Where edaphoclimatic conditions are limiting, such as in sub-Saharan Africa and northeastern Brazil, cassava often is the main carbohydrate source. On a fresh weight basis, cassava provides about 150 kcal/100 g versus 115 kcal/100 g of sweet potatoes, and on an area basis, it is competitive with grain crops in terms of calories and labor efficiency. Cassava is a good vitamin C source, containing 30–35 mg/100 g fresh weight and usually has a low fiber (1.4%) and fat (0.3%) content.

The presence of cyanogenic glucosides in all plant tissues is a major concern. Roots contain the enzyme linamarase which hydrolyzes the glucosides to prussic acid. The glucosides are water soluble, and the enzyme is inactivated at temperatures greater than 50°C; linamarin is the principal glucoside. A general but not absolute correlation exists between glucoside content and root bitterness. Glucoside levels greater than 50 mg/100 g tissue are poisonous; when greater than 100 mg/100 g, they are dangerous and can cause death. Roots of the bitter types of cassava must be treated to reduce the content of the poisonous substances before consumption. Detoxification is accomplished by soaking roots and discarding the soak water; the glucosides are highly soluble in water. Grinding or shredding the tissues and discarding the expressed liquid, and also fermentation and drying are other detoxication practices. The glucosides will decompose at temperature greater than 150°C. Glucoside content is strongly influenced by cultivar and growth environment.

The inedible peel which comprises about 20% of the root weight also is a undesirable feature, as is the relatively low-protein content, only 2–3% on a dry weight basis. Furthermore, protein quality is only fair; sulfur-containing amino acids, methionine in particular, are low. The protein content of young roots is slightly higher than in mature roots, but this attribute is not realized because of the impracticality of harvesting small roots. Occurrences of kwashiorkor disease caused by severe protein deficiency are occasionally observed in some populations that rely on cassava as the primary food staple.

In many parts of the tropics, cassava leaves are used as vegetables. They are usually harvested from sweet-type, low-glucoside plants. Yields as much as 20 t/ha are obtained where cassava is grown primarily for leaf harvest. Leaves are a good vitamin C source, contain fair amounts of pro-vitamin A, and about 30% protein on a dry weight basis. It is interesting that in Zaire, mosaic-virus-infected leaves are preferred because they are sweeter and more tender than healthy leaves. Additionally, virus-infected leaves contain less cyanogenic glucosides than noninfected leaves. In South China, leaves are used for feeding silk-worms.

Production

Cassava is an important vegetable staple because of its significant daily caloric contribution to millions of people. Essentially all cassava production occurs in developing countries, and the bulk of that production is from small farms, often having marginal soils. Cassava is very important to the rural poor, being a dependable crop even during drought. Drought tolerance and flexible harvest periods make cassava a valuable famine reserve crop.

The high bulk and rapid spoilage mandates local dissemination and consumption, and a limited volume of fresh production is sent to distant markets. However, because the dried processed product is less bulky and more durable, cassava is of some importance in international trade.

Although Brazil is the world production leader (more than 15%), the African countries collectively produce in excess of 47% of the world production. African production is largely utilized as food, whereas proportionally more of the cassava produced in South America and parts of Asia, namely Thailand, is used for animal feed and industrial products, of which a significant volume is exported.

World average yields are low because much of the production occurs in marginal agricultural conditions in the developing countries where cultivar development and agronomic and pest control practices are limited. World cassava production is shown in Table 11.2.

Crop Improvement

Future progress in cassava production relies on genetic exploitation to address the major concerns of low protein content, presence of cyanogenic glucosides, relatively rapid postharvest deterioration, disease and pest resistance, and mechanization. The Centro Internacional de Agricultura Tropical (CIAT) in Colombia and the International Institute for Tropical Agriculture (IITA) in Nigeria have active programs dedicated to cassava improvement. The governmental agency in Brazil, EMBRAPA, is also involved in developing and applying new technologies for cultivar and agronomic improvements. Additionally, Thailand, India, and Cuba have active research programs, and these organizations also function as centers for germplasm collection and conservation. Research on products derived from cassava starch and flour could open new markets and stimulate increased production. When the crop's carbohydrate-producing capability under marginal conditions is consid-

TABLE 11.2.   WORLD CASSAVA PRODUCTION, 1994

	Area (ha × 10³)	Yield (t/ha)	Production (tons × 10³)
World	15,819	9.6	152,473
Africa	9,481	7.7	72,779
North and Central America	200	4.9	984
South America	2,375	12.7	30,050
Asia	3,745	12.9	48,449
Oceania	17	12.2	211
Leading countries			
Brazil	1,838	13.1	24,009
Nigeria	2,000 [a]	10.5	21,000[a]
Zaire	2,430[a]	8.1	19,600[a]
Thailand	1,383	13.8	19,091
Indonesia	1,295	11.6	15,000
Tanzania	693	10.4	7,209
India	235[a]	22.8	5,340[a]
Ghana	607[a]	7.2	4,378[a]
China	231[a]	15.2	3,503[a]
Mozambique	908	3.6	3,294

[a]Estimated
Source: 1994 FAO Production Yearbook, Vol. 48, FAO, Rome, 1995.

ered, it lessens the concern about low protein content, which may be balanced by other sources of food.

Other species of the family Euphorbiaceae grown for vegetables in the tropics are chaya *(Cnidoscolus chayamansa)* katuk *(Sauropus androgynus),* and castor bean *(Ricinus communis)* (see Chapter 25).

SELECTED REFERENCES

Balagopalan, C., Padmaja, G., Nanda, S.K., and Moorthy, S.N. 1988. Cassava in Food, Feed, and Industry. CRC Press, Boca Raton, FL.

CIAT. Annual reports. Centro International de Agricultura Tropical/International Center for Tropical Agriculture (CIAT), Cali, Colombia.

CIP. Annual reports. Centro, International de la Papa/International Potato Center (CIP), Lima, Peru.

Cock, J.H. 1985. Cassava: New Potential for a Neglected Crop. Westview Press Inc., Boulder, CO.

Conner, D.J., and Cock, J.H. 1981. Response of cassava to water shortage. II. Canopy dynamics. Field Crops Res. *4,* 285–296.

Cooke, R.D., and de la Cruz, E.M. 1982. The changes in cyanide content of cassava *(Manihot esculenta* Crantz) tissues during plant development. J. Sci. Food Agric. *33,* 269–275.

Dahniya, M.T., Oputa, C.O., and Hahn, S.K. 1981. Effects of harvesting frequency on leaf and root yields of cassava. Exp. Agric. *17,* 91–95.

Dufour, D.L. 1988. Cyanide content of cassava *(Manihot esculenta,* Euphorbiaceae) cultivars used by Tukanoan Indians in Northwest Amazon. Econ. Bot. *42,* 255–266.

El-Sharkawy, M.A. 1993. Drought-tolerent cassava for Africa, Asia and Latin America. BioScience *43,* 441–451.

El-Sharkawy, M.A., Hernandez, A.P., and Hershey, C. 1992. Yield stability of cassava during prolonged mid-season water stress. Exp. Agric. *28,* 165–174.

Etejere, E.O., and Bhat, R.B. 1985. Traditional preparation and uses of cassava in Nigeria. Econ. Bot. *39,* 157–164.

Howeler, R.H. 1991. Long-term effect of cassava cultivation on soil productivity. Field Crops Res. *26,* 1–18.

IITA: 1990. Cassava in Tropical Africa. A Reference Manual. International Institute for Tropical Agriculture (IITA), Ibadan, Nigeria.

IITA. Annual reports. International Institute for Tropical Agriculture, Ibadan, Nigeria.

Kawano, K. 1990. Harvest index and evolution of major food crop cultivars in the tropics. Euphytica *46,* 195–202.

Lancaster, P.A., and Brooks, J.E. 1983. Cassava leaves as human food. Econ. Bot. *37,* 331–348.

Martin, F.W., and Ruberte, R.M. 1975. Edible Leaves of the Tropics. Antillian College Press, Mayaguez, Puerto Rico.

Odigboh, E.U. 1983. Cassava: Production, processing, and utilization. In Handbook of Tropical Foods. H.T. Chan Jr., ed. Marcel Decker, Inc., New York.

Oka, M., Sarakarn, S., and Limsila, J. 1989. Growth characteristics of recommended cassava cultivar, "Rayong 3" in Thailand. Japan J. Crop Sci. *58,* 390–394.

Onwueme, I.C. 1978. The Tropical Tuber Crops, Yams, Cassava, Sweet Potato, Cocoyams. John Wiley & Sons, New York.

Polson, R.A., and Spencer, D.S.C. 1991. The technology adaption process in subsistence agriculture—The case of cassava in southeastern Nigeria. Agric. Syst. *36,* 65–78.

Renvioze, B.S. 1973. The area of origin of *Manihot esculenta* as a crop plant: a review of the evidence. Econ. Bot. *26,* 352–360.

Rickard, J.E. 1985. Physiological deterioration of cassava roots. J. Sci. Food Agric. *36,* 167–176.

Wheatley, C.C. 1985. Storage of cassava roots for human consumption. In Cassava: Research, Production and Utilization. J.H. Cock and J.A. Reyes eds., U.N. Development Program and Centro Internacional de Agricultura Tropical, Cali, Colombia, pp. 673–684.

12

Yams, *Dioscorea*

Family: Dioscoreaceae

Yams are a very important starchy staple crop of tropical and subtropical agriculture. They have been almost exclusively a subsistence food for the peasant farmer, but increasingly have become more of a commercial crop and are generally preferred to cassava. The name "yam" is often erroneously used to identify several other tuberous and root vegetable crops. Although yams are not of a single species, it would be helpful if this name were used only to indicate plants of the *Dioscorea* genus.

Origin

The genus, *Dioscorea,* the largest in the Dioscoreaceae family, includes more than 500 species. The Dioscoreaceae are monocots but exhibit characteristics of both monocots and dicots, suggesting they may have been among the early Angiosperms. Some taxonomists believe there was an early separation in *Dioscorea* development into Old and New World species. This is reflected in cytological differences, where Old World (Africa and Southeast Asia) species have a basic chromosome number X = 10. New World species have X = 9 as the basic number, with most being tetraploid or hexaploid. High ploidy also occurs in Old World yams, with chromosome counts ranging from 50 to 140.

Yam cultivation appears to have developed independently in west Africa and Southeast Asia, possibly about 3000 B.C. or earlier. West African artifacts indicate yams were used as food for thousands of years. Suggested centers of domestication are Indochina, South China, West Africa, and the Caribbean/South America region.

Taxonomy

The taxonomy of the genus *Dioscorea* is sometimes confusing. For some species, morphology is not sufficiently distinctive for adequate identification, and for most cultivated species, wild forms can still be found. Although spontaneous hybridization must have contributed to the ancestry of some yams, propagation has been entirely vegetative, each "cultivar" type being a single clone. About 60 *Dioscorea* species have edible tubers. Those usually cultivated as crop plants include the following:

D. alata L.—Greater yam. Other names are water, winged, and Ubi yam. Of Southeast Asian origin but now the mostly widely distributed species. Vines twine right.* Bulbils sometimes produced. Leaves are ovate and oppositely arranged. Tubers are single or multiple, varying in size, shape and color and have a long dormancy period.

D. bulbifera L.—Potato or aerial yam. African and Asian origins. Vines twine right. Stems without hairs or spines. Leaves large and simple. Aerial tubers (bulbils) are produced in leaf axils with average weight of 600–700 g, and width and length of 4 and 8 cm, respectively. Aerial tubers, although succulent and edible, may require detoxification before consumption. Underground tubers are either absent or usually smaller than those of other *Dioscorea* spp.; they are hard, bitter, and not palatable without treatment.

D. cayenensis Lam.—Yellow yam. West African origin. Vines twine right. Vines are spiny with leaves simple, arranged opposite and cordate with pointed tip and hairless. Bulbils are not formed. Tubers vary considerably in size and shape and have a short dormancy and storage life. Requires a long growing season; truly tropical, performs best in warm temperatures with long rainy seasons, not drought tolerant.

D. dumetorum (Kunth) Pax.—Cluster, bitter, or trifoliate yam. African origin. Vines twine left.† Stems hairy with spines; leaves trifoliate; bulbils rarely produced. Tubers medium to large size and coarse, sometimes fused into a cluster. Tubers often have a high alkaloid content. After cooking, tubers harden rapidly.

D. esculenta (Lour.) Burk.—Lesser yam. Other names include potato, Asiatic, and Chinese yam. Of Indochina origin, but found in all Asiatic regions. Vines twine to the left. Thin pubescent stems are cylindrical with many spines. Leaves are few, simple, cordate, and alternate; bulbils are absent. Tubers are many and small, thin

---

*Right-hand twining rule: thumb of right hand up, fingers are in direction of vine growth.

†Left-hand twining rule: thumb of left hand up, fingers are in direction of vine growth (Fig. 12.2).

skinned, mostly cylindrical and are produced near the surface and easily harvested; some forms produce single large tubers. The dormancy period is short.

D. hispida Dennstedt—Intoxicating or starch yam. Indian and Southeast Asian origin. Vines twine left; bulbils are not produced. Tubers formed near the surface are often lobed, sometimes elongated. Some clones need to be detoxified before consumption.

D. japonica Thunb.—Jinenjo yam. Indigenous to Japan and found wild in open fields and in the mountains. Vines twine right. Although similar in cold tolerance, it differs from *D. opposita* in chromosome number, vegetative characteristics, tuber shape, and chemical composition.

D. opposita Thunb.—Chinese or cinnamon yam. Synonymous with *D. batatas.* China origin. Vines twine right; bulbils occasionally formed. Leaves are simple, oppositely arranged. Variable tuber shapes, from thin cylindrical with tendency to grow vertically downward, some also palmate to globular. Tubers do not contain sapogenins. Of the cultivated yams, Chinese yam has the best temperature adaptation and is grown in many temperate regions; more than 100,000 tons of this species is produced annually in Japan. There are three strains of *D. opposita* in Japan: naga imo, is cold resistant, produces long club-shaped tubers (Fig. 12.1), icho imo producers fan-shaped tubers, and yamato imo produces round tubers, of which some cultivars are not cold tolerant.

D. rotundata Poir.—White or Eboe Yam. Of West African origin and the major cultivated species. Considered by some as a subspecies of *D. cayenensis.* Vines spiny and twine to the right. Leaves are simple, arranged opposite, cordate with pointed tip and hairless. Bulbils are seldom produced. Produces a single usually spherical or cylindrical tuber usually in 8–10 months. Best species for making fufu.

D. trifida L.f.—Cush-Cush, Yampi, or Aja Yam. Only important cultivated species of tropical South America origin. Vines twine left; stems spineless and rectangular in cross section with flatten wings; bulbils not formed. Leaves simple with 3–5 lobes. Produces a large number of small tubers, often resembling sweet potatoes of varying flesh colors. Adapted to warm temperate regions.

Some cultivated and wild species are also used for medicinal purposes.

Botany

Yams are tropical perennials usually cultivated as annuals for their underground or aerial tubers. Plants exhibit a strong annual cycle of

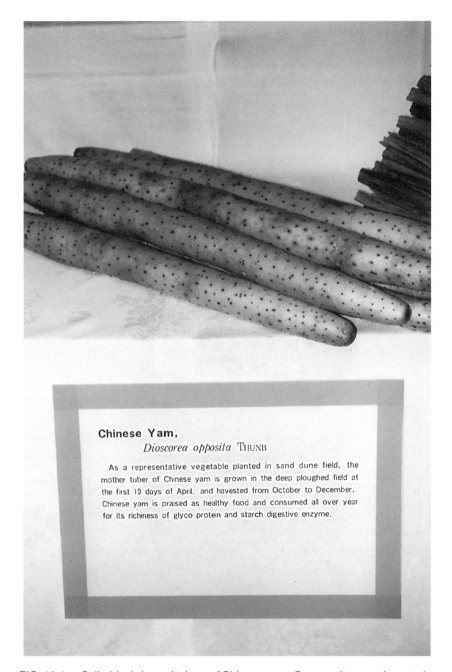

Chinese Yam,
Dioscorea opposita THUNB

As a representative vegetable planted in sand dune field. the mother tuber of Chinese yam is grown in the deep ploughed field at the first 10 days of April, and havested from October to December. Chinese yam is praised as healthy food and consumed all over year for its richiness of glyco-protein and starch digestive enzyme.

FIG. 12.1.   Cylindrical shaped tubers of Chinese yam, *D. opposita,* naga imo strain.

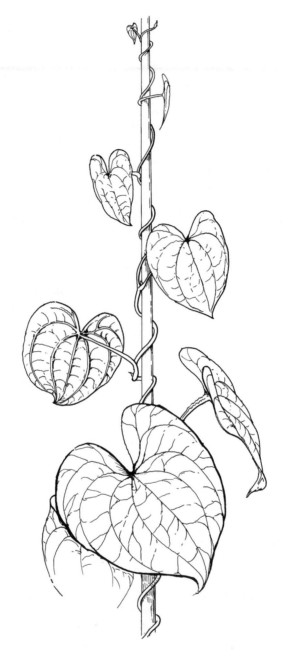

FIG. 12.2. Yam vine exhibiting right-hand twining habit. Note: arcuate veining of leaf blade.

growth and dormancy, the organ of dormancy being the tuber. The stem is usually produced annually when the rainy season begins. Initially erect, the stem climbs and twine over trees or onto supports; stems are not self-supporting. Stems are spiny or winged. Stem twining direction, either right or left, is consistent for a species. Species vary with respect to stem cross-sectional shape, presence of spines, hairs, leaf surface bloom, and winglike stem appendages. Vines typically die at the end of the rainy season or when temperatures are low.

Yams have a fibrous root system that extends mostly horizontally, with the majority of roots being relatively shallow. Adventitious roots are initiated from the cormlike stem base (primary nodal) or directly from tubers.

Borne on long petioles, leaves usually are simple and cordate, but are lobed or palmate in some species. A genus characteristic is that the primary veins of the leaf blade are arcuate (Fig. 12.2). Leaves vary in green color intensity; some occasionally produce anthocyanin pigments. Hairs are usually absent. Petioles can twist to maximize exposure of leaf lamina to light. Plants are usually dioecious, producing individual small greenish inconspicuous flowers on long racemes. Flowers, although rare, are scented and insect pollinated, often by thrips. The fruit is a dehiscent capsule with up to six seed. A seed set is rare in some species. Seed are small, light, and winged, allowing wind dispersal. Poor flowering, seed production, and frequent sterility are the probable evolutionary consequence of a long history of vegetative propagation.

All edible species develop one or more tubers, and these are renewed annually. In some nonedible species, the tuber is a perennial organ that increases in size and fiber content from year to year.

Some species produce aerial tubers known as bulbils. These are formed in the axils of newly formed leaves after plants have achieved sufficient growth, and usually continue to be formed until the end of the growing season. Bulbils are small, but otherwise similar to tubers and exhibit a wide range of shapes and colors (Fig. 12.3).

The yam is considered a tuber because it originates from the stem and resembles a stem more than it does a root. However, there are some atypical stem features such as a lack of leaf scars, buds or nodes on or near the surface, and the lack of a terminal bud. Furthermore most species exhibit strong geotropic growth. Evidence that the tuber originates from hypocotyl tissue is strong; the first meristematic activity for root, shoot, and tuber growth occurs in the area of the cormlike structure at the base of the vine. (Fig. 12.4). The primary meristem is at the distal end. As growth continues, a thin layer of meristematic

FIG. 12.3. Yam aerial tuber of *D. bulbifera* growing in the Virgin Islands. Source: Courtesy S.M.A. Crossman.

cells is left beneath the cortex of the tuber, which, with continued activity, produces parenchyma tissues. This layer also has the capability to initiate buds that develop into shoots. The major portion of the tuber is the starch-bearing parenchyma tissues which are responsible for the increase in girth. Vascular bundles pass throughout the center of the tuber characteristic of a monocot. A corky suberized skin layer that protects the tuber is formed from cork cambium.

Cultivated species produce annual underground tubers; some have tubers which are perennial. Depending on the species, singular or multiple tubers are produced. Tuber shapes also vary with species as well as the environment and may be cylindrical, branched, or lobed.

Tuber weight varies with species and environmental conditions. Some may weigh less than 1 kg, and others surpass 50 kg. Tubers weighing more than 100 kg have been reported, but most grown for food are in the range of 2–10 kg. Growth cracks caused by tuber expansion often occur in the corky layer. Interior flesh colors are white or light yellow; some yams contain anthocyanins, which impart colors ranging from pink to purple.

Tuber dormancy starts with the cessation of growth, which usually

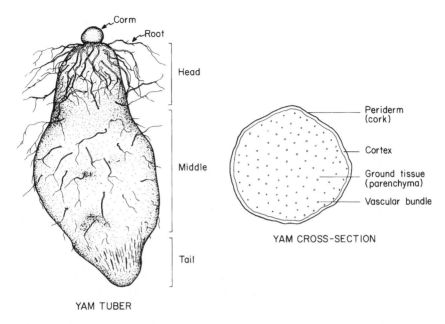

Corm

Root

Head

Middle

Tail

YAM TUBER

Periderm (cork)

Cortex

Ground tissue (parenchyma)

Vascular bundle

YAM CROSS-SECTION

FIG. 12.4. Yam tuber anatomy, *D. alata.*

correlates with the start of the dry season. By the start of the wet season, dormancy will have dissipated and new shoots develop.

Cultural Requirements

Most yams are grown in tropical and subtropical regions and are intolerant of frost. Some, like *D. opposita,* tolerate low temperatures and are grown in warm temperate areas. Optimum growth for most species occurs at about 30°C; growth is poor at temperatures averaging less than 20°C. Temperatures above 35°C are detrimental to growth. Soil mulching is frequently performed to protect against excessive heat and desiccation. Mulches improve soil texture, aeration, and fertility. Dried grasses or straw with a soil cover are often used as a mulch.

Yams are usually grown in regions of relatively well-defined wet and dry seasons. Average rainfall, about 1500 mm, is common in these yam-producing areas. Yields are reduced when rainfall is less than 600 mm, especially if the plant is under moisture stress during tuber enlargement. Vine weight and tuber yield are strongly correlated with available moisture. Tuber growth is retarded in waterlogged soils. Species differ in their response to drought; some are tolerant of low moisture.

Loose, deep, and well-drained fertile soils are required for high tuber yield. Yam root systems are typically shallow and sparse, and productivity is limited in poor soils. They are often the first crop grown in "slash and burn" practices, or because of residual nitrogen, they are likely to follow a legume crop. Compared to other root crops, yams require the most intensive management and highest soil fertility levels to obtain a high yield of good quality tubers. Nutrient utilization is similar to white potatoes, yams being responsive to nitrogen and potassium. The best results are obtained when nitrogen is uniformly supplied throughout growth. Application is most beneficial if nutrients are available at the stage of growth when the plant is no longer dependent on the tuber seed piece for nourishment. However, because the largest proportion of yam production is by subsistence farmers, supplemental fertilizers and crop care often are not provided. For good yields, the crop requires a considerable amount of labor, a primary reason why it is a relatively expensive commodity.

Yams are frequently intercropped, often with maize where the stalks have a secondary function in supporting climbing vines. However, intercropping frequently diminishes yield, which is a reason why this practice is decreasing. Stakes or poles, 2–5 m long, as well as various types of trellis provide vine support; small trees are also used for vine support (Fig. 12.5). The potential yield can be reduced drastically for some species when vines are not supported for maximum light exposure. Until vine growth is well established, weeds can easily outgrow the crop; therefore, they must be controlled or yield will be reduced.

Propagation and Planting

Small intact tubers, bulbils, or tuber portions (setts) are used for propagation. Seed tubers weighing between 100 and 150 g are usually used whole. Head pieces (crowns) of 250 g are regarded as the ideal propagating material because of rapid uniform sprouting and high productivity. Bulbils usually produce smaller tubers and often require a longer growing period before the crop is harvested. Bulbils are commonly used for *D. bulbifera* propagation, but those from other species are used less often. Cut tuber seed pieces, weighing 200–300 g, are used when small whole tubers are not available. The proximal end is preferred because it produces the best vine growth, followed by the middle section, with the poorest growth resulting from use of the distal end. Minisetts can be used to overcome the otherwise low multiplication rate of propagating materials. Minisetts are small disklike sections cut from the tuber surface. With this procedure about 40,000 setts can be produced from 1 ton of tubers.

FIG. 12.5.   Yam production on trellis in Tottori prefecture, Japan (a). Source: Courtesy D.N. Maynard. Full canopy growth of *D. esculentus* yams in Puerto Rico (b). Source: Courtesy F.W. Martin.

In some areas of West Africa, large seed pieces are preferred because it is believed that the greater food reserve provides vigorous rooting, vine growth, and high individual tuber yields. However, research has shown that seed tubers larger than 300 g offer little or no yield advantage while utilizing large amounts of otherwise edible product. A considerable amount of the crop must be reserved for propagation; as much as 3 tons of propagating material is used for 1 ha. Unfortunately, many farmers market their best yams and use inferior tubers for propagation.

Seed tubers are planted when dormant or presprouted. Once dormancy is broken, sprouting is rapid in high humidity at 25–30°C. High humidity is necessary for rootlet formation. Sprouting was shown by Onwueme (1978) to be initiated in a meristematic region within parenchyma tissues about 1 cm below the corky tuber surface, usually at the proximal (head) end of the tuber. The cells in the meristematic region produce a mass of undifferentiated cells that become organized to develop a shoot apex. The shoot apex ruptures the tuber skin as it elongates. Roots soon form where the base of the shoot apex and tuber are joined (Fig. 12.6). Sprout initiation at the proximal end occurs more readily than at the middle or distal (tail) end. Buds often are easily induced after tuber cutting because apical dominance has been eliminated.

Generally, vine cuttings are not used for propagation because production is delayed, although it is a method for increasing propagating material without using tubers. Meristem culture for virus-free propagules is possible; but because of the very slow development of plantlets, this procedure is limited. Except for breeding, the use of true seed, which is available in very few cultivars, is limited for propagation. This is because of seed dormancy, variability, poor tuber quality, and low yield.

In developing countries, considerable traditional slash and burn agriculture is practiced for yam production. Land is cleared, but some small trees may be left to provide vine supports. Farmers provide minimal tillage, although conventional tillage and cultivation may be practiced by commercial-scale producers.

Planting schedules are influenced by regional climate and are usually made at the end of the dry season. Most rely on the start of rainfall to establish new plantings. However, planting can be done during the dry season, as it allows time for seed tubers to overcome dormancy, and soon after rain, the sprouts emerge. Plantings may be on mounds, ridges, or raised beds of about 50 cm height to improve drainage and to simplify harvesting. Flat and trench cultivations are also employed.

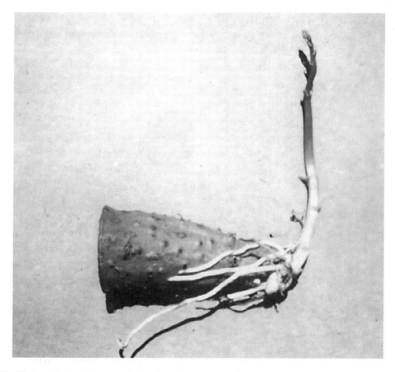

FIG. 12.6. Yam *(D. opposita)* tuber piece used for propagation showing developing sprout and roots.

These practices are less labor intensive because they can be mechanized, although they generally result in lower yields.

Unlike sweet potato or white potato where roots or stolons penetrate into the soil and later enlarge, yam tuber enlargement is initiated early, but enlargement (bulking) is a continuing process, the rate varying with species. Unless the soil is very friable, penetration is impeded because of the tuber's enlarging blunt distal portion. Therefore, a deep and loose soil, provided by tillage or mounding, is an advantage for deep penetration before tuber enlargement is initiated. Mulching is useful to protect tubers from injury during weeding operations. Propagating materials are planted about 10–15 cm deep, usually at spacings of about 1 × m, but spacings vary with the species and the size of the propagating material. Populations commonly average about 10,000 plants per hectare. High densities increase total yield but with small tuber size. However, this is a desirable practice with some species.

Growth and Development

The growth of the yam plant is strongly influenced by moisture availability. Because irrigation is not usually practiced in yam-growing countries, moisture stress severely limits growth. This is most severe at the time of tuber enlargement (bulking). Day length also has an influence; long days favoring vine growth, and short days favoring tuber induction. With some species, old plants when exposed to short days will tuberize sooner than similarly exposed young plants. Because day length in tropical regions is relatively constant, photoperiod generally is not a significant factor.

Seed tubers stored for a long time sprout earlier and tuberize sooner than those stored for a short time. Prior direct light exposure of seed tubers does not affect subsequent tuberization. Early tuberization, before production of an adequate plant canopy, is undesirable and results in small tubers and low yields.

Using seed tuber nutrients and moisture, initial shoot growth is very rapid, elongating as much as 15 cm per day. Within a month or two, the plant becomes self-supporting and in most species the initial seed piece eventually shrivels and/or rots away. In some species, the propagating tuber will continue growth or translocate its nutrients to the new tuber(s).

Stem and leaf growth is rapid and the plant reaches a full canopy about the time of tuber initiation. Foliage growth slows during tuber enlargement. New tuber formation is initially slow and follows a sigmoid curve, with tuber size rapidly increasing with time. Carbohydrates accumulate as long as growth continues. With foliage senescence, further tuber enlargement ceases and tubers become dormant. The seasonal growth of yam is illustrated in Fig. 12.7.

Harvest

In most tropical regions, harvest normally occurs after the end of the rainy season, when foliage exhibits senescence. Harvests are usually delayed as long as vines are functional. In some situations, tubers are harvested as early as 7–9 months, but 10–11 months are common. In some areas, the crop can be left until needed; the period can vary depending on species and conditions. The usual harvest procedure removes all the tubers from the soil.

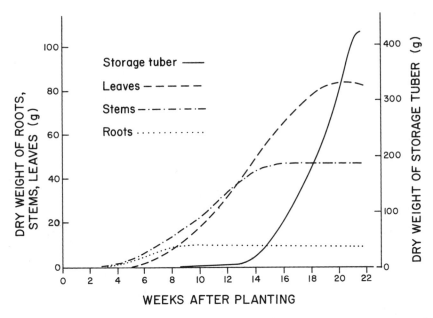

FIG. 12.7.   Seasonal growth of yam *D. rotundata*. Source: Redrawn from data of
E. Njoku et al. (1984).

When partial harvesting is performed, it generally occurs about the
midpoint of the growth cycle. In this procedure, soil is carefully removed
and some tubers or portions of large tubers are cut off below the cormlike attachment to the stem. Soil is replaced and the remaining tuber
heals and growth resumes. Additionally, new tubers may form which
can be harvested at the end of the growing season. Bulbils are harvested
when considered large enough by simply pulling them from the vines,
or collecting them after they have fallen to the ground.

Harvesting is carefully done to avoid damaging the fragile tubers. In
traditional agriculture, wood spades or digging sticks are used, making
harvests laborious and slow. Avoiding damage during handling and
storage is critical because injury leads to losses from decay (Fig. 12.8).
Variables such as tuber sizes, shapes, and depth in the soil hamper
efficient harvest mechanization. Nevertheless, in nontraditional large-
scale production, equipment is sometimes used.

Because of the many yam types and cultural and environmental
variables in regions where yams are produced, large yield variability

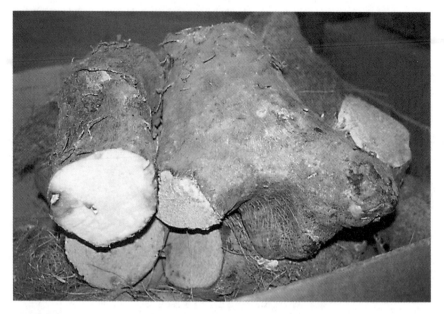

FIG. 12.8. Irregularly shaped tubers of *D. alata*. Note: the decay and discoloration of freshly cut tuber portion at the left.

occurs. About 12 t/ha is the worldwide average production, but intensive cultivation easily surpasses that level and yields exceeding 40 t/ha have been reported.

Storage

The simplest storage is to leave dormant tubers in the ground. Some species and cultivars exhibit a long dormancy period which allows tubers to be left undug. However, this period should be short to avoid sprouting when tubers are no longer dormant. Yams are also placed into a ditch or pit and either covered by soil or mulch and shaded. As with *in situ* field storage, this method is unsuitable for a long storage period.

In other storage practices, tubers are surface dried after lifting. Occasionally, they are cured by exposure for several hours to dry air in the shade to heal wounds before being carefully stacked in well-ventilated weatherproof structures or on platforms in the shade. Single layering without direct tuber-to-tuber contact is preferable to piles. Another method is to suspend tubers on twine between poles, for ventilation and rodent protection, in structures called yam barns. Adequate ventilation is very important to prevent high humidity in the area of the tubers and to remove respiratory heat. Weight loss due to respiration and desiccation can average as much as 5% per month. An optimum storage temperature is 15–16°C and 70% RH, best achieved with refrigerated cold storage and humidity management. At 10°C or less, chilling injury can occur. Under favorable conditions, some species can be stored for several months.

When first harvested, tubers are usually dormant. In this state, respiration is low and, hence, dry matter loss (shrinkage) is small. Dormancy varies with species and may last 1–3 months, sometimes longer. Once sprouting begins, respiration and moisture loss increases. Sprouting during storage can be a problem, although sprouts can be rubbed off, as is sometimes done with white potatoes. Sprout-suppressing chemicals, such as maleic hydrazide (MH), are partially effective in extending dormancy when applied to the growing plant; MH tuber soaks are effective in sprout suppression without affecting tuber quality. Although not a common treatment, gamma radiation can also reduce sprouting and weight loss.

Diseases and Pests

Bacteria
Crown gall | *Agrobacterium tumefaciens*
Tuber dry rot | *Corynebacterium* spp.

Fungi
Anthracnose | *Glomerella cingulata*
Leaf blight | *Thanatephorus cucumeris*
Leaf spots | *Cercospera* spp., also *Colletotrichum capsici* and *Gloeosporum pestis*
Neck rot | *Corticium rolfsii*
Rust | *Goplana dioscoreae*
Tuber rots | *Fusarium oxysporum, F. solani, Rhifzopus nodosus,* and species of *Armillaria, Penicillium, Armillaria, Rosellinia,* and *Sphaerostilbe*
Tuber storage soft rot | *Botryodiplodia theobromae*
Witches broom | *Phylleutypa dioscoreae*

Mosaic is the most common and most serious virus disease. Internal brown spotting virus (IBSV) is damaging to tubers during storage.

Major insect pests include the following:

Crickets	*Gymnogryllus lucens*
Greater yam beetle	*Heteroligus meles*
Lepidoptera pests	*Loxura atymnus, Tagiades gana, Theretra nessus*
Lesser yam beetle	*Heteroligus appius*
Mealybug	*Planococcus citri, Geococcus coffeae*
Yam scale	*Aspidiella hartii*
Yam shoot beetle	*Crioceris livida*
Yam weevil	*Palaeopus dioscorae*

Additional important insects pests include armyworms, termites, and grasshoppers. Nematode pests include the yam nematode *(Scutellonema bradys)*, species of root knot *(Meloidogyne)*, and lesion nematodes *(Pratylenchus* spp.). Rodents are chronic and serious pests, consuming tubers both in the soil and in storage.

Yams are poor competitors with weeds until a good leaf canopy is established. Unfortunately, a full leaf canopy is slowly achieved and hand hoeing is a principal means of weed management.

Uses and Composition

In many respects and in many countries, yams are the potatoes of the tropics, where they are an important staple food, especially for many African and Caribbean countries. Yams are also important in regions of Asia, Oceania, and tropical America. In comparison with other carbohydrate-rich root crops such as sweet potatoes and cassava, yams are usually preferred and, accordingly, they are sold at higher prices. Generally, yams are consumed locally where they are produced, and because of the crop's value and high bulk, only a small volume enters into international trade (Fig. 12.9). However, with recent population migrations to temperate regions, yams are increasingly in demand in United States and European ethnic markets.

Yams are prepared for consumption by boiling, frying, or baking. In central Africa, *D. rotundata,* in particular, is used to prepare fufu, which is made by pounding the starchy tissues into a glutinous dough. Other uses includes dried products, such as flour, flakes, or chips. Yams are generally too expensive for animal feeding or conversion into fuel alcohol. Unfortunately, substantial amounts, 10–20% of the production is lost because of decay, damage, or poor storage. Furthermore, not all of the tuber is edible; there are preparatory losses of 5–15% because

FIG. 12.9.   Roadside sale of yams in Sierra Leone. Source: Courtesy of F.W. Martin.

of the unusable thick skin and hard proximal portions and because tubers of some species are irregularly shaped. Occasionally, shoots of some wild species are boiled and eaten as greens.

Yam composition varies considerably among species. In general, tubers have a high starch content (25% fresh weight), mainly as amylopectin. Pro-vitamin A content is low, but vitamin C content varies between 5 and 15 mg/100 g. The fresh weight protein content is about 2% and the sulfur-containing amino acids are limited. There are species differences in protein content, but differences are not appreciable. However, when the daily volume of consumption is considered, yams can make a significant contribution to some diets.

The mucilaginous substance that exudes from cut tuber surfaces is mostly glycoproteins. The acrid taste of yams is attributed to the presence of tannins. Some species contain the alkaloid dioscorine ($C_{13}H_{19}O_2N$) which is water soluble and is removed by soaking and boiling. Dioscorine levels in cultivated species are generally negligible and not a problem when properly prepared. Tissues of some wild species contain high quantities and can be poisonous; however, after leaching, they can be used as a starch source. Some species contain diosgenin, a steroid sapogenin compound which is used to manufacture cortisone and other drugs.

Production

Over 95% of world yam production occurs in Africa, and more than 72% of that volume is produced in Nigeria. In recent years, world production has slightly increased, primarily because of an increase in cultivated area. Leading non-African countries producing yams were Papua New Guinea, Brazil, Haiti, Japan, and Jamaica which produced 224, 215, 212, 180, and 156 thousand tons, respectively. Although insignificant on a world scale, yam production in Central America and northern South American countries has shown recent increases. In Japan, yams are grown as an expensive specialty crop that also is believed to have some health benefits. Major yam-producing countries are shown in Table 12.1.

Crop Improvement

Overall, yam carbohydrate production is less than that of other starchy tuber and root crops, but yams, being a highly palatable and appreciated food with strong ethnic and traditional usage, are likely to continue as a major staple, especially in many African countries, Nevertheless, yam production needs improvement to continue to be

TABLE 12.1.   WORLD YAM PRODUCTION, 1994

	Area (ha × 10³)	Yield (t/ha)	Production (t × 10³)
World	3,057	9.9	30,343
Africa	2,930	9.9	29,096
North and Central America	59	7.3	430
South America	35	9.1	315
Asia	15	13.7	207
Oceania	19	15.8	294
Leading countries			
Nigeria	2,000[a]	11.0	22,000[a]
Cote Divoire	260[a]	10.8	2,824
Benin	116[a]	11.1	1,287
Ghana	229[a]	4.4	1,000[a]
Togo	48[a]	8.3	400[a]
Zaire	40[a]	8.1	322[a]
Ethiopia	65[a]	4.0	263[a]
Chad	25[a]	9.7	245[a]
Central Africa Rep.	34[a]	6.9	235[a]
Papua New Guinea	13[a]	17.8	224[a]

[a]Estimated.
Source: 1994 FAO Production Yearbook, Vol. 48, FAO, Rome, 1995.

competitive with cassava, sweet potato, and other less costly staple foods. A shift toward greater commercialization of the crop is evident and should result in reducing production costs.

Constraints to the expansion of yam production include high labor requirements, low net yields, diseases, cultural, and harvest and post-harvest management problems. Some obvious needs are for breeding and selection to obtain superior yielding, pest-resistant cultivars with improved nutritional value. Improved quality and reduction in the volume of propagating materials used are additional objectives. Alternatives for better trellising, introduction of more mechanization, and utilization of processed product are further goals. At the International Institute for Tropical Agriculture (IITA) in Nigeria and other international and national research organizations, research efforts are directed toward achievement of these and other crop improvement objectives.

SELECTED REFERENCES

Coursey, D.G. 1983. Yams. In Handbook of Tropical Foods. H.T. Chan Jr., ed. Marcel Decker Inc. New York, pp. 555–601.

Gonzalez, M.A., and Collazo de Rivera, A. 1972. Storage of fresh yam (*Dioscorea alata* L.) under controlled conditions. J. Agric. Univ. of Puerto Rico, *56,* 46–56.

Hahn, S.K., Osiru, D.S.O., Akoroda, M.O. and Otoo, J.A. 1987. Yam production and its future prospects. Outlook Agric. *16,* 105–110.

Jos, J.S., and Vehkateswaralu, T. 1978. Twining in relation to distribution among Asian yams. J. Root Crops *4,* 63–64.

Martin, F.W., and Sadik, S. 1977. Tropical Yams and Their Potential. Part 4. *Dioscorea rotundata* and *Dioscorea cayenensis,* USDA Agric. Handbook No. 502, USDA, Washington, DC.

Miege, J. and Lyonga, S.N., eds. 1982. Yams/Ignames. Clarendon Press, Oxford.

Njoku, E., Nwoke, F.I.O., Okonkwo, S.N.C., and Oyolu, C. 1984. Pattern of growth and development in *Dioscorea rotundata* Poir. Trop. Agric. (Trinidad) *61,* 17–19.

Nwoke, F.I.O., Njoki, E., and Okonkwo, S.N.C. 1984. Effect of sett size on yield of individual plants of *Dioscorea rotundata* Poir. Trop. Agric. (Trinidad) *61,* 99–101.

Onwueme, I.C. 1978. The Tropical Tuber Crops: Yams, Cassava, Sweet Potato and Coco Yams. John Wiley & Sons, New York.

Quamina, J.E., Phills, B.R., and Hill, W.A. 1982. Vine production from tuber pieces of various sizes and sections of yam *(Dioscorea alata* L.). HortScience *17,* 73.

Ramanujam, T., and Nair, S.G. 1982. Control of sprouting in edible yams *(Dioscorea* spp.). J. Root Crops *8* (1 & 2), 49–54.

Sadik, S., and Okereke, O.U. 1975. A new approach to the improvement of yam, *Dioscorea rotundata.* Nature *254,* 134–135.

13

Edible Aroids:

Family: Araceae

TARO, *Colocasia esculenta* **(L.) Schott**
TANNIA (yautia), *Xanthosoma sagittifolium* **(L.) Schott**

The history of edible aroids as food crops is ancient in both the Old and New World civilizations. In parts of the humid tropics and subtropics, the edible starchy corms and cormels, grown as a staple and subsistence crop, are an important food source for millions of people. Leaves and petioles of some species and varieties are also used as vegetable greens.

Origins

It is widely accepted that the aroids originated in swampy areas and high rainfall forests of the tropics. Low or nonacrid wild aroid domestication was likely achieved through selection in early cultivations thought to have started as long as 7000 years ago, possibly preceding rice.

Except for *Xanthosoma* species in tropical America, edible aroid genera are of Old World origin. Wild forms of *Colocasia* are found in India, north of the Bay of Bengal. From its tropical Indian origin, *Colocasia* spread eastward to China, Japan, and some Pacific islands. Westward movement also occurred to Egypt, East Africa, and the eastern Mediterranean. Further migration took place across Africa. In the post-Columbian period, taro was introduced to Caribbean and tropical American regions.

Xanthosoma sagittifolium (tannia) movement from its tropical northeastern South American origin was relatively recent. Taken to Africa during the slave trade period (16th century), tannia was rapidly adopted and now ranks behind cassava and yams in importance. Because of the resemblance to *Colocasia,* which is called "cocoyam" in some locations, tannia became known as "new cocoyam." Disease and pest resistance greatly enhanced its rapid adaptation and spread in Africa, Asia, and some Pacific islands. Cultivation of other *Xanthosoma* species are of limited importance except in certain regions of southeast Asia, India, and some south Pacific islands.

Taxonomy

Araceae is a large family of about 100 genera comprising about 1500 species. The taxonomy of most edible aroids is very confusing because species distinctions are not clear and because many local names are used interchangeably. Most production areas have their distinctive local cultivars or clones and the naming of these is often irrelevant for accurate identification. In the case of *Colocasia,* the confusion is extreme because there are hundreds of named clones. The broadest distinction appears to be that taro represents *Colocasia esculenta* and *tannia* represents *Xanthosoma sagittifolium.* Examples of different local names for these species are

Colocasia esculenta—**taro,** cocoyam. eddoe, dasheen, gabi, keladi, tari, arvi, kolkas, dalo, sato-imo

Xanthosoma sagittifolium—**tannia,** yautia, malanga, tanier, new cocoyam, belembe, maduma, gualusa

Although its taxonomy is not fully resolved, taro is considered a single polymorphic species with many botanical varieties. The cytotaxonomic background for cultivated taro varies, with the basic chromosome number believed to be $n = 14$. Most clones have $2n = 28$, others with $2n = 42$. Ploidy also varies for other edible aroids; for tannia species, the chromosome number is $2n = 26$.

Several general types are recognized in cultivated taro. Those producing one large corm and a few cormels are often called dasheen; those producing a relatively small corm and many slightly smaller cormels are called eddoe. A few other types are grown primarily for leaves and petioles.

Another distinction is that different cultivars are used for flooded and others for upland culture. Cultivar separations are based on plant size, leaf and petiole shapes and sizes, corm flesh color, flower shape, and culinary uses. Differences in acridity levels are still another feature used to distinguish cultivar or clonal differences.

Botany

Aroids are perennial herbaceous monocots. Except for *Amorphophallus* species, leaves arising from the apex of the corm are produced in a whorl with erect long petioles supporting a large broad peltate (shield) or sagittate (arrowhead) leaf blade, (Fig. 13.1). The long slender petioles are solid but have numerous air spaces permitting adaptation to flooded conditions. A common characteristic is the acrid mucilaginous juice found in all tissues.

Taro plants generally range in height from 1 to about 2 m and have long (25–50 cm), wide (12–25 cm) peltate leaf blades. Petiole attachment to the lamina occurs at about the upper middle of the leaf's undersurface, (Fig. 13.2a).

Tannia plants are generally larger than taro and occasionally grow to a height of 2 m or more. The large heart-shaped leaf blade can sometimes be almost 1 m in length. They have a prominent vein along the margin of the lamina, which is not present in taro. The petiole is similar to that of taro but is attached to the under surface at the indentation of the lamina lobes (Fig. 13.2b). Leaf turnover is continuous and the number tends to be constant as old leaves are replaced by new ones.

Taro and tannia seldom flower, and some cultivars not at all. When flowering does occur, the inflorescence arises in leaf axils, and depending on species, one or several may form. Flowers, depending on the species, are unisexual or bisexual. All are sessile and form on a spike

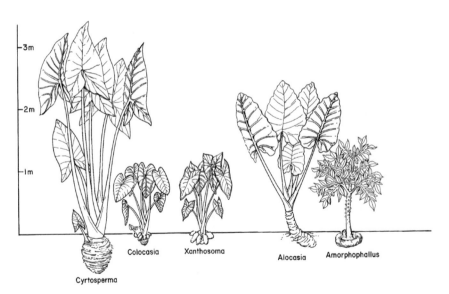

FIG. 13.1. Representative genera of the major cultivated edible aroids.

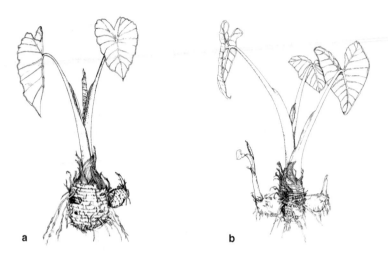

FIG. 13.2.   Taro, *Colocasia esculenta* (a) and Tannia, *Xanthosoma saggitti-folium* (b).

(spadix) enclosed in a large bract (spathe). In the unisex inflorescence, pistillate flowers are at the base of the spadix, staminate flowers at the top, and in between are a grouping of abortive sterile flowers. Insects pollinate the flowers, but fruit and seed set are rare. Fruit are berries, each containing two to five small (1 mm), oval, hard, and difficult to germinate seed.

Each species produces enlarged fleshy corms that accumulate starch. Roots developing from lower portions of the corm are adventitious, fibrous, and relatively shallow. Corms and cormels as well as leaves and petioles are edible. Corms are enlarged fleshy parts of the compressed stem base. Cormels are enlarged axillary or lateral buds, originating either from the stem or mother corm. Morphologically, both corms and cormels are stem tissue. They are usually produced underground but may develop near the surface.

Corms are usually spherical or cylindrical (Fig. 13.3). Depending on the cultivar, corm weights range from 500 g to several kg and cormels from 30 to 450 g. Large corms can achieve a length of 30 cm and width of 15 cm. Tannia corms, being less palatable, are generally fed to animals; the cormels are the portion for human consumption. For taro, both corms and cormels are consumed; small taro cormels are often used for propagation. Corm and cormel flesh color varies with cultivars; most being white, but some are cream, yellow, and sometimes

FIG. 13.3.   Taro corms, *Colocasia esculenta.*

pink. The many concentric rings observable on the surface of the corm are remnants of leaf scars. The structure of the corm or cormel consists of the relatively thick corky periderm overlaying the large volume of starch-containing parenchyma cells within which vascular bundles are scattered.

Cultural Requirements

Aroids can be grown continuously throughout the year in tropical and subtropical regions, usually in moist or flooded conditions. Favorable mean temperatures range between 21 and 27°C. A limited volume of taro production also occurs in some warm temperate regions in Japan and portions of the United States, although growth of most aroid species is arrested during periods of temperatures less than 15°C. All are shade adapted but can be grown in full sun. In traditional and home culture, partial shade is often preferred and commonly provided by intercropping with plantain, coconut, maize, and other crops. Tannia has better shade tolerance than taro and is well suited for intercropping.

High annual rainfall commonly occurs in regions where most taro and tannia are grown. Aroids obviously prefer moist soil conditions and readily available water during growth. Many taro cultivars and other aroids can be grown in flood culture. In some production regions, supplemental irrigation is provided during dry periods. Although taro tolerates some salinity, yields are depressed.

Taro plants are adapted to a wide range of soils, from clays used in paddy culture to well-drained, deep, fertile loams for upland production. Dasheen-type taro grows well in heavy soils with high moisture holding capacity and perform well under flooded conditions. For eddoe-type taro, a well-drained loamy soil is preferred, but some waterlogging can be tolerated. Eddoe types are also able to tolerate dry soil conditions.

Tannia is most productive in deep, fertile, well-drained loams and, unlike taro, does not tolerate waterlogging. However, tannia has better tolerance to low fertility and drought than taro and will tolerate periods of high rainfall if drainage is provided. Both taro and tannia can be satisfactory grown in soil with pH between 5 and 8.

Synthetic fertilizers are not normally applied in most tropical areas because of the relatively low market value of these crops. Most aroids respond to supplemental nitrogen. High soil nitrogen levels will stimulate vegetative growth and, if excessive, can delay maturity.

Propagation and Planting

Aroids are vegetatively propagated, using corms, cormels, or stem cuttings which consist of the apical portion of the corm with attached petioles. The latter are called "setts" and are usually obtained during crop harvest by cutting off 2–3 cm of the upper portion of the corm with attached petioles which are trimmed to 10–15 cm. The remaining corm portion may be used for food. In Hawaii, taro setts are called "hulis" (Fig. 13.4). Setts generally produce roots earlier and tend to yield better than corm or cormel propagation. For flooded taro culture, hulis are often the exclusive propagating material. In upland taro culture, propagules might include sprouted corm pieces, cormels, or hulis. True seed production is relatively rare, and seed are not suitable for propagation because resulting plants exhibit undesirable genetic variability and low productivity. Tissue culture has some application for cloning virus-free plants and for plant breeding.

Taro has no dormancy period and, therefore, corms and cormels are used for propagation when buds begin to sprout. A corm or corm portion with an intact apical meristem is ideal because these propagules establish rapidly and provide uniform populations. Small, 50–75 g, cormels are used intact. Corm portions other than the apical section may result in delayed sprouting and nonuniform establishment, and such propa-

FIG. 13.4.   Taro, *Colocasia esculenta,* corms with attached petioles. These may be used for propagation (huli's) or for consumption, Papeete, Tahiti.

gules should first be planted in nurseries. Following establishment, single-shoot plantlets can then be used for transplants. Plantings are made with the intention of providing one shoot per plant location; excess shoots are thinned.

For tannia propagation, head setts are frequently used. These are larger propagules than taro setts. Large corms can be used for tannia propagation because the corms of most species are not valued for consumption. However, small nonmarketable tannia cormels are also used for plantings. Although tannia cormels exhibit dormancy, it disappears after 1–3 months. Therefore, if necessary, propagating materials can be stored briefly. However, setts are more difficult to store than intact corm or cormels. Propagating practices for other edible aroid species are similar.

Corms, cormels, or setts are planted to depths of 10–15 cm. Large propagules are preferred, and about 2 tons of propagating material is utilized in planting 1 ha. Most taro plant populations average about 30 thousand per hectare, but the range can vary from 12 thousand to 100 thousand per hectare, with densities being higher in flooded culture. For tannia production, spacings are wider and populations range between 14 thousand and 20 thousand plants per hectare. Plant density and the length of the growth period influence cormel size. Large corms and cormels are produced when plant densities are low and the growing period long.

Taro Culture

In tropical climates, the time of planting for taro is mainly determined by moisture availability and is usually begun at the start of the rainy season. In temperate regions, with the availability of water, plantings can be made when growing temperatures are favorable. However, because the propagating materials are obtained from preceding crops, the time of the previous harvest often influences when a new crop is planted.

Lowland taro culture begins with fields being plowed and then flooded. A dike is erected around the field to contain the impounded water. The soil may be puddled as for paddy rice; a deep hard pan in the soil horizon can be useful for water management. Setts are transplanted, usually by hand, into moist soil and covered. Although less preferred, presprouted corms or cormels are also used. Plantings are made on a ridge or raised bed if harvesting equipment will be used; otherwise, the field surface is flat.

When newly unfolded leaves emerge, the field is again flooded. The water level is raised to 10 cm and kept at this level briefly for weed control. Lack of early weed control can result in significant yield losses. The water level is then lowered to an optimum, between 4 and 8 cm, as high water levels reduce yield. Water is allowed to flow continuously throughout most of the growth period. Interruption of flooding disrupts corm and cormel development and can depress yield. Fields are not drained except for fertilization and harvest. Allowing for evapotranspiration of about 0.5 cm per day and for percolation into the soil, an average daily water requirement is about 2–3 cm. Flooded culture usually yields more than upland rain-fed or irrigated culture, but requires a longer growing time.

For upland taro, a deep, friable, well-drained soil is preferred. In contrast to the generally more fertile soils used for flooded culture, soils used for upland culture often are of marginal fertility. After land preparation, furrows 30 cm deep and from 75 to 90 cm apart are dug for planting presprouted corms or cormels; setts are less frequently used. In-row spacings are about 40 cm. Corms and cormels are initially covered with 5–8 cm of soil; more soil is mounded later to avoid root exposure and to improve daughter corm development. Deep plantings are believed to result in higher yields than shallow plantings. Upland taro culture at low plant densities is better adapted for intercropping; maize or a legume crop are often chosen.

Tannia Culture

Tannia is grown as an upland crop because it is unable to withstand waterlogging, even though it grows well in high soil moisture. Uniformly distributed moisture, whether rain fed or irrigated, is required for good production.

Setts or presprouted corms, previously cut into pieces, each with one or more sprouts and weighing between 200 and 900 g, are used for propagation; these are two to three times larger than taro setts. Plant spacings most frequently are 90 cm within and 90 cm between rows, but equidistant spacing can range from 60 to 180 cm. Close spacings help plants compete with weeds and usually increase total yields, but corm and cormel sizes are decreased.

As with upland taro, following planting, soil is mounded repeatedly around plants to avoid root exposure, assist cormel development, and control weeds. Tannia responds to high nutrition and often is the first crop to be planted when land is cleared. Being more shade tolerant than taro, tannia is more often intercropped.

Growth and Development

For taro and tannia, it is typical for rapid vegetative growth to occur in the first 4 to 6 months after planting, during which time the maximum leaf canopy is achieved (Figs. 13.5a and 13.5c). To improve yields, farmers use larger corms or setts in order to maximize leaf growth as early as possible. Corm and cormel dry matter accumulation and leaf growth over time is shown in Figs. 13.5b and 13.5d.

Leaves develop at intervals of several to many days over a period of several months. From unfolding to senescence, the average life of a leaf early in the growth period is 40–45 days, and from 55 to 80 days for later formed leaves. Initially, each succeeding leaf increases in size until midseason, then leaf sizes begin decreasing. The photosynthetic activity of the average newly unfolded leaf increases to about the 15th day and then remains at its highest level for another 10 days, followed by a gradual decline and senescence. The leaf area index is increased with additional nitrogen and also with close spacing. Shaded plants generally have greater leaf areas that compensate for lowered photosynthetic activity. About five to six functional leaves are present during most of the growth period. A total of about 20 leaves are usually produced.

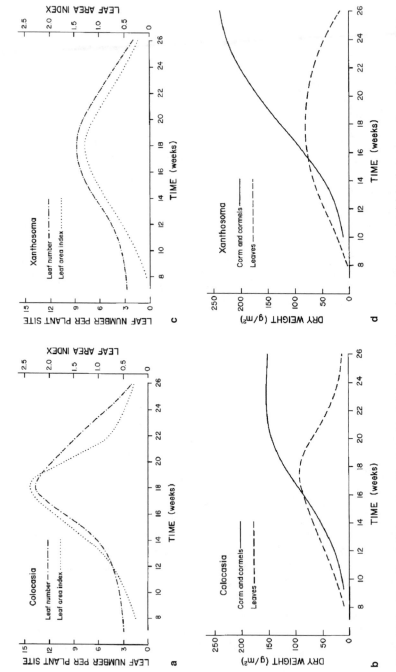

FIG. 13.5. Growth of *Colocasia* with regard to (a) leaf number, leaf area index, and (b) leaf, corm and cormel dry weight. Growth of *Xanthosoma* with regard to (c) leaf number, leaf area index, and (d) leaf, corm and cormel dry weight. Source: Redrawn from data of M.C. Igbokwe (1983).

Corm formation begins 3–5 months after planting, and enlargement continues until harvest. As plant growth continues, older leaves turn yellow and die until few are left. The upper portions of mature corms are often visible above ground. At this point of leaf senescence, corms and cormels are ready for harvest. If not harvested, growth of new foliage can begin and corm enlargement may continue. Some aroids are specifically grown for several years before being harvested.

Harvest

Harvesting of upland taro usually occurs about 8–9 months following planting, and slightly longer for tannia. Flooded taro culture requires a 12–15-month growth period. Generally, leaf yellowing is a usual indication of crop maturity, and corms are physiologically mature when sugars are at a minimum. Harvest is best performed during the dry season. For flooded taro culture, the fields are drained several weeks before harvest begins. Harvest of the upland taro crop can be delayed if necessary.

Multiple or partial harvest is not a frequent practice, as it has more disadvantages than advantages. For taro, a complete harvest is made, but for tannia, cormels can be carefully harvested, and the mother corm or smaller cormels are left in place for further growth.

Most of the world's taro and tannia crops are hand harvested, which is especially true for lowland taro. For ease of corm harvest, most often the entire plant is lifted. The main corm and cormels are severed from the roots and trimmed of petioles. Setts for propagation are taken at this time. Machinery may be used in commercial-scale production to harvest and clean corms. Yields are extremely variable within and between production regions. Taro yields are generally higher than those of tannia; occasionally ranging to 30 t/ha for taro and up to 20 t/ha for tannia. In experimental plantings, taro yields of 100 t/ha have been obtained.

Harvesting of leaves and petioles of either plant for vegetable use has little effect on corm production as long as the frequency and amount removed are minimal. Occasionally, crops are grown specifically for foliage.

Postharvest and Storage

Corms and cormels are susceptible to injury, and careful handling is important to minimize postharvest losses. Following harvest, the corms and cormels are cleaned, washed, and trimmed of petioles and attached fibrous roots, but those intended for storage are not washed. Taro corms, unlike tannia cormels, do not have a true dormancy and may sprout, a reason why storage periods are short.

Other than for short periods, most aroid storage is limited; that which occurs is usually close to production sites. In addition to in-ground storage, growers store harvested corms and cormels in underground soil covered pits. The surface is kept damp and mulched to limit moisture loss and to prevent temperature rise of the commodity. At ambient temperatures in ventilated structures, corms may be held up to 6 weeks. In refrigerated storage at 7°C and about 80–85% RH, tannia cormels can be maintained for 4 months or more. Storage at temperatures below 5°C results in chilling injury, with symptoms that appear as internal browning. Taro corms will tolerate a short storage period at 5°C.

Diseases and Pests

Some of the serious diseases affecting aroids include the following:

Bacterial leaf blight	*Xanthomonas campestris*
Bacterial soft rot	*Erwinia carotovora*
Black rot	*Ceratocystis fimbriata*
Cladosporium (brown) leaf spot	*Cladosporium colocasicola*
Cercospora leaf spot	*Cercospora chevaliegi*
Fusarium dry rot	*Fusarium solani*
Phyllosticta leaf spot	*Phyllosticta colocasiophila*
Pythium root and corm	*Pythium* spp.
Rhizopus rot	*Rhizopus stolonifer*
Sclerotium tuber rot	*Sclerotium rolfsii*
Taro leaf blight	*Phytophthora colocasiae*

Leaf blights are more prevalent in flooded taro than upland culture. Generally, *Xanthosoma* plants have greater resistance to disease and pests than those of *Colocasia*. Storage diseases are usually less important because corms and cormels are not usually stored for long period.

Vegetative propagation as presently performed is conducive to the transfer of diseases, including viruses. Dasheen Mosaic Virus (DMV), an aphid-vectored virus, causes considerable damage. In the Solomon Islands, alomae and bobone are two serious viruses. Meristem tissue clonal propagation for virus freedom is not yet feasible for general grower use.

A high population of root knot nematodes, *Meloidogyne* spp., attack both feeder roots and corms, thereby reducing yield and making corms unfit for market. Loliloli is a physiological disorder of taro, in which the loss of starch results in softening of corms.

Insect pests, although troublesome, are not usually production limiting, many being specific to one aroid species. The taro leafhopper, *Tarophagus proserpina,* is one of the more important pests and also is a virus vector. Hawk moth caterpillar, *Hypotion celerio,* armyworm, *Spodoptera litura,* and several other Lepidoptera pests cause plant damage. Beetle species of *Araeocerus, Papuana,* and *Ligyrus,* and spider mites (*Tetranychus* spp.), thrips *(Heliothrips indicus),* scale *(Aspidiotus destructor),* mealybug *(Dysmicoccus brevipes),* aphids, white fly, ants, and snails are additional pests.

Uses and Composition

Aroid corms and cormels are used as a starchy vegetable prepared by boiling and are consumed after being roasted, baked, steamed, or fried. A pastelike preparation made from cooked taro corms is known in Hawaii as poi. In Africa, a similar and also very popular product known as fufu is produced from tannia cormels. Both preparations are sometimes eaten after fermentation, which makes the product slightly acidic. Apparently, tannia fufu is preferable to preparations made from taro, which partly explains the much greater production of tannia in Africa. Although having a tropical American origin, tannia, because of easier cultivation, a greater tolerance of low soil fertility, greater disease resistance, drought tolerance, and edible qualities, has displaced much of total world taro production and is the dominant aroid crop in Africa.

Aroid corms/cormels are a low-cost, high-carbohydrate food source but nutritionally low in protein and vitamins. Aroid starch is easily digested and nonallergenic. The composition of these storage organs averages between 65 and 80% water and 20% and 25% carbohydrate (Appendix Table C). The starch content of tannia corms or cormels is generally higher than that of taro, and tannia cormels have a higher starch content than the corm. Dried corms and cormels of either species can be milled into flour. The sugar content in mature corms and cormels is low.

Young tender leaves and petioles are used as greens (Fig. 13.6). A tannialike species, *Xanthosoma braziliense* is grown exclusively for its edible foliage. Foliage of *Amorphophallus campanulatus* is also widely used in the Philippines, Malaysia, and China. Leaf protein content is about two to three times that of the corms, and vitamin C content is generally good. Blanched taro shoots, which are consumed before the leaf opens, are obtained by forcing corms in the dark at high temperatures. Acrid components, although present are usually at lower levels than in corms and are removable, usually by boiling one or more times in water.

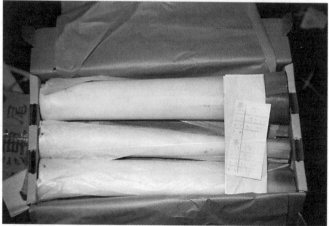

FIG. 13.6.   Bundled taro leaves displayed for market sale (top). Plastic-wrapped packaged taro petioles (bottom). Plastic film protects against desiccation.

Toxins

A common aroid feature is their acrid nature and irritant raphids. Acridity is the presence of a sharp burning or bitter taste and is associated with unidentified compounds which may be glycosides or proteins and with the presence of raphids. Raphids are tiny needlelike calcium oxalate crystals produced in specialized cells, and when consumed, puncture and irritate mouth and tongue tissues. Electron microscope studies show the crystals to be about 50 μ long with sharp barbs which make dislodgment difficult once raphids penetrate the skin. Skin irritations and swelling can also result from handling aroids. Contact with peeled taro corms generally causes more skin irritation than with tannia, although individuals differ in their sensitivity. Cooking is necessary before consumption to remove the acridity. Wild aroid species contain more of the acrid principle and are known to contain cyanogenic glycosides. However, most domesticated edible aroids have insignificant levels in either leaves or corms. Corms also contain a trypsin inhibitor which is not a problem because the inhibitor is inactivated by cooking and corms are always cooked before consumption.

MINOR AROIDS

GIANT TARO, *Alocasia macrorrhiza* (L.) G. Don

Other names: ape, ahpi (Hawaii), biga (Philippines), birah (Indonesia) also alu, kape, amu

Alocasia macrorrhiza is the main cultivated species of about 75 in the genus *Alocasia* and is primarily grown in Sri Lanka and India, presumably the center of domestication. Production is also widely distributed throughout Southeast Asia and many south Pacific island groups where it is often cultivated as a dryland crop. *Alocasia* can grow where it is too dry for *Colocasia* and has the advantage of good insect and disease resistance. Although high rainfall is beneficial, the plant is intolerant of waterlogging. Production does not occur where temperatures are less than 10°C.

Giant taro is a large, 90–270-cm tall, succulent perennial. Leaves are similar to those of tannia, but the leaf tip is inclined upward almost parallel with the petiole. Juice from a freshly cut petiole is watery and sometimes used in folk medicines. The edible, mostly aboveground, semicompressed corm is hardy, long, thick, and woody, ranging from 45 to 100 cm in length and weighing 20 kg or more. Being somewhat fibrous, giant taro is the least palatable of the edible aroids. Not being seasonal, the crop can be planted year round and harvested anytime after 12 to 18 months, but harvest may be delayed as long as 3 or 4 years. Propagation is usually with shoot tips, head setts or suckers, although small cormels remaining in the soil after harvest can sprout and serve for propagation.

Average yields are about 30 t/ha, and range from 2 to 5 kg per plant. The carbohydrate content is about 15%, with protein less than 1%.

GIANT SWAMP TARO, *Cyrtosperma chamissonis* **(Schott) Merr.**

Other names: kiha, brak

This crop, possibly originating in Indonesia, is the most tropical of all cultivated edible aroids; its growth is poor in subtropical climates. Although limited, cultivation continues to occur in low-elevation, freshwater, swampy localities in Indonesia, the Philippines, and several other Pacific islands. The plant is almost entirely grown as a subsistence or reserve food crop (Fig. 13.7).

Plant growth is tolerant of deeper water and somewhat brackish conditions than other edible aroids. Plants have a similar appearance to *Alocasia species* and also possess good insect and disease resistance. The leaves are very long, sagittate with pointed tips, and borne on long, thick, spiny petioles. Petioles are used for making mats and baskets. Plants are very large, often more than 3 m in height, and can be grown for several years in order to produce large corms. Although infrequent, some plants are grown for more than 10 years, and corms greater than 100 kg are obtained; yields average about 10 t/ha.

Planting and harvesting are not seasonal, but the usual practice is to grow plant types that mature corms in 1 or 2 years with harvests usually made at the time of flowering. Although seed are formed, they

FIG. 13.7.   Plants of *Cyrtosperma chamissonis,* growing in Ecuador. Source: Courtesy C.M. Rick, Davis, CA.

are seldom used. Instead, topped cormlets or suckers are used for propagation. Corms contain up to 29% carbohydrates, but only about 1% protein. *Cyrtosperma merkussi,* a related species, is frequently grown in the Philippines.

ELEPHANT YAM, *Amorphophallus campanulatus* (Roxburg) Blume ex Decaisne *(A. paeoniifolius)*

Other names: pongapong and tindoc (Philippines), teve (Samoa), ol ka chu (Pakistan), and kidaran (Sri Lanka)

Most *Amorphophallus* species are believed to have originated in India or Sri Lanka or possibly Southeast Asia. Plants are tropical and subtropical, with an annual vegetative and dormant cycle. Temperatures between 25°C and 35°C are preferred, as is high humidity and abundant moisture. Low temperatures and low moisture tend to induce early dormancy. New leaf growth develops after dormancy. Plants become reproductive after the third or fourth year. Leaf structure differs from some other aroids and resembles those of dicotyledonous plants, being compound segmented and umbrella-like.

The short, broad corms are harvested when plants become dormant or are permitted to enlarge for additional years. Corm size ranges from 2 kg for first-year growth to 25 kg after 3 years. Small cormlets from the base of the main corm are used for propagation. Usual plant populations are about 15,000/ha. Mature corms contain about 24% starch and about 1% protein; some clones have higher levels of protein. Consumption of young tender leaves and petioles are popular in the Philippines. In Java, two variety types are grown, *A. campanulatus* var. *hortensis* (Backer) and *A. campanulatus* var. *sylvestris,* the former recognized by smooth and the latter by rough petioles.

KONJAK, *Amorphophallus rivieri,* Durieu var. *konjak* Engler

With an Indochina–southern China origin, konjak resembles elephant yam. However, konjak is adapted to temperate conditions and grows well at 15–20°C. Plants grow to a height of 100–120 cm and tend to resemble a small tree (Fig. 13.8). The corm enlarges slowly and weighs less than those of *A. campanulatus,* about 35 g in the first year and about 600 g after 3 years. Large corms are consumed and small corms are usually used for propagation. The flattened-shaped corm is

FIG. 13.8.   Plants of konjak, *Amorphophallus rivieri,* var. *konjak* growing in Mae-bashi, Japan.

rich in mannose and mannans, which can comprise as much as 50% of dry weight, and is used to make a product called konnyaku, which is popular in Japan. The mannans give konnyaku its gellike properties; these mannans are not digestible by humans. Other processed products are pickled corms and dry chips for snack foods.

TANIER SPINACH, *Xanthosoma brasiliense* (Desf.) Engler

Other names: Tahitian taro or spinach, belembe, and calalou
Related species include X. atrovirens, X. violaceum

Tanier spinach is grown for the edible leaves (Fig. 13.9). The corms, being very small and low in starch, are used only for propagation. Tender leaves are sometimes used raw in salads, or lightly cooked. Prolonged cooking is not necessary to render the calcium oxalate crystals nonirritating, as with other crops in this genus. The protein content of fresh leaves is about 3%; the leaves also contain high levels of Ca, P, and vitamins A and C. However, much of the calcium may not be nutritionally available because of insoluble calcium oxalate.

FIG. 13.9. Plants of *Xanthosoma braziliense* grown for their foliage used as cooked greens, Londrina, Parana, Brazil.

SWEET FLAG, *Acorus calamus* L.

Sweet flag is a hardy perennial planted in bogs and grown mostly in India for vegetable and for medicinal use of the aromatic pinkish rhizomes. Plants have tall (1.5 m) grasslike leaves and are propagated by division.

AROID PRODUCTION

Most of the world's aroid production is from taro and tannia species widely grown in West Africa, Southeast Asia, and many Pacific and Caribbean islands. Production of tannia exceeds that of taro, as exhibited by its extensive production in Africa. Several minor aroids, usually grown as subsistence crops, are of limited world importance but regionally are often very important, especially when adverse weather destroys other food crops.

Aroid production occurs mostly in the warm climate developing countries. African nations produce about 61% and Asian countries about 32% of the world production Table 13.1. Climate, cultivar adaptation, diseases and pests are other factors influencing yields, but yield variability is due largely to agronomic practices that range from intense cultivation to subsistence farming. This results in yields ranging from only 1 or 2 to more than 40 t/ha. Accurate production statistics are difficult to acquire because of undocumented small and/or subsistence production and generally incomplete reporting.

Crop Improvement

Aroid improvement through breeding is likely to be slow and limited, particularly because of sexual reproduction difficulties and lack of research interests and funding. Much of the genetic gain has relied on mutants and selection for superior clones, which is likely to continue. Labor requirements in planting and harvesting are impediments to expanded production. Generally, subsistence production and small holdings preclude capital investment for mechanization and other improvements. The demand for edible aroids remains high and is a favored food in many areas. However, vegetative propagation requiring the use of a large portion of the crop for propagation, high water requirements, diseases and pests, and ineffective storages are production obstacles.

Because of their unique adaptability to wetlands or high-moisture soils and production dependability, the edible aroids will continue to

TABLE 13.1.   WORLD PRODUCTION OF TARO, TANNIA, AND OTHER AROIDS, 1994

	Area (ha × 10³)	Yield (t/ha)	Production (t × 10³)
World	982	5.9	5797
Africa	786	4.6	3586
North and Central America	2	10.2	22
South America	1	10.7	11
Asia	144	12.8	1842
Oceania	48	7.0	336
Leading countries			
China	83[a]	16.3	1360[a]
Nigeria	250[a]	5.2	1300[a]
Ghana	173[a]	7.3	1272[a]
Cote Divoire	240[a]	1.4	337
Japan	24[a]	12.9	310[a]
Papua New Guinea	33[a]	6.7	220[a]
Madagascar	20[a]	6.0	120[a]
Egypt	4[a]	31.9	115[a]
Philippines	32[a]	3.5	111[a]
Burundi	24[a]	4.3	103

[a]Estimated.
Source: 1994 FAO Production Yearbook, Vol. 48, FAO, Rome, 1995.

contribute significantly in meeting the food needs of many tropical populations. Many people rely on the production of these edible aroids for their caloric needs, and the consumption of the easily digested product can be broadened by use of its many processed forms.

SELECTED REFERENCES

Chandra, S., ed. 1984. Edible Aroids. Clarendon Press, Oxford.

Ezumah, H.C., and Plucknett, D.L. 1982. Cultural studies on taro, *Colocasia esculenta* (L.) Schott. 2: Age and moisture effects on growth and corm yield. J. Root Crops *8* (1 & 2), 17–26.

Igbokwe, M.C. 1983. Growth and development of *Colocasia* and *Xanthosoma* spp. under upland conditions. In Tropical Root Crops Production and Uses in Africa. Proc. 2nd Triennial Symposium of International Society for Tropical Root Crops, pp. 172–174.

O'Hair, S.K., and Asokan, M.P. 1986. Edible aroids: Botany and horticulture. Hortic. Rev. *8,* 43–99.

Onwueme, I.C. 1978. The Tropical Tuber Crops: Yams, Cassava, Sweet Potato and Coco Yams. John Wiley & Sons, New York.

Pardales, J.R., Jr., 1986. Characteristics of growth and development of taro (*Colocasia esculenta* (L.) Schott) under upland environment. Philippines J. Crop Sci. *11,* 209–211.

Pardales, J.R., Jr., and Dalion, S.S. 1986. Methods for rapid vegetative propagation of taro. Trop. Agric. (Trinidad) *63,* 278–280.

Plowman, T. 1969. Folk uses of new world aroids. Econ. Bot. *23,* 97–122.

Sadai, W.S. 1983. Aroid root crops: *Alocasia, Cyrtosperma* and *Amorphophallas.* In Handbook of Tropical Foods. H.T. Chan, Jr. ed. Marcel Decker Inc., New York.

Sakai, W.S., Hanson, M., and Jones, R.C. 1972. Raphids with barbs and grooves in *Xanthosoma sagittifolium* (Araceae). Science *178,* 314–315.

Valenzuela, H.R., O'Hair, S.K., and Schaffer, B. 1991. Developmental light environment and net gas exchange of cocoyam *(Xanthosoma sagittifolium).* J. ASHS *116,* 372–375.

Wang, J.K. ed. 1983. Taro: A Review of Colocasia esculenta and its Potentials. University of Hawaii Press, Honolulu.

14

Other Underground
Starchy Vegetables

It is well recognized that tuber and root crops make a huge contribution to the world's food supply. Potatoes, sweet potatoes, cassava, and yams are the major commodities produced in large volumes. In addition to edible aroids, there are many other underground starchy vegetables in certain regions that are also important for their edible carbohydrates and other nutrients.

In general, most of these starchy vegetables are produced in tropical regions and in economically developing nations. Most are not staple crops and are neither intensively nor extensively cultivated. Many are perennials which extends their usefulness as subsistence crops, and many are grown because of regional and/or ethnic preferences.

Accurate production statistics are difficult to acquire for these underground starchy vegetables because of their small volume; much of the production occurs in home gardens, for which information seldom is reported. Various estimates suggest annual world production at 5–10 million tons.

The arrangement of these other underground starchy vegetables in this chapter does not ascribe any priority about importance but is presented in alphabetical order according to botanical family. The identification of some plants can be confusing because of similar common names used for different species of similar appearance or use. For example, many plants are called arrowroot of one kind or another. Thus, *Maranta arundinacea,* perhaps the best known arrowroot, is known as West Indian arrowroot. Edible canna is known as Queensland and purple arrowroot, and species of *Tacca,* such as *T. leontopetaloides, T. pinnatifida, T. hawaiiensis,* and *T. xanthorrhiza* are known as East Indian, Polynesian, Hawaiian, and Indonesian arrowroot, respectively. Additionally, *Calathea allouia* is called Guinea arrowroot, *Curcuma angustifolia* is East Indian arrowroot, and *Curcuma pierreana* is called false arrowroot.

MONOCOTYLEDONS

EDIBLE CANNA, *Canna edulis* Ker-Gwal

Family: Cannaceae
Other names: achira, purple arrowroot, Queensland arrowroot, tous-les-mois

Edible canna is a large lilylike perennial producing branched underground starchy rhizomes. The plant is indigenous to tropical America and has a long history of cultivation. Dried rhizomes found in Peruvian excavations were dated at 2000 to 4000 years old.

The fleshy stems grow as tall as 2.5 m with green leaves which are entire, long and narrow, and with thick midribs (Fig. 14.1). At the base of the edible stem, near or slightly below the soil surface, the starchy rhizomes develop. Two plant types are known, producing either white- or purple-skinned rhizomes. Flowers, red to orange, produce three celled fruit capsules containing fertile round black seed. However, for propagation, rhizome portions are replanted or left for a ratoon crop.

Plants are adapted to moist soils. The best growing temperature is between 25°C and 28°C; plants are sensitive to higher temperatures and drought. An important attribute of canna is its adaptability to altitudes as high as 2600 m and to low temperatures where other starchy crops such as cassava are less productive.

The crop can be harvested as needed about 8 months after planting, although more time is usually allowed for higher yields, which range between 12 and 25 t/ha. Some of the largest rhizomes are 60 cm long and weigh over 25 kg. Harvest delay tends to increase rhizome fiber content.

Baking is a preferred preparation method; rhizomes can be, but seldom are, eaten raw. Rhizomes contain about 25% of a unique and easily digested starch consisting of the largest known starch grains, about 100 times larger than those of aroid starch. Production is concentrated in South America, especially in Peru and Colombia, although canna is produced on a relatively intensive scale for industrial starch in Australia. In Vietnam, the highly valued starch is used for noodle production, and as a luxury food elsewhere in east Asia.

Other edible species of *Canna* are known as brick canna, *C. discolor*

FIG. 14.1.   Canna plant, *Canna edulis,* growing in China.

Mexican canna *(C. glouca),* iris canna *(C. iridiflora),* broad-leaved canna *(C. latifolia),* Inca canna *(C. languinosa),* and Andean canna *(C. paniculata).*

CHUFA, TIGER NUT, *Cyperus esculentus* L. var. *sativus* Boeck.
YELLOW NUTGRASS, *Cyperus esculentus* L.

Family: Cyperaceae

Chufa is a perennial plant with grasslike stems and leaves; it is found throughout the tropics and in many warm temperate areas. Although better known as a common noxious weed, chufa is cultivated for the small, 1–2 cm, fleshy, almond-flavored underground tubers. Chufa can be propagated by seed and also with rhizomes or tubers. Tuber initiation is a response to shortening day lengths. Most of the oblong tubers are found grouped below the plant, each at the tip of a stringlike rhizome. Although most tubers sprout the first year, some persist without sprouting until the second year.

Tubers mature in about 3–4 months and contain about 25% starch, 20% fat, and 5% protein on a dry weight basis; the fiber content is fairly high. Cultivated and noncultivated tubers are a common food in many parts of north Africa. In cultivated plantings, tubers are larger and yields of more than 10 t/ha are achievable. The tubers are used as a raw or baked vegetable. In Spain, tubers are used in a beverage called "horchata de chufa." They are also used roasted as a coffee substitute and sometimes for livestock feeds.

A closely related species is purple nutgrass, *C. rotundas,* and although tubers are similar, they are seldom eaten because of bitterness.

LILY BULBS, *Lilium* L. species

Family: Liliaceae

Cultivated bulbs of several *Lilium* species, many of which are native to Japan and China, are popular foods in the Orient (Fig. 14.2). Nonbitter or only slightly bitter species are used as a boiled or steamed vegetable. Propagation is with bulbils or scalings planted in the fall, and the crop is dug and harvested the following fall after leaves have turned

FIG. 14.2. Lilly bulbs, *Lillium* spp., grown in Japan.

completely yellow. Production usually occurs in cool climates, as this tends to improve edible qualities of the bulb. Bulbs contain about 17% starch, some sugars, and about 2% protein; these amounts can vary. Occasionally, bulbs are dried before consumption in order to increase the proportion of starch.

Some of the more popular species are *Lilium leichtlinii* var. *maximowiczii*, known as devil or tiger lily, and *L. lancifolium (L. tigrinum)*, also known as tiger lily. Cultivars of *L. auratum* known as mountain, gold banded, or golden ray lily are also popular, as is the old cultivar, Hong Kong, of the species *L. brownii* var. *viridulum*. Hemerocallis species such as *H. fulva* (day lily) and *H. lilioasphodelus* (lemon lily) are also grown, but mainly for their large edible flowers.

ARROWROOT, *Maranta arundinacea L.*

Family: Marantaceae
Other names: West Indian arrowroot, true arrowroot, jamachipeke

Arrowroot, a tropical perennial, is indigenous to northern South America and possibly the West Indies. A related perennial species, *M. ruiziana,* is also grown in the western Amazon region.

The name arrowroot is believe to have developed because rhizomes were used to treat poison arrow wounds. Because the shape of the leaf resembles an arrowhead is another possible explanation (Fig. 14.3). Arrowroot also appears to be a generic term applied to several similar species eaten after cooking or from which flour is made.

Arrowroot grows to a height of 1–1.5 m and is shallow rooted with deep-penetrating fibrous and thick rhizomes. Leaves are lanceolate, about 20 cm long and 10 cm wide. White flowers develop in 3 months, but seed is seldom set.

Propagation is made with small pieces of rhizome having several buds. Plantings are usually spaced at 40 × 80 cm. Growth is optimum at high temperatures and high humidity. To be productive, sandy soils and abundant uniform moisture are required. Toward the end of the growing season, foliage turns yellow and rhizomes are harvested. Rhizomes, from 20 to 40 cm long, are enclosed in scaly bracts; generally, yields are between 4 and 10 t/ha, occasionally higher. Plantings can be productive for 5–7 years, from regrowth of rhizome portions left after harvest.

Two cultivar types are commonly grown. The "Creole" form has good keeping qualities but the long and thin rhizomes are difficult to harvest.

FIG. 14.3.   Arrowroot, *Maranta arundinacea,* growing in Puerto Rico.

The "Banana" form has shorter and thicker, easily dug rhizomes; because of its poor keeping qualities, they should be consumed within a short period. Rhizomes are reported to contain more than 12% starch on a dry weight basis with a protein content of about 1–2%. The primary usefulness of arrowroot is for the highly digestible starch which has smooth consistency and high viscosity features.

LEREN, BAMBOO TUBER, CALATHEA, *Calathea allouia* (Aubl.) Lindl.

Family: Marantaceae
Other local names: allouis, topi, tambu, aria, Guinea arrowroot, and
sweet corn tuber

Leren, grown for the tuberous roots, is a perennial plant endemic to tropical America, namely Brazil and the Amazon region. Cultivation occurs in areas of the Caribbean, in India, and throughout many high-temperature and moist regions of Southeast Asia and Indonesia. A requirement for abundant and uniform moisture, and drought sensitivity, limits wider cultivation. Plants are shade tolerant but are more productive in full light.

Mature plants reach a height between 60 and 120 cm and exhibit dense clustered growth of multibranched rhizomes, some of which can become fleshy, and others become pseudostems with elongated leaves (Fig. 14.4). Leaf blades are simple, oval shaped, and 30–60 cm long. The edible, elongated, spherical, stubby tuberous roots, 2–4 cm wide and up to 8 cm long, are formed near the end of the fibrous roots.

Green or pale yellow flowers usually do not set seed; hence, propagation is with offsets or presprouted short rhizome sections. Plantings are made at the start of the rainy season with harvest scheduled for the dry season, and the tuberous roots are harvested after 10–12 months.

FIG. 14.4.   Leren, *Calathea allouia,* growing in Puerto Rico.

Adequate and uniform moisture availability strongly influences yields, which range from 2 to 12 t/ha. During the dry season, the plant is leafless and dormant. Leren can be stored at room temperature for several months with little desiccation or quality loss. Low-temperature storage results in chilling injury.

A strong feature is its crisp texture, retained even if overcooked. Tuberous roots are eaten like the white potato. Boiling is the principal preparation method, but removal of the thin barklike skin is difficult. The flesh is white and resembles sweet corn in taste; sometimes, it can have a slightly bitter aftertaste. Protein content on a dry weight basis is as much as 6%, with starch content between 13% and 15%. Young flowers of a related species, *C. violacea,* are used for food in Brazil.

EAST INDIAN ARROWROOT or TACCA, *Tacca leontopetaloides* (L.) Kuntze
POLYNESIAN ARROWROOT, *Tacca pinnatifida* (L.) Kuntze

Family: Taccaceae
Other names: piam, mokmok

Tacca is a perennial grown for its starchy underground tubers. Endemic to Southeast Asia, the plant is also cultivated in other parts of tropical Asia and Oceania. Its importance is diminishing, being displaced by other more easily produced starchy vegetables.

Resembling an aroid, the leaves are entire and large with long petioles. Plants are deciduous and become dormant in winter. Many yellow-green to purple flowers are formed on a tall scape. Fruit are berries with many ovoid and flattened seed. Seed can be used in propagation; however, tubers or stem divisions are usually used.

The potatolike tubers are about 15–20 cm long, weigh less than 1 kg and mature in 8–10 months. Other *Tacca* cultigens produce smaller tubers. The white-fleshed tubers are acrid and inedible when raw and, therefore, require considerable soaking and/or boiling to remove the bitter principle, taccalin. The recovered starch is used in many food preparations, a common preparation is for poi-like dishes. On a dry weight basis, tuber starch content averages about 25% and protein about 5%.

SHOTI or ZEDOARY, *Curcuma zedoaria* (Christm.) Roscoe

Family: Zingiberaceae

Shoti, native to northeast India, is cultivated in India, Sri Lanka, and throughout Southeast Asia. The perennial tillering plant, which has some tolerance to poor drainage, produces erect stems arising from the top of lateral rhizomes. The green leaves have purple-brown veins. Fleshy rhizomes become tuberous and are used for their easily digestible arrowrootlike starch, reportedly about 12% on a fresh weight basis. Yields of 10 t/ha are produced. Young buds and shoots are used in salads.

DICOTYLEDONS

ARRACACHA, *Arracacia xanthorrhiza* E.N. Bancroft *(A. esculenta)*

Family: Apiaceae (Umbelliferae)
Other names: apio, Peruvian carrot, Peruvian parsnip, mandioquina-salsa, zanahoria, racacha

Arracacha, a native perennial of the Andean highlands, continues to be an important and often preferred staple food in much of South America. Wild types have been found in Peru, and considerable diversity is found in Ecuador.

The short-day behavior and frost sensitivity limits the spread of arracacha to other areas. Plants are not suited to lowland production, but good productivity is obtained when grown at elevations between 1000 and 2500 m. Favorable growing temperatures range between 14°C and 21°C, whereas temperatures in excess of 25°C suppress growth. Ideal soils are light, well drained, and slightly acid, (pH 5–6). Loose soils favor lateral root enlargement and smoothness. Plant water use is high (600–1000 mm) and moisture should be uniformly available. Surprisingly, nutritional requirements are greater for phosphorus than nitrogen. Nitrogen applications are minimized to avoid luxurious foliar growth which is competitive to root enlargement. Plants are susceptible to spider mites, nematodes, and viruses.

Arracacha in some ways is similar to carrot and celery. Enlarged, fleshy, lateral roots resemble white carrots, with yellowish flesh, and sometimes the ring of the cambium tissue is purple (Fig. 14.5a). Produc-

FIG. 14.5. Arracacha, *Arracacia xanthorrhiza,* roots (a) and (b) plants, Brazil.

tive plants develop several to as many as 10 lateral carrot-sized roots clustered around the main rootstock. Root length typically ranges from 5 to 25 cm with diameters from 2 to 6 cm. Yields range between 5 and 15 t/ha.

Dark green, purple-tinged petioles resemble celery and are eaten raw or cooked (Fig. 14.5b). Petioles are tied together for blanching. Leaves are deeply divided and may reach heights of about 1 m. Small yellow or purple flowers are borne on loose umbels. If permitted, seed may be set; however, they are not used for propagation.

Arracacha is propagated with offshoots that form as a crown at the base of the primary stem. Interestingly, a grower practice of lightly slashing the base of offshoots after removal is done to stimulate better and uniform shoot and lateral root development. After a 2–3-day period to heal the wounds, these are transplanted. Plantings are usually started at the beginning of the rainy season. Most planting populations range from 25 thousand to 30 thousand plants per hectare. Short-day conditions are required for good root production, although cultivars vary in this response. Three cultivar types are recognized according to root color; these are blanca (white), amarilla (yellow), and morada (purple).

Harvest typically precedes flowering, but to enhance root enlargement, farmers remove flowers as they occur. Furthermore, once flowering begins, root quality diminishes. Arracacha is normally harvested by hand digging or plowing out the roots. At a sacrifice in yield, harvest can begin as early as 4–5 months. In the more normal situation, production takes more time, 10–14 months of growth to produce a mature crop and high yield. Delayed harvesting results in fibrous and off-flavored roots. Because it costs less to produce, arracacha frequently replaces potatoes in many diets. Roots have a short postharvest life; hence, they are infrequently stored. If stored, they are placed at 3–5°C, during which time, some starch will convert to sugar.

Roots must be cooked to be edible and are commonly prepared fried, boiled, or baked. They have an unusual taste, resembling a mixture of celery, carrot, and parsnips. Some describe the taste as roasted chestnuts combined with celery and carrot flavors. Starch content varies from 10% to 25% with starch granules that are small, similar to cassava. The starch is easily digested and, therefore, often prepared for infants and the elderly.

SQUAW ROOT, *Perideridia gairdneri* (Hook & Arn.) Mathias (*P. oregana* (S. Wats.) Mathias)

Family: Apiaceae
Other names: yampah, epos root

Squaw root grows wild in the western United States (southern Oregon, northern California). Related species grow in the western United States and British Columbia. Plants are perennial and prefer moist soils. Enlarged storage roots are collected in the summer and autumn by local people and are eaten raw or cooked. They are also dried for use during winter. Epos served as an food for natives and early settlers in these regions. The starchy dark-skinned roots resemble nuts with a white flesh.

JERUSALEM ARTICHOKE, *Helianthus tuberosus L.*

Family: Asteraceae (Compositae)
Other names: girasole, topinambour, sunchoke

Jerusalem artichoke, although a perennial plant, is cropped as an annual. Endemic to North America, plants can be found growing wild from north central United States, adjoining regions of Canada, and the eastern coast of the United States. Native American Indians were utilizing the tubers long before European colonization (Fig. 14.6a). Early in the 17th century, Jerusalem artichoke was introduced to Europe where its establishment was rapid and it achieved some importance as a vegetable. However, its primary use in Europe reverted to feeding livestock.

The crop is called Jerusalem artichoke, but it is not an artichoke nor did it originate in Jerusalem. The name probably was derived from the Italian word, *girasole*. Tubers taste somewhat like the globe artichoke, *Cynara scolymus.*

Shoots originating from buds on the tuber grow rapidly to produce tall stems and considerable foliage. Vigorous plants are as tall as 4 m (Fig. 14.6b). A well-branched but shallow fibrous root system develops from the stem bases. Rough textured leaves are long ovate to ovate oblong shaped with serrate–dentate margins. Early growth is often vigorous enough to outgrow and shade weed competitors. Typical compositelike flower heads, 4–8 cm in diameter, are yellow. Seeds are produced but are not used for propagation because resulting plants are not uniform for plant type.

FIG. 14.6. Tubers of Jerusalem artichoke, *Helianthus tuberosus,* (a) and (b) plants.

Plants are adapted to temperate regions requiring 4–5 months of frost-free growth. The favorable temperature range is between 18°C and 26°C; higher temperatures tend to depress yield. Plants are killed by frost.

Because of its low monetary value, Jerusalem artichoke is often relegated to grow in poor soils, with minimum nutrition, moisture, or care. Although having a reputation of performing well in adverse conditions, in reality, yields are low when growing conditions are poor. Provided fertile, well-drained sandy loam soils, good cultural care, and nutrition, the crop is capable of very high yields. Plants are sensitive to poorly drained soils, especially during tuber development. Water consumption is fairly high at 100 cm or more per crop-year.

For propagation, uniform whole tuber or tuber pieces weighing 60–90 g are planted about 5–10 cm deep. Tuber buds exhibit apical dominance, and freshly harvested tubers pass through a rest period before sprouting can occur. Raised beds, even if slight, improve subsequent tuber crop development and harvesting. During early plant growth, soil is moved toward the base of stems, with the result that tubers formed will develop about 10–25 cm below the surface. Commonly used row spacings are 1–1.5 m between rows and 50 cm within rows.

Decreasing day lengths initiate tuberization concurrent with translocation of foliar and stem carbohydrates to developing tubers. Tuberization causes a cessation of vegetative growth and foliage; some begin to senesce. Flowers sometimes are removed to avoid seed development competing with tuber enlargement. Plantings should be made as early as possible in the growing season in order to maximize the period of vegetative growth before tuberization. Short growing periods limit foliage growth and carbohydrate accumulation and thus result in low yields.

The fleshy tubers are variable in size and shape and are highly branched with irregular thin-skinned surfaces that make harvesting and soil removal difficult, especially in fine textured soils. Tuber colors vary from white to yellow, and sometimes are red or purple. The fleshy interior is white with a crispy and watery texture.

Tubers are harvested in the late fall when stems are completely senescent or killed by cold temperatures. Stems are cut away to facilitate harvesting. Tubers are difficult to harvest because they are branched and many break apart and are not recovered. Yields vary considerably; about 15–20 t/ha is common. However, under favorable conditions, yields of 40 t/ha are achieved. Because the thin-skinned tubers bruise easily, careful handling during harvest and postharvest is necessary. Even undamaged tubers are subject to rapid desiccation.

Ground storage is feasible where soils do not freeze and are well drained. However, when the rest period is broken, sprouting can occur. Harvested tubers can be stored for several months when held at 0–2°C and at 95% RH. Tubers held at low temperatures taste sweet and should be reconditioned at warm temperatures for several days before consumption to reduce the sugar content.

A unique feature of Jerusalem artichoke is the production of inulin, a polymer of fructose, which averages about 15% of the fresh weight and is the main storage carbohydrate. Tubers are a suitable food for diabetics because this carbohydrate form can be utilized by persons with glucose metabolic dysfunction. Tubers are cooked much like potatoes and can also be eaten raw in salads. Considerable use is made of tubers and foliage for animal feeding. Jerusalem artichokes are also used for sugar extraction and fuel alcohol production.

A crop similar to Jerusalem artichoke is Indian breadroot, *Psoralea esculenta* Pursh (Fabaceae), also known as prairie potato. It is a noncultivated North American native plant that produces tuberous roots having a dry weight starch content greater than 60%.

YACON, *Polymnia sonchifolia* Poepp. & Endl. (*P. edulis* Weddell.)

Family: Asteraceae
(Compositae)
Other names: jiquama, jacon, apple of the earth

Wild forms of yacon can be found in the Andean regions of Colombia, Ecuador, and Peru. In these areas, the crop is cultivated at high elevations, but not much above 3000 m. The perennial plant is cold tolerant, and when aboveground portions are damaged or killed by freezes, reestablishment occurs by regrowth from underground portions. Plants also have some high-temperature tolerance. Yacon has the potential for cultivation in semitropical and some temperate climates because of the broad adaptation, day-neutral characteristic and relatively easy culture even in poor soils. The crop is grown as an annual for its foliage and as a perennial for the underground tuberous roots.

Plants grow as tall as 1.5 m, with stems that are hairy, green with purple markings, and bear yellow or orange flowers which are formed on lower axillary stems. The tops become senescent and die after flowering. Offsets from the base of aboveground stems and stem cuttings are used for propagation. Developing tuberous roots, usually four or more, form as a cluster, each about 15 cm in length. These appear as if fused to the swollen stem and are enlarged portions of basal stems and storage root tissues. Shapes vary from spindlelike to round, and the tuberous roots usually can be easily separated. The thin skins are tan or a purplish brown color. The flesh can be white, pale yellow, or orange and is of variable sweetness.

The tuberous roots are normally harvested after a long growing period of more than 200 days. These typically weigh from 200 to 500 g, some as much as 2 kg. Yacon yields are relatively high, frequently more than 20 t/ha. After harvest, they can be stored fairly well for several weeks if kept cool (5°C) and in the dark.

The crisp textured tuberous roots are juicy and frequently eaten raw in snacks and salads and also are boiled or baked. Sun-drying preservation is used to extend usefulness; exposure to sunlight is believed to improve flavor and sugar concentration. The skin must be peeled because it is resinous. Most taste sweet because of the fructose content, although this is not a uniform trait. By selection, a greater uniformity of sweet types can be achieved. The moisture content is between 70% and 85%; almost 65% of the dry weight is sugar. As with some other Asteraceae species, carbohydrate is stored as inulin, a fructose polymer, in an abundant and very pure form. The tuberous roots contain about 6–7% protein and are high in potassium. Stems and leaves, most often fed to animals, can be used as a cooked vegetable or eaten raw.

ULLUCO, *Ullucus tuberosus* Lozano

Family: Basellaceae
Other names: Olluco, melloco, papalisa

Ulluco, a tuber-producing Andean native crop, has been cultivated for more than 4000 years in Peru, Bolivia, Ecuador, and northern Argentina. Wild forms can still be found in these countries. The crop is most popular in Andean regions, although tubers are marketed in many parts of South America. Production in Peru alone exceeds 20,000 ha. Ulluco and other Andean tubers, oca *(Oxalis tuberosa)* and anu *(Tropaeolum tuberosum),* were considered for introduction to Europe, particularly after the potato blight epidemic, but with disappointing results. The main obstacle for successful cultivation being the short-day-length requirement of the introduced plants.

The crop possesses good cold tolerance, necessary at the high elevations (2500–4000 m) where it is grown. Plants have high tolerance to drought as well as to high soil moisture and are relatively productive on marginal soils. Tuber growth is poor in hot climates. Overall adaptability is comparable to that of the potato, *Solanum tuberosum*.

Ulluco is a succulent plant with forms varying from prostrate to bushy to semiclimbing vines. Tubers are brightly multicolored, almost ornamental, with inconspicuous surface buds and a thin soft waxy skin (Fig. 14.7). The white to light yellow fleshy texture remains crisp even after cooking; some describe its flavor as nutlike. Tubers are elongated; some are strongly curved and small (4–7 cm). In regional markets, small size is preferred.

Tubers are initiated during short days and enlarge at the end of the stolons. Stolons form along the stem, and a common cultural practice places soil high against the stem to encourage additional stolon development and, hence, more tubers.

Propagation is vegetative, using small whole tubers. Greenish yellow to reddish flowers rarely produce seed; however, fertile seed have been obtained experimentally. Ulluco is frequently intercropped with potato, mashua, and oca, although pure stand culture is typical when in rotation with potato. Ulluco has relatively few disease or pest problems other than a virus complex that is present in almost all material. Although plants are productive even when infected, meristem tip culture can be used to produce virus-free planting stock. Nematodes can also cause crop losses.

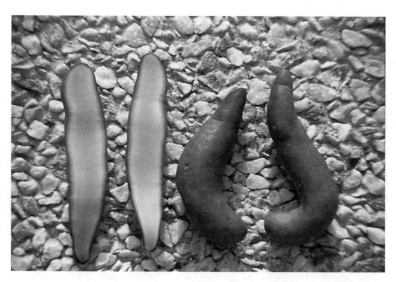

FIG. 14.7.   Ulluco tubers, *Ullucus tuberosa,* Cali, Colombia.

Tubers mature in 6–8 months. Average yields between 5 and 9 t/ha are less than white potato yields grown in the same area. Ulluco tubers store well at cool temperatures and are kept in the dark to avoid greening.

Starchy tubers are used for their thickening property in soups and stews. They can be dehydrated in a product like "chuño" to store for years. Edible foliage contains about 12% protein on a dry weight basis and has mucilaginous qualities resembling malabar spinach, *Basella alba*. The mucilaginous material is also present in tubers, but this can be washed away before cooking.

MACA, *Lepidium meyenii* Walp.

Family: Brassicaceae (Cruciferae)
Other names: maka, Peruvian ginseng

With the exception of some high-altitude-adapted potatoes, maca is unique in being grown in the very high elevation (3800–4200 m) and harsh environment of the "puna" region of Peru. The labor-intensive cultivation of this annual plant, most often for subsistence cropping, is at least 3000 years old. At high elevations, the plant tolerates intense sunlight and exhibits good growth at temperatures averaging between 5°C and 10°C and has good resistance to cold and frosts. Some believe that maca tends to exhaust the soils in these Andean highland regions and that several years of fallow are required before the land is again cultivated.

The plant's growth habit is matlike, with leaves forming a prostrate rosette (Fig. 14.8). The long, deeply scalloped foliage resembles cress and is reported to be edible. Flowers are white, self-fertile, and produce a two-celled silique. Thought to be a biennial, maca probably is a long-lived annual. Propagation is with seed, the production of which requires a special culture. The tuberous hypocotyl is harvested and then replanted, often after a short storage period, to resume foliage growth and to produce seed. The seed are generally broadcasted and trampled into the soil by sheep. Flowering appears photoperiod insensitive, and low-temperature vernalization may not be required.

The turnip-shaped tuberous hypocotyl-root axis is the edible organ. These are about 8 cm in diameter, narrowing into a tap root, and are found in an array of colors from creamy yellow to purple or almost black, yellow being preferred; the interior flesh is white or creamy yellow.

Harvest occurs 7–11 months or more after planting and usually at

FIG. 14.8.   Maca plant and tuber, *Lepidium meyenii.*

the beginning of the dry season. Yields are relatively low, about 3 t/ha. The tuberous roots are lifted and are allowed to sun dry. When dried, they become brown, sweet, and aromatic, developing a musky tangy taste and a butterscotch smell. Dried tuberous roots have a long storage life, often more than a year, but the quality diminishes with time. In addition to baking and roasting as vegetables, the tuberous roots are used for desserts and in various drinks. On a dry weight basis, maca is high in sugar and starch and contains about 10% protein. However, the tuberous roots also contain alkaloids, tannins, saponin, and glucosinolates, most not yet identified. Maca is also used medicinally, and folklore relates that its consumption improves human and animal fertility.

BUFFALO GOURD, *Cucurbita foetidissima* HBK
Family: Cucurbitaceae

Buffalo gourd is a long-lived monoecious perennial endemic to the southwestern United States and northwestern Mexico. The crop is able to grow in arid conditions, being extremely drought resistant due to a large and deep root system with stored food reserves. Plants are sensitive to poorly drained soils, but otherwise grow fairly well even on arid and nutritionally poor soils. Plants require a long period of warm dry weather for good growth, but at temperatures above 40°C, vine growth is restricted.

New plantings are propagated by adventitious rooting from vine runners. Vine growth resumption from established plants typically begins after the last spring frost and growth continues until the first frost in the fall. Initially, large fleshy and deeply penetrating roots develop; some are 4 or 5 m long. Roots continue enlargement (thickening) each year, many attaining a weight of 30 kg after two seasons and as much as 50 kg after 3 or 4 years. Yields in excess of 30 t/ha have been reported. Storage roots are very high in starch, often as high as 60% of the dry weight, with a starch composition similar to cassava. Roots also contain glycosides; however, these are easily leached by soaking in water.

Small, inedible, round, yellow, hard-shelled fruit, about 5–7 cm in diameter, are produced, as many as 100 from a single plant. Seed are white and flat, about 1 cm long by 0.5 cm wide. Native peoples have used the mature seeds for food for centuries. Seeds contain about 30–35% protein and about 30% oil; seed yields as much as 1 t/ha are obtained.

Presently, most buffalo gourd production is feral, but the plant is presently undergoing domestication because of its potential for starch production and as a high-protein, edible seed oil crop.

APOIS, *Apois americana* Medikus

Family: Fabaceae (Leguminosae)

Apois presently is mostly a noncultivated perennial vine-like plant endemic to streams and rivers from southeastern Canada to southeastern United States. The white-fleshed, starchy tubers have been used by natives in these regions for centuries; the seed are also edible.

Succulent vines varying in length from 1 to 6 m produce white, pink, purplish, or brownish red flowers. Seed pods are 5–13 cm long. Plants produce rhizomes from which in stringlike fashion several tan to black-skinned, white-fleshed tubers are formed. Additionally, tubers may be clumped or widely spaced. Tuber sizes vary from 3 to 10 cm in length and from 2 to 5 cm in width. Accessions from different regions show wide variations in tuber numbers, shape, and development; some plants do not form tubers. Yield vary from negligible to almost 4 kg per plant.

Propagation is with seed or tubers, either or which result in large plant variations. Early harvested tubers exhibit different levels of dormancy, apparently influenced by the position along the rhizome. Dormancy is broken when tubers are exposed to low temperatures for several months. Seed production and germination is also variable; pods often are without seed, and pod dehiscence is common.

In its native habitat, apois can be found in waterlogged and acid soils, but growth and yields are better in well-drained soils with a pH between 5 and 7. Better growth occurs when plants are trellised. Tubers can be harvested at any time, although yield increases until frost kills the tops. Tubers can be left in the ground until harvested in the spring.

Some tubers contain as much as 45% carbohydrate and 15% crude protein on a dry weight basis. Protease inhibitors are present but are inactivated by cooking. Tubers have a taste resembling a combination of boiled peanut and Irish potato; the texture is mealier than that of potato. Recent interest to domesticate the plant as a food crop involves selection and breeding. Most accessions are diploid ($2n = 22$), although triploids are also found.

AHIPA, *Pachyrhizus ahipa* (Wedd.) Parodi
POTATO BEAN, *P. tuberosus* (Lam.) Sprengel

Family: Fabaceae (Leguminosae)
Other names: ajipa ashipa, frijol chuncho, Central American yam
   bean

No wild forms of ahipa have been found; its likely origin is the southwestern Amazon basin near Bolivia or northern Argentina where it was domesticated at least 2000 years ago. Production and consumption remains largely regional in Bolivia and Peru, although the potential for expansion to other areas is high.

The potato bean has a wide adaptation, growing from sea level to 3000 m in tropical, subtropical, and some temperate regions in either wet or dry environments. A warm climate and moderate rainfall is the most favorable environment for growth. Although grown at high elevations, plants are frost sensitive. However, although having some similarity to yam bean *(Pachyrhizus erosus)* some potato bean cultivars are day neutral with respect to flowering and initiation of root enlargement.

Plants are seed propagated. Growth of the semierect, 3–6-m-long trailing vines is rapid; *P. ahipa* vines are shorter than those of *P. tuberosus.* Leaves are entire and rhomboical ovate shaped. Flowering occurs 2–3 months after planting; lavender or white flowers produce pods with black kidney-shaped seed. Often, growers remove developing flowers to enhance root enlargement.

The below-ground, swollen tuberlike root is the edible organ. Immature pods are not eaten because of irritant hairs and toxicity; all other plant parts are toxic. When ready for harvesting, the typical root is elongated, about 15 cm long, and tapers at each end. Tuberous roots are harvested as early as 5–6 months and weigh between 500 and 800 g. The thin, pale, yellow skin does little to prevent desiccation; thus, they are commonly left in the ground until needed. Roots of *P. ahipa* are smaller and mature earlier than those of *P. tuberosus.* Potato bean roots have a succulent, crispy white interior, and cut surfaces are slow to discolor. They are mainly eaten raw for salads, but when cooked, the texture remains crisp.

The Asian yam bean, *P. angulatus,* is another edible root crop. Confusion as to the identification of various yam bean species is not unusual because both common and species names are frequently interchanged.

YAM BEAN or JICAMA, *Pachyrhizus erosus* (L.) Urban

Family: Fabaceae (Leguminosae)

The Information about this crop is found in Chapter 22.

KUDZU, *Pueraria lobata* (Willd.) Ohwi *(P. thunbergiana, Pachyrhizus trilobus)*

Family: Fabaceae (Leguminosae)
Other names: fankot, ko (China), kuzu (Japan)

Kudzu is a perennial, hairy, climbing vine commonly grown in Japan and China for the accumulated starch in the enlarged tap root. Kudzu vines are leafy and vigorous, growing as long as 8 m, a characteristic useful for fodder or cover crop production. In some areas, kudzu has escaped to become a persistent weed. A related tropical plant, *Pueraria phaseoloides* (Roxb.) Benth., known as Indian kudzu, also a perennial, is even more vigorous and widely used for fodder and pasturing.

Kudzu leaves are large and rhomboidal shaped. Flowers are purple and produce hairy, oblong, narrow pods about 8 cm long. Propagation is usually with seed, but in some climates, few seeds are set and plants are vegetatively propagated from stem cuttings. Plants are usually staked to support the vines.

FIG. 14.9.   Kudzu storage root, *Pueraria lobata.*

At harvest, some enlarged roots can be 1 m in length, weigh as much as 35 kg, and contain over 25% starch. However, it is the young roots that are used as a baked or boiled food (Fig. 14.9). The starch has some qualities like arrowroot and is similarly used in food preparations. A potential use of kudzu starch is for ethanol production.

AFRICAN YAM BEAN, *Sphenostylis stenocarpa* Hochst. ex A. Rich) Harms. *(Dolichos stenocarpa)*

Family: Fabaceae (Leguminosae)

The African yam bean possibly originated in Ethiopia, although wild forms are known to occur in West Africa and Zaire. The plant is grown in many tropical African locations. In some localities, wild and domesticated plants are cultivated. The plant is a climbing vine of about 2 m length having trifoliate leaves and flowers that are either purple or pink; some have yellow or greenish white petals. Propagation is with seed or stem cuttings.

Edible portions are the small spindle-shaped underground storage root that looks like sweet potato and tastes much like white potato, and the mature seed. Roots are harvested after 8 months; the flesh is white and watery with a fresh weight starch content of 20% and about 3.5% protein, and are much admired for their flavor.

In central Africa, the storage root is the primary crop portion, whereas in West Africa it is the seed. The plants cultivated in West Africa produce large (10–12 mm) seed compared to the seed (2–3 mm) produced by plants grown for the storage root. A disadvantage of the seed is that they require a long cooking period before consumption.

FALSE YAM, *Icacina senegalensis* A. Jess.

Family: Icacinaceae

A subsistence crop, false yam is an infrequently cultivated perennial shrub of west and central African origin, grown for the tuberous hypocotyl. Erect shoots, which grow up to 90 cm, arise from the fleshy underground organ. Plants have ovate to obovate-shaped leaves about 5–10 cm long that become leathery and dark green when mature. White inconspicuous flowers produce bright red fruit that contain a single seed covered with a white pulp.

The principal vegetable value of this plant is the large turniplike hypocotyl, which can achieve a length and width of about 40 and 25 cm, respectively. These contain 10–15% starch, but of an edible quality that is better suited for industrial use. Additionally, the white flesh contains a bitter-tasting resinous compound.

COLEUS POTATOES

Family: Lamiaceae (Labiatae)

COLEUS or CHINESE POTATO, *Coleus parviflorus* **(C. tuberosus)**
SUDAN or HAUSA POTATO, *Solenostemon rotundifolius* **(Poir.)**
J.K. Morton *(Coleus rotundifolius)*
LIVINGSTON, KAFFIR, or HAUSA POTATO, **Plectranthus**
esculentus N.E. Br.

Common and/or regional names for these species often result in confusion as to their identity, as all are frequently called coleus or Hausa potato.

Coleus or Chinese potato, *Coleus parviflorus,* is a crop of some importance in southeastern India, the probable center of origin and where

FIG. 14.10.   Coleus potato, *Coleus parviflorus,* Kuala Lumpur, Malaysia.

it is commonly called kourka (Fig. 14.10). The crop is also grown in other dry tropical and subtropical regions of Asia and Africa.

The plant is an annual with simple leaves having serrate margins that are oppositely attached to succulent, nearly square-shaped stems. Although flowering occurs, no seed is produced. For propagation, seed tubers are first used to produce stem cuttings, which are then transplanted to establish the main crop. Cuttings, 10–15 cm long, are planted on raised beds at densities of 7–10 plants/m^2.

The clusters of white-fleshed, small, round, brown or black starchy tubers that form at the base of the primary stem are the main crop. Leaves can be used for flavoring. Tubers mature about 4–6 months after planting; harvest should not be delayed, as quality declines with overmaturity. Yields vary from 5 to 19 t/ha. Tubers store well in cool well-ventilated conditions. They have an aromatic flavor, contain about 11% carbohydrate and 5% protein, and are eaten raw or cooked.

Sudan or Hausa potato, *Solenostemon rotundifolius,* is sometimes also called coleus potato. This annual is grown in Southeast Asia, India, Sri Lanka, and Africa for its clustered growth of dark brown potatolike tubers that developed at the base of the main stem. The culture and productivity is similar to *Coleus parviflorus* with good growth in areas of high rainfall, cool nights, and well-drained sandy soils. Edible tubers mature in 6–8 months and have a carbohydrate content of about 20% and protein about 1%.

The Livingston potato, also called Kaffir or Hausa potato, *Plectranthus esculentus,* is a deciduous perennial that grows to about 75 cm in height with long trailing branches that readily root at the nodes. The leaves are pubescent and sessile, and the yellow flowers bloom after foliage drop.

The Livingston potato is grown in tropical Africa. It is propagated with plantlets taken from the tubers. Tubers mature in 7–8 months following planting; yields average 5–6 t/ha. The thick, yellow-skinned, branched, tuberous roots are eaten after cooking; the taste is similar to white potato.

The aromatic shoots of *C. amboinicus* Lour., known as Spanish thyme or Indian borage, also have culinary use in parts of tropical America. *Coleus barbatus (C. forskohlii)* is another aromatic perennial plant producing tuberous roots that are pickled and eaten as a condiment in India.

CHINESE ARTICHOKE, *Stachys affinis* Bunge (*S. tuberifera, S. sieboldii*)

Family: Lamiaceae (Labiatae)
Other names: chiyorogi, Japanese artichoke, knotroot, kon loh

A native of the Far East, Chinese artichoke is a perennial herb grown for the somewhat cylindrical tubers (Fig. 14.11). These are formed at the tips of underground stems. Some underground stem tips also

FIG. 14.11.   Tubers of Chinese artichoke, *Stachys affinis.*

emerge from the soil and grow above ground. Aboveground stems have the characteristic square-shaped stem of the Lamiaceae family. Nettle-like in appearance, these plants grow to a height of 30–45 cm. Petiolate roughly ovate leaves form in opposite pairs on the stems. Flowers are white or pink. The slender white tubers, also used for propagation, are 5–8 cm long and about 1–2 cm in diameter and have distinctive constricted internodal segments resembling a string of beads. Rather than starch, much of the carbohydrate (about 60%) is the tetrasaccharide stachyose.

Tubers are dug, washed, and eaten after cooking, much like white potatoes. Being thin skinned, tubers readily lose moisture and are easily bruised; therefore, they are commonly stored in damp sand. The tubers, usually pickled, are a popular food in China and Japan and are also eaten fried or fresh in salads.

MAUKA, *Mirabilis expansa* (Ruiz & Paron) Standley

Family: Nyctaginaceae
Other names: yuca inca, chago, miso, arracacha de toro

Mauka is a little known, high-altitude tuber crop. Wild ancestors are found in Andean mountain regions. Plants have good cold tolerance and are able to adjust to wet as well as arid conditions, but are heat sensitive. Although perennial, it is cultivated as an annual but requires at least a year to reach harvest maturity in that mountainous environment.

The low, compact plants produce succulent foliage. Edible parts are tuberous portions of lower stems and upper root. Flowers are purple or white. Although propagation is possible with seed, stem portions or offshoots are used.

With developing maturity, the edible, thickened, fleshy axis forms at or just below the surface. The tuberous portion ranges in length from 20 to more than 30 cm, usually with a diameter of 5–8 cm. The flesh is white to yellow in color. These tuberous forms are clustered at stem bases. They can be left in the ground a second year for further enlargement; yields up to 20 t/ha are obtained. The freshly harvested tuberous portions of some cultivars have a strong astringent taste which can be reduced by exposure to sunlight. Sun exposure and drying also aid in the conversion of starch to sugar. Mauka is used boiled and often mixed with other foods. On a dry weight basis, carbohydrates comprise about 85% of the tuber and protein about 7%. Leaves are also eaten in salads and stews.

OCA, *Oxalis tuberosa* Molina

Family: Oxalidaceae
Other names: papa roja (Mexico), truffette acide (France)
  knollen-sauerklee (German), New Zealand yam, apilla, ibia

Oca is a ancient crop cultivated by pre-Columbian people and presently by farmers in the Andean highlands. To these people, oca is nearly as important as the potato, *Solanum tuberosum*. In these mountainous environments, oca yields could be twice that of the potato. Peru leads the world in production with over 20,000 ha. In addition to production in the Andean region, commercial production occurs in Mexico, Central America, and the south island of New Zealand, where, unfortunately, they are incorrectly called "yams."

The compact, bushy growth of this perennial is between 20 and 30 cm in height. Tolerant of harsh climates, oca is grown at elevations up to 4000 m, at relatively low temperatures and under a wide range of rainfall. Although freezing kills the foliage, the plant will usually regrow. Oca is heat sensitive, and at temperatures greater than 28°C, growth is arrested. Leaves have purple coloration and are hairy. On the whole, flowering is abundant and viable seeds develop. However, propagation is by use of whole tubers. Plantings are made at the start of the rainy season. Oca is often interplanted with other tuber crops or with legumes or grains. Oca is also a commodity frequently considered in rotations with potatoes in the high Andes.

Like the potato, edible tubers develop on underground stolons, and tuberization is initiated by short day lengths. Tuberization is also influenced by temperature and usually occurs about 4 months after planting, with maturation 2–3 months later. Tubers are small and elongated, about 6–8 cm in length, and appear wrinkled with white to bright red surfaces and firm white flesh (Fig. 14.12). There are sweet and bitter cultivars due to variations of the sugar and oxalic acid content. Some, usually the white-colored tubers, appear to have a slightly higher oxalic acid content.

Tubers can be eaten immediately after harvest, but tuber flavor is better when exposed to the sun in order to reduce bitterness. Oca is consumed after boiling, baking, frying, or pickling and also can be eaten

FIG. 14.12. Oca tubers, *Oxalis tuberosa*. Source: Courtesy M. Holle.

uncooked. Leaves and young shoots are eaten in salads or as cooked vegetables. Bitter types are usually converted into dried forms or made into "chuño de oca," also called "kaya" or "okaya." Oca chuño is reported to taste better than chuño made from white potato. Oca chuño is made by first soaking the tubers for several weeks in running water, then subjecting them to alternating outdoor freezing and drying conditions. The taste is reported to be similar to dried figs. Tubers are also processed by canning.

Oca storage characteristics are relatively good if properly cured and moisture loss is minimized. Nutritional qualities are comparable to potato. On a dry weight basis, about 85% of the tuber is carbohydrate; the starch is of good quality.

PERUVIAN POTATO, *Solanum hygrothermicum*

Family: Solanaceae
Other names: urahji, kurahji, cachariqui, and moshaki

Peruvian potato is cultivated in the warm humid eastern lowlands of Peru. Plants and tubers look like potato *(Solanum tuberosum),* and, like the cultivated potato, plants are tetraploids. Tubers are small, oval, or somewhat cylindrical. The skin is yellowish white, and the shallow eyes are surrounded by a purplish pigment. Tubers are reported to have a short rest period.

MASHUA, *Tropaeolum tuberosum* Ruiz & Pavon

Family: Tropaeolaceae
Other names: añu, isaño (Peru), nabos, cubios (Colombia)

Mashua, grown for its tubers, is endemic to the Andean region and also of ancient domestication. About 5000 ha are grown in Peru, but production outside the Andean region is relatively small. The crop is well adapted to growing at high elevations (2500–4000 m), and is often intercropped with other Andean crops such as oca, ulluco, or potato. Plants are tolerant of low temperatures and frosts and have a short day length requirement for tuberization.

A semiprostrate perennial with vigorous sprawling growth that is capable of twining, mashua resembles garden nasturtium, *Tropaeolum majus*. Orange to deep red, bisexual, edible flowers are produced profusely and the fruit contain many seed. Seed are not used for propagation because of resulting plant variability; instead, small tubers are used.

Because tubers develop near the surface, soil is moved toward the base of the plant to assure tuber development below the surface. Harvest occurs about 6–8 months after planting. About the size of a small potato and somewhat cone shaped, tubers are attractively colored, often in combinations of white, yellow, and red; the flesh is yellow.

Tubers quickly lose moisture, but if stored at low temperatures and in the dark, they will keep well for several weeks. They can also be left in the ground and harvested as needed. Because of their strong pungency, tubers are not eaten raw. The sharp mustardlike taste is due to the presence of glucosinolates which disappear after exposing tubers to 4–5 sunny days, or by cooking. Tubers are boiled and used in soups and in other preparations. Tender young leaves are used as a potherb. Nutritional values of mashua are comparable to those of other root and tuber crops.

SELECTED REFERENCES

Bemis, W.P., Curtis, L.D., Weber, C.W., and Berry, J.W. 1978. The feral buffalo gourd, *Cucurbita foetidissima*. Econ. Bot. *32*, 87–95.

Bemis, W.P., Berry, J.W., Weber, C.W., and Whitaker, T.W. 1978. The buffalo gourd: A new potential horticultural crop. HortScience *13*, 235–240.

Blackman, W.J., and Reynolds, B.D. 1986. The crop potential of *Apois americana*—Preliminary evaluations. HortScience *21*, 1334–1336.

Chubey, B.B., and Dorrell, D.G. 1983. The effect of fall and spring harvesting on the sugar content of Jerusalem Artichoke tubers. Can. J. Plant Sci. *63*, 1111–1113.

DeVeaux, J., and Schultz, E.B., Jr. 1985. Development of buffalo gourd as a semiarid land starch and oil crop. Econ. Bot. *39*, 454–472.

Erdman, M.D., and Erdman, B.S. 1984. Arrowroot *(Maranta arundinacea)*, food, feed, fuel, and fiber resource. Econ. Bot. *38*, 332–341.

Erdman, M.D., Phatak, S.C., and Hall, H.S. 1985. Potential for production of arrowroot in the Southern United States. J. ASHS *110*, 403–406.

Hodge, W.H. 1954. The edible Arracacha—A little known root crop of the Andes. Econ. Bot. *8*, 195–221.

Imai, K., and Ichihashi, T. 1986. Studies on dry matter production of edible canna (*Canna edulis* Ker.). Japan J. Crop Sci. *55*, 360–366.

Kaldy, M.S., Johnson, A., and Wilson, D.B. 1980. Nutritive value of Indian bread-root, Squaw-root, and Jerusalem artichoke. Econ. Bot. *34* (4), 352–357.

King, S.R., and Gershoff, S.N. 1987. Nutritional evaluation of three underexploited Andean tubers: *Oxalis tuberosa* (Oxalidaceae), *Ullucus tuberosus* (Basellaceae), and *Tripaeolum tuberosum* (Tropaeolaceae). Econ. Bot. *41*, 503–511.

Leon, J. 1964. The "Maca" *(Lepidium meyenii)*, a little known food plant of Peru. Econ. Bot. *18*, 122–127.

Martin, F.W., and Cabanillas, E. 1976. Leren *(Calathea allouia)*, a little known tuberous root crop of the Caribbean. Econ. Bot. *30*, 249–256.

Mokady, S., and Dolev, A. 1970. Nutritional evaluation of tubers of *Cyperus esculentus* L. J. Sci. Food Agric. *21*, 211–214.

National Research Council. 1989. Mashua. In Lost Crops of the Incas: Little Known Plants of the Andes with Promise for Worldwide Cultivation. National Academy Press, Washington, DC. pp. 66–74.

Nelson, J.M., Scheerens, J.C., Berry, J.W., and Bemis, W.P. 1983. Effect of plant population and planting date on root and starch production of Buffalo gourd as an annual. J. ASHS *108*, 198–201.

Potter, D. 1996. African Yam bean. Personal communication.

Reynolds, B.D., Blackman, W.J., Wickremesinhe, E., Wells, M.H., and Constantin, R.J. 1990. Domestication of *Apois americana*. In Advances in New Crops. Proc. 1st National Symposium for New Crops, Indianapolis, Indiana. J. Janick and J.E. Simon, eds. Timber Press, Portland, OR, pp. 436–442.

Rousi, A., Jokela, P., Kalliola, R., Pietila, L., Salo, J., and Yli-Rekola, M. 1989. Morphological variation among clones of ulloco (*Ullucus tuberosa,* Basellaceae) collected in Southern Peru. Econ. Bot. *43*(1), 58–72.

Rousi, A., Saro, J., Kalliola, R., Jokela, P., Pietila, L., and Yli-Rekola, M. 1989. Variation patterns in ulluco. Acta Hort. *182*, 145–152.

Sperling, C.R., and King, S.R. 1990. Andean tuber crops: Worldwide potential. In Advances in New Crops. Proc. 1st National. Symposium New Crops: Research, Development, Economics, Indianapolis, Indiana. J. Janick and J.E. Simon, eds. Timber Press, Portland, OR, pp. 428–435.

Ugent, D., Pozorski, S., and Pozorski, T. 1984. New evidence for ancient cultivation of *Canna edulis* in Peru. Econ. Bot. *38*(4), 417–432.

Velayudhan, K.C., Amalraj, V.A., and Muralidharan, V.K. 1988. Studies on the growth of *Coleus parviflorus* Benth. J. Root Crops *14*, 71–72.

Zardiru, E. 1991. Ethriobotanical notes on Yacon. Econ. Bot. *45*, 72–85.

15

Sweet Corn, *Zea mays* L.

Family: Poaceae (Gramineae)

Sweet corn, a form of maize, is an important and popular vegetable, especially in the United States where about 250,000 ha are grown annually. North America dominates world sweet corn production. The popularity of sweet corn has greatly increased and an apparent trend is for further expansion of production in European and Asian countries, especially Japan and China.

Overall, maize is a very important food grain for humans and animals and has a multitude of food and nonfood uses. The crop is commonly called corn in the United States, Canada, and Australia. Corn was a generic term for a grain or cereal, for example, wheat in England and oats in Ireland and Scotland. Thus, it is not surprising that the name corn was applied to the new grain species settlers found in the Americas. Nevertheless, most of the world's population recognize *Zea mays* as maize. The relatively broad growth adaptation of maize permits its extensive cultivation throughout much of the world. In terms of production volume, maize is preceded only by wheat and rice.

Origin and Taxonomy

Maize is an ancient crop, as revealed from cob remnants dated to about 5000 B.C. found at the Tehuacan cave excavations in Mexico. Domestication is estimated to have begun about that time.

Two major genera of Poaceae (Graminaceae) native to the Americas are *Zea* and *Tripsacum*. The progenitor of corn is generally accepted to be teosinte, *Zea mays* var. *mexicana* (Fig. 15.1). Present maize plants can pollinate teosinte, and hybrid seed are produced, but resulting progenies are not useful. One theory suggests that some now extinct form of pod corn was the ancestor and that teosinte is a mutant of the pod corn. However, wild types of pod corn cannot be found.

Sweet corn is believed to have resulted as a mutation of grain corn

FIG. 15.1.   Teosinte, *Zea mays* var. *mexicana,* the likely progenitor of corn.

identified at the *Su1* locus of chromosome 4, (*Su1/Su1* = grain corn, *su1/su1* = sweet corn). Such mutations undoubtedly occurred often. Apparently, sweet corn was grown in pre-Columbian times as a source of alcohol for ritual purposes. However, a broader use of *su1* corn was ignored for a long time, most likely because the wrinkled seed had less starch and also stored poorly compared to *Su1* corn. Commercial sweet corn cultivation began about 200 years ago in the United States.

Botany

Sweet corn is a herbaceous monocot and warm season annual. Plants are monoecious with male flowers borne as the terminal inflorescence (tassels) on the main stem (axis or culm) and female flowers borne separately as a lateral inflorescence (ear) developing at a leaf axil. Plants produce one to several ears. Occasionally, staminate flowers form at the end of ears, and pistillate flowers form in tassels.

After germination, the initial primary root provides early establishment. Whorls of secondary roots develop at the basal nodes of the stem and grow laterally. It is these relatively shallow, profusely branched-adventitious-roots that nourish the plant. Aerial buttress roots provide additional anchorage and some absorption. These develop at above-ground compressed basal nodes and are unbranched until entering the soil.

The rigid stem ranges in height from 1.5 to 2.5 m and is enveloped by alternating leaf sheaths originating from each node. Stem nodes are prominent. The leaf sheath forms at the node and closely encloses a length of the main stem, often over the next node. At the ligule, each leaf sheath then angles away from the stem as a long, broad, curved leaf. The ligule fits tightly around the stem at the top of the sheath. Leaf blades are alternate and grasslike in appearance. The long leaves are fairly uniform in width and have a prominent midrib, with many small veins parallel to the length of the leaf. It is not uncommon for seedlings before they are 20 cm tall to have already initiated all their leaf buds and the terminal inflorescence.

Branching (suckering) commonly occurs at the base of the stem. Suckers are secondary shoots or tillers developing in the axils of lower leaves near the soil surface. Ears formed on the secondary shoots are late to develop and are seldom productive. The merits of early sucker removal are controversial; the consensus is that removal requires considerable effort for which there is no yield benefit. Large genetic variation exists for the tendency to branch; preferred cultivars are predominately single stalked.

Inflorescence

The male inflorescence is a loose terminal panicle, (tassel) consisting of the spikelike central axis and lateral branches (Fig. 15.2). The central axis usually has four or more rows of paired spikelets; the lateral branches usually have two rows. Each spikelet pair consists of a sessile and a pedicelled flower. Tassel florets contain stamens and a rudimentary pistil that degenerates early, although under some conditions, the pistil may develop.

When male flowers are mature, anthesis starts near the center of the tassel spike and then proceeds up and down. Anthers are exerted first from the upper spikelet, soon followed by anthers from the sessile spikelet, resulting in an extended period of pollen shedding. Pollen exits through a pore at the anther tip. It is estimated that about 25,000 pollen grains are produced for each style (silk).

FIG. 15.2. Male inflorescence (tassel) of sweet corn.

Pollen shed is influenced by temperature, air movement, and cultivar, and may be completed in 3–10 days. Pollen shed begins before the pistillate stigmas of flowers emerge (silking), thus favoring cross-pollination. Pollen dispersal is by wind and gravity. Field plantings should be made to take advantage of wind patterns that might improve pollen dispersal. Pollen grains rapidly desiccate and lose viability, especially when temperatures exceed 30°C and relative humidity is low. Even in good conditions, viability can be lost in 3–4 hours.

The pistillate inflorescence develops at the terminus of a lateral ear shoot originating from the leaf axil, usually near the center of the main stem. The lateral shoot is very short because of short internodes. From each node of the lateral shoot, a leaf is produced. Because of the closeness of the nodes, the leaves overlap and thereby form the "husk" that encloses the developing ear (Fig. 15.3). For some cultivars, tassel development appears to influence ear shoot development.

Female flowers occur as paired spikelets on the central axis of the lateral shoot, which is referred to as the cob or rachis. Pistillate spikelets occur in paired rows on the cob; and depending on the cultivar, this ranges from 8 to 20 rows. Each spikelet has two florets of which the lower floret usually aborts. After fertilization, paired rows of kernels develop. In some cultivars, both florets of the spikelets are functional, and when abortion of the lower florets does not occur, the paired-rows

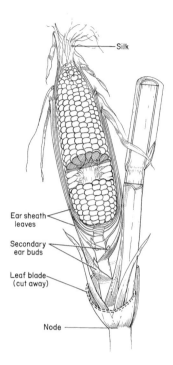

FIG. 15.3. Female inflorescence (ear) of sweet corn.

configuration of kernels is obliterated. A very long style (silk) grows from the simple pistil. Pistillate florets also contain stamen primordia, but they are arrested early in their development.

The first silk originate from the basal pistils of the ear, with one silk for each potential kernel. Silk usually emerge 1–3 days after anther dehiscence begins and are receptive when emerging from the ear husk. Complete emergence of all silk take from 2 to 7 days, depending on temperature and plant vigor. Practically all of the kernel set occurs 3–5 days after the first silk emerge. High temperatures during pollen shed and silk emergence are detrimental because of possible pollen desiccation. Pollination can occur within a wide temperature range, the optimum is about 30°C. For many cultivars, temperatures above 36°C, under hot dry winds, or when the plant is under moisture stress will result in poor pollination with poor "ear fill" as a consequence.

The pollen tube of the germinated pollen grain grows down the full length of the silk and enters the embryo sac, where it bursts, releasing two sperm nuclei. One fuses with the egg nucleus, forming the zygote

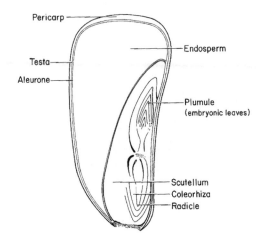

FIG. 15.4.   Structure of sweet corn kernel (caryopsis).

and restoring diploidy. The other fuses with one of the two polar nuclei, which in turn fuses with the second polar nuclei, resulting in the triploid endosperm. Of the many pollen grains which germinate and transverse the silk, only one enters the embryo sac. The distance the pollen tube transverses in growing through the silk is roughly equivalent to 1500 times the diameter of the pollen grain.

Seed

The mature single-seeded fruit is called a caryopsis (Fig. 15.4). It is a flattened grain with either a convex or indented top and pointed base. It consists of the endosperm surrounding the embryo, the aleurone layer, and the surrounding layer of pericarp tissues. The pericarp layer, which is maternal tissue, is thin in sweet corn, about 5 cell layers thick, compared to more than 20 layers in popcorn. The endosperm may be about 85% of the kernel weight and is the food source for the embryo during germination. Kernel colors usually are white or yellow; some cultivars have mixtures of white and yellow kernels on the same ear. However, kernel color in other corn types, such as those grown for ornamental use, can range from red to dark blue-black.

Different kernel types of *Zea mays,* largely identified by kernel appearance are as follows:

Sweet corn, *sacharata** or *rugosa*—kernel endosperm initially accu-

*The italicized names are not botanically recognized but have been used in other publications to identify different maize kernel types.

mulates sugars, but with increasing maturity, starch accumulates. Mature kernels are wrinkled, somewhat translucent. Pericarp tissue ranges from thin to thick. Plant height is short to intermediate; cobs are small to intermediate.

Popcorn, *everta*—characterized by a hard endosperm and very thick pericarp, and small kernels that are often pointed. Trapped moisture within kernels when heated becomes steam and explosively bursts the seed coat and exposes the expanded endosperm. Kernels have a high ratio of starch to sugar. Plants and ears are generally small.

Dent corn, *indentata*—kernel top surface becomes indented when the central core of soft starch shrinks during drying more than the surrounding hard endosperm. Plants are tall; ears are short and thick.

Flint corn, *indurata*—large hard kernels have a broad, smooth, rounded, top surface and thick pericarp. The endosperm has a high ratio of hard to soft starch with a small amount of soft floury opaque tissue near the center of the kernel. Ears are long and slender. Although sugar accumulation is well below that of sweet corn, immature ears are occasionally used as a vegetable, called "roasting ears."

Flour corn, *amylacea*—kernels have a greater proportion of soft floury (amylose) starch in the endosperm. Because of the thin pericarp, kernels are easily ground; the flour is used to make tortillas. Kernel endosperm does not have an opaque appearance.

Waxy corn, *ceretina*—kernel composition is almost entirely amylopectin starch, differing from other corn having differing ratios of amylopectin to amylose starch.

Sweet Corn Kernel Characteristics

The kernel endosperm is where sugar and starch are stored. The primary endosperm sugar is sucrose with some glucose, fructose, and maltose. The major components of endosperm starch are amylose and amylopectin. These are usually present in a ratio of 1: 3 in grain corn, but in sweet corn cultivars, the amount of starch is less and its composition differs (Fig. 15.5).

The *Su1* gene for starchy kernels is homozygous dominant *(Su1/ Su1)* in grain corn, whereas in sweet corn, it is homozygous recessive, *su1/su1*. Grain corn endosperms with the grain corn dominant *Su1* gene store considerably more starch than sugars. Sweet corn with the recessive *su1* (sugary) gene accumulate more sugar than grain corn. The *su1* gene results in a preferential accumulation, on a dry weight basis, of sugar (15%) and water-soluble polysaccharides (35% phytoglycogen) in endosperm tissues. Kernel phytoglycogen content is important for the creamy texture.

FIG. 15.5. Comparison of kernel appearance of sweet corn (left) and grain corn (right). Note: the dented kernels of sweet corn result because of a lower starch content and a high ratio of amylopection to amylose compared to the grain corn kernels. Source: Courtesy V.L. Guzman.

An additional effect of the *su1* gene is slowing sugar to starch conversion. Starch content in sweet corn slowly increases with maturity, but after about 20 days tends to remain constant, whereas in grain corn, starch continues to increase and reaches much higher levels, up to 75% of dry weight. Because sweet corn kernels contain less starch, they are wrinkled and somewhat translucent when dry.

Many recently introduced cultivars have recessive genes such as sugary enhancer *(se1)* and shrunken-2 *(sh2)* that act with modifiers yet to be identified to further improve endosperm sweetness. The sugary enhancer, *se1* gene, is a recessive modifier of the sugary gene *su1* and significantly increases kernel sugars, and thereby allows for a longer harvest period with less sugar loss. In *se1* cultivars sugars are increased without reducing phytoglycogen content. The sugar to starch conversion of *se1* cultivars occurs at the same rate as for normal *su1* types, but having more sugar initially, *se1* types remain sweet longer.

The *sh2* gene produces the highest sugar content (as much as 50% of kernel dry weight), especially that of sucrose, but results in reduced phytoglycogen. The rate of sugar conversion to starch is less than that of normal *su1* sweet corn. As with *se1* types, having more sugar initially enables *sh2* kernels to remain sweet for longer postharvest periods.

The benefits derived from these endosperm mutant genes are not all positive and are associated with some undesirable characteristics. Normal *su1* sweet corns have a thin pericarp and a creamy kernel texture. The *se1* endosperm types have the thinnest and most tender pericarps and are intermediate with regard to creamy texture (Table 15.1).

Although capable of retaining sweetness for a long period, the *sh2* gene has been associated with tough pericarps and a lack of creaminess. However, kernel tenderness is not closely associated with the endosperm mutations.

Seed of *sh2* cultivars have very little starch reserves; therefore, the mature seed exhibit a pronounced shrunken endosperm. Expectedly, seed emergence and early seedling growth are weak and require extra cultural care until established. Seed of *sh2* and *se1* cultivars should be planted shallow and into warm soils to accelerate germination and emergence. Fungicidal treatment is common to protect the vulnerable seed from diseases. Additionally, *su1, sh2,* and *se1* cultivars require isolation from contaminating pollination by cultivars of different endosperm characteristics to avoid possible starchy kernel development (Table 15.2).

Pollen from one cultivar can influence the kernel characteristics of another cultivar; this condition is known as xenia. For example, grain corn pollen fertilizing sweet corn cultivars will result in high starch kernels and may also affect kernel color (Table 15.3).

Because different endosperm types create a problem of unwanted cross-pollination, physical isolation and other means to avoid pollen

TABLE 15.1.   CHARACTERISTICS OF MUTANT ENDOSPERM SWEET CORN CULTIVARS

Mutant endosperm type	Sweetness (days)[a]	Approximate sugar conc.(%)[b]	Endosperm texture	Pericarp texture
su1	Sweet (1–2)	8–18	Creamy	Tender
se1	Very sweet (4)	15–40	Creamy	Very tender
sh2	Extremely sweet (10)	20–50	Less creamy	Moderately tender–tough

[a]Number of days sweetness can be retained during low (0–5°C) temperature and high humidity (95%) conditions.
[b]Approximate percent sugar concentration at 22 days after pollination.

TABLE 15.2.   KERNEL CHARACTERISTICS AS AFFECTED BY POLLEN AND
KERNEL GENOTYPE INTERACTIONS

Kernel genotype	Pollen genotype			
	Su1	*su1*	*se1*	*sh2*
	Grain corn	Sweet corn		
Su1 Grain corn	Starchy	Starchy	Starchy	Starchy
su1 Sweet corn	Starchy	Sweet	Segregates 1 : 2 : 1	Less starchy
se1 Sweet corn	Starchy	*su1* level of sweetness	Sweet	Starchy
sh2 Sweet corn	Starchy	Starchy	Starchy	Sweet

contamination are necessary. For crop as well as seed production, cultivars are grouped according to endosperm characteristics and are grown so as to avoid cross-pollination (see Tables 15.2 and 15.3). Different groups should not be planted closer than 75 m and should be more than 200 m apart when grown for seed.

The *sh2* sweet corn cultivars are affected more by inadequate isolation than *se1* types. As an alternative to isolation by distance, cultivars can also be isolated by scheduling plantings with flowering periods that differ by at least 14 days or by locating cultivars upwind from those that would be subject to undesired pollination.

Additional genes affecting endosperm qualities are brittle-1 *(bt1)*, brittle-2 *(bt2)*, amylose extender *(ae1)*, dull *(du1)*, and waxy *(wx1)*. The *sh2* gene blocks starch synthesis by affecting the enzyme ADPG pyrophosphorylase. Some cultivars with endosperm genotypes *bt1* and *bt2* are better suited for growth in the high-temperature tropics. The *ae1* (amylose extender) gene increases the ratio of amylose to amylopectin starches in the endosperm. The *wx1* mutant results in almost 100% production of amylopectins. The *du1* gene affects the ratio of sugar to starch in the endosperm. Sweet corn cultivars may contain more than one of the mutant endosperm genes.

TABLE 15.3.   POLLEN EFFECT (XENIA) ON SWEET CORN KERNEL COLOR

Pollen source	Expected genotype kernel color	Result
Yellow	White	Some yellow kernels among white
Mixed (W/Y)	White	Few yellow kernels among white
Yellow	White/yellow	White and yellow, with more yellow than white kernels
White	Yellow	Yellow kernels, no effect
White	White/yellow	White and yellow, no effect

Culture

Sweet corn has relatively good adaptation to a broad range of frost-free climates being cultivated at latitudes as far as 50° from the equator. However, sweet corn is not well adapted to the wet tropics.

Almost always sown directly, seed is planted at depths of 3–5 cm. Common sweet corn spacings average about 20–25 cm within, and from 75 to 90 cm between rows. Hill plantings consisting of several stalks are spaced wider. Low plant densities improve two-ear-per-stalk development capability. Fungicidal seed treatment is recommended, especially when planting in cool and/or wet soils. Hybrid seed use is extensive because of greater vigor and productivity.

Seed emergence is optimum between 21°C and 27°C and is very slow or fails at soil temperatures below 10°C. Following emergence, seedling and plant growth can be satisfactory in a range from 10°C to 40°C, but is best between 21°C and 30°C. Low temperatures have less effect during the seedling stage, but thereafter temperatures should be higher for good growth. Low temperatures significantly limit growth, especially after tasseling is initiated. Thinning, if necessary, should be performed before seedlings are 20 cm tall.

Warm days as well as warm nights improve overall growth, and although warm temperatures are ideal for vegetative and ear growth, moderate temperatures are optimum for carbohydrate accumulation. Figure 15.6 shows the various phases of sweet corn growth. Generally, wherever grain corn grows, sweet corn can be grown, because less time is required to reach its harvest stage.

Flowering and crop development are influenced by day length and temperature; plants flower sooner under short days. Many tropical cultivars will not flower in temperate regions until the day length decreases to less than 13 or 12 h. Under long days, these tropical types remain vegetative and occasionally can reach heights of 5–6 m before tasseling. However, very short (8 h) days and temperatures less than 20°C also delay flowering. When grown in short-day temperate conditions, tropical cultivars tend to flower early. Because of insufficient vegetative growth to support ear and kernel development, yields are poor. Long days are more advantageous for the production of a greater number of leaves and greater carbohydrate production. Accordingly, plant breeders have developed cultivars that are adapted to differing photoperiods.

Productivity and earliness generally benefit from warm temperatures. Some cultivars can be harvested in a minimum of 70 days, whereas late-maturing cultivars require more than 110 days. Plants

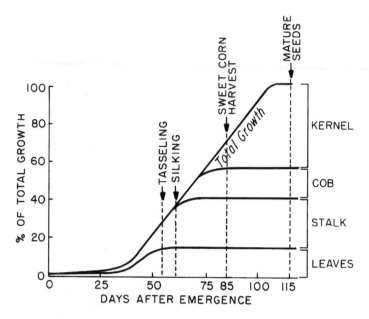

FIG. 15.6.   Growth stages of sweet corn.

usually have some field tolerance to short periods of low-temperature exposure but are sensitive to frost.

Soils, Moisture and Nutrition

Sweet corn is successfully grown in a wide range of soil types. Clay loams are preferred because of their high soil moisture holding capability. The crop is sensitive to acidic soils and grows best at a pH range between 6.0 and 6.8 and has some tolerance to saline conditions.

The crop has a high-moisture demand, typically ranging from 500 to 700 mm per season. Moisture stress is most critical during silk formation and ear filling. A short-term soil water deficit can often be tolerated with little effect on kernel development. However, continued deficit after pollination significantly decreases kernel dry matter. Under these conditions, kernel growth is partially maintained by remobilization of stored assimilates in the stem. Overall, plants have fair drought resistance but are sensitive to poor drainage and intolerant to waterlogging.

Sweet corn is responsive to high levels of fertilization. For high yields, supplemental nutrition is usually required. In temperate climates, an interesting association of a nitrogen-fixing bacteria, *Azospirillum brasilense* in the root rhizosphere has occasionally shown increased plant weight and nitrogen content.

Harvest and Postharvest

Sweet corn harvests generally occur about 18–24 days after pollination and is usually evident by the external appearance of dried silks, tightness of husk leaves, and firmness of the ear when grasped. Harvests occurs when kernels are still immature, at the milk, and before the early dough stage. At this point, kernels contain about 72–76% moisture. Cultivars with *sh2* endosperm are harvested at the higher moisture content. Corn processed as "cream style" is allowed to reach the early dough stage in order to obtain starchier kernels. An additional 30 days is required to produce mature seed (Fig. 15.6). Harvest as well as planting schedules are frequently planned using the accumulated heat unit procedure, where the base and upper temperature thresholds for sweet corn are 10°C and 30°C, respectively.

The ear is removed by snapping downward away from but without breaking the main stalk. Leaving the main stalk intact allows the remaining ear(s) to continue development without much disruption for harvest a few days later. Harvest machinery is extensively used for the processing crop and increasingly for fresh market. Machine harvest physical damage is not as evident in the processed product as it can be for fresh market, because processing occurs within hours of harvest. Ear husk leaves offer some protection from damage, but they also respire and withdraw moisture from kernels. Uniformity of ear location (orientation and height above ground) becomes an important factor in facilitating both hand and machine harvest efficiencies (Fig. 15.7).

Early morning or night harvesting helps reduce field heat and time and energy for postharvest cooling. Properly administered hydrocooling and vacuum cooling are effective methods for reducing product temperatures and respiration rates. When vacuum cooled, husk leaves should be wet, so that moisture loss during that procedure is from the husk and not the kernels. Sweet corn respiration is high and sugar-to-starch conversion can be rapid. Maintaining quality requires low temperatures and high humidity at all handling steps. Rapid cooling and holding temperatures near 0°C are essential to minimize respiratory losses. Properly cooled sweet corn can be held in good condition for a week or more if temperatures are low (0°C) and relative humidity high. Long shanks and the long outer green leaves (flag leaves) at the end of the ear can be trimmed because they extract moisture from the kernels.

a   b

FIG. 15.7.   Field hand harvest and packing of sweet corn with mechanical aid (a) and mechanized harvest (b) in Florida. Source: Courtesy V.L. Guzman.

Diseases and Other Pests

Anthracnose	*Colletotrichum graminicola*
Bacterial soft rot	*Pseudomonas syringae,* other *Pseudomonas* spp., *Sclerospora graminicola*
Common rust	*Puccinia sorghi*
Common smut	*Ustilago maydis*
Downey mildews	*Peronosclerospora* spp. and *Sclerophthora* spp.
Head smut	*Sphacelotheca reiliana*
Northern corn leaf blight	*Exserohilum turcicum*
Seed damping off	*Pythium* and *Fusarium* spp.
Southern corn leaf blight	*Bipolaris maydis (Helminthosporium maydis)*
Southern rust	*Puccinia polysora*
Stalk rot	*Diplodia maydis, Fusarium* spp.
Stewart's bacterial wilt	*Erwinia stewartii*
Tropical rust	*Physopella zeae (Angiospora zeae)*
Yellow leaf blight	*Mycosphaerella zea-maydis*

Southern corn leaf blight during the 1970–1971 epidemic was not a serious problem in hybrid sweet corn production because the cytoplasmic male sterility (CMS) used differed from the susceptible form used for other maize hybrids.

Virus diseases
  Cucumber mosaic virus (CMV)
  Maize chlorotic dwarf virus (MCDV)
  Maize dwarf mosaic virus (MDMV)
  Maize mosaic virus (MMV)
  Maize rough dwarf virus (MRDV)
  Maize streak virus (MSV)
  Sugarcane mosaic virus (SCMV)
Virus-like diseases
  Corn bush stunt (mycoplasma, CBSM)
  Corn stunt (spiroplasm, CSS)
  Maize wallaby ear (leafhopper toxin reaction)

Nematode pest include stem and bulb nematodes (*Ditylenchus* spp) and stubby root nematodes (*Paratrichodorus* spp). Additionally, species of lesion, dagger, lance, and needle nematodes also attack sweet corn.

Insect pests include the following:

Armyworms	*Spodoptera frugiperda* and *Pseudaletia unipuncta*
Chinch bug	*Blissus leucopterus*
Corn earworm	*Heliothis zea*
Corn leaf aphid	*Rhopalosiphum maidis*
Cutworms	*Agrotis* and *Feltia* spp.
European corn borer	*Ostrinia nubilalis*
Flea beetles	*Systena* spp.
Grasshoppers	*Melanoplus* spp.
Rootworms	*Diabrotica* spp.
Seed corn maggot	*Hylemya platura*
Spider mite	*Tetranychus urticae*
Stalk borer	*Papaipema nebris*
Tarnish plant bug	*Lygus lineolaris*
Thrips	*Anaphothrips obscurus*
Variegated cutworm	*Peridroma saucia*
Wireworms	*Melanotus* spp.

Maysin, a flavone glycoside found in corn silk, has been shown to have antibiotic activity to corn earworm larvae; genotypes vary in silk maysin content.

Effective early weed management during the seedling stage usually permits plants to outgrow and compete with most weeds.

FIG. 15.8. Sweet corn ear infected by corn ear smut fungus, *Ustilago maydis.* At an early stage, the fungal fruiting bodies are used much like mushrooms, especially in Mexico.

Use and Composition

Sweet corn is eaten fresh, usually after being cooked. A considerable volume is processed by canning and freezing of kernels after removal from the cob; corn on the cob is also processed by freezing. Sweet corn provides diversity to meals because of its texture and flavor, in addition to significant nutrition. Sweet corn is a very good source of proteins and lipids. Many sweet corn cultivars are high in pro-vitamin A (cryptoxanthin), a carotenoid pigment. Corn smut fungus *Ustilago maydis* (Fig. 15.8), is a frequent sweet corn production problem; however, the mushroom-like fruiting bodies are consumed as a delicacy, especially in Mexico.

Crop Improvement

Most of the future improvements in sweet corn are expected to rely on plant breeding in which genetic engineering will play a major role. Grain corn is a worldwide major crop that commands extensive research devoted to its improvement, and from which sweet corn improvement has and will continue to benefit. Research can provide a greater understanding of starch and protein synthesis pathways, and perhaps genetically modify respiration to further extend postharvest quality.

A further advantage would be gained if storage protein (zeins) content were increased. Corn is low in the essential amino acid, lysine; new cultivars with high lysine content are being developed. Other breeding programs directed to improved disease and pest resistances and for better cold adaptation offer further advances. Improved cold adaptation would permit earlier plantings in many areas presently restricted because of short growing seasons. Similarly, better adaptation to the humid tropics would further increase production capabilities.

An important goal is to have the benefits of the mutant gene expression without the unfavorable association of poor seed emergence and performance characteristics. Additional objectives would include improvement of plant and ear characteristics such as stronger stalks, fewer suckers, shorter cob shanks, smaller diameter cobs of increased length and kernel number, and cob husks that fully enclose the tip of the ear to reduce pest entry.

SELECTED REFERENCES

Andrew, R.H. 1982. Factors influencing early seedling vigor of shrunken-2-maize. Crop Sci. *22,* 263–266.

Beadle, G.W. 1981. Origin of corn: Pollen evidence. Science *213,* 890–892.

Boyer, C.D., and Shannon, J.C. 1984. The use of endosperm genes in sweet corn improvement. Plant Breeding Rev. *5,* 139–161.

Cantrell, R.G., and Geadelmann, J.L. 1981. Contribution of husk leaves to maize grain yield. Crop Sci. *21,* 544–546.

Carey, E.E., Rhodes, A.M., and Dickinson, D.B. 1982. Postharvest levels of sugars and sorbitol in sugary enhancer *(su se)* and sugary *(su Se)* maize. HortScience *17,* 241–242.

Courter, J.W., Rhodes, A.M., Garwood, D.L., and Mosely, P.R. 1988. Classification of vegetable corns. HortScience *23,* 449–450.

Galinat, W.C. 1985. The missing links between teosinte and maize: A review. Maydica *30,* 137–160.

Galinat, W.C. 1988. *The origin of corn.* In Corn and Corn Improvement—Agronomy Monograph No. 18, 3rd ed., pp. 1–31. Madison, WI: ASA–CSSA–SSSA.

Galinat, W.C. 1992. Maize: Gift from America's first peoples. In Chillies to Chocolate: Foods the Americas Gave the World. N. Foster and L.S. Cordell, eds. University of Arizona Press, Tucson, pp. 47–60.

Gerber, J.M., and Caplan, L.A. 1989. Priming sh2 sweet corn seed for improved emergence. HortScience 24, 854.

Gilmore, E.C., Jr., and Rogers, J.S. 1958. Heat units as a method of measuring maturity in corn. Agron. J. 50, 611–615.

Goodman, M.M. 1988. History and evolution of maize. CRC Crit. Rev. Plant Sci. 7, 197–220.

Herrero, M.P., and Johnson, R.R. 1980. High temperature stress and pollen viability of maize. Crop Sci. 20, 796–800.

Quattar, S., Jones, R.J., and Crookston, R.K. 1987. Effect of water deficit during grain filling on the pattern of maize kernel growth and development. Crop Sci. 27, 726–730.

Sears, P.B. 1982. Fossil maize pollen in Mexico. Science 216, 932–934.

Shurtleff, M.C., Holdeman, Q., Horne, C.W., Kommedahl, T., Martinson, C.A., Nelson, R.R., Schiefle, G.C., Weihing, J.L., Wikinson, D.R., Worf, G.L., Wysong, D.S., Smith, H.E., and Muller, G.J. 1973. A Compendium of Corn Diseases. American Phytopathological Society, St. Paul, MN.

Tracy, W.F. 1993. Sweet corn Zea mays L. In Genetic Improvement of Vegetable Crops. G. Kalloo and B.O. Bergh, eds. Pergamon Press, Oxford, pp. 777–807.

Tracy, W.F., and Galinat, W.C. 1987. Thickness and cell layer number of the pericarp of sweet corn and some of its relatives. HortScience 22, 645–646.

Waters, L., Jr., Burrows, R.L., Bennett, M.A., and Schoenecker, J. 1990. Seed moisture and transplant management techniques influence sweet corn stand establishment, growth, development and yields. J. ASHS 115, 888–892.

16

Plantain, Starchy Banana, Breadfruit, and Jackfruit

PLANTAINS and STARCHY BANANA, Usually triploids comprised of *Musa acuminata* Colla and *Musa balbisiana* Colla germplasm

Family: Musaceae

Plantains and starchy bananas are important food staples for more than 50 million people in many tropical countries. They are one of the least expensive source of caloric food to produce and are commonly available throughout the year. The compatibility of plantains for inter-crop cultivation is a decided advantage for their extensive use in subsistence farming.

The distinction between plantain and the dessert banana is fairly clear, but it is less well-defined between plantains and starchy bananas and the names are often interchanged (see Taxonomy section). The best known distinction is that plantain and starchy bananas are eaten cooked, even when ripe, and thus are commonly considered together as "cooking bananas."

Plantain and starchy banana fruit are unpalatable unless cooked; when eaten raw, they have a strong and lasting astringency. Even after cooking, the fruit pulp usually remains relatively firm. Ripe dessert banana fruit are sugary and usually eaten without cooking. A major compositional difference between plantains and starchy bananas compared to dessert banana fruit is the amount and rate at which the accumulated starch is converted into sugars as fruit ripens. Plantains and starchy bananas are consumed when very little of the starch contained in the pulp has been converted, whereas dessert bananas are consumed when the conversion is already well advanced. In comparison, the total solids content of ripe bananas has about 80% sugar and less than 5% starch, whereas mature plantains and starchy bananas contain about 66% sugar and 17% starch. Dessert bananas are starchy when immature and can be used as a cooked vegetable much like plantains and starchy bananas but are typically permitted to ripen, thereby allowing for the conversion of starch to sugar.

Origin and Domestication

Indochina and Southeast Asia are presumed centers of diversity and domestication of wild *Musa* species, although wild types range from India to Thailand and are also found in New Guinea and Australia. Most botanists consider Malaysia the area of origin for *M. acuminata;* fringe areas in India and the Philippines are believed to be areas of origin of *M. balbisiana.* The geographer Carl Sauer suggested that some species of *Musa* were probably among the first plants domesticated. Various types of bananas have been grown in Southeast Asia for thousands of years. Bananas were reported to be grown in India about 500 B.C. and in West Africa about A.D. 500. *Musa* species were observed by European explores of West African coasts in the late 15th century. The word "banana" possibly originated from the local languages in Sierra Leone or Nigeria. European traders visiting such localities may have adopted the word "banana" from the sound-alike "bana." Eastward movement from Southeast Asia brought bananas to Pacific island areas. Plantains and bananas in the Americas are believed to have been transferred from western Africa by European traders.

It is likely that the corm and inner stalk tissues of wild plants were first used as food, as the extremely seedy fruit were essentially inedible because of many stone-hard seed and relatively little pulp. These wild forms evolved toward greater edibility, presumably by human selection for plants producing parthenocarpic fruit.

Taxonomy

The family Musaceae consists of two genera, *Ensete* and *Musa.* Because of continued use of previously incorrectly applied nomenclature, there is some confusion about their taxonomy. *Ensete,* with six to seven species, originated in Asia but has also existed in a wild state in Africa for centuries. *Ensete ventricosum* (synonyms *E. edule* and *Musa edule*) is commonly known as Abyssinian banana and to a limited extent is cultivated in Ethiopia. Although all parts of the plant are eaten, the starchy corm and inner stalk tissue are the plant portions principally used as food.

The genus *Musa* has about 40 species, *M. acuminata* and *M. balbisiana* being the most important. All edible bananas and plantains are derived from the two wild types of diploid *M. acuminata* and *M. balbisiana* with the chromosome composition of $2n = 22$. Although all *Musa* wild species are diploids, not all have the same chromosome number, some having $2n = 14$, $2n = 18$, or $2n = 20$. Nearly all cultivated plantains, starchy bananas, and dessert bananas are polyploids, most being triploids, derived by hybridization within and between *M. acuminata* and *M. balbisiana*. Geneticists identify the basic genome of *M. acuminata* as AA, from which triploids (AAA) occurred through natural polyploidization. The AAA grouping is horticulturally significant because two major dessert banana clones, Gros Michel and Cavendish, are derivatives of this group. Cavendish clones are still grown extensively as a commercially export crop.

The *M. balbisiana* genome is identified as BB. No diploid form of *M. balbisiana* have edible or parthenocarpic fruit and no BBB triploids are known. *M. balbisiana* is found in areas of the Indian subcontinent where discrete wet and dry seasons occur and is generally believed to be a hardier and more drought-tolerant species. Wild *M. balbisiana* has not been found in the humid tropics.

It is proposed that through human transfer of *M. acuminata* diploid and triploid clones to other areas resulted in natural hybridizations with *M. balbisiana* diploids. This produced hybrids with genomic compositions of AB, AAB, and ABB. Hybridization between *M. acuminata* and *M. balbisiana* is easily accomplished in nature, and hybrids with various characteristics likely would have occurred at different times and locations. The parthenocarpic and better edibility of *M. acuminata* when combined with hardiness of *M. balbisiana* permitted plantain and starchy banana, and dessert banana production in many less humid tropical areas. In addition to hardiness, *M. balbisiana* also contributed greater disease resistance, fruit starchiness, and acidity.

The plantain subgroup resulting from crossings of *M. acuminata* × *M. balbisiana* has the genomic identity of AAB and ABB. The AAB plantain group is abundant in west Africa, where considerable diversity occurs. It is also abundant in parts of tropical America but is uncommon in Southeast Asia. On the other hand, ABB clones are numerous in India and parts of Southeast Asia, and essentially all are starchy bananas, although incorrectly sometimes called plantains. Starchy bananas are often preferred to plantains because they have better vigor and culinary characteristics.

Parthenocarpic triploid clones are more productive than diploids, possibly attributable to the absence of seed. These clones have greater vigor and larger fruit and are thus cultivated preferentially to tetraploids and diploid clones. Diploids, often seeded, are grown least of all and are of little economic importance. Nevertheless, some seed-bearing diploid clones of *M. acuminata* are still grown in Southeast Asia. Tetraploids, such as AAAB, AABB, and ABBB also occur, but few ABBB clones have any local importance. AAAA plants are not found naturally but have been produced by breeders.

Botany

Plantains and starchy bananas are large perennial herbaceous monocots grown mainly for their starchy parthenocarpic fruit. The initial roots of the suckers are short lived and are quickly replaced by many adventitious roots that form from the lower part and sides of the underground stem (corm) tissues. As the corm continues to enlarge, additional advantageous roots and root laterals are formed; thus, the presence of several hundred adventitious roots is not unusual. Most roots are located in the upper 30 cm of soil. They are relatively small and thin, although some have a lateral matlike spread of several meters which enhance nutrient absorption and anchorage. Roots are continuously produced, replacing those that have died as well as increasing in total number. Individual roots have a life span between 4 and 6 months.

The corm is a compressed subterranean stem with many closely spaced internodes. The apical portion of the corm contains meristematic tissues from which develop foliage, roots, and suckers. During growth, the corm also enlarges and terminates its growth in producing an aerial stem and inflorescence. The corm of a mature plant may range from 15 to more than 30 cm in width. Axillary buds at upper portions or sides sprout and develop as suckers, which may be used for propagation. Apical dominance restricts the number of buds that sprout as suckers, although pruning practices also influence how many suckers are initiated.

Leaves are initiated in a spiral succession inward from the outer portion of the flatten apical meristem. This results in a pseudostem, a trunklike structure consisting of closely rolled leaf sheaths often mistaken for a stem. Pseudostem length is a cultivar characteristic but is also influenced by environment.

Each leaf consists of a sheath, petiole, midrib, and leaf blade (lamina). The first leaves are nearly bladeless, but the lamina size increases in width and length with succeeding leaves until flower initiation; further leaf initiation ceases. Subsequent emerging leaves are smaller. Much of the leaf's development occurs within the pseudostem. There, it is formed as a coiled double-rolled cylinder. The right half of the leaf blade is rolled upon itself and the left portion is rolled over the right half and midrib. It is in this configuration that the leaf moves by elongation of the sheath and petiole upward through and exits vertically at the top of the pseudostem. After, emergence, the coiled leaf unfolds within a week while continuing the remainder of its expansion. Leaf position changes with age, and from its nearly vertical position, the leaf gradually bends to and beyond the horizontal positions and droops when senescent (Fig. 16.1).

Veins within leaf blades are unbranched and parallel to each other, but perpendicular to the midrib. Leaf blades have little resistance to transverse tearing, often caused by strong winds; this arrangement enables the vascular connection of veins to the midrib to remain intact, and leaf functions are not significantly impaired. Thus, torn and ragged foliage may not be as damaged as its appearance indicates.

The inflorescence appears when the apical meristem elongates and pushes upward through the pseudostem. Depending on cultivar, the inflorescence is usually formed after the production of 30 or more leaves. Usually, the floral stem requires about a month from the time of initiation to emergence from the top of the pseudostem and are several to more than 7 m in length.

The inflorescence bears both female and male flowers on the aerial stem extending through and largely supported by the pseudostem. The initially erect inflorescence usually becomes pendent because of its own increasing weight. Although varying among cultivars, geotropism and the presence or absence of the male flower bud are also believed to influence inflorescence position, so that some are not fully pendent.

Flowers are formed as nodal clusters; the first basal nodes produce

FIG. 16.1. Plantain with pendent bunch of developing fruit. Note: the maiden sucker at left of pseudostem and the sword sucker at the right.

female flowers. A cluster of female flowers is called the "hand," and the individual flowers, "fingers." The entire inflorescence, often called "bunch," will have several to 10 or more spirally arranged hands, each containing from 5 to 20 fingers (Fig. 16.2). Bunch size, number of hands and fingers, and their size and shape are cultivar characteristics, and variations among cultivars are considerable.

Female and male flowers appear identical initially but change with further growth. Female flowers have an elongated ovary comprised of three fused carpels. Because of the usual male and female flower sterility, ovules do not develop, the fruit being a unfertilized berry. Flowers produce nectar and can attract insects, birds, and bats, which are pollinators of wild *Musa* species.

After initiation of several basal clusters (hands) of female flowers, intermediate nodes of the inflorescence form clusters of hermaphroditic flowers, and male flowers are initiated at distal nodes. Both of the latter flower forms generally degenerate. At the terminus of the inflorescence is the male apical bud, which for many cultivars can continue producing male flowers until the bunch is harvested.

Fruit of the AAB plantain group are long and curved, whereas those

FIG. 16.2. Examples of fruit clusters, generally called "hands". The individual fruit are called " fingers". O.L'ewai is a triploid French type plantain female parent, Calcutta 4 is the diploid male parent, and 597/77 is the vigorous tetraploid hybrid. Source: Courtesy Rodomiro Ortiz and Dirk Vuylsteke, IITA, Nigeria.

of the ABB starchy banana group are short, angular, thick, and almost straight. At first, fruit may be green or various shades of red, but the color changes as fruit ripen, becoming yellow, brown, and/or black; the interior pulp is white or yellow.

Propagation

Plantain propagation is exclusively vegetative. Materials used are plant suckers identified as sword, maiden, peepers, and water suckers which arise as shoots from buds on the corm (Fig. 16.1). Sword suckers, the most preferred propagules, are large shoots about a half to a meter long and are formed from the lower axillary buds of the corm. They have narrow leaves and are usually planted intact with a large portion of the mother corm and roots attached, sometimes the foliage is trimmed. Roots of sword and other kinds of suckers are usually lightly trimmed before planting. Maiden suckers are larger, and older sprouts (5–8 months) have tall and broad leaves; these usually have already developed a fairly large size corm. Before planting, the foliage is cut back to remove diseased, damaged, or insect-infested leaf sheaths. Suckers and corms can be disinfected for nematode control by a hot-water (53–55°C) treatment. The immersion period is usually about 20 min. but can vary depending on the size and volume of the propagules being treated. Peepers are recently emerged shoots having few leaves; and because of the small size and lack of stored food reserves, they are usually rejected for propagation. Water suckers have a narrow pseudostem, broad leaves, and small attachment to the mother corm. These are also avoided because they produce weak plants and low yields.

In addition to obtaining propagation material, pruning is performed to selectively remove unwanted suckers that would compete with the primary plant. Plantains generally are not pruned until 2 or 3 months after fruit harvest because, unlike dessert bananas, sucker formation is inhibited until the fruit is harvested. Propagating material is also obtained from nurseries where large numbers of sword and maiden suckers are produced. The practice of destroying the apical meristem of the mother plant allows axillary buds to sprout. Normally, about 5 usable suckers can be harvested from each nursery plant but with intensive management, as many at 8–10 can be obtained. Suckers taken from existing plants cause some injury to the corm and mat roots of the mother plant; therefore, it is preferred that suckers be obtained from nursery plantings. Suckers should be planted soon after harvesting, because they have a short postharvest life and do not store well.

Also used for propagation are corms or corm portions, called "bits." Those containing several viable axillary buds are trimmed of aerial growth before planting. Large corms can be split into two or three portions, depending on initial size and the number and location of axillary buds.

Shoot tip meristem tissue culture is a useful and an expanding alternative to obtain large numbers of propagules rapidly. Its use is more costly than traditional propagation and is presently, only feasible for intensively cultivated plantations. Tissue culture has the advantage of providing material free of disease, and the general uniformity of propagules allows for better scheduling of plantings than those from nursery-produced materials. Nevertheless, tissue culture propagation has shown increases of somoclonal variation of about 5%. True seed use is essentially limited to plant breeding activities.

Sucker and corm propagules are planted in deep holes about 40–60 cm in diameter and depth. A mound of loose fertile top soil mixed with some preplant fertilizer is usually first placed into the opening. The sucker or corm is then placed on the mound and covered with soil so that the neck of the sucker is at or just below the surface. For corms, about 15 cm of soil should cover the top. An adequate depth of planting is important because shallow plantings make an already shallow root system susceptible to upheaval by winds. Planting on raised mounds or beds may be necessary if drainage is poor.

Plant spacing varies with clones. Early maturing, small "False Horn"-type plantains are usually spaced 2×2 m apart, large "Medium French" types about 2.5×2 m, and the larger "Giant French" types at 3×2 m. Such spacings result in about 2500, 2000, and 1500 plants per hectare, respectively. To provide appropriate populations, various configurations are used in commercial plantations that allow for equipment movement. Plant densities are adjusted accordingly where intercropping is practiced (Fig. 16.3). Spacings for home garden production are frequently wider, as much as 3–6 m between individual plants.

The use of dwarf types is likely to continue to increase preferentially to that of large plant types because of the ease of cultivation and because fruit size is not adversely affected. Although a greater number of bunches are produced in high-density plantings, fruit development is delayed and smaller bunches and fruit are produced. However, yields are comparable to low-density plantings. The minimum fruit size for most commercial production is about 270 g. High-density plantations are more representative of intensive production and require more management and usually have a shorter productive life span.

In the humid tropics, planting can be started almost anytime, but

FIG. 16.3. Interplanting within a plantain plantation, a frequent cultural practice. The woody legume plant is being used as a cover crop. Plantain yields are improved when this cover crop is grown on poor soils in Nigeria before planting plantains. Source: Courtesy Rodomiro Ortiz and Dirk Vuylsteke, IITA, Nigeria.

in tropical areas with seasonal rainfall, plantings are scheduled just before the rainy season starts. In scheduling, consideration is given to avoid drought conditions at flowering. The ability to irrigate is an important factor affecting planting schedules. Time of planting is also influenced by market opportunities as well as by climate, and sometimes these factors are not compatible. By using different ages and sizes of propagating materials, harvests can be scheduled for continuous production. Production schedules can also be adjusted by using different clones. For example, the "False Horn" plantain types flower in about 10 months and are harvested about 2 months later. The "Medium French" types flower in 12 month and are harvested at 15 months, whereas those of the "Giant French" flower at 15 months and are harvested after 18 months.

After the first production cycle, subsequent growth can be continued by ratoon cropping, with suckers emerging from the mother plants of the preceding crop. However, in succeeding years following the initial planting, flowering becomes less synchronized, and reestablishment may be appropriate, especially if weeds, disease, and other factors contributing to yield decline exist.

Culture

The best plantain growth occurs in the humid tropical lowlands, but crops are cultivated in many frost-free locations, within 30° of the equator in both wet or dry climates. Optimum growth occurs at 27–28°C. Temperatures much below 20°C or above 38°C are growth restricting, although plants are able to survive a wider range of temperatures. Depending on length of exposure, temperatures of 10°C can cause chilling injury to plants, and temperatures greater than 45°C are also damaging. On sloping land, plantings are made to take advantage of maximum sun exposure. Although not photoperiod responsive, plants require high light intensity for maximum growth, although fair growth occurs even when plants are partially shaded. Dry matter accumulation is most favorable at 24–28°C, whereas temperatures of 32–34°C increase the rate of new leaf appearance.

The best soils for production are deep, well-drained loams with high fertility, organic matter content, and water holding capacity. Soil type often can be of greater importance in site selection than climate, although in some subtropical regions, climate may be the primary consideration. Gently sloped land or hillsides are often selected to provide good drainage. Many tropical areas receive high amounts of rainfall, and because plantains are sensitive to waterlogging, good drainage is important; sometimes field drains are installed. For optimum root development, light textured soils and those with high organic matter are preferred. Because most of the adventitious fibrous roots are located in the upper 50 cm of soil, shallow soils can and often are used. Having poor root systems, plantains are easily blown over by strong winds; accordingly, dwarf clones are frequently grown because of greater resistance to wind damage. In some situations, windbreaks are used to reduce wind force, but a disadvantage is that windbreaks create shade and occupy field space. Plantains are sensitive to salinity. The preferred soil pH ranges between 5.0 and 7.5; a pH less than 5.0 limits productivity. Lime should be applied to correct high acidity.

In many tropical areas and especially for subsistence farmers, nutrients for the crop are largely dependent on existing soil organic matter. Because plantains require relatively good soils with adequate organic matter, newly cleared forest land is often used. Because nutrition is most critical in early plant growth, supplemental organic and/or synthetic fertilizer is commonly applied at planting. In nutrient-poor soils, fertilization must be at high levels and frequently supplied. A major portion of the fertilizer for a cropping cycle is commonly applied at the time of planting because subsequent applications are difficult to apply without damaging the shallow root system. Increments of nitrogen fertilizers usually are applied several times during the production cycle. Supplemental nitrogen can be effectively supplied with irrigation. The highest demand for potassium occurs at flowering.

Mulching is known to significantly increase yield and is utilized as much as possible. Mulches are beneficial because they restrict weed growth and soil and nutrient runoff, stabilize soil temperatures, and reduce evaporative losses from the soil surface.

Access to a uniform supply of moisture is very important; an average of 150–200 mm precipitation per month is usually adequate for a cropping cycle; 25 mm weekly is considered minimal. In most plantain production regions, annual precipitation ranges from 1000 to 2000 mm. Dry periods, especially during flowering, depress productivity; in such situations, supplemental irrigation is desirable. Good drainage and well-aerated soils are important prerequisites, because plantains are easily damaged by flooded conditions. Flood irrigation is widely used, but because of nutrient leaching, the practice is being replaced by overhead or low-level sprinkler irrigation; drip irrigation use is also rapidly increasing.

In subsistence cultivation, more often than not, plantings have a short productive life primarily due to minimal culture or neglect. The fairly large diversity of plantain and starchy banana clones existing in subsistence farming often require different cultural practices than those of the more uniform clones used in intensive monoculture. In some areas, a bush–fallow system is followed, where plantains are grown for several years and then the land is fallowed before being used for other crops. Rotations, especially with crops having soil and nutrient improvement characteristics, are recommended. In commercial plantings, the economic productive life ranges from four to eight crop cycles, each varying from 12 to 18 months. Plantains are often grown indefinitely on household compounds, especially where organic wastes can be frequently provided to supplement soil nutrients. Intercropping is widely practiced by subsistence and commercial producers, especially throughout Africa. Cassava, yams, maize, okra, aroids, beans, and groundnuts are crops commonly intercropped.

Weed control is achieved largely by hand cultivation, especially in traditional culture. Control is most critical during early growth when plants are least competitive. When a full canopy is achieved, weeds are fairly well suppressed. Mulching with leaves or other organic matter is an effective and widely used practice for weed control. In plantation-scale production, chemical herbicides are more likely to be used.

Sucker removal is a common cultural practice to reduce competition with the mother plant and to maintain desired plant density, as well as to provide a source of propagating material. Leaf prunings are made to remove diseased and senescent foliage or to prevent foliage from scaring the developing fruit. The male flower bud may be removed to lessen the weight of the developing bunch and/or harvested for vegetable use.

Growth and Development

The rate of leaf initiation and leaf size varies with specific clones and growing conditions. The number of leaves that develop and persist, and the size of surface areas are important components of the cropping cycle. For mature plants, the maximum number of expanded leaves at any one time will range from about 10 to as many as 15, with a number of developing nonexpanded leaves. Leaf growth is rapid, and elongation of about 10 cm per day is not unusual. A new leaf is produced about every 7–10 days, and during that time the oldest leaf usually becomes senescent. The leaf area index (LAI) reaches a maximum value about the time flowering is initiated and then slowly declines. Plantains are generally smaller plants and have smaller leaf surfaces than dessert bananas. Although the plantain leaf canopy is impressive with the total leaf surface area exceeding 10 m^2, it is much less than that of many dessert banana cultivars.

The inflorescence usually emerges after 30–40 leaves are produced and is fairly well developed before it exits the pseudostem; for some cultivars, as many as 50 leaves are produced. At inflorescence initiation, further leaf production ceases and the size of recently formed leaves decreases. With the emergence of the inflorescence, further leaf emergence ceases (Fig. 16.4). Inflorescence size is largely determined by the size of the apical meristem at the time of initiation. At flowering and during fruit development, propping or tying to support the inflorescence is usually necessary.

Flowering and fruit development periods vary for different clones; fruit maturation requires from 2 to 4 months after flowering. Only one inflorescence (bunch) develops per corm, followed by senescence of the stalk; although, after harvest, suckers continue to develop and grow unless removed. The total life of the pseudostem may be as little as 12 or as long as 24 months, the length of time being influenced by clonal characteristics, temperature, and growing conditions.

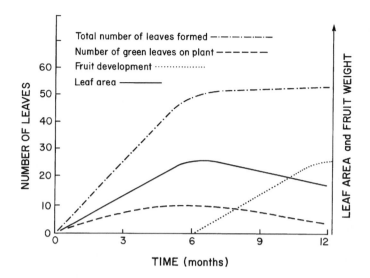

FIG. 16.4. Growth pattern of an early maturing plantain cultivar.

Diseases and Pests

Dessert bananas seem to have a much greater and broader susceptibility to diseases than plantains. Nevertheless, plantains share a number of diseases that seriously affect production; some important field
and postharvest diseases are as follows:

Anthracnose crown rot	*Colletotrichum musae*
Black sigatoka (black leaf streak)	*Mycosphaerella fijiensis* var. *diformis*
Chloridium leaf speck	*Chloridium musae (Cladosporium musae)*
Cigar end rot	*Trachysphaera fructigena, Verticillium dahliae*
Cordana leaf spot	*Cordana musae*
Deightoniella leaf spot	*Deightoniella torulosa*
Enset bacterial wilt	*Xanthomonas campestris* pv. *musacearum*
Fusarium wilt (Panama disease)	*Fusarium oxysorum* f. sp. *cubense*
Head rot	*Erwinia chyrsanthemi* pv. *paradisica*
Moko	*Pseudomonas solanacearum*

Pitting disease	*Pyricularia grisea*
Stalk rot	*Sclerotinia sclerotiorum, Verti-cellium* spp.
Stem end rot	*Ceratocystis paradoxa*
Yellow sigatoka	*Mycosphaerella musicola*

Various strains of *P. solanacearum,* in addition to causing moko disease, are believed responsible for diseases known as blood disease and bugtok (tapurog) disease. Bunchy top virus and a strain of cucumber mosaic virus, each being aphid transmitted, are serious diseases affecting plantains. An important stress-induced physiological disease results in failure of heart leaf unfurling disorder and leaf engorgement. This condition, exhibited by shortened internodes, interferes in bunch development.

Some disease control is achieved through the use of disease-free propagating materials and other cultural management that avoid and/ or limit disease incidence, and with chemical pesticides. Pesticides have fairly wide usage in intensive commercial production, and to some extent in subsistence farming. Variable effectiveness and a greater occurrence of pesticide resistance make it obvious that the search for genetic resistance must be expanded.

The banana (borer) weevil, *Cosmopolites sordidus,* is considered the principal insect pest. Species of *Temnoschoita* and other weevils also cause damage or destruction of pseudostem tissue similar to that of the banana weevil. The locust, *Zonocerns variegatus,* as well as various aphids, thrips, and spider mites are other important pests. The banana aphid, *Pentalonia nigronervosa,* is responsible for virus transmission. Caterpillar feeding of *Plusia chalcites* and *Sibine apicalis* cause serious crop losses. Additional Lepidopterous insects are *Caligo eurilochus, Ceramidia viridis,* and *Opsiphanes tamarindi.*

Often more serious than insect or diseases is the devastating damage caused by root burrowing nematodes. The lesion root nematode, *Radopholus similis,* is a major pest causing serious crop losses due to lodging or collapse of pseudostems. Other nematodes damaging roots and corm tissues are *Pratylenchus coffeae, Helicotylenchus multicinctus,* and species of *Meloidogyne,* often grouped as lesion, spiral, and root knot nematodes, respectively. Nematode control is somewhat limited because the semipermanent nature of the crop limits nematicide treatments. Nematicides are commonly used in intensive cultivation, but rotations, fallow, and flooding are other cultural control practices.

In newly cleared forests in humid zones of Africa, yields of plantains commonly decline in the second or third crop cycle. This is a frequent occurrence and is due to a number of factors that may include (1) rapid reduction of organic matter, (2) rapid deterioration of soil fertility because of exhaustion or erosion, (3) increased nematode populations, (4) banana weevil infestation, (5) high mat growth, and (6) weeds. High mat growth causes the crown of the corm to rise above soil level and roots fail to adequately penetrate and exploit the soil. Weeds compete with the crop for nutrients and moisture and can greatly suppress yield. These factors limit the ability to sustain production over a large number of cropping cycles. Declining yields are somewhat less common in starchy banana production than with plantains.

Harvest and Postharvest

Visual inspection of fruit fullness and angularity of fruit cross-sectional shape are commonly used criteria to determine harvest time. The peel of immature fruit exhibits some angularity, which becomes circular with maturation. In harvesting plantains, workers using machetes first slash through part of the pseudostem so that the weight of the bunch bends the cut pseudostem down, making the bunch more accessible and easy to cut from the floral stem. Generally, this procedure is not necessary with dwarf cultivars. Because pseudostem growth is already terminated with inflorescence development, slashing is not a detrimental practice, and clearing of "spent" pseudostems is necessary for preparation for the crop cycle that follows.

In commercial-scale production, harvested bunches are cut into hands and field packed. "Dehanding", the practice of dividing the bunch into hands, helps to reduce physical damage and repeated handling. In order to further minimize physical damage, the hands may be wrapped or bagged in polyethylene film and placed in padded boxes. To avoid the latex from cut surfaces staining fruit, they may be washed; sometimes, fruit are given a protective fungicide dip. The earlier practice of handling bunches intact during transport and distribution, whether to repacking sheds or directly to consumers, has largely ceased for commercial producers. However, subsistence and small-scale farmers still transport harvested whole bunches from the fields to homesteads and local markets (Fig. 16.5).

In general, plantain yield variability is enormous, ranging from less than 5 t/ha for subsistence farmers to as much as 50 t/ha in intensive cultivation. Plantain yields are much lower than those of bananas. Cultivars of the False Horn and French types in West Africa yield from 10 to 30 t/ha for subsistence and intensive production, respectively. Plantain yields vary among clonal types, although they may be fairly

FIG. 16.5.   Roadside marketing of plantain and cooking bananas in Honduras.
Source: Courtesy Phil Rowe, FHIA, Honduras.

comparable when calculated as yield per hectare per year. For example, in Table 16.1 the theoretical yields from intensive cultivation of three clonal types after adjustment for space and length of growth period is nearly equivalent.

Nonuniformity of fruit development and the ability to accurately predict crop maturity are important harvest variables. The fruit is harvested at a preclimacteric stage, with further maturation continuing during the postharvest stage. At the preclimacteric stage, almost all

TABLE 16.1.   PROJECTED PRODUCTIVITY OF THREE DIFFERENT INTENSIVELY CULTIVATED PLANTAIN CLONES

	Clones		
	Giant French	Medium French	False Horn
Bunch weight (kg)	30	20	12
Population (per ha)	1600	2000	2500
Yield (t/ha)	48	40	31
Growth (months)	18	15	12
Yield (t/ha/year)	32	32	31

Source: Du Montcel (1987).

the carbohydrate is in the form of starch. During subsequent ripening, carbohydrate is progressively changed into reducing and nonreducing sugars. Initially, sucrose is the dominant sugar, but later glucose and fructose predominate. In plantains, carbohydrate breakdown is not complete, whereas in dessert bananas, starch breakdown is usually completed during ripening. Plantains do not require and do not receive ethylene treatment for ripening. In contrast, dessert bananas, especially those for export, commonly receive ethylene or acetylene (100 ppm for 12–24 h) treatment to stimulate early and uniform ripening.

Major harvest and postharvest problems mostly relate to the susceptibility of fruit to physical damage and decay from pathogenic fungi. Unfortunately, plantains do not receive the level of postharvest attention or handling care equivalent to that provided dessert bananas.

Freshly harvested plantains can be refrigerated at 13°C to delay natural ripening. Shelf life is prolonged when plantains are packaged in sealed polyethylene, preferably with an ethylene-absorbing material. Avoiding low relative humidity is very important because that tends to shorten the preclimacteric period and increase moisture loss. Various coating materials effectively retard fruit ripening and water loss but are not widely used because of the low economic value of plantains.

Being chilling sensitive, fruit should not be held at temperatures below 12°C. Because cold storage facilities are rare in countries producing plantains, chilling injury is not frequently observed. Under controlled-atmosphere storage (O_2 at 1–3% and CO_2 at 5–10%), lower temperatures are permissible. Hypobaric storage at one-half atmospheric pressure can also extend shelf life. At temperatures greater than 25–30°C, the proportion of starch that converts to sugars increases; the amount varies among clones.

Production

Although Food and Agriculture Organization (FAO) statistics indicate that the volume of reported banana production was much larger (about 52 million versus 28 million tons) than that of plantains, it is generally believed that perhaps as much as half of the total banana production is used as a cooked vegetable. Plantains and starchy bananas are clearly a crop almost exclusively grown in developing countries. The volume of homestead or backyard plantain and banana production is immense, especially in many African countries, and this volume is not well identified in production statistics. Furthermore, plantains are frequently intercropped, and such joint land use makes accurate production statistics difficult to obtain and interpret. It is

TABLE 16.2. WORLD PRODUCTION OF PLANTAINS AND STARCHY BANANAS AND BANANAS, 1994

	Production tons × 10³		
	Plantains and starchy bananas		Bananas
World	28,744		52,584
Africa	20,776		6,686
North and Central America	1,643		7,988
South America	5,561		14,397
Asia	760		21,503
Oceania	5		1,608
Leading countries			
Uganda	8,613	India	7,900[a]
Colombia	2,970[a]	Brazil	6,022
Rwanda	2,600[a]	Ecuador	4,715
Zaire	2,300[a]	Philippines	3,250[a]
Nigeria	1,447[a]	China	3,200[a]
Ghana	1,322[a]	Indonesia	2,300[a]
Cote Divoire	1,300[a]	Colombia	2,000[a]
Ecuador	978[a]	Costa Rica	1,932[a]
Peru	877	Mexico	1,700[a]
Cameroon	860[a]	Thailand	1,658[a]

Note: Some country reports do not distinguish between plantains or starchy bananas and bananas.
[a]Estimated.
Source: 1994 FAO Production Yearbook, Vol. 48, FAO, Rome 1995.

estimated that 85% of all plantains and starchy bananas are grown by peasant farmers and on small holdings for local consumption and trade. Exportation has increased from almost nothing to about 12 million tons annually. Africa produces more than 70% of the world's total plantain and starchy bananas, with Uganda singly producing about 30% (see Table 16.2).

Fresh uncooked consumption is the widely recognized major use of bananas and large volumes of fruit are exported throughout the world. On the other hand, the bulk of plantain and starchy banana production is directed toward local or regional consumption. Unlike dessert bananas, few plantains and starchy bananas are shipped great distances, not so much because of postharvest concerns but because of the limited demand from areas outside tropical regions. However, changing demographics have resulted in more plantains entering international trade. Since the 1960s, exports have markedly increased in response to demand expressed by ethnic migrant populations in importing countries.

Uses and Composition

As a staple and subsistence crop, individual consumption of one or two plantain fruit per meal is common. Being an inexpensive source of calories has unfortunately identified plantains as a low status food in many countries. Besides providing a very good source of readily digestible carbohydrate (about 30% of edible portion), plantains have high palatability and satiety values, provide a good level of pro-vitamin A, about 20 mg/100 g fresh weight of vitamin C, and fair amounts of B vitamins. They are a high source of potassium, although low in iron and sodium content. Protein and fat content is about 1% and 0.3% of the edible portion, respectively. A comparison of the nutritional contribution of plantain with other important starchy vegetable is reflected in favor of plantain (Table 8.1).

Fruit pulp astringency is due to the presence of tannin compounds that cooking removes. The peel, which is not consumed, constitutes about 40% of the fresh weight. Plantains are prepared in many ways: toasted, boiled, baked, or fried as a vegetable, and used in soups and gruel. Plantains are also processed into chips, flakes, and flour. In Rwanda, Burundi, and parts of Uganda and Zaire, much of the starchy banana produced is used for wine and beer production.

The large terminal male flower buds (Fig. 16.6) also have vegetable usage, although preparation requires several changes of boiling water

FIG. 16.6.   Terminal male bud showing male inflorencence. The bud is used as a cooked vegetable.

in order to remove astringency. Fleshy corms are eaten as a starchy food in New Caledonia and other areas; some clones are grown expressly for this purpose. In India, the inner sheaths of the pseudostem are also used as a vegetable. Additionally, pseudostems are often used for livestock feed and as mulch in plantain groves. Foliage is also used for mulching and have a fair number of uses as wrapping and padding materials.

Future Improvement

Because of its important food status, the level of plantain and starchy banana production is likely to continue. Research agencies have active breeding efforts to improve productivity, disease, insect, and nematode resistance and to broaden site adaptability. The possibility is good for successful hybridization within plantains and between plantains and other bananas at international research centers, such as the International Institute of Tropical Agriculture (IITA) and the International Network for the Improvement of Banana and Plantain (INIBAP). These and other agencies are stressing sustainable production of plantains and starchy bananas for subsistence farmers in the tropics because this is viewed as a high food priority and important income source. Substantial germplasm resources have been collected and are maintained by institutions in Honduras, Brazil, Burundi, the Philippines, India, and Malaysia.

Progress is being achieved in disease, insect pest, and nematode resistance and for cultural management likely to increase productivity. Increases in intensive production can make plantains available beyond local markets. Although the commodity is not well recognized outside the tropics, postharvest practices are already adequate for distant marketing of plantains and export production. This is important for the economies of developing countries.

BREADFRUIT, *Artocarpus altilis* (Parkins.) Fosb.

Family: Moraceae
Other names: Arbre a pain (F.), fruta de pan (S.), ulu (Polynesia)

Breadfruit, so named because of its texture and taste similarity to bread, is a native of Malaysia and is grown throughout the humid tropics. Although a starchy vegetable of some importance, and a staple in some locations, production is declining in the presently major areas of Indonesia, Polynesia, and the Caribbean. Breadfruit grows best at warm, 21–33°C, temperatures and in the lowland tropics, where annual rainfall ranges from 1500 to 2500 mm. Some clones are known to have fair drought, salt spray, and soil salinity tolerance. Cultivation is best in well-drained soils and in wind-sheltered areas.

The breadfruit is a tall (12–25 m) evergreen tree in the humid tropics, with large (50 cm), glossy, and pinnately lobed leaves. Dwarf cultivars (4–6 m) have been identified; however, these exhibit considerable variation in leaf size and shape. Seedless and seeded types occur, these being identified as *A. altilis* var. *apyrena* and *A. altilis* var. *seminifera,* respectively. The latter, sometimes called breadnut, is wind pollinated. Seedless cultivars are propagated by root cuttings, or rooted adventitious shoots; micropropagation procedures can also be used. For seeded cultivars, either vegetative methods or seed are used for propagation. Seed are prone to desiccation and should be planted as soon as extracted from the fruit.

Trees are monoecious and begin bearing fruit about 2–6 years after planting. Staminate flowers are borne on an elongated inflorescence; on the same branch, an oval-to-globular inflorescence bears the pistillate flowers. The structure of the syncarpus fruit consists of the clustered aggregation of as many as 2000 smaller fruit fused together and embedded in the fleshy perianth. The fruit surface, usually very rough, has a pronounced dissected appearance, although some cultivars have a relatively smooth surface. Fruit mature about 60–90 days after anthesis and are between 10 and 20 cm in diameter, 20–30 cm in length, and weigh 1–4 kg. Large mature trees can produce as many as 700 fruit. Production tends to be cyclic, with higher yields occurring during the summer in contrast to the winter. Harvests are normally made when fruit acquire a brownish green color and the pulp is still white and mealy. Following removal of the yellowish green outer rind, the pale yellow or whitish flesh is firm and starchy, having the consistency of white potato.

Consumption usually occurs soon after harvest because postharvest life is very short. In ambient conditions, fruit will soften within 2–3 days and often deteriorate in less than 5 days. Retarding softening is important for improving market prospects. Some extension of shelf life is obtained with low temperature (12–13°C) and 90% RH; however, below 12°C, chilling injury occurs. Controlled-atmosphere storage and use of various films or waxes have extended shelf life for 2–3 weeks. Some of the breadfruit production is processed by canning or dehydration.

Breadfruit is usually cooked after removal of the skin and central core. It can also be used without removal of the skin. Either sliced or whole, it may be boiled, baked, or fried. Seed are boiled or roasted for a snack food. The pulp, which is high in starch, is sometimes made into flour. Seed are a better source of protein than the pulp.

JACKFRUIT, *Artocarpus heterophyllus* Lam.

Family: Moraceae
Other names: Jak, jaca

Jackfruit, a native of southern India, is closely related to breadfruit. Growth is best in lowland tropical climates, and the plant tolerates slightly lower temperatures than breadfruit. Like breadfruit, well-drained soils are preferred. Other similarities are its tall tree growth and late fruiting, which begins about 4–6 years after planting.

Jackfruit is dissimilar to breadfruit in several ways. The leaves are relatively small and entire, so that the tree has a different appearance. The syncarpus fruit are oblong and very large, some as long as 70 cm and as wide as 40 cm, and weighing more than 25 kg. However, most fruit are not that large and weigh about 8–10 kg (Fig. 16.7). Another major difference is the cauliflorous bearing habit, whereby fruit are borne singly or in bunches attached directly to the tree trunk by a short stout twig; all fruit are seeded.

Large trees can produce 150–250 fruit in a year. Fruit reach full maturity after 100–120 days. The edible portion is the fleshy and juicy yellowish pulp (perianth tissue) surrounding individual gelatinous covered seeds. The pulp, which has a sweet and acid taste, comprises about 30% of the fruit, the seed about 5%. There are two types of jackfruit: one with a soft and melting pulp, another having a firm fleshy pulp. Fruit shapes are mostly oblong with a surface of numerous short, sharp, hexagonal fleshy spines.

Propagation from seed is more common than by vegetative means even though resulting plants are highly heterozygous. The relatively large seed rapidly lose viability soon after extraction, often within a month, and therefore should be planted without delay. Seed are sown directly in the field because transplanted seedlings establish poorly, mainly due to injury of the delicate taproot. Results of propagation by budding or grafting have been variable, but success has been obtained by stooling, air layering, and stem cuttings; *in vitro* propagation has also been reported to be satisfactory.

Like breadfruit, jackfruit is highly perishable and chilling sensitive, but it can be maintained after harvest for several days at about 12°C.

FIG. 16.7. Jackfruit, *Artocarpus heterophyllus*. The immature fruit are used as a starchy vegetable, Los Banos, Philippines.

It is best that they be utilized without delay. Fruit composition is about 75% water, almost 25% carbohydrates, and contains little protein. In the immature stage, the pulp has vegetable uses similar to those of breadfruit.

The seed, prepared boiled, baked, or roasted, taste like chestnuts. They contain about 6% protein and nearly 40% carbohydrates. In Java, young flower clusters are eaten with syrup and agar-agar. Seed of other related species like *A. ovata* and *A. blancoi* are eaten in the Philippines. Other related species having minor vegetable use are *A. hypargyrea* and *A. hirsuta*.

SELECTED REFERENCES

Barker, W.G. 1969. Growth and development of the banana plant: Gross leaf emergence. Ann. Bot. *33*, 523–535.

Brantjes, N.B.M. 1981. Nectar and the pollination of breadfruit, *Artocarpus altilis* (Moraceae). Acta Bot. Neerlandica *30,* 345–352.

Chatterjee, B.K., and Mukherjee, S.K. 1981. Effect of invigoration and wounding in the rooting of cuttings of jackfruit (*Artocarpus heterophyllus* Lam.) Indian J. Hort. *38* (½), 1–3.

Du Montcel, H.T. 1987. Plantain Bananas. Macmillan, Hampshire, England.

Gawel, N., and Jarret, R.L. 1991. Cytoplasmic genetic diversity in bananas and plantains. Euphyptica *52,* 19–23.

Gowen, S., ed. 1994. Bananas and Plantains. Chapman & Hall, London.

Graham, H.D., and Negron de Bravo, E. 1981. Composition of the breadfruit. J. Food Sci. *46,* 535–539.

Krikorian, A.D., and Cronauer, S.S. 1984. Aseptic culture techniques for banana and plantain improvement. Econ. Bot. *38,* 322–331.

Maharaj, R., and Sankat, C.R. 1990. The shelf-life of breadfruit stored under ambient and refrigerated conditions. Acta Hortic. *269,* 411–424.

Marriott, J., and Lancaster, P.A. 1983. Bananas and plantains. In Handbook of Tropical Foods. H.T. Chan, ed. pp. Marcel Dekker, Inc. New York. pp. 85–143.

Rowe, P. 1984. Breeding bananas and plantains. Plant Breeding Rev. *2,* 135–155.

Sauer, J.D. 1993. Historical Geography of Crop Plants: A Selected Roster. CRC Press, Boca Raton, FL.

Simmonds, N.W. 1987. Classification and breeding of bananas. In Banana and Plantain Breeding Strategies. Proc. International Workshop. G.J. Persley and E.A. De Langhe, eds. Cairns, Australia, pp. 69–73.

Stover, R.H., and Buddenhagen, I.W. 1986. Banana breeding: polyploidy, disease resistance and productivity. Fruits *41,* 175–191.

Stover, R.H., and Simmonds, N.W. 1987. Bananas, 3rd ed. Halsted, New York.

Swennen, R., and De Langhe, E.A. 1985. Growth parameters of yield of plantain (*Musa* c. AAB). Ann. Bot. *56,* 197–204.

Thomas, C.A. 1980. Jackfruit, *Artocarpus heterophyllus* (Moraceae), as source of food and income. Econ. Bot. *34,* 154–159.

Part **B**

Vegetables Consisting of Succulent Roots, Bulbs, Leaves, and Fruits

17

Alliums

Family: Alliaceae (Amaryllidaceae)

Allium is a large and diverse genus of about 500 species. Table 17.1 lists the important cultivated species and their common names. Onion, the most important, is grown from tropical to subartic regions. Regional preferences result in other Alliums sharing the popularity of onion, such as, garlic in Korea, leek in western Europe, and Japanese bunching onion in China and Japan.

Vegetable Alliums are largely Asian in origin with important species found across the continent, stretching from eastern China to the Mediterranean regions. Major areas of genetic diversity are Afghanistan, Iran, and western Pakistan. The Mediterranean basin is considered a secondary center. Many wild Alliums are also found in other areas.

TABLE 17.1. BOTANICAL GROUPING of CULTIVATED VEGETABLE ALLIUMS

Species	Common name
A. cepa var. *cepa*	Onion
A. cepa var. *aggregatum*	Multiplier onion
A. cepa var. *solaninum*	Potato onion
A. cepa var. *perutile*	Ever-ready onion
A. cepa var. *ascalonicum*	Shallot
A. cepa var. *viviparum*	Topset onion[a]
A. cepa var. *bulbiferum*	Tree onion[a]
A. cepa var. *proliferum*	Egyptian topset onion[a]
A. cepa × *A. fistulosum*	Wakegi onion[b]
A. fistulosum	Japanese bunching or Welsh onion
A. ampeloprasum var. *porrum*	Leek
A. ampeloprasum var. *aegyptiacum*	Kurrat
A. ampeloprasum var. *holmense*	Great-headed garlic
A. ampeloprasum var. *sectivum*	Pearl onion
A. chinense	Rakkyo
A. sativum var. *sativum*	Garlic
A. schoenoprasum	Chives
A. tuberosum	Chinese chives

[a]Hybrids of *A. cepa* (onion) × *A. fistulosum.*
[b]A hybrid of shallotlike *A. cepa* var. *caespitosum* × *A. fistulosum.*

For example, *A. amplectens* and *A. anceps* in northern California were gathered and used for food and flavoring by natives of that region.

The taste and odor characteristics of the Alliums are their major attribute. Other features are the umbel inflorescence, flowers with nectaries, a three-chambered ovary, and a basic chromosome number of 8 for the cultivated species. Differences in flowering habit, floral morphology, leaves, scapes, storage organs, and flavor are useful for species identification (Table 17.2 and Fig. 17.1).

Allium taxonomy has undergone periodic revisions and that probably will continue. The Alliums are now included in the family Alliaceae rather than Amaryllidaceae, or as previously classified in the Liliaceae.

Allium Flavor and Lachrymator Compounds

The major flavor of Alliums results from the activity of the enzyme, alliinase, acting on certain sulfur-containing compounds (*s*-alkyl cysteine sulfoxides) when tissues are broken or crushed. The volatile flavor compounds in onions are mainly propyl disulfide and methyl propyl disulfide. The tear- or lachrymator-inducing compound is thiopropanal sulfoxide. Volatile flavor components and the lachrymator are liberated

TABLE 17.2.   CHARACTERISTICS USEFUL IN IDENTIFYING VEGETABLE ALLIUMS

Allium	Diploid chromosome number	Usual edible portions	Usual flower color	Bulbs formed	Bulbils in inflorescence
Onion and shallot	16	Bulbs, foliage leaf bases and foliage blades	White, green striped	Yes	Absent in most cvs
Garlic	16	Cloves	Lavender to pale green and white	Yes	Very common
Leek and kurrat	32	Pseudostem	White to purple	No	Sometimes
Great-headed garlic	48	Cloves few and large	White to purple	Yes	Usually not
Japanese bunching onion	16	Pseudostem, foliage leaf bases and leaf blades	Pale yellow to white	No	Absent in most cvs
Chive	16, 24, or 32	Foliage leaf blades and leaf bases	Purple or rose, rarely white	No	Rarely
Rakkyo	16, 24, or 32	Bulbs, swollen foliage leaf bases	Rose-purple	Yes	No
Chinese chives	32	Foliage leaves, scape and flower buds	White	No	No

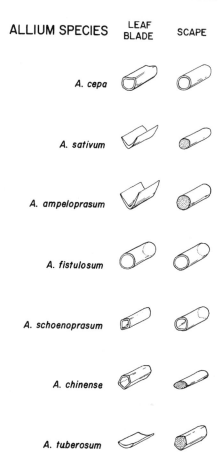

FIG. 17.1.   Leaf and scape cross-sectional characteristics useful for identification of *Allium* species.

by the same enzyme. Table 17.3 presents the active radicals of some sulfur compounds involved in the volatile flavors detectable in different Alliums. For example, onion, welsh onion, chives, and leek share similar levels of propyl disulfide; therefore, it is logical that they have taste similarities.

Gas chromatography and mass spectrometer analyses are useful in identifying the various flavor compounds; biotechnology techniques using flavor compound analyses are useful to accurately identify taxonomic and some other relationships among *Allium* species.

Yellow and red onions and shallots contain high levels of dietary

282   ALLIUMS

TABLE 17.3.   DETECTABLE VOLATILE FLAVOR PRECURSOR COMPOUNDS OF DIFFERENT ALLIUMS

Crop	Radical of sulfur compounds[a]					
	Me$_2$	Me–Pr	Me–Al	Pr$_2$	Pr–Al	Al$_2$
Common onion	+	+	+\–	+++++	+	–
Japanese bunching onion	+	++	+\–	++++	+	–
Leek	–	++++	+\–	+++	+\–	–
Chives	+	++	+\–	++++	+\–	–
Garlic	+\–	+\–	++	–	–	++++
Great-headed garlic	+\–	+\–	++	–	+	++++
Chinese chives	+++++	–	+	–	–	–
Rakkyo	+++++	+	+\–	–	–	–

[a]Me = methyl, Pr = propyl, Al = allyl. +++++ = very high, ++++ = high, +++ = medium, ++ = low, + = very low, +\– = very low to none, – = none.

flavonols, quercetin, and quercetin glycosides, (about 200–1000 mg/kg). These compounds have antibacterial and antifungal properties and are claimed to also exhibit anticarcinogenic activities as well as anticoagulant properties, and other health benefits. In plants, these compounds form part of a defense mechanism to prevent microbial infection.

ONION, *Allium cepa* var. *cepa L.*

Origin and Domestication

Wild forms of onion are not known, although central Asia is generally regarded as a center of domestication. Although production is widely distributed, most are concentrated in the Northern Hemisphere, with many cultivars specifically adapted to the various ecological habitats. The wet tropics and much of Southeast Asia are areas where onion production is limited. Unfavorable climate and handling conditions make shallot production a preferred choice to onion in these areas.

Onions have been cultivated for more than 4000 years for food and flavor and also for health and religious purposes. Its therapeutic benefits, whether real or imagined, are highly regarded in many areas. Onions were cultivated in India about 600 B.C. The Greeks and Romans were using onions as early as 400–300 B.C. Introduction into northern Europe occurred about 500 A.D., at the start of the Middle Ages. Figure 17.2 shows some of the diversity among different onion types.

Botany

Onion is an herbaceous biennial monocot cultivated as an annual. Each leaf consists of a blade and sheath, the blade may or may not be

FIG. 17.2. Example of some of the diversity among different onion types. Source: Courtesy National Garden Bureau, Downers Grove, Illinois.

distinctive. The sheath develops to encircle the growing point and forms a tube that encloses younger leaves and the shoot apex. Young leaves grow up through the center of the sheath of the preceding leaf (Fig. 17.3). Collectively, the grouping of these sheaths comprise the pseudo-stem. Leaves are initiated alternately and opposite each other. They arise from the short, compressed, disklike stem which continues to increase in diameter and, with maturation, resembles an inverted cone. The leaf blades are tubular, slightly flattened on the adaxial side, and, although hollow, are closed at the tip. Where the leaf blade and sheath join is a pore through which the succeeding leaf blade emerges. Each succeeding leaf is larger than the preceding leaves until bulbing is initiated. At that time, newly formed leaves become progressively shorter and then bladeless.

Thus, the onion bulb consists of a vegetative stem axis and the bases of the concentric storage and vegetative leaves. The onion skin is formed from the dry paperlike outermost leaf scales that lose their fleshiness during bulbing. The next layer of leaf sheaths are called false scales and are fleshy with leaf blades. Further inward are the true scales of fleshy storage leaf sheaths that are bladeless, and at the bulb center are the primordial sprout leaves. A transverse slice through the bulb clearly reveals the ringlike structure (Fig. 17.4). At maturity, a typical

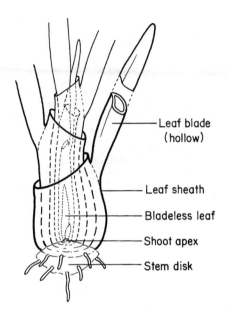

FIG. 17.3. Onion foliage development: younger leaves developing within the leaf sheaths of older leaves, the youngest leaf being bladeless.

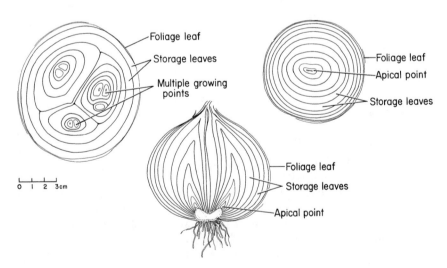

FIG. 17.4. Longitudinal and cross sections of an onion bulb. Onions with a single center are preferred to those with multiple growing points, especially for the preparation of onion rings.

bulb will usually have two dry skins enclosing three to five swollen sheaths from bladed leaves. These enclose three to five swollen bladeless leaves (scales), within which are enclosed four or five bladed leaf initials (sprout leaves). Branching occurs when a lateral bud, present in each leaf axil, sprouts and produces leaf scales. This results in multiple centered bulbs.

Major bulb features are uniformity of shape, size, and skin color. Shapes range from spherical to nearly cylindrical and include flat and conelike bulbs. Size variation is considerable as is skin color which may be white, yellow, brown, red, or purple. Other features such as pungency and dry matter are important. Each of these features is genetically determined but can be altered by environmental conditions.

Onion roots are shallow, most occur within 15–20 cm of the surface, and seldom extend horizontally beyond 50 cm. Seedlings initially produce a primary root; otherwise, all roots are adventitious. Roots are initiated from the stem at the base of the leaves and grow downward through the stem disk to emerge. Onion roots are short lived, being continuously produced. Roots rarely branch, rarely have root hairs, or rarely increase in diameter. As old roots die, new adventitious roots, about three to four, are formed each week. During early growth, the number of active roots increases, but as the bulb matures, roots die at a more rapid rate than they are formed. Late-developing roots sometimes can be seen exiting through the bases of earlier formed leaves.

The terminal inflorescence develops from the ringlike apical meristem. Scapes, one to several, generally elongate well above the leaves and range in height from 30 to more than 100 cm. The scape is the stem internode between the spathe and the last foliage leaf. At first, the scape is solid but, by differential growth, becomes thin walled and hollow. The onion scape has a characteristic bulge at the lower third of its length. The number of scapes that develop depends on the number of sprouted lateral buds.

A spherical umbel is borne on each scape and can range from 2 to 15 cm in diameter (Fig. 17.5). During early development, the inflorescence is initially enveloped by a spathe. The umbel is an aggregate of many flowers at various stages of development, usually there are 200–600 small individual flowers, but can range from 50 to more than 1000. The flowering period may last 4 or more weeks; individual flowers are fertile for a week. Infrequently, bulbils are produced at the top of the inflorescence. Flowers are perfect, having six white petals, six stamens, and a three carpel pistil. Protandry (pollen shedding before the stigma is receptive) promotes out-crossing and a reliance on insect pollinators.

FIG. 17.5.   Onion inflorescence (umbel).

Flowers have nectaries, an attractant to pollinating insects, usually honey bees.

Onion seed mature about 45 days after anthesis. Seed are black, irregularly shaped, and relatively small, about 250 weigh 1 g. Seed lose viability rapidly unless stored under optimal conditions of 0°C and low RH (relative humidity). Under the high temperature and humidity of tropical conditions, viability may be less than a year.

Climate, Soils, Moisture, and Nutrition

Onions are a cool season crop that have some frost tolerance but are best adapted to a temperature range between 13°C and 24°C. Optimum temperatures for early seedling growth are between 23°C and 27°C; growth is slowed at temperatures above 30°C. Acclimated plants are able to tolerate some freezing temperatures.

Soils used for onions range from light sands to heavy clay loams. Peat soils or sandy soils, if irrigation is available, are preferred and often used. Adequate moisture is critical for uniform seedling emergence. Soils with high water holding capacity are better able to provide moisture to the shallow rooting system but must also drain well to be suitable. Growth is retarded when available soil moisture is low, but onions are also sensitive to a high water table or waterlogging. Uniform moisture availability, about 400–800 mm per crop, is conducive to large bulb size and high yields. Favorable soil pH is about 6.5–8.0 in mineral soils and about 5.8 in peat.

An adequate and uniform nitrogen supply is essential for productive plant growth, bulb yield, and quality, and is also important for seed production. To better accommodate the shallow root system, fertilizer banding near the plant is usually preferred to broadcast applications. It is also advantageous that nitrogen be applied in frequent small increments rather than in large quantities at any one time. Late or high-nitrogen applications should be avoided because continued late vegetative growth can result in thick necks, multiple centers, and even bulb splitting. High levels of P and K, more than for most vegetables, are needed for rapid growth and high yield. Copper, manganese, and zinc deficiencies may occur in high-pH peat soils, whereas molybdenum deficiency can result at low pH. Onions are sensitive to salinity.

Propagation (Seed)

Direct sowing of seed is commonly practiced in the United States and many countries with advanced cultural practices, and although some equipment is required, it can be the least costly method for bulb and especially for green onion production. Advancements in planting equipment to handle primed or pregerminated seed have improved seedling establishment. Seed are planted 1–3 cm deep at a rate of 2–4 kg/ha. Evenly distributed seed and spacings improve uniformity of plant growth and bulb size. A firm rather than a friable seed bed is preferred, as that tends to enhance capillary movement of soil moisture to the imbibing seed.

The mechanism of onion seed germination is quite different than that for most vegetable seeds. Elongation of the cotyledon base pushes the embryonic root–shoot axis out of the seed, but the tip of the cotyledon remains in the seed while continuing to absorb nutrients from the endosperm. The relatively long cotyledon appears like a bent joint (called the knee stage) while pushing up through the soil. The joint eventually straightens out, pulling the end of the cotyledon up and away from the soil, and usually free of the seed coat. Soil crusting is often a problem for onion seedlings, especially when they lack vigor.

Although seed germinate over a wide temperature range, 0–32°C, germination is most rapid between 21°C and 27°C. At these temperatures, seedlings usually emerge in 6–8 days. Emergence is slow at low temperatures and sometimes requires as long as a month. During such conditions, nonuniform emergence and establishment often result.

Uniform and adequate stand establishment is very important, as thinning is difficult and not practical nor is filling-in with transplants. Nonuniform stands affect plant size and bulb yield. Seed are often graded and pelleted to improve uniformity of placement and spacing by planting equipment. Insecticide and fungicide seed treatments can also be made to improve plant establishment.

Determining sowing densities rely on cultivar germination character-istics and field conditions. Planting arrangements vary; in some situa-tions, single row plantings are made, and in others, multiple rows are used. Generally, wide spacings favor more vigorous vegetative growth which may delay bulbing; narrow spacings tend to reduce bulb size. Rapid bulb growth tends to produce elongated shapes, whereas slow and long growth periods tend to increase bulb diameter.

A compromise is often made between yield and bulb size. High densi-ties may result in a greater total yield and early maturity, but with many small bulbs resulting. Conversely, low densities may result in less total yield and late maturity, but with large bulbs. Precision plant-ers make spacings more uniform. Nevertheless, when less precise seed drills are used, resulting plants seem to adjust fairly well to the less uniform spacing and plant-to-plant competition. About 40–80 plants/ m^2 is an objective for most bulb crops; however, the range can vary from 25 to 100 per m^2. Plant densities for dehydration or pickling onions range from 100 to 200 per m^2, and from 200 to 400 per m^2 for green onion production. Densities of 1600 per m^2 are used for the production of small pickling (cocktail) onions.

Propagation (Transplants)

Worldwide, most onion production relies on transplant propagation. Transplanting can be a totally manual procedure or can involve special-ized equipment. In areas where growth periods are too short to permit production from direct seedling, transplants or sets are used to over-come that limitation. In addition to earliness, transplants frequently provide higher yields. They are used in the spring to expedite planting schedules, such as when closely following previous crops.

Transplants are produced in the field, in glasshouses, or in sheltered plastic tunnels. The most favorable growth is achieved when day and night temperatures of 17°C and 10°C, respectively, are provided. Seed-ing rates of 80–100 kg/ha provide high populations. Transplants are suitable for use after 8–12 weeks of growth and/or when stem diameters are 3–4 mm, and most often are handled as bare rooted seedlings. With regard to plant schedules where the crop is to be overwintered, transplant size is an important consideration. For that situation, transplants should be large enough to survive winter conditions, but not too large as to be sensitive to vernalization and therefore early bolting. Herbicides are a valuable asset to production because they allow use of much narrower row spacings than might be used if only mechanical cultivation were available.

Advances in transplant production and handling technology have greatly contributed to mechanization. Although a more costly procedure, the use of multiple cell tray containers for transplant production has increased. In some situations, multiseeding of individual cells or soil blocks to provide a greater number of seedlings per unit area is increasingly practiced.

When transplants are used for bulb production, field populations are usually considerably less than those of directly seeded plantings. It is difficult to justify using transplants for green onion production because of the higher labor and costs required for the high populations involved.

Propagation (Sets)

Onion sets are another means of propagation, (Fig. 17.6). Sets are small bulbs, their growth having been intentionally arrested for the purpose of resuming growth at a later period. Short-day cultivars are commonly grown during long-day conditions to produce sets for propagation. Their principal purpose is earliness or to accommodate short growing seasons. To produce sets, plantings of the selected cultivar are scheduled so that appropriate day length conditions induce early bulbing. Plantings for onion set production are most often begun in the spring, with harvest occurring during the summer. High plant densities, 1000–1300 plants/m^2, provide strong plant competition and thus limit bulb size. Set yields in excess of 20 t/ha can be achieved.

Ideally, sets have diameters between 15 and 20 mm, and each weighs 2–3 g. Those greater than 25 mm are likely to be sensitive to low-temperature induction and, thus, prematurely bolt after growth resumes. The use of minisets is a recent development. These sets are less than 10 mm in diameter and can be handled by some seed drills much like seed. They are produced in less time but also tend to produce small bulbs. However, resulting bulb uniformity may be affected because of improper orientation of sets during planting. Although they can be handled much like sets, bulbils (topsets) are infrequently used for propagation.

A high level of nutrition is important to support adequate early growth during set production, but excessive nitrogen can cause luxuri-

FIG. 17.6. Onion propagation: sets as well as transplants and seed are used. Source: Courtesy National Garden Bureau, Downers Grove, Illinois.

ous foliage, which can result in thick necks that are more difficult to dry and prone to disease. Prior to harvesting sets, the plants are deprived of moisture. Sets are ready to harvest when leaf tops have dried. Tops are mowed off or left intact if small and dry. The plants are undercut and lifted from the soil, windrowed, and further dried in the field for several days, protected from moisture and direct sunlight.

With optimum conditions, good quality sets can be stored as long as 6–8 months. Sets are stored at 0–5°C and 60% RH or at 20–30°C and 60% RH in ventilated crates or net bags. Temperatures greater than 20°C help reduce bolting tendencies, but also reduce storage life and weight. Storage at 28–30°C also results in earlier bulbing than storage at 20°C. Between 5°C and 20°C and relative humidities greater than 75%, sets are likely to sprout, develop roots, and/or decay. Therefore, storage temperature options are less than 5°C or greater than 20°C.

Propagation for Seed Production

Two methods used to produce onion seed are bulb to seed and seed to seed. The first requires planting of mature nondormant (mother) bulbs that have been stored long enough at a low temperature to induce bolting or have been planted in the autumn and allowed to grow through winter to induce flowering. Exposure of bulbs to bolting induction could be experienced either in the field or during storage. Generally, bulbs of long-day cultivars are harvested in the fall and stored at low temperatures for spring planting. The resulting seed crop is harvested in the late summer or early fall. Short-day cultivars are grown during winter and bulbs are harvested in the spring. These are stored during summer, often in ambient conditions, for replanting during the fall; seed harvest occurs in late spring or early summer.

For the seed-to-seed method, plantings are made at the appropriate time, usually late summer, for the specific cultivar to obtain good plant size. Plants are vernalized in the field, bolt, and produce seed. The advantage of the second method is that it requires less labor and cost, as bulbs are not lifted or stored. Seed-to-seed yields can be as high or higher than bulb-to-seed yields. The quality of seed from the bulb-to-seed method is better than from the seed-to-seed procedure. This is because it is possible to exercise additional selection and rogue bulbs for off-color and type before planting for seed production. However, excellent quality can be maintained in seed-to-seed production, provided stock seed is grown from bulbs that are rigorously monitored for genetic and other qualities.

Growth and Development

As seedlings become established and grow, new foliage and roots continue to be produced, along with a slight elongation and widening of the compressed stem. At first, successive leaves tend to be longer and have wider leaf bases. Leaves and roots continue to be produced at a relatively uniform rate, although with bulbing, leaf growth changes so that leaves tend to be shorter and smaller and change shape in becoming bladeless (Figs. 17.3 and 17.7). The foliage growth pattern is also altered by bolting.

Bulbing

Bulbing is a change in leaf morphology initiated when sufficient exposure to a critical day length is exceeded, although temperature has an influence. Each cultivar has a critical day length for bulbing induction. The duration of light exposure is most important, and the exposure process is cumulative. A brief exposure to the appropriate day length stimulus is not adequate to permit bulbing to proceed. When cultivars reach their critical day length before adequate vegetative

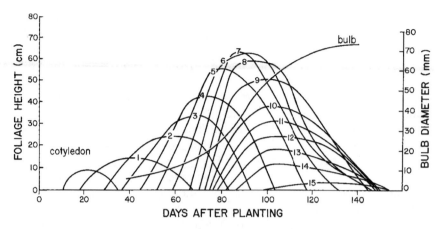

FIG. 17.7. Growth of consecutive leaves of an onion throughout its development from seed emergence to dry bulb; numbers represent sequence of leaves formed.

growth is achieved, resulting bulbs will be small. Cultivars that require long days to bulb will not bulb when grown during short days.

Onions are identified as short-, intermediate-, or long-day cultivars. Short-day cultivars bulb when day length is equal to or greater than 11–13 h. Intermediate cultivars bulb in response to day lengths equal to or greater than 13–14 h, and long-day cultivars bulb in response to day lengths of 14 h or longer. These designations have a positive correlation with latitude (Fig. 6.2). For bulb production, short-day plants are usually grown at less than 30° latitude, intermediate between 30° and 38°, and those grown at latitudes greater than 38° are long-day types. Actually all cultivars are long-day plants for the bulbing response, because they bulb in response to increasing rather than decreasing day length.

Induction of bulbing causes the mobilization of food reserves into the leaf bases, resulting in an enlargement that forms the storage structure called the bulb. Photosynthate partitioning differs with various growth phases. During development prior to bulbing, leaf blade growth is greater than that of leaf sheaths, but at early bulbing, leaf sheath growth accelerates compared to leaf blade growth. As bulbing advances, inner scale or bladeless leaf growth becomes dominant.

The general effects of day length and temperature on onion bulbing and bolting are presented in Table 17.4.

A site-specific example of the influence of day length (latitude), tem-

TABLE 17.4.   DAY LENGTH AND TEMPERATURE EFFECT ON ONION BULBING
AND FLOWERING

Temp.	Day length	
	Short days (11 h)	Long days (15 h)
21°C	No bulbing, no floral initiation	Rapid bulbing, no floral initiation
	No emergence of previously formed initials	Previously formed initials destroyed
10°C	No bulbing, floral initiation; slow bolting	When bulbing, floral initials formed can emerge
		When not bulbing, floral initiation; rapid bolting

Source: Adapted from Brewster (1977).

perature, and planting dates for short-, intermediate-, and long-day onions is presented in Fig. 17.8.

In example I of Fig. 17.8, short-day (12 h) cultivars were planted on the first of April, May, June, July, August, and September. For each planting, seedlings grew during day lengths of 13 h or more. Under these conditions, seedlings received the long-day stimulus and prematurely produce very small bulbs. In example II, short-day cultivars were planted with emergence about the first of October. Seedlings grew in the fall while temperatures were mild and day lengths were less than 12 h. Such plants usually grow sufficiently large before low temperatures (less than 10°C) in December, January, and early February cause vernalization. With active growth resumption during warmer April temperatures, the seed stalks elongate to the disadvantage of bulb development. In example III, seedlings of short-day cultivars planted about the first of November, December, January, or February will likely still be in a juvenile stage during the period when vernalization temperatures occur. Because these plants will not be vernalized, they continue to grow vegetatively until day lengths of 12 h or more occur during late March, when bulbing is induced. The potential size of the mature bulb depends on plant size at the time of induction. In example IV, seedlings of short-day cultivars emerging about the first of March will, after a relatively short growing period, receive the bulbing stimulus about the end of March. As little vegetative growth will have occurred prior to the time the critical day length was reached, the bulbs produced will be small. In principle, the response for intermediate- and long-day onion cultivars would be similar.

Day length has the most predictable influence on bulbing; the influence of temperature is less and tends to vary. Therefore, scheduling planting periods and cultivar selection rely heavily on the day length

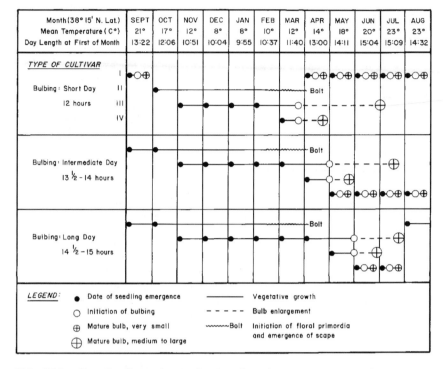

FIG. 17.8.   Growth effects due to day length and temperature on onion types at 38° N latitude (Davis, CA). For further explanation of examples I, II, III, and IV, see the text.

requirement. However, temperature also interacts in the bulbing process. Bulbing and maturation occur earlier and faster under long days and higher temperatures. Low temperature will not prevent, but may delay, bulbing. Once bulbing is initiated, temperature becomes very important and is the main factor influencing foliage growth and bulb enlargement. There are situations when certain cultivars will not bulb at low temperatures when the appropriate day length is experienced, but will bulb at warm temperatures.

In the tropics, temperature tends to be as important as day length for bulbing. For example, the bulbing response at the critical day length will be shortened at high temperatures, but temperatures greater than 40°C retard bulbing.

Bulb yield is highly dependent on the leaf area developed prior to bulbing. An ideal situation for high yield is to have 70–90% of the shoot dry weight translocated into the bulb. The rate of bulb growth and maturation is also influenced by nutrition, moisture supply, plant competition, and light intensity and quality. Large and old plants are more responsive to bulbing than small and young plants when day length requirements are met. However, given a strong photoperiodic stimulus, even a one-leaf seedling can bulb. Nitrogen deficiency occurring near the critical day length tends to accelerate the initiation of bulbing. On the other hand, even at the critical day length, excessive nitrogen can delay the onset of bulbing. Low-moisture stress and competition with plants or weeds can also accelerate the response. Under inductive day length conditions, high light intensities increase bulbing. Far-red light promotes initiation; red light prevents or can reverse initiation.

Onions growing at less than their critical day length requirement continue new leaf development but do not bulb. Use is made of this response in growing green onions because bulbing is undesirable for this commodity. Accordingly, long-day, usually white skinned, cultivars are commonly used for green onion production.

Short-day cultivars grown during long-day conditions bulb early and produce small bulbs because of inadequate plant growth prior to bulbing. Short-day cultivars are grown during long-day conditions to produce sets for propagation, or for small bulbs required for specific uses and some processed products.

During bulb growth, lateral buds can produce multiple apices within the bulb, each surrounded with bladeless storage leaves. Single center bulbs are highly desirable, especially for processing as fried onion rings. The tendency for lateral bud development is cultivar related, but it is also subject to influences such as length of growth period, plant spacing, nutritional level, and even herbicides. Important onion bulb features are earliness and uniformity of size, shape, and color. Also important is flavor (pungency), dry matter, and storage life. Additional desirable features are intact and attractive skins, thick leaf scales (rings), single centered bulbs, thin necks, and resistance to early bolting and to disease and insect pests. Each of these characteristics is genetically influenced but can be modified by environment and cultural practices.

Bolting

Bolting is the formation of a seedstalk and associated inflorescence. Cultivars differ greatly in their response to low temperature and the duration of exposure needed for bolting. A period of exposure to 5°–10°C for 1–2 months is adequate for vernalization of many cultivars. For some cultivars, temperatures between 10°C and 15°C are adequate to stimulate bolting (Fig. 17.9). A reversion to high temperature can partly

FIG. 17.9. Effect of different bulb storage temperature on subsequent onion bolting for seed production.

nullify the cumulative cold inductive effect. Rapid and vigorous bulbing can suppress seed stalk emergence even if already initiated. However, it is possible to have bulbs and seed stalks developing at the same time.

Plant size is important regarding induction response. Once beyond the juvenile seedling stage, low temperatures are conducive to bolting, and large plants are more responsive than small plants. Plants with less than four or five leaves or "neck" diameters less than 6 mm are usually considered to be in the juvenile stage and are not responsive. Onion sets with less than 16-mm bulb diameters are also less responsive.

Plants grown for seed production should achieve ample growth before vernalization in order to maximize plant growth and bulb size and thereby subsequent seed stalk development. To get the vegetative growth needed for high seed yields, plantings are made in mid to late summer so that large plants are developed before inductive temperatures occur in the fall and winter. Normally, bulb onions, intended for seed production, require a cool dormant period during which time floral primordia are initiated. Storage temperatures for bulbs or sets can influence sensitivity to bolting. Storage at either 0°C or 25°C are less conducive to bolting than temperatures in between. The subsequent seed stalk development is enhanced when produced from large plants and bulbs. The number of flower stalks per plant depends on the number of lateral bud shoot apices, and large bulbs have more. However, when plants are grown directly from seed, usually only one seed stem is formed. A disadvantage of bolting sensitive cultivars is that the progeny from the seed produced may, in some conditions, prematurely bolt and, hence, be of little value as bulb onions.

Harvesting

As bulbs enlarge and photosynthate is transferred from foliage leaf blades into storage leaves, the foliage becomes senescent. Varying with cultivars, this stage usually occurs 80–170 days after planting. In anticipation of harvest, a usual practice is to allow the soil to dry about 2–3 weeks before harvest occurs. Bulb onions are commonly harvested when about 50–80% of the tops have collapsed (Fig. 17.10a). In situations when onions are harvested in very warm and humid weather, a delay in harvest beyond 80% top fall sometimes favors an increased incidence of scale infections, particularly the mold *Aspergillis*. When harvested during cool weather, such infections are less frequent. Onions harvested at full top senescence tend to have a shorter storage life. Thus, the optimum time for harvest is a compromise between increasing bulb weight and a possible decrease of postharvest quality and storage ability.

Ideally, when tops have fully collapsed, the proportion of total biomass produced and harvested as bulbs will be about 90%, the remainder being dried tops. When onions are harvested while tops are erect and fleshy, bulb yield is reduced, and the potential for postharvest and

a   b

FIG. 17.10. Example of total onion foliage collapse (a). Ideally, the crop should be harvested when about 80% of the tops have fallen. Example of well cured onions with dried tops ready for trimming (b), source: Courtesy National Garden Bureau, Downers Grove, Illinois.

storage problems increase. Such onions have a high moisture content and a relatively short postharvest life. In the case with more mature onions, as the foliage dries, the pseudostem (neck) shrinks, and with good closure, the incidence of disease is reduced and storage potential improved (Fig. 17.10b).

During harvest, much of the foliage, whether dried or not, is flailed or cut away; the bulbs are undercut and lifted from the soil. Sometimes undercutting precedes harvest to accelerate senescence. In many regions, bulbs are hand harvested and final trimming of tops and roots is also performed by hand. Onions for dehydration or other processed purposes are more likely to be machine harvested because their per unit value is less than fresh market bulb crops and because possible bulb damage is better tolerated. However, for early season marketing advantages, bulbs are sometimes harvested while the foliage is still erect. In the tropics, new foliage growth often is continuous; thus, bulbs are harvested with tops still green. The world average yield of bulb onions is about 16 t/ha, but with excellent growth and management, yields of 50–60 t/ha can be obtained.

Green onions are undercut, pulled, and tied into bunches after size sorting and removal of decayed, aged, or damaged leaves. Roots are freed of attached soil and trimmed; plants are usually washed before packing.

Bulb Curing

Following bulb harvest, curing is necessary to improve postharvest handling characteristics and to limit entry of rot-causing organisms into still fleshy pseudostems or injured tissues. Curing is also performed to enhance the formation of well-colored intact outer skins. When weather permits, bulbs are cured in the field.

For field and ambient temperature curing, bulbs are gathered into windrows to air dry or placed into ventilated boxes or bags for several days to a week. During this period, while encouraging air circulation, the onions are protected from moisture and sunburning (scalding). Sunburn can damage and cause the loss of outer scale leaves, and when severe, it can damage inner scales. Any loss of the outer skin tissues reduces bulb appearance, market value, and ability to further protect bulbs from injury and desiccation. Curing is also accomplished with forced circulation of warm (30°C), low-humidity air through bins or piles of onions placed on slatted floors for 12–24 h; deep piles should be avoided. Following that treatment, temperatures are lowered because continued high-temperature drying results in dark coloring of bulb skins; best skin color develops at 24–32°C. During curing, onions can lose as much as 5% of initial harvest weight.

Bulb Storage

Storage extends the availability of bulbs over long periods. Storage disadvantages are dry matter and moisture loss. Other possible losses include decay, sprouting, and rooting. Most bulb shrinkage is due to respiration. During storage, translocation of carbohydrates occurs via the stem plate from the outermost succulent swollen scales to inner scales. The outermost succulent scale gradually desiccates, becoming a dry protective scale that help reduce water loss from inner succulent scales. This process can continue, resulting in an increase in the number of dry outer scales and, in turn, a decrease of an equal number of succulent scales, along with a concomitant decrease in bulb diameter. Onion respiration rates are generally low, but, as expected, do increase with elevated temperature. Respiratory heat must be removed by ventilation or refrigeration. Relative humidity has a large influence on storage life; sometimes its influence is greater than that of temperature.

Atmospheres with elevated CO_2 and reduced O_2 are known to extend onion storage life. Careful handling, trimming, avoidance of large and high piles, and direct contact with moisture are important. Deep piles can exert sufficient compaction pressure to distort bulb shape.

Mature onion bulbs store best at or near 0°C and at 65–70% RH; surprisingly, storage at temperatures between 25°C and 35°C is also satisfactory. At the higher temperatures, bulbs of some cultivars can be stored for 3–6 months without sprouting, but once removed from these storage conditions, the tendency for sprouting is high. A general correlation exists between long-day types, high solids, and long storage characteristics, whereas short-day types, low in solids, tend to a have short postharvest life. Efforts are in progress to improve characteristics of postharvest storage life in the tropics and high-temperature regions.

Green onions should be held at 0°C and 95% RH. Under such conditions they can be maintained in good condition for 10–20 days, whereas at 5°C, storage life may be limited to 1 week.

Bulb Rest and Dormancy

At the approach of harvest maturity, onion bulbs enter a state of rest for a period that may continue for 4–9 weeks. During this natural dormancy, even though apical meristems are active, such bulbs will not sprout or continue visual growth because of inhibitors synthesized in green leaves that had been translocated to the bulb. The inhibitors are gradually destroyed with time. This is an oversimplification of the physiology of dormancy, but it is important to have normal senescence of foliage in order to improve storage life and reduce early sprouting. The change from rest to dormancy is gradual and highly dependent on cultivar genotype. When dormant and held at optimal storage temperatures, bulbs do not sprout. Once dormancy is passed, with favorable temperature and humidity provided, root emergence occurs, followed by the appearance of leaf shoots.

Bulb Sprouting and Sprout Prevention

Temperature has the most influence on sprouting, which is inhibited at 0°C and at about 65% RH, and also at high temperatures (30°C). Sprouting is optimum at 10–15°C (Fig. 17.11). Although not practical, continuous removal of newly formed roots tends to delay sprout emergence.

For chemical inhibition of sprouting, maleic hydrazide (MH) is applied when about one-third of the tops have fallen prior to harvest. The chemical is absorbed by the remaining green tissues and translocated to meristems where mitosis is inhibited. MH is usually applied at 2500 ppm at a rate of 500 liters of water per hectare. This provides a threshold

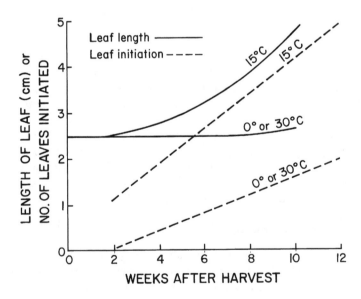

FIG. 17.11.   Leaf length and number of leaves initiated by onion bulbs in different storage conditions. Source: Redrawn from Abdulla and Mann (1963).

of about 20 ppm in the center shoot of the bulb, which is the concentration needed to suppress sprouting. When applied too soon, foliage injury occurs; when too late, foliar absorption is insufficient to be effective. To improve absorption, applications are made when most of the foliage is still green and when dew is not present. Bulbs intended for propagation for seed production should not be treated with sprout inhibitors. Onions treated with MH and held at a temperature between –2°C and 0°C and 65–70% RH can be stored for as long as 6–7 months without sprouting. Gamma irradiation also will inhibit bulb rooting and sprouting, as will controlled-atmosphere storage at reduced oxygen levels.

Allium Diseases and Other Pests

Diseases

Bacterial leaf streak	*Pseudomonas viridiflava*
Bacterial soft rot	*Erwinia carotovora* subsp. *carotovora*
Black mold	*Aspergillus niger*
Black stalk rot	*Pleospora herbarum*
Blue mold	*Penicillium* spp.
Botrytis brown stain	*Botrytis cinerea*
Cercospera leaf spot	*Cercospera duddiae*
Charcoal rot	*Macrophomina phaseolina*
Cladosporium leaf blotch	*Cladosporium allii-cepae*
Collar or neck rot	*Botrytis allii*
Botrytis leaf blight	*Botrytis squamosa*
Botrytis rot of garlic	*Botrytis porri*
Damping off	*Pythium* spp., *Fusarium* spp., *Rhizoctonia solani*
Downy mildew	*Peronospora destructor*
Fusarium basal rot	*Fusarium oxysporum* f. sp. *cepae*
Fusarium basal rot, garlic	*Fusarium culmorum*
Leaf rot (blast)	*Sclerotinia squamosa*
Pink root	*Phoma terrestris (Pyrenochaeta terrestris)*
Powdery mildew	*Leveillula taurica*
Purple blotch (scald)	*Alternaria porri*
Rust	*Puccinia porri*
Sclerotinia rot	*Sclerotinia sclerotiorum*
Slippery skin	*Pseudomonas gladioli* pv. *alliicola*
Smudge	*Colletotrichum circinans*

Smut	*Urocystis colchici*
Sour skin	*Pseudomonas cepacia*
Southern blight	*Sclerotium rolfsii*
Stemphylium leaf blight and stalk rot	*Stemphylium vesicarium*
Twister (Anthracnose)	*Glomerella cingulata*
White rot	*Sclerotium cepivorum*
White tip	*Phytopthora porri*
Xanthomonas blight	*Xanthomonas campestris*

Virus and viruslike agents and (vectors)
Aster yellows, microplasma (leafhopper)
Garlic mosaic virus (aphid)
Leek yellow stripe virus, LYSV (aphid)
Onion yellow dwarf virus, OYDV (aphid)
Shallot latent virus, SLV (aphid)

Insects

Armyworm	*Spodoptera exigua*
Bulb flies	*Eumerus* spp.
Cutworms	*Agrotus* spp.
Green peach aphid	*Myzus persicae*
Leafminers	*Liriomyza* spp.
Leek moth	*Acrolepiopsis assectella*
Lepidoptera larva	*Lepidoptera* spp.
Onion fly	*Delia antiqua*
Seed corn maggot	*Delia platura*
Shallot aphid	*Myzus ascalonicus*
Thrips	*Thrips tabaci, T. palmi*
Western flower thrips	*Frankliniella moultoni*
Wireworm	*Agriotes* spp.

Various Lepidoptera species also are chronic pests. A common pest is the red spider mite *(Tetranchyus urticae)*, another is the garlic mite, *Aceria tulipae.*

Nematodes

Burrowing	*Radopholus* spp.
Lance and dagger	*Longidorus* spp.
Lesion	*Pratylenchus* spp.
Reniform	*Rotylenchus reniformis*
Rootknot	*Meloidogyne* spp.
Stem and bulb nematode	*Ditylenchus dipsaci*
Stubby root	*Trichodorus* spp.

Weed management is especially critical during early seedling growth because of slow emergence and growth of onion seedlings and those of most other Alliums. Therefore, considerable time is required before young plants achieve sufficient leaf area to shade and compete with weeds. Several herbicides are effectively used, and such use is likely to continue. Nevertheless, hand and mechanical cultivation and crop rotations continue as important control procedures.

Production

Onion production in Asia accounts for more than 56% of the world's volume; European production at about 18% follows. The major onion-producing countries, China and India, collectively account for more than 27% of total production (Table 17.5).

Uses and Composition

Onions are an almost indispensable food in nearly all kitchens. The main use is as cooked vegetables, and significant amounts are consumed raw. Foliage of immature plants, known as green or salad onions, are also used raw in salads and cooked in many dishes. Bulb size, shape, color, pungency, and dry matter content are important marketing fea-

TABLE 17.5.   WORLD PRODUCTION OF DRY ONIONS, 1994

	Area (ha × 10³)	Yield (t/ha)	Production (t × 10³)
World	2,023	16.1	32,546
Africa	182	14.1	2,560
North and Central America	88	35.7	3,136
South America	138	16.4	2,259
Asia	1,232	14.9	18,330
Europe	379	15.7	5,945
Oceania	5	41.9	216
Leading countries			
China	266[a]	17.4	4,629[a]
India	390[a]	11.0	4,300[a]
USA	65	44.1	2,859
Turkey	100[a]	20.0	2,000
Iran	39[a]	36.6	1,435[a]
Japan	32[a]	44.4	1,400[a]
Korea, Rep.	17[a]	60.4	1,051[a]
Brazil	81	12.6	1,024
Spain	27	37.1	1,017
Pakistan	70	14.2	987

[a]Estimated.
Source: 1994 FAO Production Yearbook, Vol. 48, FAO, Rome, 1995

tures and are very much subject to regional preferences. Other features that makes onion a popular vegetable are its excellent processing characteristics as canned, pickled, frozen, dehydrated, and flavoring products. Bulb onions for dehydration are usually well cured and if taken from cold storage should be held briefly at ambient temperatures or warmed to 25–30°C. This is a conditioning procedure that reduces discoloration that may occur in dehydrated onion products.

The cultivar has the most influence on pungency, but temperature and soil types also have some effect. In tropical regions, strong pungency is preferred, whereas in many temperate regions, less pungency is preferable.

Soluble solids are an important component for onion storage life and processing quality. This characteristic varies greatly among cultivars. Those developed specifically for dehydration have high-soluble solids, some exceed 18–20%, whereas other cultivars have as little as 5%. Low-soluble solids are typical of short-day, low-pungency cultivars usually grown for fresh consumption. High-solids tend to be associated with cultivars having high pungency and long storage capabilities.

Starch is absent in onion and other Alliums. The carbohydrates present are mainly sucrose, glucose, fructose, and fructosan, a fructose polymer. Protein, fat, and fiber contents are low. In addition to flavor, onions make a significant contribution to human diets in terms of caloric energy and nutrients. Green onion tops have a high pro-vitamin A and fair vitamin C content. Several medicinal attributes such as blood thinning and diuretic functions are assigned to onions and other *Allium* species.

OTHER ALLIUM CEPA *CROPS*

SHALLOT, *Allium cepa* L. var. *ascalonicum* (*A. cepa* var. *aggregatum*)

Not grown to the extent of onions, shallots are widely grown as a popular green onion and also for the bulbs in many tropical regions. They are a highly appreciated vegetable in northern European countries, France in particular, where production exceeds domestic use.

Tops and bulbs of shallot are similar to but smaller than onions. Another difference is that the pear-shaped, reddish-brown-skinned bulbs are clustered at the base of the plant. Such clusters may contain a few to as many as 15 bulbs. The bulb clusters develop because of rapid formation of lateral buds in the bulb. In turn, these bulbs produce additional clusters. Bulbs in clusters are variable as to shape and size (Fig. 17.12). Well developed bulbs are about 5 cm in diameter.

Plant variability is high; some readily flower, produce seed, and are interfertile with onion; others rarely flower. When first initiated, the scape is solid, but when fully elongated, it is hollow and 60–70 cm tall. Because of heterozygosity, seedlings are unlike the parents and, therefore, the crop is usually propagated by bulbs.

Because shallots have a rest period similar to onion, bulbs are commonly stored for several months before planting. It is important that healthy virus-free bulbs be used for propagation and that they be planted shallow. Plant populations are commonly 300,000/ha. Seed use results in low yields because only one bulb is produced. Improvements in seed have been promising but are still inadequate for propagation use, although some use is made in transplant production.

Harvest occurs after leaves have wilted; depending on growing conditions, this varies from 60 to 100 days. Bulbs are easily hand pulled and permitted to dry before being cleaned and tied in bunches. An alternative is to cut away the dry leaves and individually market bulbs or bulb clusters. Shallot bulbs are most appreciated for their subtle delicate flavor and have similar usage as onions.

FIG. 17.12.   Shallot bulbs, *Allium cepa* var. *ascalonicum*.

MULTIPLIER ONION, *Allium cepa* L. var. *aggregatum*
POTATO ONION, *Allium cepa* L. var. *solaninum*

Multiplier onion plants are characterized by broad hollow leaves and the production of a cluster of flattened bulbs, slightly larger than shallot bulbs. Usually four or five bulbs are formed, and sometimes many more. These are enclosed within the leaf sheath of the mother bulb until the bulbs are well developed. When the sheath is torn, the daughter bulbs become separated.

Flowering is sporadic, and when it occurs, few viable seed are produced. Propagation is by division of bulb clusters. Mature bulbs are used for the same purposes as shallots, and tops are used as green onions.

EVER-READY ONION, *Allium cepa* L. var. *perutile*

A perennial plant, the ever-ready onion exhibits vigorous vegetative growth and produces many, 10 or more, small slender reddish skinned bulbs. The bulbs show no dormancy. Plants rarely flower, and those produced are sterile, but when flowering does occur, the scape is short (40–50 cm) and the umbel small. Propagation is by division using bulbs which are usually too small for food use. The narrow long (1 × 30–50 cm) hollow leaves are used as green onion.

TOPSET ONION, *Allium cepa* L. var. *viviparum*
TREE ONION, *Allium cepa* L. var. *bulbiferum*
EGYPTIAN TOPSET ONION, *Allium cepa* L. var. *proliferum*

These names are applied to plants best recognized for topset or bulbil production, tillering, and cold-temperature hardiness. Although morphologically similar, plants of this grouping of *A. cepa* are sometimes given botanical variety designation. For example, topset, tree, and Egyptian topset onion are identified as *A. cepa* var. *viviparum,* var. *bulbiferum,* and var. *proliferum,* respectively. Some researchers believe the Egyptian topset onion was derived from hybridization between *A. cepa* and *A. fistulosum.*

Flowers formed are usually sterile; bulbils produced in the inflorescence are used for propagation (Fig. 17.13). Bulbils that develop at the top of the scape can sprout, producing miniature onion plants. Because of this characteristic, in Japan these plants are called *kitsune negi,* meaning foxy or mysterious onion. They are often cultivated in the

FIG. 17.13.   Topset onion plant, *Allium cepa* var. *viviparum.*

Orient, mostly for the foliage, although bulbils are also eaten. These types of *A. cepa* are mostly grown in home gardens and are of limited commercial importance.

WAKEGI ONION *Allium cepa* L. var. *caespitosum* Mak. × *Allium fistulosum* Araki (Syn., *A. wakegi*)

Wakegi, also known as turfed stone leek, is cultivated for its foliage in far-eastern Asia, principally in China, western Japan, South Korea, the Ryukyu islands, and Taiwan. Plants are intermediate in size between Japanese bunching onion *(A. fistulosum)* and shallot *(A. cepa* var. *ascalonicum).* Wakegi is an allodiploid believed to have resulted from a chance hybridization between Japanese bunching onion and shallot in ancient gardens many years ago.

The crop is propagated by bulbs, as no viable seed are produced.

Plants grow in clumps much like shallots, and daughter bulbs are produced. There are two major ecotypes, each probably domesticated separately. The Japanese-type clones, grown in Japan and South Korea, are slow growing during winter, grow rapidly in the spring, and become dormant in the summer. The Southern-type clones are grown in the Ryukyu Islands and Taiwan. Although growth is affected by cold, these continue growth during the winter. Daughter bulb formation occurs under long days and warm temperatures and is earlier than the Japanese type.

GARLIC, *Allium sativum* L.

Origin and Taxonomy

Central Asia is believed to be a possible center of origin for garlic, and *A. longicuspis,* endemic to central Asia, is possibly the wild ancestral type. Interestingly, *A. longicuspis* produces viable seed, whereas garlic usually does not. However, some clones of *A. sativum* do produce a few viable seed; this is being pursued for the possibility of seed-propagated cropping. Recent isozyme evidence suggests that a separate species distinction from *A. longicuspis* is not appropriate.

There is evidence that garlic was used in Egypt before 2000 B.C., and in China and India for more than 1000 years. European traders facilitated further distribution so that garlic became and continues to be an important vegetable commodity throughout the world for flavoring many foods. Although probably one-tenth of the consumption of onion, garlic is the second most widely used *Allium.* Garlic production occurs in most countries and its cultivation ranges from the equator to about the 50th latitude.

Allium sativum var. *ophioscorodon* is a top-setting garlic known as hardneck, ophio, or stick garlic. The *ophioscorodon* types usually produce a seed stalk with the top-bearing small aerial bulbils. These types generally produce smaller bulbs, with fewer and more uniform cloves in a fairly well organized arrangement compared to the many random size and less organized cloves of the larger bulbs of *A. sativum.* One *A. ophioscorodon* type known as Rocambole is sometimes called snake or serpent garlic because of its coiled scape. *Allium canadense,* a unrelated wild garliclike plant, is a native of North America.

The variability observed among the many different clones of *A. sativum* are due to mutations providing opportunities through natural and human selection for adaptation to various growing environments.

Botany

Although somewhat resembling onion in growth and appearance, garlic differs in having solid, flattened, V-shaped, longitudinally folded leaf blades with a keellike lower surface (Fig. 17.1). Another difference is that garlic foliage leaf bases do not store food; only the bladeless storage leaf of the clove performs this function. The disklike stem is very short, and the adventitious root system, although similar, is somewhat more extensive than onion.

Garlic scapes are straight and solid, but vary in height because of differences among clones and growing conditions. The umbel inflorescence of garlic is subspherical and usually containing only bulbils, or bulbils and flowers which do not or rarely set seed. When infrequently formed flowers do occur, they are lavender and usually wither and abort. However, bulbils, varying with cultivar, can form on the scape. With some bolting-type garlics, the inflorescence may not be evident, and although bulbils may be produced, they occur within the pseudostem slightly above the bulb.

Long believed to be sterile, fertile flowers of *A. sativum* have been reported to occur in central Asia and several other locations. Seed production is highly clone-specific and environmentally dependent. Additionally, for seed to be produced, topsets must be removed. Seed propagation of garlic would have a huge impact on the present management of clove storage and planting, and also that of virus and nematode control.

Precise classification of garlic forms is incomplete. The identification of the many clones is mostly based on growth and bulbing response to temperature and day length, cold hardiness, length of bulb dormancy, and some morphological features. Not all garlic clones produce an inflorescence, a feature used for subspecies classification with bolting types identified as *A. sativum* var. *sagitatum,* and nonbolting as *A. sativum* var. *vulgarae.* The former are further separated according to flower stem and bulb-forming characteristics. The latter nonbolting types are divided into ecotypes with regard to bulb maturity and other morphological features. Bolting tendency is very variable among clones; with some, bolting is fairly reliable, and with others, bolting seldom occurs. Bolting is uncommon in tropical production.

The garlic bulb consists of a grouping of sessile lateral bulbs (cloves) which develop from axillary buds of young leaves in the central axis of the plant. The number of cloves formed is quite variable and range from 1 to more than 25 (Fig. 17.14). Most growers consider an average of 8–10 cloves ideal, although 15–20 are commonly produced. Cloves

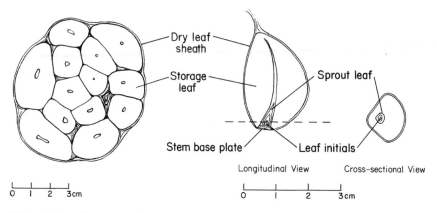

FIG. 17.14.   Cross section of garlic bulb; longitudinal and cross sections of a garlic clove.

are usually ovoid to ellipse-oblong; each clove consist of two mature leaves. One is a paper-thin protective cylindrical sheath that encloses a single second thickened storage leaf that contains a small central vegetative bud. The storage leaf, accounting for most of the clove size, is fleshy and bladeless. The vegetative bud consists of a rudimentary stem, a bladeless sprout leaf, and four to six dormant foliage leaf initials (Fig. 17.14). When conditions are favorable for growth, and after completion of the rest period, the vegetative bud will sprout and grow.

The bulb consists of several cloves, each originating from a lateral bud which is enclosed by the sheaths of the third and subsequent foliage leaves of the mother clove. Younger leaves emerge inside older ones of the pseudostem. The pseudostem is formed by the sheathing bases of successive leaves. The plant's outermost older leaves develop into the smooth parchmentlike protective sheath surrounding the developing bulb. The uppermost apical meristem either forms a flower stalk (scape) or the final leaf. With maturity, an abscission layer is formed between the base of each clove and its attachment to the plant's initial stem plate.

New cloves develop from lateral buds in leaf axils of foliage leaves of the clove used for propagation; the first clove usually develops near the center of the plant, and subsequent cloves continue to originate from near the center of the plant.

Cloves are initiated preferentially from buds in axils of young inner foliage leaves. Each bud primordia forms from two to six growing points, each of which can develop into a lateral bud, and subsequently into a clove. Irregularly shaped bulbs are caused by cloves formed in leaf axils further from the center, often as a result of an excessive period of low-temperature exposure. Insufficient exposure to low temperatures results in no bulb formation; this often occurs when garlic is grown in the tropics.

Culture and Propagation

Garlic plants are very hardy and tolerate low, even some freezing temperatures. In temperate regions, garlic is planted in the fall or early spring and allowed to grow under cool conditions in order to acquire the chilling (vernalization) required for bulbing.

Following planting and during the process of sprouting, the sprout leaf of the vegetative bud elongates and foliage leaves are produced. Subsequently, lateral buds are formed in leaf axils of some of the youngest leaves. These lateral buds have the potential to develop cloves within the bulb, each clove containing a dormant vegetative bud.

Garlic should not be planted following any preceding *Allium* plantings. Although grown in numerous soil types, garlic plants prefer those which are well drained. The plant has a relatively shallow and limited root system; therefore, soils should be maintained near field capacity during most of the growing period. Plants are easily stressed by insufficient moisture and also by waterlogging. Garlic has a moderate to high fertilizer requirement, with banding a preferred application method. Excess nitrogen promotes unwanted secondary growth.

The high level of floral sterility necessitates vegetative propagation with cloves from bulbs preferably having been stored at low temperatures for several months. This period is usually adequate to eliminate postharvest bulb dormancy. Bulbils are seldom used because they produce small plants, hence small bulbs. Low-temperature storage of cloves is particularly important for plants that are grown in the subtropics or at temperatures above 10°C that will not provide sufficient chilling in the field for subsequent bulbing.

In temperate regions where garlic is grown, the crop is generally planted in the fall or early winter and bulbs are harvested during the summer. Early-maturing cultivars are planted during October and November and harvested from April to June; late varieties are planted about November until January for harvest May to September. In areas of very low temperatures, crops are spring planted and harvested in late summer or fall.

For planting, the bulb is separated into cloves; the small central cloves are usually not used because they produce small plants, and therefore small bulbs with few cloves. The cloves are planted, about 3–5 cm deep; they should be planted right side up with the base down. Improperly placed cloves result in emergence variability.

Leaves are formed from the central meristem of the stem tissue. Upon sprouting, the sprout leaf elongates and exits through the pore at the tip of the storage leaf; this is followed by the first foliage leaf that emerges through the pore at the tip of the sprout leaf. The first formed leaf is the protective leaf that ensheathes the developing storage leaves.

To obtain large-sized bulbs, 25–40 cloves/m^2 are planted, but for high yields, plant densities are increased to about 60 or 70 cloves/m^2. Approximately 1 ton of cloves are used to plant 1 ha. Individual clove size greatly influences yield potential; large cloves consistently outyield small ones.

Mosaic virus and stem and bulb nematode *(Ditylenchus dipsaci)* infestations of propagating materials are major problems. Most garlic planting stock is contaminated with mosaic virus. Meristem tissue culture can produce virus-free and nematode-free propagules, and the procedure has been applied on a commercial scale. Virus-free cloves are more productive, but their vigorous growth may sometimes delay plant maturity compared to infected cloves. Plants produced by virus-free propagation usually become infected during growth. Thus, cloves from these plants are no longer virus free and tissue culture must be repeated to again obtain virus-free planting stock.

For maximum yields, garlic should be planted as early as possible. Fall plantings usually produce higher yields than winter or spring plantings. In the tropics, plantings can be made anytime. Development is dependent on prevailing environment, normally 8–11 months in temperate regions and 3–4 months in the tropics.

Bulbing

Bulbing is a two-stage process. The first is the formation of axillary buds in the axils of several leaves, usually among the youngest. The second is the transformation of the buds into storage leaves (cloves) when the primary (apical) bud elongates. The aggregation of the cloves results in the multicloved bulb. This is the usual sequence for garlic bulb development.

Duration and level of low-temperature storage of planting stock is important regarding bulb induction. Extensive low-temperature storage increases the possibility of producing single-clove bulbs or early, small bulbs. Single-clove bulbs or those with few cloves occur because of precocious bulbing, or when produced by small plants such as those propagated from small cloves or from bulbils (topsets). In such plants, only an apical bud or few lateral buds are formed. The amount and period of cold should only be sufficient to break apical dominance. It should not be too cold or too long, which can initiate apical bulbing (single-clove) development. Another result is the development of lateral buds in outer leaves which results in rough or irregularly shaped bulbs. The number of cloves commonly formed during bulbing can range from 1 to more than 25, about 15 are usually produced.

Low field temperature exposure of cloves during development and/ or when dormant in storage is required to initiate bulbing. Temperatures less than 15°C are usually adequate for many cultivars, although some require a period of exposure to temperatures less than 4°C. Normally, cold storage of bulbs (cloves) meets the low-temperature requirement for most cultivars.

For some cultivars, long-term storage at temperatures less than 4°C can result in secondary clove initiation at the shoot apex. Development of secondary clove initiation is also subject to temperature and photoperiod experienced after planting. Axillary bud induction during bulb storage is minimized or can be reversed by a change from a low temperature to higher temperatures.

Vigorous clove sprouting and plant growth tend to be promoted by long-term, low-temperature storage. In some situations, long photoperiod and high temperature soon after planting can promote bulbing before axillary branching occurs, which often results in the production of small bulbs with few cloves.

Once bulbing is initiated, warm temperatures and long days tend to advance bulb enlargement; cultivars differ in the response to temperature and day length. For example, "California Late" is day length sensitive, but the early-maturing "Creole" is day length insensitive. However, many cultivars, especially those grown in temperate regions, must first be exposed to low temperatures in order to respond to these conditions. Cloves stored at temperatures greater than 25°C or those obtained from plants grown at temperatures above 25°C may not bulb or bolt, and vegetative growth may continue indefinitely.

Garlic usually does not bulb or bulbs poorly in the warm and short-day conditions of the lowland tropics. However, some cultivars can produce bulbs when grown at the cool temperatures of high-elevation areas, although they tend to be small and of inferior quality. When temperate-region cultivars are grown in the tropics, all leaves formed are likely to be foliage leaves; therefore, bulbing or bolting does not occur.

Temperate interacts with photoperiod during bulb enlargement. Low and prolonged temperature exposure can shorten the day length requirement. Thus, a variety grown at 20°C may require a 16-h day, but when grown at 15°C, a 12-h day may be sufficient. Response under these conditions normally requires an exposure of about 2 months. Storage leaf growth (clove) is optimal at 17–26°C.

Bolting

Bolting induction does not occur in storage, but differentiation is promoted after planting by relatively cool temperatures and short day length. For some cultivars, low field temperatures during growth may be adequate to initiate bolting. Small plant size, low moisture, and nutritional stress also limit inflorescence development. However, bulbing tends to suppress scape elongation. The extent of scape development determines if the inflorescence is fully or incompletely developed and varies among clones ranging from rudimentary to well-extended scapes having topsets and, occasionally, flowers; flowers are seldom fertile. Storage at very low (–2 to 2°C) temperatures will optimize flower induction after planting.

After scape emergence, warm temperatures and long days enhance its elongation and topset development. For normal multicloved bulb production, inflorescence initiation should occur before storage leaf differentiation. For many temperate-region cultivars, inflorescence induction appears to be a prerequisite for normal multiclove bulbing. Perhaps this occurs because inflorescence development weakens apical dominance, resulting in axillary bud differentiation into storage leaves that form cloves. In nonbolting cultivars, axillary branching should occur before storage leaf differentiation. Once that occurs, increased temperatures and day lengths promote storage leaf differentiation and growth.

Harvest and Storage

Ideally, plants should achieve adequate growth before bulbing commences, so that the foliage is capable of producing large bulbs and high yields. Maturation occurs in mid to late summer, as indicated by drying leaves and falling tops.

Garlic is usually harvested after tops have fallen and are well dried. The soil should also be dry; therefore, irrigation should be stopped at least 3 weeks before harvest. Bulbs are loosened by undercutting and are hand pulled or machine lifted and placed in shallow windrows for further drying. Bulbs are covered to prevent sunburning or exposure to moisture; any direct contact with moisture should be avoided. Cover may be the dry plant leaves, or other dry and shade-providing materials. Bulbs often are placed into slatted containers or netted sacks that allows air movement. Providing good ventilation is important to facilitate drying. Usually after a week to 10 days, bulbs should be sufficiently dried to be trimmed of roots and/or leaves and are ready to be marketed or stored. In production areas where rainfall or high humidity may occur, indoor drying may be necessary. In some situations, the dried garlic tops are not removed but are used for bunching and/or braiding bulbs together.

Cloves become dormant as bulbs approach maturity; the length of dormancy varies with clones and is most intense at full maturity. Thereafter, dormancy declines and usually disappears after several weeks, although it can persist up to 2 months. For immature bulbs, dormancy is readily broken by holding at 35°C, whereas with fully mature bulbs, dormancy is rapidly depleted during 5–10°C storage. Bulbs intended for consumption are held at ambient temperatures and can remain in good condition for several months. However, for extended storage, they should be held at 0°C and at 60% RH. Well-dried mature bulbs stored at –2°C can be maintained in good condition for as long as 8 months. When no longer dormant, cloves readily sprout at temperatures between 5°C and 10°C. However, like onion, storage at 25°C can prevent sprouting, but results in high shrinkage. Garlic should not be stored at greater than 70% RH. Treatments with maleic hydrazide or gamma irradiation are sometimes used to extend storage life. Such treated cloves should not be used for propagation.

Bulbs for planting stock use are similarly stored but are usually conditioned at 5–10°C for a few days before planting. This is critical, because without conditioning, cloves held at temperatures less than 5°C may result in early maturation of small and/or rough bulbs, whereas those held at temperatures above 18°C may exhibit delayed sprouting.

Production

Asia leads the world in garlic production with more than 82% of the total world supply, with China producing 63% of the world supply; China's garlic production exceeds its dry onion production, (Table 17.6). On a per capita basis, the Republic of Korea consumes the most garlic.

Use and Composition

Garlic is primarily grown for its cloves used mostly as a food flavoring condiment. Green tops and also blanched tops are eaten fresh and

TABLE 17.6.   WORLD GARLIC PRODUCTION, 1994

	Area (ha × 10³)	Yield (t/ha)	Production (t × 10³)
World	813	9.7	7914
Africa	21	15.5	326
North & Central America	23	13.1	299
South America	35	5.8	203
Asia	617	10.6	6526
Europe	118	4.7	559
Oceania	<1	5.8	1
Leading countries			
China	372[a]	13.4	4986[a]
Korea, Rep.	43[a]	11.1	476[a]
India	89[a]	4.2	370[a]
Egypt	8[a]	31.9	255[a]
United States	12	19.1	224
Spain	31	6.5	199
Indonesia	25[a]	6.1	153[a]
Thailand	24[a]	4.5	110[a]
Brazil	17	5.3	92
Turkey	11[a]	8.1	92

[a]Estimated.
Source: 1994 FAO Production Yearbook, Vol. 48, FAO, Rome, 1995

cooked in ways similar to those for green onions, especially in tropical areas. Consumption of immature bulbs for salad use is also popular. Considerable amounts of garlic, especially in North America, are processed as dehydrated chips, flakes, granules, and powder. Garlic is also appreciated by many people for its many medicinal attributes, perceived and actual. Medical researchers have studied garlic for its possible role in reducing the incidence of atherosclerosis and its anticoagulant and anticarcinogenic effects.

Garlic flavor is due to a group of sulfur-containing compounds. The dominant flavor compound is alliin (S-allylcysteine sulphoxide), which actually is odorless until broken down to allicin (diallyl disulphide) after tissue rupture.

LEEK, *Allium ampeloprasum* L. var. *porrum (A. porrum)*

Wild forms of leek have not been found, but leek probably is a cultigen of wild *A. ampeloprasum* relatives that were domesticated in the eastern Mediterranean, where other species of the Ampeloprasum group are found. Leeks have a long history of cultivation as documented in biblical writings, and now are cultivated throughout the world.

Within the Ampeloprasum group, besides leek are subgroups such as kurrat, and great-headed garlic that have different morphological forms and also vary in ploidy. Leek, a tetraploid and although usually an outbreeder, is self-compatible and also compatible with other members of the Ampeloprasum group of similar ploidy. Bulb-forming members of the Ampeloprasum group such as great-headed garlic and pearl onion are hexaploids.

Leeks are robust plants larger than onions with solid, linear, flat, and longitudinal keeled leaf blades that are "V" shape in cross section. Leaves are alternate in opposite rows, erect, and also rise progressively one above the other (Fig. 17.15). The middle to upper portions of well-developed outer leaves tend to turn downward, and young inner leaves are upright. Plant heights vary between 40 and 75 cm. Leaves are initiated from the apical meristem of the suppressed disklike basal stem. New leaves are formed from the ringlike meristem inside each previous leaf sheath. Each meristem produces a leaf as a tubelike

FIG. 17.15.   Leek plant, *Allium ampeloprasum* var. *porrum*.

sheath which may be from 5 to more than 50 cm in length. With subsequent development and elongation, the leaf is modified above the sheath portion, as a linear solid leaf blade. Young leaves emerge from within the previous leaf sheath, and collectively, the leaf sheaths produce the pseudostem.

Leeks are broadly adapted to different soil and climatic conditions and are able to extract moisture and nutrients from soils because of the profuse, although relatively shallow, adventitious root system concentrated near the base of the stem. Best temperatures for vegetative growth are between 20°C and 25°C. Being day length insensitive and not having a rest period also improves and broadens their adaptation. They have better cold tolerance than onions and commonly are successfully overwintered. Having no definite harvest period and long storage life are other attributes which make leek a popular and versatile vegetable.

Under most temperature conditions, leeks do not form a noticeable bulb. However, under long-day conditions, food reserves can accumulate within the leaf sheath bases giving the appearance of a thick-necked bulb. Temperatures between 12°C and 18°C tend to enhance leaf base swelling, whereas temperatures above or below this range do not.

Pseudostem characteristics vary with cultivars and somewhat with cultural practices. Some cultivars produce short, thick pseudostems; for others, pseudostems are long and thin. Preference is given for uniformly long white pseudostems. Short pseudostem length appears to be associated with winter hardiness. Plant growth is continuous, and with age, the outer leaves become unpalatable. However, although the outer encircling leaves lose, fleshiness, they do not dry to the parchmentlike scales characteristic of onions.

Leeks are vernalized by low temperature, and when grown continuously at or less than 15°C, many cultivars will bolt; some bolt even at 21°C. Short days in association with low temperature meet vernalization requirements. A long photoperiod after vernalization tends to accelerate flower initiation. Advanced plant size and age also increase susceptibility to bolting. Temperatures above 18°C may cause devernalization. Lateral bulbs are sometimes produced in leaf axils during long-day conditions following flowering; some cultivars produce these bulbs more profusely than others. However, these bulbs are seldom used for propagation. Flowering rarely occurs in the tropics.

The scape is round and solid and grows through the center of the pseudostem; scape lengths vary between 40 and more than 100 cm. From 6 to 12 cm in diameter, the inflorescence contain many flowers that are light purple or white. Seed resulting from selfed and/or insect-aided cross-pollination are black and small; about 350 weigh 1 g. Bulbils (topsets) can form in umbels; flower removal tends to stimulate their formation.

The usual propagation of leek is with seed-produced transplants. Leeks are slow to develop and have a long growth period that rules against direct field sowing. Seed germination also is slow, and overall development is promoted by seed priming and/or pregermination prior to sowing. Germination is acceptable to good at soil temperatures from 11°C to 23°C; above 27°C, it is sharply reduced. Seed are planted shallow, about or less than 1 cm deep. Seedlings are grown for 2 to as long as 3 months before being transplanted. Bulbils can also be used for propagation.

Transplants, 5–6 mm in stem diameter, are planted about 10–15 cm or deeper into a shallow trench or furrow in order to assist subsequent blanching of the base of the developing pseudostem. The length of the blanched portion of the pseudostem influences its market value, thus growers strive to maximize that length. This is achieved by deep planting so that more of the pseudostem develops below the soil surface, and by hilling soil against the plant as it grows.

In place of individual transplants, the use of multiseeded module or soil block transplants providing two to four plants within the cell or block are increasingly used. It has been found that closely spaced plants are able to adjust with relatively little interference to final pseudostem size or shape. Usual plant populations are about 25–40 plants/m².

A growth period from 120 to 150 days after transplanting is usually required in order to obtain high yields. Leek differ from onion or garlic bulbs because they are harvested while still growing vegetatively. Harvesting is done by uprooting the plants which requires considerable effort because of the extensive interlocking mat of roots. Harvest machinery have made the process less difficult. In addition to field removal, washing and trimming is laborious. Washing is very important because soil is commonly present in the upper portion of the pseudostem. Roots are trimmed close to the base of the pseudostem, outer dried and/or damaged leaves are removed, and remaining leaves are cut back to a length suitable for market presentation. The pseudostem typically represents about 40% of total plant weight. Yields range from 25 to 50 t/ha.

When maintained at 0°C and 95% RH, leeks can have a postharvest life of about 2 months. Controlled atmosphere (10% CO_2 + 1% O_2) can further extend the storage period. Leeks are held or stored upright to avoid geotropic curvature of pseudostems.

Leeks are a popular year-round vegetable in northern and western Europe. They are used extensively in soups and stews because of the mild flavor and somewhat mucilaginous character; they are also cooked separately and used fresh for salads. Northwestern European countries are major producers and consumers. Worldwide production occurs on about 50,000 ha.

Prei anak, also a type of leek, is a perennial plant grown in southeast Asian countries primarily for its green pseudostem, which resembles but is less prominent than that of European leeks. A major plant characteristic is its profuse tillering. The tillers, which are not eaten, are used for propagation.

KURRAT, *Allium ampeloprasum* L. var. *aegyptiacum (A. kurrat)*

Kurrat plants are tetraploids and similar in appearance to leek. However, kurrat plants are smaller, and their pseudostem exhibits negligible growth. Propagation is by seed, usually sown in the fall or winter. The plant is cropped mainly for its leaves, which can be harvested several to many times each year. Foliage is used fresh and for seasoning other foods. The crop is popular in Egypt and other eastern Mediterranean areas; production elsewhere is minor.

GREAT-HEADED GARLIC/ELEPHANT GARLIC, *Allium ampeloprasum* L. var. *holmense*

Great-headed garlic leaves are very flat and resemble those of leek, but differ in producing large garliclike bulbs. Umbels, similar to leek are produced on round solid scapes. Plants are hexaploids, and although many cultivars will flower, flowers are either sterile or seed are not viable; thus, the crop is bulb propagated. Bulbils are formed but are not used for propagation because of their slow development (Fig. 17.16).

Plants that flower usually form a cluster of several, 2–10, bladeless storage leaves, similar to garlic cloves, although individual cloves are much larger. Plants that do not flower typically produce a single large clove. Single-clove bulbs, called "rounds," have the highest market value, and bulbs with few cloves are valued higher than multiple-cloved bulbs. Growers often break off the flower scape when first evident to encourage development of larger but fewer cloves. Only the best quality cloves should be used for propagation. Some growers, to minimize use of edible product for propagation, use small cloves and those not suitable for marketing. This is an unwise practice that usually results in poor quality and small bulb production.

FIG. 17.16.   Bulbils of great-headed garlic, *Allium ampeloprasum* var. *holmense;* these can be used for propagation.

Harvests are made before foliage becomes fully senescent, otherwise the outer bulb sheath breaks or tears, which reduces market value. Yields up to 24 t/ha have been obtained. World production is not extensive; the crop is often viewed as a specialty vegetable having taste and flavor intermediate to garlic and onion.

PEARL ONION, *Allium ampeloprasum* L. var. *sectivum*

Pearl onion plants differ from leek in forming bulbs; the pseudostem is essentially absent. These winter-hardy plants are grown for the small, almost spherical, white-skinned daughter bulbs which form in clusters at the base. They do not have the structure of an onion bulb, instead they are similar to the single storage leaf of a garlic clove. Rarely developed white flowers produce seed which are fertile. Nevertheless, the crop is usually vegetatively propagated with bulbs.

Pickling is the primary method of preparation. Small onion bulbs similarly prepared are sometimes erroneously called pearl onions.

JAPANESE BUNCHING ONION/WELSH ONION, *Allium fistulosum* L.

The ancestry of *A. fistulosum* is unknown; it is not found in the wild state. Central and northwest China are thought to be the possible center of origin. That possibility is suggested by the presence of *A. altaicum* that is found and occasionally gathered for food in Mongolia and Siberia. The name "welsh" is attributed to the German meaning for "foreign," and first used to identify the plant when introduced into Germany.

Evidence exists that cultivation occurred earlier than 300 B.C. in China, and about A.D. 500 in Japan. Japanese bunching onion is the principal crop used as greens for centuries in China, Japan, and Korea where it continues to be of major importance. Japanese annual production from about 25,000 hectares is almost 600,000 tons. Bunching onion production is most important in China and exceeds common onion production. In Japan and the Republic of Korea, bunching onion production is about 50% of the bulb onion tonnage. Overall, bunching onion remains mostly an east Asian crop; there is little large-scale production elsewhere.

Plants are adapted to a very wide climatic range, being grown in high-latitude cold regions of northern China and Japan and frequently overwintered. They also tolerate hot and humid conditions in Southeast Asian regions. Plants are very susceptible to waterlogging, but when drainage is provided, heavy and frequent rains are tolerated. Water consumption is fairly high, yet plants do tolerate low moisture stress very well. Nutritional needs are high; the preferred soil pH is neutral, but a higher pH can be tolerated.

To accommodate the many growing environments, growers have access to numerous cultivars. Although perennial, the plant is mostly grown as an annual and sometimes as a biennial for its green tops and blanched leeklike pseudostem, some of which have a length in excess of 25 cm.

The hollow tubelike vigorous leaf growth is circular in cross section; all leaves have blades. Basal portions of bulb sheaths are storage tissues from which lateral buds produce tillers. They can be harvested or used for propagation. In tillering, short rhizomes temporarily connect adjacent tillering shoots. These shoot clusters eventually separate from adjacent rhizomes and appear as individual plants.

Unlike common onion, plants do not form well-developed bulbs. Bulbing is induced by day lengths greater than 12 h and advanced by temperatures above 20°C. However, bulbing is indistinct; those produced are thin and oblong, and seldom more than 10 cm long. Lateral bulbs are few or essentially absent.

For many cultivars, flowering is most often induced by temperatures less than 13°C and short days, but a minimum plant size is also critical for flower induction; considerable cultivar variation exists. Flowering is uncommon in the tropics. The scape is round, hollow, and 40–75 cm tall, and does not have the typical bulge of onion scapes. The umbel, from 3 to 7 cm in diameter, is not spherical. A distinguishing feature is the order of opening of the pale yellow flowers, which begins at the top of the umbel and proceeds toward the base. This is in contrast to most other Alliums except chives.

Japanese bunching onion is self-compatible, but flowers are commonly out-crossed. Crosses of *A. fistulosum* with different shallot forms of *A. cepa* resulted in the cultivars: Beltsville Bunching, Louisiana Evergreen, Wakegi, and Delta Giant. The first two cultivars are nonbulbing green salad types; the latter two are bulb-forming types.

In the Orient, bunching onion receives considerable attention and many cultivars are available for the different latitudes and climatic niches where the crop is grown. More than 100 cultivars are grown in Japan. Cultivars are identified according to growing season, winter hardiness, blanching characteristics, degree of tillering, and whether pseudostems or tops are used. For example, major cultivar types, Kaga and Kujou, are known for their growth and dormancy adaptation to the coolest and warmest growing environments in Japan. Senju is a type with an intermediate adaptation. Kaga and Kujou types are usually grown for pseudostem production, whereas Senju types are tiller producers. Nebuka is a term that also identifies cultivars producing long pseudostem. Hanegi refers to cultivars generally grown for tiller production.

In temperate regions, Japanese bunching onions are usually grown as seed-propagated transplants. Some cultivars produce bulbils, but these are not used because they often produce smaller plants, although they are suitable for green-leaf production. When plantings are directly seeded, they generally are intended for green-leaf production. In tropical regions, basal tillers are the preferred method of propagation; in some situations, ratoon propagation is practiced. Ratooning is more suitable for green-leaf rather than pseudostem production.

Raised-bed culture is common, and for the pseudostem crop, friable soil is important because that will facilitate soil blanching. Pseudostem-type cultivars are transplanted to reduce the long growth period and to maximize blanching because transplants can be placed deeper into soil than seed. Transplants are planted into shallow furrows about 15 cm deep, and as with leek production, further blanching of leaf sheaths is achieved by continually mounding soil around the lower leaf bases as plants grow.

Field populations differ according to whether cropping is for green-leaf or pseudostem purposes, and range between 400,000 and 500,000 plants per hectare.

Green leaves are harvested 2–3 months after transplanting, but a longer growth period is required before the harvest of the blanched pseudostems can occur. The period may be 6 to more than 9 months. In Japan, harvest mechanization is used. Nevertheless, harvest is labor intensive, considering the need to uproot, trim, clean, and bundle pseudostems and/or green leaves for market presentation.

Pseudostem lengths range from 15 to 50 cm and from 1 to 3.5 cm in width. The blanched portion of the pseudostem is an important quality parameter and should exceed one-fourth to one-third of the total length (Fig. 17.17).

This important vegetable has many food uses. It is consumed raw and cooked in numerous salads and other dishes. Young seedling plants are used for some special recipes. Relatively little is processed, although a portion of the production is prepared for dehydration. Additionally, many therapeutic benefits are ascribed to its consumption. The flavor of

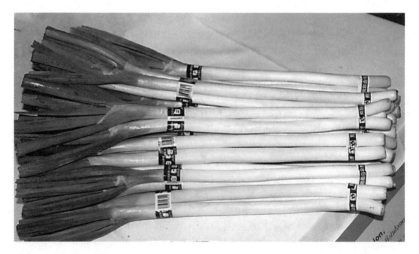

FIG. 17.17. Market-prepared Japanese bunching onion, *Allium fistulosum,* in Japan.

Japanese bunching onion is similar to onion. Green tissues of Japanese bunching onion provide good levels of pro-vitamin A and vitamin C. The eating quality is improved during low temperatures because sugars as well as protein content increase.

CHIVES, *Allium schoenoprasum* L.

Chives have a extremely wide distribution across the Northern Hemisphere, but their specific origin is unknown. Plants are perennial, cold hardy, and also tolerant of high temperatures. Optimum growing temperatures are between 17°C and 25°C. Chives grow well in a moist habitat, but not in waterlogged soils.

Chives are grown expressly for their narrow tubelike leaves; leaf lengths are from 15 to 50 cm. An axillary bud develops and forms a side shoot after every two or three leaves have formed, and thus plants develop clusters of shoots. Plants tiller and increase as clumps; and with continuing growth, a field population of disconnected clumps results that can make row distinction almost disappear. Poorly developed ovoid to oblong bulbs, about 1–3 cm long, form in clusters at the base of the plant but are usually inconspicuous. Lateral bulbs, few to many, are also formed.

Floral induction apparently is dependent on low temperatures and short-day lengths. However, small plants, those without adequate storage reserves, are less responsive to induction. Flowering usually does not occur if temperatures are above 18°C.

The umbel inflorescence that forms in the spring has a round and hollow scape, 20–60 cm tall, that bears purple and sometimes white or pink flowers. Flowers open first at the top of the umbel much like the Japanese bunching onion. Chives are generally an out-crosser, and flowers are insect pollinated, but selfing frequently occurs.

At the end of the growing season with decreasing temperatures and shortening day length, leaves become senescent. As carbohydrates are translocated into bulbs and roots, plants enter a rest stage of 4, 5, or more weeks, which is broken naturally by the end of winter. This dormancy can also be removed by soaking bulbs in warm water for 8–16 h beginning at 40°C and allowing for normal cooling of the water. The treatment possibly leaches growth inhibitors and/or results from an increase of the metabolic rate. Subjecting plants to 43–44°C for 36–48 h will also remove dormancy, as will prolonged low-temperature exposure.

Propagation is usually with seed, although propagation by division are also performed. Plantings, which may be cultivated for 3–4 years, are initially made in multiple rows at very high densities. Bulbs are sometimes used in forcing glasshouse winter crops. An accurate prediction of plant entry into or exit from the rest state is important to obtain uniform forcing results.

Multiple harvests are commonly made by hand cutting leaves about 2–3 cm above the soil surface; leaves are typically handled as bundles. Harvest in the first year is often limited to two or three cuttings. After plants have built up their food reserves, more frequent harvests can be made. Yields as much as 7 or 8 t/ha are obtainable from six to seven harvests each year. Only leaves are eaten, as the bulbs are too small.

Chives are known for their delicate flavor and are used in salads, for flavoring and garnish. A considerable volume is dehydrated, some by a freeze-dry process.

RAKKYO, *Allium chinense* G. Don (*A. bakeri*)

Other name: Ch'iao t'ou

Rakkyo an important crop in the Orient is believed to have originated in central and eastern China. Rakkyo and ch'iao t'ou are the common Japanese and Chinese names. Rakkyo, of which there are numerous strains, is a popular vegetable known to be cultivated since the 9th century in Japan where annual production exceeds 35,000 tons; Chinese production is similarly large.

Rakkyo is a perennial plant cultivated as a biennial. However, the productive growth cycle is commonly completed within a calendar year. A temperate-zone crop, rakkyo performs best at latitudes between 30° and 40°. Bulb formation and flowering are promoted by long days. Optimum temperatures for growth and bulbing are between 15°C and 25°C. Plants have good drought tolerance, require well-drained soils, but not necessarily fertile soils. Actually, high plant nutrition tends to produce large and soft bulbs, which are less preferred than those small and firm.

Leaf growth varies from 30 to 50 cm in length. The hollow leaves are sharply angular, having a "D" or pentangular-shaped cross section. The clumped growth habit and narrow leaves resembles chives.

Foliage leaves have sheathing bases, and bulbs are formed by the basal thickening of four to five concentric leaf sheaths. Normally elliptical in shape and in cross section, the rakkyo bulb does resemble the common onion bulb. However, unlike onions, all leaves have blades, even the very short ones. Fully developed bulbs are 3–5 cm long and 0.7–1.5 cm wide; although some cultivars produce bulbs with diameters as large as 2.5 cm.

Most rakkyo cultivars are autotetraploids, and when plants flower, viable seed is not set. Therefore, propagation relies on the use of bulbs. Bulbs obtained from the preceding crops are stored before planting for 1–2 months to overcome the physiological rest period that accompanies maturation. Each propagule has the potential to produce 5–15 new bulbs, but numbers and size varies with cultivars.

Plantings are usually made in the late fall. For most production, bulbs are planted 8–10 cm apart within rows and at 40–60 cm between rows. High-density plantings are made at equidistant spacings of 15 cm. During growth, soil is mounded against the leaf bases in order to blanch the bulbs.

Soon after planting, the growing point rapidly divides. From the many lateral buds, clusters of sprouting leaves are formed. These continue to grow and divide during winter and spring, and by the summer, the leaf bases will have thickened to form new bulbs, which are dormant. At that time, the foliage will have withered and died, and from each original bulb, a cluster of new bulbs held together at their base by a common stem will have resulted. Of the many strains of rakkyo, some are preferentially grown to produce small bulbs, others for large bulbs.

The crop is harvested once a year, usually during the late summer after leaf tops become senescent and die. Harvests may occur as early as 6–8 weeks after planting. Hand harvesting is common, during which the bulbs are cleaned of attached soil and bundled for fresh market, or after top removal, delivered in bulk for pickle processing (Fig. 17.18). A portion of harvest is retained for propagation. Yields of 15–20 t/ha have been reported in Japan.

In mid to late summer in temperate regions, if the plant is not harvested, it produces a solid scape, 40–60 cm in height, and flowers. The scape is oval in cross section and bears reddish purple flowers. Further shoot and bulb growth follows scape development but occurs to the side of the scape so that new leaf growth does not ensheathe the scape. When grown in the tropics, plants are small, short, and seldom flower.

The highest consumption of rakkyo is in Japan and China. Although bulbs are eaten raw and cooked, they are mostly consumed as sweet–sour pickles prepared in vinegar with sugar. For preferred quality

FIG. 17.18. Rakkyo bulbs, *Allium chinense.*

pickles, small dense bulbs are used. To achieve that quality, the crop will sometimes be grown for 1 year without harvest in order to produce a larger number of small bulbs in the second year of production. Rakkyo has been used to prevent thrombosis by inhibiting blood platelet aggregation, which is also ascribed to other Alliums.

CHINESE CHIVES, *Allium tuberosum* Rottler ex Sprengel

Other names: Kau tsoi, nira

The origin of this species is not clear but is thought to be China, where most of the production occurs. Kau tsoi and nira are Chinese and Japanese names, respectively, for this plant. Total world annual production is estimated to be greater than 75,000 tons. Plants are perennials and tend to grow in clumps and spread by tillering growth of lateral buds initiated on well-developed and persistent rhizomes that connect the clumps. Plants are out-breeders, with a high frequency of apomictic progeny. Propagation is usually with seed, although the rhizomes can be divided and transplanted, usually appearing as a cluster of attached plants. Various plant spacings used generally result in about 12–20 plant clusters per m².

The primary product is the 5–10-mm-wide, solid, slightly keeled green leaves (Fig. 17.19a). Leaf length, dark green color, flavor, and tenderness are quality parameters. Bulbs are poorly formed and of little consequence. The rhizome essentially replaces the bulb as the storage structure. White flowers, few in number, are borne on an erect, solid, somewhat angular-shaped scape.

Short day lengths and low temperatures induce dormancy and stop leaf growth. Dormancy is removed with increasing day lengths, or exposure to long periods of low temperature. Long days are required for flowering; the critical period varies with cultivars, and for some cultivars, low-temperature exposure accelerates flowering. Day-neutral cultivars are also grown. After a sufficient number of leaves have formed, the inflorescence is initiated from the rhizome; several are formed annually by each rhizome.

Optimum leaf growth occurs at 20°C, and depending on temperature,

a

b

FIG. 17.19.   Chinese chives plant, *Allium tuberosum* (a) and bunched flower stalks (scapes) as prepared for marketing in Taiwan.

three to nine harvests can be made annually; plants can be harvested for several years. Leaves are harvested when about 20 cm in length. Young scapes and flower buds are also eaten. Scapes are harvested when 30–40 cm in height and while flower buds are still green and closed (Fig. 17.19b). Some cultivars are grown expressly for scape production; some for ornamental use.

During late summer and early fall, multiple harvests at about 15–20-day intervals, are made of the foliage and/or developed scapes. Leaves and scapes are often blanched by covering with clay tile pipes or in opaque tunnels covered with rice straw mats supported with bamboo stakes. The green, or blanched, etiolated leaves are cut and tied into small bunches for marketing. Scapes are similarly handled; flower buds may or may not be detached from the scapes. A common cropping sequence is to harvest two or three cuttings as green leaves, two or three as blanched leaves, and conclude the annual production by a harvest of scapes with or without flower buds.

Chinese chives have a high respiration rate; to retain quality, a good postharvest handling procedure is necessary.

Crop Improvement

As with many crops, improvements in the productivity and quality of vegetable Alliums will be accomplished largely via advances in plant breeding and cultural management. Tissue culture is useful for haploid breeding or to overcome ploidy difficulties in species such as leek. The use of true seed rather than clonal propagation may provide yet unknown benefits for garlic production. Greater disease resistance and improvement in handling and storage will limit crop losses.

An advantage of crops such as onion and garlic is their relatively long storage ability which allow for year-round supplies. Favorable transport characteristics permit wide distribution and trade.

F_1 hybrid cultivar introductions are widely used and such use should continue to expand. Although generally more productive than open pollinated cultivars, there is a danger of erosion of valuable and adaptive genes of landrace material. Agencies, such as the International Board for Plant Genetic Resources, encourage genetic conservation and exchanges to minimize loss of such germplasm.

Several interesting interspecific hybrids, including onion–garlic and onion–leek, have recently been produced. They may lead to the development of new commodities having different production and flavor characteristics.

SELECTED REFERENCES

Abdalla, A.A., and Mann, L.K. 1963. Bulb development in the onion (*Allium cepa* L.) and the effect of storage temperature on bulb rest. Hilgardia *35,* 85–112.

Bertaud, D.S. 1988. Effects of chilling duration, photoperiod, and temperature on floral initiation and development in sprouted and unsprouted onion bulbs. In Fourth Eucarpia Allium Symposium, Warwick, U.K., pp. 254–261, United Kingdom.

Brewster, J.L. 1977. The physiology of the onion. Hort. Abst. *47,* 102–112.

Brewster, J.L. 1994. Onions and Other Vegetable Alliums. CAB International, Wallingford, U.K.

Brewster, J.L., and Rabinowitch, H.D., eds. 1990. Onions and Allied Crops. Vol. III: Biochemistry, Food Science and Minor Crops. CRC Press, Boca Raton, FL.

Currah, L. 1985. Review of three onion improvement schemes in the tropics. Trop. Agric. (Trinidad) *62,* 131.

Currah, L. 1986. Leek breeding: A review. J. Hort. Sci. *61,* 407–415.

El-Okch, I, Abdel-Kader, A.S., Waiky, J.A., and El-Kholly, A.F. 1971. Comparative effects of gamma irradiation and Maleic Hydrazide on storage of garlic. J. ASHS *96,* 637–640.

Etoh, T., and Nakamura, N. 1988. Comparison of the peroxidase isozymes between fertile and sterile clones of garlic. In Fourth Eucarpia Allium Symposium, Warwick, U.K., pp. 115–121.

Folster, E., and Krug, H. 1989. Influence of the environment on growth and development of chives (*Allium schoenoprasum* L.). II. Breaking of the rest period and forcing. Sci. Hortic. *7,* 213.

Hamon, N.W. 1987. Garlic and the genus *Allium*. Can. Pharmaceut. J. *120,* 340–342, 344.

Jones, H.A., and Clarke, A. 1943. Inheritance of male sterility in the onion and the production of hybrid seed. Proc. ASHS *43,* 189.

Jones, H.A., and Mann, L.K. 1963. Onions and Their Allies: Botany, Cultivation and Utilization. Leonard Hill Books, London Interscience Publ., New York.

Krug, H., and Folster, E. 1976. Influence of the environment on growth and development of chives (*Allium schoenoprasum* L.). I. Induction of the rest period. Sci. Hortic. *4,* 211.

Lancaster, J.E., and Boland, M.J. 1990. Flavor biochemistry. In Onions and Allied Crops Vol. 3. H.D. Rabinowitch and J.L. Brewster, eds. CRC Press, Boca Raton, FL, pp. 33–72.

Mann, L.K., and Stearn, W.T. 1960. Rakkyo or ch'iao t'ou [*Allium chinense* G. Don, Syn. *A. bakeri* (Regal)], a little known vegetable crop. Econ. Bot. *14,* 69–83.

Matsubara, S., and Kimura, L. 1991. Changes in ABA content during bulbing and dormancy and in vitro bulbing in onion plant. J. Japan. Soc. Hort. Sci. *59,* 757–762.

Pike, L.M. 1986. Onion breeding. In Breeding Vegetable Crops, M.J. Bassett, ed. AVI Publ. Co., Westport, CT pp. 357–394.

Pooler, M.R., and Simon, P.W. 1993. Garlic flowering in response to clone, photoperiod, growth temperature and cold storage. HortScience *28,* 1085–1086.

Pooler, M.R., and Simon, P.W. 1994. True seed production in garlic. Sex Plant Reprod. *7,* 282–286.

Rabinowitch, H.D., and Brewster, J.L., eds. 1990. Onions and Allied Crops. Vol. I: Botany, Physiology and Genetics. CRC Press, Boca Raton, FL.

Rabinowitch, H.D., and Brewster, J.L., eds. 1990. Onions and Allied Crops. Vol. II: Agronomy, Biotic Interactions, Pathology and Crop Protection. CRC Press, Boca Raton, FL.

Rahim, M.A., and Fordham, R. 1988. Effect of storage temperature on the initiation and development of garlic cloves (*Allium sativum* L.). Sci. Hort. *37,* 25–38.

Rickard, P.C., and Wickens, R. 1977. Effect of pre-harvest treatments on the yield, storage characteristics and keeping qualities of dry bulb onions. Exp. Hortic. *29,* 52–57.

Saghir, A.R., Mann, L.K., Bernhard, R.A., and Jacobsen, J.V. 1964. Determination of aliphatic, mono- and disulfides of *Allium* by gas chromotography and their distribution in common food species. Proc. ASHS *84,* 386–398.

Walkey, D.G.A., Webb, M.J.W., Bolland, C.J., and Miller, A. 1987. Production of virus-free garlic (*Allium sativum* L.) and shallot (*A. ascalonicum* L.) by meristem tip culture. J. Hort. Sci. *62,* 211–220.

18

Lettuce and Other Composite Vegetables

Family: Asteraceae (Compositae)

The Asteraceae is a widely distributed and very large family of about 800 genera and possibly 20,000 species, of which relatively few are cultivated. The previous family name of Compositae was representative of the composite of many florets typically found in the compact involucrate head. What may appear as a single flower is actually the composite of many individual flowers. Another characteristic of many Asteraceae species is the milklike latex contained in stems and other tissues. Dried latex has some narcotic properties and has been used as a sedative. Lactucarium, the dried latex produced from *Lactuca virosa* has drug use.

Principal Vegetable Species and Common Name

Arctium lappa	Edible burdock
Artemista lactiflora	White mugwort
Chrysanthemum coronarium	Garland chrysanthemum/ shungiku
Cichorium endivia	Endive/escarole
Cichorium intybus	Chicory, leaf and witloof
Crassocephalum crepidodes	Sierra Leone bologi
Cynara cardunculus	Cardoon
Cynara scolymus	Globe artichoke
Gynura bicolor	Gynura
Helianthus tuberosus	Jerusalem artichoke*
Lactuca sativa	Lettuce
Petasites japonicus	Butterbur/fuki
Polymnia edulis	Polymnia*
Polymnia sonchifolia	Yacon*

*See Chapter 14.

Scorzonera hispanica	Scorzonera
Spilanthes oleracea	Paragrass
Taraxacum officinale	Dandelion
Tragopogon porrifolius	Salsify

LETTUCE, *Lactuca sativa* L.

Lettuce is the world's most used salad crop. It is a major and extensively grown cool season vegetable best adapted to temperate locations. In some countries, lettuce consumption is large enough to make a significant nutritional contribution. World lettuce production is estimated at about 3 million tons and grown on more than 300,000 ha.

Origin

Lactuca sativa, the only domesticated *Lactuca* species, is native to the eastern Mediterranean basin. Evidence from ancient Egyptian tomb paintings indicated that a nonheading form was grown as early as 4500 B.C. Its early use probably was medicinal and for edible seed oil. A landrace lettuce, known as USDA Plant Introduction 251245, is used for the seed oil.

Wild types of lettuce often have prickly leaves and stems, are nonheading and bitter tasting, and contain an abundance of latex. Domestication probably emphasized absence of spines, slow bolting, large nonshattering seed, less latex, and less tissue bitterness. Additional changes include: reduced suckering, broad and large leaves, and a heading habit. Heading lettuce is a relatively recent cultivated crop, being first described as cabbage lettuce in 1543.

Taxonomy

Species of *Lactuca* have basic chromosome numbers of 8, 9, and 17, the $n = 17$ species may be natural amphidiploids. *L. sativa* is one of four interfertile species having nine pairs of chromosomes. Others are *L. serriola, L. virosa,* and *L. saligna. L. sativa* is not found wild and is believed to have been derived from *L. serriola,* either alone or in combination with one of the other species. *L. sativa* and *L. serriola* are cross-fertile, crosses are easily made, and progeny share some characteristics. However, genetic barriers often make crossings difficult with other wild types of *Lactuca,* and when achieved, resulting progeny are few and often sterile. Nevertheless, some cultivated lettuces have been derived from *L. sativa* × *L. virosa* and *L. sativa* × *L. saligna* crosses.

Wild-type germplasm will likely continue to be used, especially for the transfer of disease resistances.

L. serriola exhibits a tendency for rosette growth, but is nonheading. Seed are small; leaves are narrow and spiny, with entire or toothed margins. *L. saligna* differs from *L. serriola* in having very narrow toothed leaves and bearing sessile flowers on a spike, compared to pedicelled flowers on panicles. The blue-green-leafed *L. virosa* plants exhibit annual as well as biennial growth habit. Plants show a strong rosette growth habit with panicles closely resembling *L. serriola*.

Botany

Lettuce is an annual polymorphic plant, especially with regard to foliage characteristics. Plants rapidly develop a deep taproot accompanied with thickening and extensive development of largely horizontal lateral roots. Even though the tap root can penetrate to almost 1 m, lateral roots near the surface are responsible for most of the moisture and nutrient absorption.

Leaves, often many and usually nearly sessile, are spirally arranged in a dense rosette. Considerable diversity occurs in color, shape, texture, and leaf margins between different forms. Glabrous leaves can be smooth, savoy, or crumpled. Leaf margins may be lobed, smooth, or finely divided, and leaf colors can vary from light to dark green; some cultivars have red or purple coloration. Interior leaves of leafy cultivars tend to be lighter in color, whereas those of heading types are blanched.

For most lettuce types, except stem lettuce, the cylindrical stem is short and compressed. Upon bolting, the stem elongates, becoming tall and branched. The inflorescence is a dense corymbose panicle of numerous capitula, each consisting of 10–25 florets which are self-pollinating, although insect pollination occasionally occurs. Fertilization occurs in as little as 3–6 h after pollination.

Flowering may continue over a period of 1–2 months. However, all florets in the same flower head open only once for a brief time in the morning. Thus, seed develop simultaneously in the same flower head, each floret producing a single-seeded dry fruit called an achene. Seed are prone to shattering and are small, ribbed, and topped with pappus hair; about 1000 weigh 1 g. Variations in seed color range between white and shades of yellow, brown, gray, and black. Most newly harvested seed exhibit a short postharvest dormancy due to a water-soluble inhibitor that is dissipated with time or leaching. Some cultivars exhibit photochemical dormancy, and all have varying levels of thermodormancy.

Lettuce Morphological Types

Some lettuce types are sufficiently distinctive and have been given botanical variety designations such as the following:

Crisphead and butterhead	*L. sativa* var. *capitata*
Cos (Romaine)	*L. sativa* var. *longifolia*
Loose leaf (Leaf)	*L. sativa* var. *crispa*
Stem	*L. sativa* var. *asparagina*
Latin	*L. sativa*

Four generally recognized morphological forms of lettuce (shown in Fig. 18.1) are crisphead, butterhead, cos, and loose leaf. Stem and Latin lettuce are other forms.

Crisphead forms are commonly referred to as "iceberg" or head lettuce. Following early rosette development, additional leaf growth begins overlapping each other and eventually entrap newly formed young

FIG. 18.1.   Lettuce, *Lactuca sativa*, types: (a) crisphead (iceberg); (b) butterhead; (c) cos (Romaine); (d) loose leaf. Source: (a) Courtesy Frank Zink, (c) Courtesy National Garden Bureau, Downers Grove, Illinois.

leaves. Continued expansion of the entrapped leaves increases head density; heads usually are nearly spherical. Heads can become very firm, and with increasing enlargement, the head may burst. Overmature foliage becomes bitter-tasting. The tightly folded interior leaves are rugose, brittle, and crispy. Outer leaves usually are dark green, inner leaves being progressively lighter in color. When harvested, field-grown plants usually weigh between 700 and 1000 g. Glasshouse-produced lettuces are generally much smaller. Good shipping and shelf-life characteristics are important attributes of crisphead lettuces.

Butterhead cultivars, sometimes called cabbage lettuce, are extensively grown; for many consumers they are preferred because of their delicate flavor and quality. Plants are smaller, slightly flattened, and produce a less compact head than crisphead types; leaves are broad, crumpled, and tender, with a soft oily texture. Two major types are produced, one being day neutral with large fairly firm heads. The other, short-day type, produces small, less firm heads and is commonly grown in protective shelters. Both types are easily bruised, so shipping and shelf life are poor. Batavia-type cultivars have features intermediate between crisphead and butterhead plants. These cultivars are grown outdoors as well as in protective structures.

Cos cultivars, also known as romaine, have elongated, coarse, crispy textured leaves with prominent broad midveins. The long relatively narrow leaves tend to grow upright and may loosely overlap each other, but do not form a head. Postharvest features are similar to crisphead types.

Among loose-leaf cultivars, considerable variation in leaf size, margins, color and texture occur. Each develop-leaves as a tight rosette cluster. Some have crispy tender foliage, some have smooth leaves, and others are intermediate. Postharvest handling is more critical with these lettuces because of leaf tenderness; the shelf life, while better than butterheads, is fairly short, even with refrigeration and good handling.

Stem lettuce is also known as celtuce or asparagus lettuce (Fig. 18.2). Plants are nonheading and are produced mainly for the erect thickened edible stem which may be as long as 30–40 cm. Before consumption, stems are peeled and the soft translucent green core is used as a raw or cooked vegetable; uncooked, the taste resembles cucumber. Except for the youngest foliage, leaves are not palatable because of their high latex content and bitterness. Stem lettuce is popular in China and Egypt, but elsewhere production is rather limited.

Latin lettuce cultivars generally produce a rosette of loose, elongated, and soft cos-like leaves that occasionally form a partly closed head. Leaves somewhat resemble butterhead leaf texture and are shorter

FIG. 18.2. Stem lettuce, *Lactuca sativa* var. *asparagina,* a popular vegetable in China.

than those of cos-type lettuces. This crop has some popularity in Mediterranean regions and to some extent also in Argentina and Chile.

Indian lettuce, *L. indica,* is native to Asia, domesticated and grown in China, Japan, and several Southeast Asian countries. Plants are perennial and produce a tall (1 m) tillering growth. Partial harvesting of the radial rosette of leaves begins about 60 days after planting and harvests generally continue until flowering; flowers are yellow. The succulent leaves are sessile and oblong to lanceolate in shape. They are used either raw or cooked for salads. Plants with anthocyanin-containing foliage are also grown.

Climatic Requirements

Moderate temperatures are ideal for high-quality lettuce production; a regime of 20°C days and 10°C nights is optimum. Temperatures greater than 30°C usually stunt growth, are conducive to bolting, and result in bitterness and loose head formation in heading types. Leafy-type lettuces generally are better adapted to a wider range of temperatures than heading types.

Crop maturation is strongly temperature dependent and harvests can occur in as few as 60 days with warm weather, whereas winter-grown crisphead cultivars may require more than 120 days. Nonheading cultivars mature more rapidly than crisphead lettuces. When properly temperature acclimated, some cultivars tolerate light frost, although frost may cause surface blistering of epidermal tissues of the outermost leaves. Seedlings are more tolerant of low temperature than mature plants.

Generally, high light intensity and long days increase growth rate and hasten leaf area development by producing broader leaves, resulting in advanced head formation. However, under long days, some lettuce cultivars are induced to bolt; the tendency is strongly accelerated by high temperatures. Cultivars vary as to day length sensitivity, but through selective breeding, most present-day cultivars are day neutral.

Soils, Moisture, Nutrition

Lettuces are grown in a wide range of soil types. Soils having good moisture retention qualities with adequate drainage, such as sandy loams or organic soils, are preferred. Lettuce is sensitive to soil compaction and acidity. In mineral soils, pH should be above 5.5; a range from 6 to 8 is most satisfactory. Seedlings have a low tolerance to salinity, whereas older plants are more tolerant.

Uniform and well-formed smooth seed beds, whether flat or raised, are important for uniform seedling establishment and growth. Raised beds, commonly 15–25 cm in height, provide drainage and aeration that help reduce disease incidences. Soil crusting interferes with seedling emergence and thereby affects uniformity and plant populations. Crust formation can be minimized by irrigations that do not exceed infiltration rates. Phosphoric or sulfuric acid applications to the seed line are sometimes used to prevent soil crusting. Shallow cultivations are also made to break crusts. Mulches help in preventing crust formation.

The relatively shallow lateral root system makes lettuce plants susceptible to moisture stress. Growth is optimized by a uniform moisture supply, and prolonged soil saturation should be avoided. About 400 mm of water, well distributed during growth, is usually adequate for most lettuce crop production. Supplemental irrigation is relied upon to meet plant needs in arid production areas and in other locations when rainfall is inadequate. Irrigation methods include flooding, subbing, furrow, sprinkle, and drip. The use of drip irrigation has markedly increased because it has been shown to significantly improve water use efficiency and productivity.

The relatively small and shallow root system also requires an easily accessible nutrient supply. Timing of fertilizer applications is very important to obtain efficient nutrient use. The nitrogen source is important, and a mixture consisting of nitrate and ammoniacal nitrogen is considered to be better than either alone because the mixture provides both rapid and sustained growth. Phosphorus is very important for early plant development. Because about 75% of plant biomass is produced in the last few weeks before harvest, the supplied fertilizer is more effective when most of it is applied prior to the accelerated growth. Plant nutrients introduced via drip irrigation systems permit frequent low rate applications and result in efficient and effective fertilizer use.

Cultural Practices

Lettuce seed is used directly for field sowing or for the production of transplants. Germination is optimum at 24–25°C; the rate of germination diminishes with lower temperatures, down to about 2°C. Between 25°C and 30°C germination is not reliable because of high-temperature-induced dormancy. Thermodormancy can be avoided by not sowing during high temperatures. Temperatures near the seed zone can be lowered by evaporative cooling from moist soil surfaces. Scheduling irrigations late in the day or during the night when temperatures are lower is also effective. The critical point is to avoid seed imbibition during the high-temperature period. Gibberellic acid seed treatment and pregermination are other alternatives that can be used. Seed priming treatments with polyethylene glycol (PEG) or other osmoticants are also effective in reducing susceptibility to thermodormancy. Seed, primed in an aerated 1% solution of K_3PO_4 for 6–9 h, exhibit improved high-temperature germination compared to nonprimed seed; germination sometimes can be achieved even at 35°C. Seed priming is also used to improve uniformity of germination and can be combined with light exposure that is required for germination of some lettuce cultivars (see Chapter 6).

Seed are generally planted shallow to ensure that oxygen and light will not be limiting. Precision planters that provide accurate seed placement are increasingly used. Being small and irregularly shaped, lettuce seed are difficult to singularize for use in precision seeders. Therefore, to facilitate handling, the seed are coated to form pellets of uniform size and shape. Seed coating materials commonly are combinations of either clay, diatomaceous earth, talc, or similar products with binders applied to form a uniform layer of coating material around individual seed. Other materials and treatments may be included prior to preparing pelleted seed. These can include priming or other proprietary germination enhancing treatments and seed protectant products. A minimum thickness of the coating is usually preferred because that allows for rapid moisture imbibition and better oxygen access. Pregerminated seed (radicle not protruding) can be pelleted but must be planted without delay.

Precision planters utilize various mechanisms to separate and individually sow either pelleted or untreated seed. Many precision planters are capable of sowing more than 25 seed per second.

Even with precision planters, an excess of seed is generally sown to assure adequate stand establishment. Surplus seedlings are later thinned to provide the desired population. Sowing at 5-cm intervals in the row requires about 375,000 seed per hectare. Final plant spacings range from 25 to 40 cm in the row and from 40 to 75 cm between rows.

A common practice in the western United States lettuce production areas is the planting of twin rows on the level surface of a raised bed. The height of the bed is between 15 and 25 cm, and the distance from the center of one bed to the adjoining bed is 1 m. A seed row is planted near each edge (shoulder) of the bed; the twin rows are about 30 or 35 cm apart (Fig. 18.3). Final plant population usually ranges from 60 thousand to 75 thousand per hectare. Lettuce, especially crisphead types, if spaced too closely may become misshapen and lose market value. Therefore, precise spacings during seeding and after thinning are important. Thinning is usually done by hand and often combined with hoeing for weed control; machine thinning is infrequently prac-

FIG. 18.3. Typical twin-row raised bed culture of head lettuce in the southwestern United States.

ticed. Depending on seedling development, thinning is done as soon as 3 weeks after sowing and is seldom delayed beyond 6 weeks.

Transplanting is a common practice for many lettuce producers. Improved technology and equipment have greatly increased efficiencies in providing uniform, high quality, and rapid establishment of transplants. Transplants, often produced by specialized growers, are usually between 4 and 6 weeks old when planted. If too young, they are difficult to handle, whereas older plants are easier to handle but are slower in resuming growth. Transplanting machines, some highly automated, are used in many countries. Nevertheless, considerable quantities of lettuce continue to be hand transplanted.

Diseases and Pests of Lettuce and Other Asteraceae Vegetables

Bacterial

Bacterial leaf spot	*Xanthomonas campestris* pv. *vitians*
Corky root	*Rhizomonas suberifaciens*
Soft rots	*Erwinia carotovora* subsp. *carotovora, Pseudomonas marginalis, P. viridilivida*
Varnish spot	*Pseudomonas cichorii*

Fungal

Anthracnose	*Microdochium panattonianum (Marssoninia panattoniana)*
Artichoke black rot	*Ascochyta cynarae*
Artichoke powdery mildew	*Leveillula taurica*
Artichoke ramularia	*Ramularia cynarae*
Artichoke root rot	*Rosellinia necatrix*
Botrytis rot/Gray mold	*Botrytis cinerea*
Bottom rot	*Rhizoctonia solani*
Cercospora leaf spot	*Cercospora longissima*
Damping off	*Pythium* spp., *Rhizoctonia solani*
Downy mildew	*Bremia lactucae*
Fusarium wilt	*Fusarium* spp.
Jerusalem artichoke tuber rot	*Fusarium* spp.
Jerusalem rust	*Puccinia helianthi*
Lettuce drop	*Sclerotinia minor, S. sclerotiorum*
Powdery mildew	*Erysiphe cichoracearum*

Rust	*Puccinia* spp.
Salsify leaf spot	*Alternaria tenuis*
Salsify white rust	*Albugo tragopogonis*
Septoria leaf spot	*Septoria lactucae*
Stemphilium leaf spot	*Stemphilium botryosum* f. sp. *lactcae. Alternaria porri* f. sp. *cichorii*

Viruses

Artichoke Mottle Crinkle Virus (AMCV)
Artichoke Curly Dwarf Virus (ACDV)
Beet Western Yellows Virus (BWYV)
Beet Yellow Stunt Virus (BYSV)
Bidens Mottle Virus
Big Vein Virus (assumed a virus)
Broad Bean Wilt Virus (BBWV)
Cucumber Mosaic Virus (CMV)
Lettuce Infectious Yellows Virus (LIYV)
Lettuce Mosaic Virus (LMV)
Lettuce Necrotic Yellows Virus (LNYV)
Sow Thistle Yellow Vein Virus (SYVV)
Tomato Spotted Wilt Virus (TSWV)
Turnip Mosaic Virus (TuMV)

Aster yellows is a phytoplasm (mycoplasma) transmitted by the six-spotted leafhopper, *Macrosteles fascifrons*. The big vein viruslike agent is transmitted by the fungus, *Olpidium brassicae*.

Production and certification of seed that is free of lettuce mosaic virus (LMV) greatly limits potential disease and crop losses. Methods for the detection of LMV infected seed from sampled seed lots include (1) leaf symptom expression within the population of 30,000 seedlings grown from a sample of the seed lot, (2) the reaction of LMV-sensitive plants to extracts of ground-up seed obtained from a 30,000-seed sample of the seed lot, and (3) a less intensive method tests the seed sample for antibody response in an enzyme-linked immunosorbent assay (ELISA).

Resistance has been incorporated into many cultivars for diseases such as downy mildew, LMV, big vein, corky root, and anthracnose.

Physiological Disorders and Associated Factors

Tipburn is associated with rapid growth and restricted transpiration during abrupt temperature increase, and reduced calcium transport to rapidly growing new tissues. The disorder tends to be more common in greenhouse production.

Russet spotting is a result of ethylene exposure.

Bolting is usually caused by high temperatures during growth, some cultivars bolt following prolonged long-day exposure.

Pink rib occurs more frequently with older plants and is accentuated by high transit and storage temperatures and low O_2; specific causes are uncertain.

Rib blight—cause unknown.

Brown stain is associated with interaction of high CO_2, low O_2, and certain sensitive cultivars.

Internal rib necrosis is associated with the interaction of LMV infection and certain cultivars.

Glassiness can occur when restricted transpiration occurs under conditions of high humidity and high soil moisture.

Cultivars vary in their tolerance to the above disorders.

Nematodes

Needle	*Longidorus* spp.
Root knot	*Meloidogyne* spp.
Stem and bulb	*Ditylenchus* spp.
Stubby root	*Paratrichodorus* spp.

Insects and mites

Alfalfa looper	*Autographa californica*
Armyworm	*Pseudaletia unipuncta*
Artichoke plume moth	*Platyptilia carduidactyla*
Artichoke thistle aphid	*Brachycaudus cardui*
Aster leafhopper	*Macrosteles fascifrons*
Beet armyworm	*Spodoptera exigua*
Black cutworm	*Agrotis ipsilon*
Cabbage looper	*Trichoplusia ni*
Chrysanthemum leafminer	*Phytomyza syngenesiae*
Chrysanthemum root fly	*Psila nigricornis*
Corn ear worm	*Heliothis zea*
Cribrate weevil	*Otiorhychus cribricollis*
Cricket	*Gryllus* spp.
Cucumber beetle	*Diabrotica undecimpunctata*
Darkling beetles	*Blapstinus* spp.
Flea beetles	*Phyllotreta* spp.
Garden symphylan	*Scutigerella immaculata*
Greenhouse whitefly	*Trialeurodes vaporariorum*
Green peach aphid	*Myzus persicae*
Granulate cutworm	*Feltia subterranea*

Leafminers	*Liriomyza* spp.
Lettuce aphid	*Nasonoria ribisnigi*
Lettuce root aphid	*Pemphigus bursarius*
Lygus bug	*Lygus* spp.
Painted lady moth	*Vanessa cardui*
Potato aphid	*Macrosiphum euphorbiae*
Saltmarsh caterpillar	*Estigmene acrea*
Silverleaf whitefly	*Bemisia argentifolii*
Springtails	*Papirivs maculosus*
Sweet potato whitefly	*Bemisia tabaci*
Thrips	*Thrips tabaci*
Tobacco ear worm	*Heliothis virescens*
Two-spotted mite	*Tetranychus urticae*
Variegated cutworm	*Peridroma saucia*
Western flower thrips	*Franklinella occidentalis*
Wireworms	*Limonius* spp.
Yellow striped armyworm	*Spodoptera ornithogalli*

Birds occasionally become a serious impediment to establishing plant stands. Rodents, snails, namely the brown garden snail *(Helix aspera)* and slugs such as the gray garden slug *(Agriolimax reticulatus),* black European slug *(Arion ater),* and also *Milax gagates* and *Deroceras reticulatum* also damage seedlings.

Lettuce plants are poor competitors with most weeds and unless controlled, yields are reduced. Several effective selective herbicides are widely used; even with such use, some hand hoeing is often required.

Harvesting

Lettuce vegetative growth and development is highly responsive to temperature, and growth rate increases with increasing temperatures. Depending on lettuce type and cultivar, the time from planting to harvest varies greatly; judgment of harvest maturity is, therefore, somewhat subjective. Premature harvest results in low yield, and late harvest results in reduced quality. Plants often do not develop uniformly, and multiple harvests may be required. A major goal for many commercial producers is to obtain a single-cut harvest.

Lettuce is hand harvested and packaged. Large-scale producers use skilled harvest crews whose experience and organization make harvesting rapid and efficient. Fully mechanized field harvesting has been thoroughly investigated in the United States and elsewhere. At this time, the economic feasibility has not been demonstrated. The need for selective harvesting with minimal damage and the high cost of developing and manufacturing such harvest equipment are limitations for full mechanization. The hand-harvesting procedure consists of identifying well-developed plants, cutting through the stem at ground level, trimming senescent and unusable leaves, and packing into containers (Fig. 18.4). Direct field packing is a common procedure in the United States. However, many variations of labor-assist equipment, such as conveyors and mobile packing machines are used. Presently, in major U.S. lettuce-production areas, compact mobile packing machines are also used. These units facilitate film wrapping or bagging of trimmed lettuce heads and are also used for traditional nonwrapped lettuce packing into cardboard containers. Field packing has several advantages versus shed packing, namely less handling, reduced damage, minimized waste disposal, and not having to acquire and maintain permanent packing facilities.

Lettuce harvested for processing are placed into large bulk container

FIG. 18.4.   Hand harvest and field packing of head lettuce in central coastal California.

of about 1 m^3 for transport to processing facilities. At such facilities, the lettuce is (lightly processed) by being precut or shredded, washed and centrifuged dry for marketing as a "value-added" product. Such lettuce, packaged alone or mixed with other salad items, has exhibited substantial market expansion. Lettuce produced for value-added processing has less restrictive size or head shape requirements. Thus nonselective "cut all" one-time harvests are common, and for this purpose, mechanized harvest equipment may become feasible.

Because of the many types of lettuces grown and harvested at different maturity states, it is difficult to generalize about yields. Thus, yields range from 50 to a potential of more than 70 t/ha.

Postharvest

To retain high field quality, lettuce should be quickly transported from the field to cooling facilities. For long shelf life, cooling to remove field heat and reduce respiration is important. Vacuum cooling is an excellent method for quickly lowering product temperature. Its introduction was largely responsible for permitting in-field packing in place of shed packing and icing practices. Where vacuum-cooling facilities are not available, forced-cold-air cooling, although less rapid, is effective. Hydrocooling is effective for nonheading lettuces but is not used for head lettuce. Reliance on good postharvest handling and temperature management throughout marketing is critical for extending shelf life.

Lettuce held at temperatures of 1–2°C and high humidity (90–95% RH) can be maintained in good condition for 2–3 weeks. Ethylene exposure should be avoided at all times, as even minute quantities can cause early senescence, russet spotting, and significant quality loss.

Uses and Composition

Most lettuces are eaten uncooked and are popular salad vegetables because of their color, texture, and flavor contribution to meals. Often the major component of vegetable salads, lettuce has a high water and low carbohydrate and protein content. Nevertheless, because of the volume annually consumed (about 12 kg per capita in the United States) lettuce is recognized for its nutritional contribution of minerals, pro-vitamin A, vitamin C, fiber, and bulk. Cos- and leaf-type lettuces provide more pro-vitamin A because of a greater proportion of green leaves compared to crisphead types.

Protected Culture

A significant volume of lettuce is produced in protective structures such as glasshouses and plastic tunnels (Fig. 18.5). Such structures

FIG. 18.5. Glasshouse production of butterhead lettuce near Versailles, France.

provide environment management and are especially useful during periods when outdoor growth is limited by low temperatures. With a managed temperature growth environment, opportunities for off-season production are increased. However, that advantage is partly offset by high production costs attributed mainly to the structure and heating energy.

A suggested temperature regime for lettuce early seedling growth is 13°C night and 16°C day. When temperatures exceed 21°C, structures usually are ventilated. During the rosette stage, a 10°C night and 13°C day regime is recommended with ventilation when temperatures are above 18°C. To minimize heating costs, a lower-temperature regime can be 7°C night and 13°C day with ventilation at 16°C or above. In addition to directly increasing air temperatures within the structure, soil heating with underground hot-water pipes is sometimes practiced. In hydroponic production, heating the circulating nutrient solutions helps to advance growth.

Light is a commonly limiting factor in most protective production. Unfortunately, light supplementation is often more expensive than supplying heat. However, when light is adequate, additional heating and CO_2 enrichment can be provided to further increase growth rate. Some lettuce cultivars have been bred specifically for good performance in low light as well as low-temperature conditions.

Sheltered lettuce production relies on transplant usage, because it is not economical to use the expensive space in protective structures for starting plants when transplants can be produced more efficiently elsewhere. Specialized rooms that provide optimum light and growing conditions and accommodate a high population of seedlings are sometimes used.

Transplants are usually set out after 4 weeks of development. Plants are narrowly spaced in order to maximize use of available space and commonly average about 20 cm between plants. They can be planted directly into the soil, which is usually sterilized to destroy disease, pests, and weeds. Sterilization is achieved with chemical fumigants or steam. Transplants are also planted and grown in soil bags, rockwool slabs, and straw bales, and also used in various modification of hydroponic procedures that include gravel and nutrient film technique (NFT) culture.

Pest and disease management is a major concern for protective cultivation; good sanitation and other problem-preventative practices are required for successful production. The integration of biological and chemical pesticide practices along with nutrient management is critical for effective production. This is especially important for hydroponic or other nonsoil cultivation because contamination can occur more readily in these systems. Other cultural procedures that provide proper moisture application, relative humidity management, and appropriate ventilation help to minimize problems.

Harvest maturity varies with cultivar and seasons and may range from as few as 30 days after transplanting to as long as 85 days. Plant weights also vary greatly, ranging from 100 to 400 g. Continuing growth to achieve higher plant weight becomes progressively less profitable, and usually it is more expedient to begin a new cropping cycle. Lettuce types most frequently grown in protective structures are butterhead cultivars, although leaf, crisphead, and batavia types are also produced.

ENDIVE, *Cichorium endivia* L.
CHICORY, *Cichorium intybus* L.

Endive and chicory are native to the Mediterranean region, and it is known that they were cultivated in ancient Egypt. Except for root chicories, cultivars of each species are grown for the foliage used in salads, and as cooked potherbs. Foliage, both green and red, is strongly flavored and exhibit different levels of bitterness that are savory and delicate to some, and harsh and unpalatable to others. Old and dark-colored leaves are more bitter than young or pale foliage. Blanching is commonly practiced to reduce bitterness and to increase leaf tenderness.

Endives are hardy annual or biennial plants, initially producing a dense rosette of leaves on a short compressed stem. The foliage resembles that of some loose-leaf lettuces, but tends to be prostrate. Two main cultivar types are grown. The escarole type produces broad coarse crumpled leaves, whereas the endive type produces narrow, deeply cut, curled leaves (Fig. 18.6). Upon bolting, the stem elongates and branches. The inflorescence bears many capitula, usually having pale blue, but sometimes white, flowers. Flowers are self-pollinating, and each floret produces an individual seed; about 800 seed weigh 1 g.

Favorable growing temperatures are between 15°C and 18°C with harvest maturity resulting about 70–100 days after planting. Endives are slightly more heat tolerant than chicories, and both are more tolerant of extreme temperatures than lettuce. High temperatures and long days are conducive to bolting for annual cultivars; high light intensity tends to accelerate bolting. Vernalization is required for biennials. Plants are frequently blanched 1–3 weeks before harvest by tying leaves together or using various coverings to exclude light. Most cultural, spacing, harvest, and handling practices are similar to those for lettuce.

Chicories are deep-rooted perennial plants grown as annuals or bien-

FIG. 18.6. Finely cut/curled leaves of one form of endive (left) and a heading form of chicory known as radicchio (right).

nials; the latter are harvested with the intention for forcing new growth at a later time.

Initial growth results in the production of a dense rosette of leaves. Leaf forms exhibit greater variation than those of endive. Principal cultivar types include leafy salad, forcing, and root types. The array of leafy salad types is numerous. They vary from nonheading "radichetta" and "asparagus" types with narrow, deeply notched leaves, to heading broad relatively smooth leaf "sugar loaf" and cabbagelike "radicchio" types (Fig. 18.6). Foliage colors range from light green to dark purple. In general, smooth leaf forms are usually cooked; the crispy leaf types usually are eaten fresh in salads. Chicory cultivars like those of endive are harvested about 60–90 days after planting. The postharvest life of both chicory and endive is about 2 weeks when held at 1°C and 95% RH.

Some chicories are grown for the high carbohydrate content of the fleshy enlarged root. Root chicory foliage is usually considered too coarse for salad use. Roots are processed after drying for use as a coffee adulterant or substitute. Magdeburg and Brunswick are major root chicory cultivars. The high content of fermentable carbohydrates, as much as 20% soluble solids in some root chicory cultivars, is of interest for fuel alcohol production. Production practices for root chicories are similar to those used for sugar beets, whereas the production of leafy salad types is similar to that for lettuce.

After vernalization, multiple-branched seed stems produce light blue flowers that are cross-pollinated by insects. Seed have the same appearance and size as those of endive.

Witloof Chicory

An important forcing type of chicory is known as witloof (a Flemish word for white leaf), also as Belgian or French endive. Second-year growth of this biennial is forced in order to produce the enlargement of the apical bud situated on the highly compressed stem. The bud is harvested before significant stem elongation or loss of compaction occurs. In France, the enlarged apical bud is called a *chicon* (Fig. 18.7a). The harvested chicons are eaten raw in salads and as cooked vegetables. Chicons range in length from 10 to 15 cm and in width from 4 to 7 cm.

Uniform storage roots are necessary to effectively produce the forced crop. The first-year growth in the field produces a rosette of foliage and a storage taproot. Good storage root development is necessary for adequate accumulation of food reserves for the subsequent forcing crop. At the conclusion of seasonal growth, usually before the plants are uprooted, the foliage is mowed or hand cut 2–3 cm above the growing point (apical meristem). In some operations, mechanical harvesting

FIG. 18.7. Witloof chicory, *Cichorim intybus:* (a) chicons packaged for marketing in the Netherlands, (b) field forcing near Cambari, France, (c) hydroponic forcing in France.

equipment is used to undercut and lift out roots while cutting the foliage. Plants that exhibit bolting are discarded. Excessive foliage growth, whether due to high fertility or other causes, should be avoided. Production of uniform, well-sized roots and the identification of root maturity relative to forcing performance are the most difficult aspects of forced chicory production.

After harvest, roots are graded and trimmed. This operation can be performed before or after roots are placed into storage; before is preferable, to avoid storage of unusable roots. Optimum root lengths are between 15 and 18 cm with crown diameters of 3–5 cm. With larger-diameter roots, axillary buds tend to produce suckers that compete with apical buds, and marketable chicons are not obtained. Undersized roots lack sufficient carbohydrate reserves to produce chicons large enough for marketing. Unbranched taproots are preferred because they are easier to handle.

Roots are stored at 0°C with high relative humidity to achieve vernalization and to maintain root turgor and quality by minimizing respiration and desiccation. The storage roots should not be allowed to wilt; those with some attached soil actually store better. Cultivars vary with regard to vernalization requirements and the length of appropriate low-temperature storage can vary from a few weeks to several months. For forcing, roots are removed from storage and placed into a warm environment that supports new feeder root initiation and chicon growth.

Witloof Forcing Practices

Soil-Grown Culture   Previously, the procedure for forcing involved digging a long, wide, shallow trench in the soil. Into the trench a layer of fermenting manure is placed and then covered with a layer of soil. The storage roots are then placed on top. Prior to placement, roots are trimmed to a uniform length and then placed in an upright position against each other, so that root crowns are at the same level. The stacked roots are covered with about 20 cm of soil into which the enlarging chicons can expand. Heat from the fermenting manure and supplied moisture promote the growth of secondary roots and the apical bud.

An alternative method of bottom heating is provided by warm water or air through pipes beneath the soil. The forcing trench usually is outdoors but could be in a shed (Fig. 18.7b). For either method, the growing area is insulated to minimize temperature fluctuations.

Soil cover is used to exclude light and to maintain compactness of expanding chicons. Other covering materials such as heavy carpets are used. Their advantage is that they do not contaminate the chicons with soil and require less labor.

The harvest procedure for traditional cultivation involves removing the soil or other covering. The roots are lifted from the soil and the chicons easily hand snapped from the storage roots. Chicons are trimmed, sized, and quality graded and packaged.

Hydroponic (Soilless) Culture   During the late 1960s, plant breeders introduced cultivars that produced compact chicons without the required soil cover compression. This freed growers from trench and soil cover culture. Roots could be placed in trays for forcing, and trays could be stacked above each other, effectively multiplying productive surface areas (Fig. 18.7c). Use of hydroponic procedures eliminates soil contamination of the growing chicon. The subsequent introduction of hybrid cultivars further improved crop uniformity, quality, and yields.

Most current witloof production is performed in forcing structures that: exclude light, provide heating and/or cooling, allow ventilation, and are designed for hydroponic culture. Hydroponic tray cultivation procedures also rely on close placement of storage roots. Roots are placed into shallow trays that are about 20 cm deep and usually provide a surface area of about 1.2 m² and are handled and moved as necessary with forklift equipment. Trays are stacked upon each other, but with space allowed between for the expanding chicons. Water, often with added plant nutrients, is supplied by continuous recirculation from tray to tray or by periodic application and drainage with replenishment when necessary. Aeration of the water in contact with the roots is important for healthy and productive growth.

A very important aspect of forcing is temperature management. Growth will occur at temperatures as low as 5°C, but very slowly. Therefore, most indoor production is carried out with air temperatures of about 15°C. Ideally, root temperature is maintained about two or three degrees above ambient air temperatures, which is achieved by heating the hydroponic solution. High temperature results in the development of less compact and more elongated chicons, whereas low temperature has an opposite effect. Depending on temperature and cultivar characteristics, the period from initiation of forcing to harvest may require as little as 20 days but can extend beyond 30 days. Because of better temperature control, hydroponic production tends to be more predictable with regard to forcing and harvest schedules.

When sufficient size is achieved, storage roots and the attached chicons are removed from the trays. The chicon is snapped from the root, trimmed of unusable leaves, graded for size and quality, and packaged. Quality is judged by size, compactness, and shape. In the absence of light, the chicons are blanched to a white or light cream color. The presence of green foliage is considered a defect and such leaves are removed. A few cultivars are grown that produce red (anthocyanin) chicons. Yields ranging from 40 to 80 kg can be obtained from 100 kg of forced roots. Individual chicons weigh 60–80 g. Postharvest shelf life is best at 0°C, 95% RH, and in the absence of light.

GLOBE ARTICHOKE, *Cynara scolymus* L.

Origin

Globe artichokes are thistlelike herbaceous perennials grown mainly for the edible bracts and receptacle of immature *capitula,* commonly called "buds." Fleshy petioles and offshoots are also eaten. The crop is an important vegetable in the Mediterranean basin and north African countries, where it originated and where its wild relatives, *C. syriaca, C. sibthorpiana,* and *C. cornigera,* are found. The word artichoke is derived from the Arabic word meaning "barbs of the earth."

Botany

Vegetative growth consists of a rosette of large, deeply lobed or divided pubescent grayish green leaves attached to a compressed stem (Fig. 18.8). When cultivated as a perennial, plant renewal occurs each growing season from offshoots. These arise as offshoot (sucker) growth from axillary buds at the base of the stem and sometimes from slightly below the soil surface; each is capable of developing its own adventitious roots. Initially, the root system is mostly fibrous, but during the first year of growth, the largest roots thicken and become more of a fleshy storage organ. Seed-propagated plants develop a conspicuous, thick, fleshy taproot.

Axillary buds are formed which are the progenitors of offshoots for the following growing season. Because of apical dominance, not all of the many axillary buds at the stem base develop offshoots, and if too

FIG. 18.8.   Field production of globe artichokes, *Cynara scolymus,* near Castroville, California.

many appear, the excess offshoots are removed. Each vegetative off-shoot also has the capability of producing a flower-bearing stem. Floral stem induction is influenced by temperature and photoperiod. Cultivars differ in their low-temperature and day length requirements; while some require long days to initiate flowering, others appear day length neutral.

Flower primordia are formed at the apex of the main stem and branched axillary stems. Following initiation, the floral stem elongates rapidly and the apical or primary bud is the first to appear above the rosette of leaves. The flower-bearing stem is erect and can grow to a height well above 1 m. Secondary, tertiary, and higher-order buds develop on branching stems arising from leaf axis of the primary stem. The primary terminal bud achieves the largest size, with size decreasing sequentially for secondary, tertiary, and higher-order flower buds. The terminal bud exhibits apical dominance over subsequent bud development, which is diminished when the bud is removed. Secondary buds in turn have dominance over higher-order buds. The results of nonuniform bud development are protracted harvests.

The flower bud is comprised of many florets crowded onto a fleshy receptacle and surrounded by a whorl of multiple rows of bracts. The basal portions of the bracts are thick and fleshy and upper portions are progressively thinner. Outermost bracts are the largest, most fibrous, and least palatable. Inner bracts progressively decrease in size but increase in tenderness. The inner tender bracts and tender portions of the receptacle is sometimes called the "heart." When flower bud growth is continued, the many florets at the upper surface of the receptacle will further develop. The stage of development is conspicuous by the increasing length of *pappus* hairs. Well-developed florets have a slender corolla tube through which the style grows and the stigma protrudes. Corolla colors are usually blue or reddish purple. Flowering behavior is protandrous, thus favoring cross-pollination, as stigma surfaces are not receptive at the time anthers shed pollen. Seed are thick achenes topped with a prominent pappus, which collectively are sometimes referred to as the "choke." About 20 seed weigh 1 g.

Cultivars and clones are mostly identified by the shape of the flower buds and bracts, and also color and presence of spines. Shapes are conical, elongated, globed, or flattened spheres. Bracts may be wide, narrow, long or short, and thin or fleshy. Bud and bract colors are solid or mixed shades of green or reddish purple. Depending on cultivar or clone, bract tips may or may not have a spine; spines vary from minute to threatening. Spines can also appear on stems and leaves.

Cultural Requirements

Globe artichoke plants are adapted to a wide temperature range for growth, but day temperatures of 15–18°C and night temperatures of 10–12°C are optimum for quality flower bud development. Locations providing cool to moderate day and low night temperatures extend the period of floral induction and thereby lengthen the production period. Along with cool growing temperatures, high relative humidity enhances bud quality. Plants are tolerant of high temperatures above 30°C. However, high temperatures tend to decrease edible flower bud quality because of the rapid development of floral organs and fiber in bract and other tissues. Growth is slow at temperatures less than 10°C, and all tissues are frost sensitive, especially leaves and buds. Accordingly, production sites are often located near large bodies of water that provide temperature moderation to limit or prevent frost.

Artichokes are more productive in loams or clay loams than in sandy coarse textured soils. A preferred pH range is between 6 and 8; plants are moderately sensitive to salinity. Although somewhat drought tolerant, artichokes are intolerant of prolonged waterlogging. For the perennial cultivation method, seasonal moisture consumption is about 800–900 mm. Relatively little product biomass in the form of immature flower buds is removed during harvest, with the bulk of vegetative growth left or returned to the field. As a result of such nutrient recycling, supplemental nutrient demands usually are not high.

Propagation

Vegetative propagation is used extensively because seed propagation results in undesirable plant variability and performance. Plant breeders have improved the quality and performance of some seed-propagated cultivars, and it is likely that the exclusiveness of vegetative propagation will greatly diminish. Materials used for vegetative propagation include divisions (rooted basal stem portions), suckers (rooted offshoots), and ovoli (enlarged detached axillary buds). Propagating materials are generally obtained from established plantings, often from those destined for replacement because of declining or low productivity and disease. Rather than starting new plantings with such material, it is prudent to use uniform and healthy propagating material, especially when fields are intended for 5 or more years of cropping. Tissue culture procedures are occasionally used for nursery production of disease-free plantlets. Improved production and uniformity of seed-produced plants increase the feasibility of annual cultivation. Seed propagation can reduce the cost of high plant populations and also provide initially disease-free plants.

Culture and Harvest

Most plant spacings are either 1×1.2 m or equidistant at 1 m apart and provide about 7000–10,000 plants per hectare. In the United States, wider spacings result in populations from 2500 to 3500 plants per hectare. Wider spacings are used to produce large-sized buds, a premium in U.S. markets. Although some European markets also prefer large buds, many prefer buds that weigh from 100 to 200 g, in contrast to those of 300 g and higher.

In established plantings, a number of offshoots will develop at the site of the initial plant, with each offshoot essentially an independent plant. The initial field spacings influence the number of offshoots formed and how many are permitted to develop. In European cultivation, a plant site may support between one and two flowering stems at different stages of development, whereas in the United States, five or six flower stems are allowed to develop at a plant site. As growth proceeds, buds are harvested and the spent senescent stems are replaced with newly developing flower-bearing stems. Growers sometimes physically remove the flower stem after bud harvest because it is believed that removal stimulates offshoot and flower stem development. In addition to the terminal bud, each floral stem usually has the potential to produce two to five secondary, five to eight tertiary, and many fourth-order flower buds.

However, not all the buds produced are harvested or marketable. In general, the total number of buds harvested per plant ranges from 5 to 8. Failure to achieve suitable fresh market size is usually the reason many late-developing buds are not harvested; off-types, disease, and pest damage are other reasons. Nevertheless, small buds are not always a production loss, because they are often used for processing; sometimes, these are harvested preferentially for processing. Whereas all sizes of artichoke buds can be used for processing, small buds yield proportionally more edible product than large buds.

Flower bud production can occur throughout the year, but can be managed to adapt to climatic conditions and/or market needs. A common procedure for scheduling production is to prevent or retard growth until desired. A method for managing growth is that of "cut-back," the removal of aboveground vegetation. Plants are cut back to about 5 cm below soil level. The cut foliage is finely chopped and incorporated or buried in the field to reduce insect infestations. Sometimes it is burned or removed for animal feeding. The cut-back plants are not killed, but the practice stimulates development of new shoots which emerge soon after growth resumes. How rapidly growth resumes can be managed by withholding or providing moisture. Determining when plant cut-back occurs and when moisture is supplied enables producers to schedule production periods.

Applications of 25–50 ppm of gibberellic acid (GA_3 and/or GA_{4+7}) to the foliage can accelerate flower bud development. This treatment can advance bud harvest about 5–7 weeks. Whereas the practice does not increase total production, it shifts production to an earlier period and, thus, may permit off-season production.

For optimum edible quality, flower buds are harvested when the floret primordia on the receptacle are rudimentary, and before significant bract spreading and tissue fiber occur. Harvesting, normally by hand labor, involves cutting through the stem below the base of the flower bud. Depending on market requirements, buds are harvested with stems 10–15 cm in length, sometimes more, and sometimes short enough to be flush with the base of the bud. Because of the sequential nature of bud development and plant-to-plant variability, concentrated harvests are infrequent. This makes mechanical harvesting of artichoke not feasible.

Harvest frequency is strongly influenced by temperature. To avoid harvest of overdeveloped buds during warm periods, harvests are made every 4 or 5 days. During cool temperatures, the interval may be 2 weeks or longer. Some growing regions have a short harvest period; for others, production may extend to 8 or 9 months, or even year-round, and involve about 25 or more harvests. Yields vary greatly, ranging from 5 to more than 20 t/ha. Some seed-propagated plantings tend to produce a high concentration of uniform buds which can make harvesting more efficient, and with high-density populations, mechanical harvesting may be feasible.

Production

The European countries in the Mediterranean region are world leaders, collectively accounting for almost 80% of world total production, with Italy and Spain producing more than 45% and 26%, respectively; see Table 18.1. This volume does not reflect the trend of a slow but steady decrease in overall production. Although an important vegetable commodity in southern Europe and parts of the Mediterranean region, artichoke is generally considered a minor crop and specialty vegetable elsewhere.

TABLE 18.1. WORLD PRODUCTION OF GLOBE ARTICHOKE, 1994

	Area (ha × 10³)	Yield (t/ha)	Production (t × 10³)
World	110	11.2	1234
Africa	8	10.0	82
North and Central America	4	14.2	51
South America	6	14.6	93
Asia	2	11.2	25
Europe	90	10.9	984
Leading countries			
Italy	51	11.0	561
Spain	24	13.9	326
Argentina	4[a]	19.5	74[a]
France	13	5.3	68
USA	3	14.6	50
Morocco	3	12.4	34
Egypt	2[a]	12.5	25[a]
Greece	2[a]	10.4	24[a]
Chile	2	7.5	17
Tunisia	2[a]	7.4	14[a]

[a]Estimated.
Source: 1994 FAO Production Yearbook, Vol. 48, FAO, Rome 1995.

Postharvest and Uses

Artichoke flower buds require careful handling to avoid bruising and desiccation. Following harvest, buds are cooled as soon as possible. Hydrocooling is effective for both cooling and cleaning. When storage or holding is required, conditions providing temperatures of 0–2°C and 95% RH can maintain good quality for as long as 10 days. Frosts can injure artichoke buds, but unless the damage is severe, it is mainly aesthetic. In the United States, an industry promotional program effectively markets frost-affected buds, claiming that they are "winter kissed" and therefore taste better.

Most artichoke production is consumed fresh. Artichokes can be, but are usually not, eaten raw. They are, however, cooked in a wide variety of dishes, and also processed by canning in marinade, vinegar, or brine as well as by freezing. Extracts from artichoke tissues are prepared for tonics and for other medicinal and nonfood uses.

CARDOON, *Cynara cardunculus* L.

Similar to artichoke in many ways, cardoon is cultivated on a much smaller scale. The edible product is the fleshy basal portions of the large thick leafy petioles. Propagation is with seed, as uniformity of floral bud development is not a concern nor is floral development desirable. The basal rosette of leaf stalks are blanched by tying leaves

FIG. 18.9. Cardoon, *Cynara cardunculus,* as prepared for marketing.

together several weeks before harvest. This makes the interior tissues tender and reduces bitterness. Harvests are made about 120–150 days after planting. The bunched leaf stalks are cut at the soil surface and trimmed to a length of 45–60 cm (Fig. 18.9). The harvested product, in appearance, has a slight resemblance to stalk celery, and tastes like artichoke when cooked.

SALSIFY, *Tragopogon porrifolius* L.

Salsify is a seed-propagated, slow-growing, hardy biennial native to the southern European Mediterranean region. From the appearance of its foliage, the plant acquired the name of "goat's beard," and from the taste of the root, "vegetable oyster."

Foliage growth develops as a rosette from the compressed stem. The grasslike leaves are long, linear, keeled, glaucous, and tightly clasped about the base. Following vernalization, the elongated, well-branched

FIG. 18.10.   Salsify root, *Tragopogon porrifolius.*

flower stem produces blue or purple perfect flowers that form within a single head (capitula) upon a fleshy hollow peduncle. All flowers in the head open at the same time and only once. The seed is a long achene with a noticeable break; about 70–90 seed weigh 1 g.

The taproot and a small portion of the hypocotyl are edible. This fleshy axis is 20–35 cm long and gradually tapered (Fig. 18.10). Some lateral roots tend to develop, but are usually small. However, these can interfere with harvest as well as food preparation. Deep, sandy, loam soils are preferred for easier root removal at harvest. Plant spacings are usually 10–15 cm between plants and 60–90 cm between rows.

The exterior color is cream or buff; the flesh is white and bleeds latex when cut. In appearance, the salsify root resembles parsnip. It is hardy and often overwintered. The usual growing season is 6 or more months. Low temperatures at maturity tend to improve flavor by converting stored starch to sugar.

SCORZONERA, *Scorzonera hispanica* L.

Other names: Black salsify, Spanish salsify

Scorzonera is endemic to southern Europe, possibly Spain, and is grown for its long, fleshy, black-skinned taproot. Taproots usually exhibit little or no branching. Roots are harvested when about 30–35 cm long, although lengths as long as 75 cm are attained. They are slightly tapered, with crown widths of 2–4 cm; the exterior is black and the interior is white. Taproots are eaten after peeling and cooking. Young blanched leaves are also prepared as a vegetable potherb. Although a perennial, scorzonera is grown as an annual. The rosette of alternative leaves are mostly lanceolate and may be entire and dentate. Scorzonera differs from salsify in having yellow flowers. Plant growth and taproot enlargement is slow, and sometimes the growth period before harvest may be as long as 200 days.

EDIBLE BURDOCK, *Arctium lappa* L. var. *edule*

Other name: Gobo

Edible burdock is a temperate-region biennial or short-lived perennial native of northeast China and Siberia; uncultivated forms are found across Asia and Europe. A relative is the weedy common burdock, *A. minus*. Plants are mainly grown for the light brown, long, narrow taproots that are eaten after scraping and cooking (Fig. 18.11). Root lengths range from 30 to 60 cm and sometimes have a length of 100 cm; root diameters usually are between 2 and 4 cm. The interior flesh

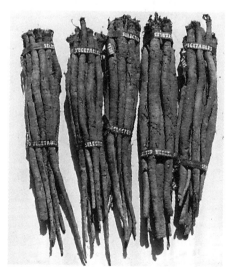

FIG. 18.11. Edible burdock roots, *Arctium lappa.*

is white but quickly darkens when exposed to air. Roots are somewhat fibrous and have a firm texture, and supposedly have some medicinal properties.

The rosette of large dark green leaves are heart shaped and slightly puckered with purplish veins, the undersides exhibiting short white hairs. The petioles are ridged and pubescent; these are also edible when peeled and cooked. Plants are tolerant of high temperature and can be grown in the subtropics. However, optimum growth occurs at moderate (20–25°C) temperatures. Interestingly, although foliage is damaged by cold temperatures, roots can tolerate some freezing temperatures and therefore can be overwintered or stored in the ground. Low temperatures and long days are conducive to bolting. Flowers are purplish red.

Burdock is seed propagated, requiring exposure to light for germination. Plantings in friable and deep, well-drained, light textured soils are preferred. The resulting long root growth makes hand harvesting difficult. Plants are frequently grown on raised beds to facilitate removal of roots. Large-scale operations utilize especially made deep plowing equipment to lift the roots. The growth period to obtain suitable size is about 10 months.

DANDELION, *Taraxacum officinale* Wiggers

Dandelion is a perennial plant that frequently occurs as a weedy pest of the Northern Hemisphere; its specific origin is unknown. Dandelion has been naturalized in Europe and central temperate Asia, and although a weed to many, the plant also is a highly regarded vegetable green for raw use in salads, or as a cooked potherb. Dandelion leaves have a high content of pro-vitamin A and vitamin C. Roots and flowers are also edible.

Cultivars bred for vegetable usage have little resemblance to weedy forms, having been selected for greater leafiness and less bitterness. Leaves are often blanched to further reduce the tart and bitter taste. Latex is contained throughout all plants parts. Leaf growth is a dense rosette varying in leaf length and shape, which range from narrow to broad with margins entire or divided. Flower heads are yellow and the seed produced have a conspicuous white pappus.

Plants are seed propagated, either sown in place or transplanted. Divisions are also used for propagation. A limited amount of production is forced during low-temperature periods in protective structures or tunnels. Usual field spacings are 10–15 cm × 50 cm; preferred soils are sandy loams.

Plants are harvested much like spinach, being cut just below the stem, with several individual plants tied together as a bunch, or the leaves cut free of the stem and bunched or handled in bulk.

GARLAND CHRYSANTHEMUM/CHOP SUEY GREENS, *Chrysanthemum coronarium* L.

Other name: Shungiku

This hardy leafy annual is probably native to the Mediterranean region. The well-branched plant usually grows to a height of 30–70 cm and up to 1 m when flowering. Young shoot tips and foliage are cooked, often rinsed in fresh water, before consumption. The crop is a popular cooked green in Japan, China, and other Asian countries. Daisylike flower heads are yellow or yellowish white and are also eaten. All plant parts have aromatic flavor qualities, being most pronounced in older foliage.

Garland chrysanthemum is commonly grown during cooler periods, and temperatures above 25°C are detrimental to leaf quality. The three major plant types can be characterized by leaf size and shape; leaf features exhibit some correlation to climatic adaptation (Fig. 18.12). Those with small and finely dissected foliage tend to have good low-

FIG. 18.12.   Garland chrysanthemum foliage, *Chrysanthemum coronarium*.

and high-temperature tolerance, but are low yielding. Plant types with intermediate leaf size and leaf margin divisions are also adapted to warm and cool temperatures, are high yielding, and are the most popular. Large size and shallow-lobed leaf types are high yielding, but only adapted to warm temperatures. Generally, short days and cool temperatures tend to retard stem length, delay flowering, and promote increased leaf area.

Propagation is with seed. The harvest method, whether by the single removal of intact plants or partial leaf removal, determines plant spacings. Plants are usually harvested 30–40 days after sowing when shoots are still tender, and before reaching full height and flowering. Growers often remove early flowers in order to prolong vegetative growth.

A leafy plant very popular in China that resembles garland chrysanthemum is *C. spatiosum,* also cultivated for its young edible leaves. *C. morifolium,* known as shokuyo giku or garnish chrysanthemum, is grown for its edible ligulate flowers, used as a cooked vegetable or as fresh garnish in Japan. *C. morifolium* is probably a native of China, and cultivars grown for the edible flower heads are selections of the common florist chrysanthemum. Cultivars are identified by flower color variation which may be yellow, pink, red, purple, or white. Another related species is *C. balsamita,* also known as costmary or mint geranium.

BUTTERBUR/FUKI, *Petasites japonicus* (Sieb. & Zucc.) Maxim.

A perennial dioecious plant, butterbur is appreciated for its succulent petioles, although leaf blades and flower buds are also edible (Fig. 18.13). Favorable growing temperatures range from 10°C to 23°C, optimum being 18–20°C; foliage is killed by frost. Most crops are grown in the spring or fall, as growth is usually slowed by summer temperatures.

Plant height is about 1 m, and the leaves are broad, almost spherical or kidney shaped with the upper surface pubescent, the lower wooly. Petioles are long and valued according to their length, which range from 20 to 80 cm. The inflorescence contains white, usually sterile, ray flowers surrounded by long green bracts. Cultivated forms of butterbur are triploids; wild forms are fertile diploids, but triploids also occur.

Propagation is by vegetative division using the thin 10–30 cm-long rhizomes produced by the plant; usually three to five rhizomes are produced. When planted, the rhizomes are commonly spaced about 30–50 cm apart. Most plantings are maintained for 5–6 years before replacement, although plants can live considerably longer and often

FIG. 18.13. Butterbur plant, *Petasites japonicus,* growing in Japan.

revert to weeds; therefore, they are widely found throughout eastern Asia. In some situations, plants are grown as annual crops.

Harvests are made as petioles achieve acceptable lengths. Before consumption, petioles are parboiled to remove astringency and to make peeling the epidermis easier. The crop is most important in Japan where a range of cultivars exists that differ by size and maturity periods. Examples such as cultivars Aichi early fuki, Akita fuki, and Mizu fuki permit harvesting from February to May. Aichi early fuki is the earliest producer and is grown in the spring and also cultivated for winter forcing in plastic tunnels or houses. Mizu fuki is a late producer and small plant, and Akita fuki is a large plant usually produced for processing. Akita fuki also has the best cool season adaptation. Butterbur is also widely grown in China and Korea.

GYNURA, *Gynura bicolor* DC.

Gynura is a perennial plant native to the Old World tropics. It is a popular crop grown in China for the young shoots and leaves which are used as a vegetable or in soup preparation (Fig. 18.14). The plant is recognized by the dark green upper surfaces and the magenta or

FIG. 18.14.   Seedling plants of gynura, *Gynura bicolor,* growing in China.

purple lower surfaces of leaves and petioles, and also for its orange-colored daisylike flowers. The erect stems grow to a height of about 1 m.

SIERRA LEONE BOLOGI, *Crassocephalum crepidodes* (Benth.) S. Moore

This native west African climbing perennial sometimes also identified as *Senecio biafrae,* is grown for its nearly triangular-shaped succulent leaves which are eaten as a potherb. Propagation is with cuttings.

WHITE MUGWORT, *Artemisia lactifolia* Wallich ex DC.

This erect perennial has a smooth, grooved stem and grows to a height of 1.5 m. Leaves are long and pinnatified into ovate–lancelote lobed segments. The flowers are white. Plants are grown for the aromatic qualities of the foliage, and cultivation is popular in China. *A. indica* Willd. is another form of mugwort.

COSMOS, *Cosmos caudatus* Kunth

Cosmos is characterized by a turpentinelike smell and its strong-tasting edible foliage. The plant is primarily a home garden vegetable and infrequently found in markets. Related species are widely grown as ornamental flowers.

SOWTHISTLE, *Sonchus oleraceus* L.

The plant is a leafy potherb, more often gathered from weedy growth, but occasionally it is cultivated. Plants are rich in latex and become very bitter with age.

PARAGRASS, *Spilanthes oleracea* L.

Paragrass is a potherb and salad green consumed for its sharp tasting tops and leaves especially in southeast Asian countries. Plants are annuals or short lived perennials and generally not cultivated; even when marketed they are gathered from natural populations. Leaves are broadly ovate, about 6–8 cm long with wavy margins; flowers are greenish yellow. Related weedy species are *S. paniculata* and *S. iabadicensis*.

SELECTED REFERENCES

Basnizki, Y., Goldschmit E., Luria, Y., Itach, M., Berg, Z., and Galili, D. 1986. Effect of acidified GA_3 spray on yield of globe artichoke (*Cynara scolymus* L.). Hassadeh *15*, 1814–1817.

Bianco, V.V., and Pimpini, F., eds. 1990. *Orticoltura*. Patron Editore, Bologna, Italy.

Borthwick, H.A., Hendricks, S.B., Toole, E.H., and Toole, V.K., 1954. Action of light on lettuce-seed germination. Bot. Gaz. *115*, 205–225.

DeVos, N.E. 1992. Artichoke production in California. HortTechnology *2*(4), 438–444.

Endo, M., and Iwasa, S. 1982. Characteristics and classification of edible and garnish chrysanthemum (*Chrysanthemum morifolium* Ramat.). J. Japan. Soc. Hort. Sci. *51*, 177–186.

Nelson, J.M., and Sharples, G.C., 1986. Emergence at high temperature and seedling growth following pre-treatment of lettuce seeds with fusicoccin and other growth regulators. J. ASHS *111*, 484–487.

Price, K.R., DuPont, M.S., Shepherd, R., Chan, H. W-S., and Fenwick, G.R. 1990. Relationship between the chemical and sensory properties of exotic salad crops—Colored lettuce (*Lactuca sativa*) and chicory (*Cichorium intybus*). J. Sci. Food Agric. *53*, 185–192.

Robinson, R.W., McCreight, J.D., and Ryder, E.J. 1983. The genes of lettuce and closely related species. Plant Breeding Rev. *1*, 267–293.

Ryder, E.J. 1979. Leafy Salad Vegetables. Chapman & Hall, New York.

Ryder, E.J. 1986. Lettuce breeding. In Breeding Vegetable Crops. M.J. Bassett, ed. Chapman & Hall, New York, pp. 433–474.

Ryder, E.J., and Whitaker, T.W. 1980. The lettuce industry in California: A quarter century of change, 1954–1979. Hortic. Rev. *2*, 164–207.

Snyder, M.J., Welch, N.C., and Rubatzky, V.E. 1971. Influence of gibberellin on time of bud development in globe artichoke. HortScience *6*, 484–485.

Takahaski, B. 1994. Burdock. In Horticulture in Japan. Organizing Comm. XXIVth International Horticultural Congress, eds. Asakura Publ. Co., Ltd., Tokyo, p. 104.

Valdes, V.M., Bradford, K.J., and Mayberry, K.S. 1985. Alleviation of thermodormancy in coated seed by priming. HortScience *20*, 1112–1114.

Whitaker, T.W., Ryder, E.J., Rubatzky, V.E., and Vail, P. 1974. Lettuce Production in the United States. USDA Agric. Handbook 221. USDA, Washington, DC.

Zink, F.W., and Yamaguchi. M., 1962. Studies on the growth rate and nutrient absorption of head lettuce. Hilgardia *32*, 471–500.

Zohary, D., and Basnizki, Y. 1975. The cultivated artichoke, *Cynara scolymus,* its probable wild ancestors. Econ. Bot. *29*, 233–235.

19

Cole Crops, Other *Brassica,* and Crucifer Vegetables

Family: Brassicaceae (Cruciferae)

Commonly known as the mustard family, the Brassicaceae consists of more than 300 genera and 3000 species. Included are important annual and biennial vegetables as well as valuable oil seed and ornamental crops distributed throughout the world. Most are grown in temperate regions, and some are even grown in subarctic climates. Many Brassicaceae crops commonly known as crucifers are widely recognized for their contribution to human nutrition and for other possible healthful benefits. Recent research suggests that some crucifers may have cancer-preventive attributes.

Pungency is a common family characteristic. *Brassica* and many other Brassicaceae genera contain glycosinolate compounds that are converted by the enzyme, myrosinase, to give bitter-tasting and goitrogenic substances such as isothiocyanates, thiocyanates, nitriles, and goitrin. These compounds, while contributing to flavor and odor, have been reported to interfere with thyroxine production and cause goiter (thyroid enlargement). Through selection and breeding, glycosinolate levels have been greatly reduced in cultivated species when compared to levels existing in primitive forms.

Brassica is the most important genus, with about 40 species, several of which are important annual and biennial foliage and root vegetables. Some *Brassica* species also include important oil seed, animal feed, and cover crops. Additional vegetable crucifers that are not brassicas are also discussed in this chapter. However, discussion of Japanese horseradish and watercress are included in Chapter 26 and maca in Chapter 14.

Brassica Taxonomy

The majority of cultivated vegetable brassicas occur in six species. three, *B. nigra, B. oleracea,* and *B. rapa (B. campestris),* are monogenomic with 8, 9 and 10 pairs of chromosomes, respectively. Their genomic composition are customarily identified as B, C, and A, respectively. The other three are amphidiploid species *B. carinata, B. juncea,* and *B. napus* and, are identified as BC, AB, and AC, respectively, and in the same order have 17, 18, and 19 pairs of chromosomes. The latter three amphidiploid species are believed to have evolved naturally. Amphidiploids are diploid for two genomes, each from a different species. Evidence supporting this development is provided from artificially created amphidiploids that strongly resemble the natural forms with which they can easily intercross. The interrelationship of these six species was described by the geneticist U in 1935. Figure 19.1 shows how *B. carinata* would have originated from hybridization of *B. nigra* with *B. oleracea, B. juncea* from hybridization of *B. nigra* with *B. rapa,* and *B. napus* from hybridization of *B. rapa* and *B. oleracea.*

Brassica taxonomy is complex and still not fully resolved, and common names do not necessarily reflect their species association. Often crops producing large succulent leaves are called cabbages, those with enlarged storage roots are called turnips, those producing succulent foliage and spice seeds are called mustards, and those producing seed oil are called rapes. Table 19.1 provides a list of important vegetable *Brassica* species.

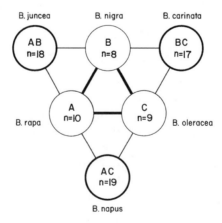

FIG. 19.1.   Genomic interrelationships among *Brassica* species as proposed by U in 1935.

TABLE 19.1. *Brassica* CLASSIFICATION

Species	Botanical group[a]	Cytoplasm designation and basic chromosome number	Common name[a]
Monogenomic species			
nigra		B (8)	Black mustard
oleracea		C (9)	
	Acephala		Kale/collard[c]
	Alboglabra		Chinese broccoli/Chinese kale/kailan
	Botrytis		Cauliflower/heading broccoli
	Capitata		Cabbage[d]
	Gemmiferae		Brussels sprouts
	Gongylodes		Kohlrabi
	Italica		Sprouting broccoli/calabrese
rapa		A (10)	
	Chinensis		Chinese mustard/pak choi
	Japonica[b]		Mizuna/mibuna
	Oleifera		Turnip rape
	Parachinensis		Choi sum/false pak choi
	Pekinensis		Chinese cabbage/pe tsai/celery cabbage
	Perviridis		Tendergreen/spinach mustard
	Narinosa		Chinese flat cabbage
	Rapifera		Turnip
	Ruvo		Broccoli raab/rapini
	Septiceps		Seven top turnip/Italian turnip
Amphidiploid species			
carinata		BC (17)	Ethiopian/Abyssinian mustard
juncea		AB (18)	Mustard (various forms)[e]
napus		AC (19)	
	Napobrassica		Rutabaga/swede turnip
	Oleifera		Oil seed rape/canola
	Pabularia		Siberian kale/Hanover salad

[a]Agreement is not uniform as to the classification of different forms of *B. oleracea* as either botanical varieties or as botanical groups. We include both variety and group.
[b]Possibly a result of introgression with *B. juncea.*
[c]Within this botanical group are various plants sometimes identified as varieties of the Acephala group. Examples are Portuguese kale/cabbage as var. *costata,* marrow stem kale as var. *medullosa,* 1000 headed kale as var. *millecapitata,* tree/Jersey kale as var. *palmifolia,* and collard as var. *plana* or var. *salellica.*
[d]In some references, further subdivision of cabbage forms may be seen as forma *alba,* f. *rubra,* and f. *sabauda* for white, red, and savoy cabbages, respectively. Many regard such identification as superficial and confusing.
[e]These mustards, although widely known as Indian, brown, or yellow mustards have many morphologically different forms that are cultivated. Morphological differences include plants with large, small, broad, narrow, smooth, and curled leaves, as well as broad, long petioles, some green, some white, swollen stems, and plants with tillering characteristics and some tuberous rooted forms. Botanical varietylike nomenclature (usually somewhat descriptive) is often ascribed to each form. However, even for similar forms, nomenclature frequently differs between and within different localities.
To avoid adding to existing confusion, the *Brassica* listing is limited to those species and subdivisions recognized by most taxonomists.

COLE CROPS, forms of *Brassica oleracea*

This polymorphic species includes vegetables collectively referred to as cole crops. The word "cole" is possibly from the Middle English, *col*, initially derived from Anglo Saxon, *cal, cawl*, or *cawel*, or from Old Norse *kal;* both also possibly originating from the Latin *caulis*, meaning stem or stalk. Current English usage is also thought to be derived from the German *kohl*, or from sound-alike words from other languages. The Ancient Greeks referred to cole crops as *kaulion*, meaning stem. *Brassica* is from the Celtic word *bresic*, meaning cabbage.

Origin and Domestication

Early Greek and Roman literature mention the medicinal properties of these plants, and it is surmised that early forms were cultivated for their perceived medicinal attributes. Mention of cole crops was also made in various herbals, and other writings during the Middle Ages, as their cultivation spread across Europe.

Various cole crop forms are thought to have evolved via mutation, through self-adaptation, and selection from human involvement. Many forms of the species are self-incompatible, and therefore cross-fertilization probably expanded opportunities for introducing variation. Present Mediterranean wild species exhibit annual growth habits. Biennial selectivity must have occurred with domestication and adaptation to northwestern and northern European environments. Along with selection for winter hardiness, it is also probable that selection reduced bitterness.

Wild kales and nonheading cabbages were likely the first domesticated. Initially, medicinal use probably was more important than food. It was reported that these plants were used to relieve gout, diarrhea, deafness, and headache; cabbage juice was a prescribed remedy for mushroom poisoning. The present firm-headed cabbages are descendants of the wild nonheading *Brassica oleracea* var. *sylvestris* and possibly other wild species such as *B. cretica, B. insularis,* and *B. rupestris.* Wild forms have been found near the coasts of Britain, the Bay of Biscay, and in regions surrounding the Mediterranean Sea. Most presently known forms of *B. oleracea* were not well documented until the 16th century. First important in Europe, cabbage soon became a crop of worldwide usage.

Sprouting broccoli and cauliflower are thought to have been domesticated in the Mediterranean region and possibly as localized as Cyprus or Crete. The first written description of cauliflower appeared in 1544.

Sprouting broccoli, which possibly preceded cauliflower in domestication, only recently has become a popular vegetable beyond its Mediterranean origin, although cauliflower production beyond the Mediterranean region occurred earlier than sprouting broccoli. It was only during the latter half of this century that sprouting broccoli (calabrese), introduced into the United States from Italy, became a popular vegetable. Its popularity in the United States spread to other parts of the world; in fact, its reintroduction and appreciation in Europe is very recent. The rate and magnitude of sprouting broccoli production throughout the world has been very rapid; and although total production is still less than cauliflower, their positions are likely to change. Heat-tolerant tropical forms of cauliflower were grown in India during the last 200 years. These were developed by selection from the United Kingdom and other European sources.

The first written description of brussels sprouts was about 1587. The crop has had limited distribution and/or production beyond northern Europe. Kohlrabi, first described in ancient Rome, was domesticated in northern Europe about the 15th century. Chinese broccoli differs from other cole crops because its domestication apparently occurred in Southeast Asia where many landraces are grown. Although once thought to be a native of China, recent DNA evidence suggests that Chinese broccoli is closely related to Portuguese cabbages and kales and may have been introduced to Asia by Portuguese traders.

Although polymorphic, cole crops are grouped together because of taxonomic similarities. It is difficult to believe that so many forms with such a diverse morphology are related, but all readily intercross and share similar cultural requirements and pest susceptibilities.

Botany and Morphology

Cole crops are annual and biennial herbaceous dicots; the biennials forms are commonly grown as annuals. As young seedlings, the various cole crops are hard to distinguish, but soon each develop recognizable characteristics. Cultivated forms of *B. oleracea* are illustrated in Fig. 19.2.

Root systems are moderately shallow; the taproot, although not prominent, readily branches and has many fibrous roots, mostly concentrated within 30–35 cm of the surface. Characteristically, leaves are thick, somewhat leathery, glaucous, smooth, and some are pubescent. Notable exceptions of smooth foliage are kale and savoy cabbages. Leaves are usually alternate, petiolate, oblong, and simple; some are deeply pinnatified. The typical inflorescence is an elongated raceme

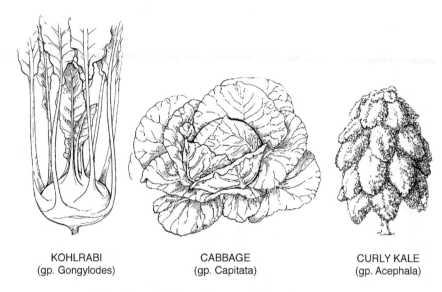

| KOHLRABI | CABBAGE | CURLY KALE |
| (gp. Gongylodes) | (gp. Capitata) | (gp. Acephala) |

FIG. 19.2. Cultivated forms of *Brassica oleracea.*

with many small flowers formed at terminal portions. Brassicaceae flowers are perfect and generally characterized by four petals standing opposite each other in a square or crucifix pattern. Also typical are the six stamens, of which two are shorter; the ovary is two celled. Most flowers are yellow or pale yellow and sometimes white. Insect pollinators are readily attracted to the abundant flowers and nectar. Small, round, dark brown seed form in a podlike fruit called a *silique.* The thin, 3–5-mm-wide pods are 50–100 mm long and often dehisce when seed mature. Seed usually mature 50–90 days after fertilization. Most seed have a short postharvest dormancy; the inhibitors are easily leached or dissipated within a month or two. For most cole crops, cold temperature exposure (vernalization) is required for flowering (Table 19.2). Exceptions are some cultivars of sprouting broccoli and tropical cauliflower having annual growth habits.

CABBAGE, *Brassicca oleracea* L. var. *capitata* L. (Group Capitata)

A cabbage head is best described as a single, large terminal bud comprised of tightly overlapping leaves attached to and enclosing most

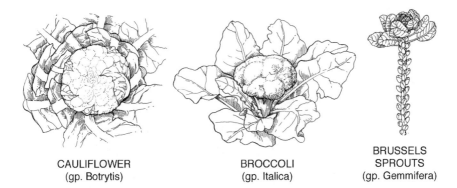

CAULIFLOWER
(gp. Botrytis)

BROCCOLI
(gp. Italica)

BRUSSELS
SPROUTS
(gp. Gemmifera)

FIG. 19.2.   *Continued*

of the unbranched short stem. Plant height commonly ranges between 40 and 60 cm. For most cultivars, early leaf growth is elongated and fairly prostrate. Subsequent leaves are progressively shorter, broader, and more erect, and begin to overlap younger leaves. Continued leaf formation and growth beneath the overlapping leaves results in increasing the density of the developing head. Coincidental with leaf growth, the stem also continues to slowly elongate and thicken. Continued interior head growth beyond the mature (firm) stage can result in head bursting. Important commodity variables are head size, density, shape, color, leaf texture, and maturity period. Shapes range from pointed ellipsoids to flattened drumlike heads, with spherical or nearly spherical forms preferred. Foliage color, either with or without a waxy surface

TABLE 19.2.   COLE CROP VERNALIZATION REQUIREMENTS FOR PRODUCTION OF HARVESTED PORTION AND FLOWERING

Crop	Harvested portion	Flowering
Broccoli, early type	No	No
Broccoli, late type	Yes	Yes
Brussels sprouts	No	Yes
Cabbage	No	Yes
Cabbage, tropical type	No	No
Cauliflower, early type	No	Yes
Cauliflower, late type	Yes	Yes
Cauliflower, tropical type	No	No
Chinese broccoli	No	No
Collards	No	Yes
Kale	No	Yes
Kohlrabi	No	Yes

FIG. 19.3. Savoy cabbage, *Brassica* oleracea, gp. Capitata. Source: Courtesy National Garden Bureau, Los Altos, California.

bloom, varies from light green to dark blue-green and also reddish purple. Leaf texture may be smooth or savoyed (Fig. 19.3).

PORTUGUESE CABBAGE, *Brassica oleracea* L. var. *costata* DC. (Group Capitata)

Portuguese cabbage is generally considered a botanical variety of the Capitata group. The plant closely resembles a loose-headed cabbage with leaves having prominent and succulent main ribs. Plants have some resemblance to kale and are believed to have originated from hybridization of cabbage and kale forms with domestication probably occurring in Portugal or Spain. Variability is extensive among the mostly landrace cultivars. Important crop features are the extent of heading (many do not head), leaf tenderness, and maturity periods.

CAULIFLOWER, *Brassica oleracea* L. var. *botrytis* L. (Group Botrytis)

The edible plant portion is commonly called a *"curd"* or head. The curd in the early types consists of tightly clustered undifferentiated, usually white, shoot apices formed upon thick, hypertrophied, repeatedly branched, fleshy terminal portions of the short thick stem. Chlorophyll is typically absent in the curd tissues. Following vernalization curds of strongly biennial, late-maturing cultivars produce shoot apices with differentiated floral primordia. Floral primordia become evident usually after the curd has passed its prime market quality stage. Flower buds are formed in the axils of the elongated branches of the curd. Tropical cauliflowers are most similar to summer types and will produce curds at relatively high temperatures and readily flower after relatively little low-temperature exposure.

Slow elongation associated with rapid thickening of lateral branches result in a short, thick, and compressed dome-shaped curd for most cultivars; some produce pointed and/or pyramidal shaped curds. Continued growth of the curd results in elongation of its many branches, causing spreading and a loss of compactness and shape.

Plant heights are variable; most cultivars are 50–80 cm tall. Leaves are usually upright and oblong, and longer and narrower than those of cabbage. Foliage colors are grayish to blue-green with a waxy bloom, and with leaf margins either smooth or curly. Small inner leaves initially envelop and protect the curd from discoloration due to sunlight. However, as curd size increases, the innermost leaves become reflexed and are less able to overlap and cover the head. The extent of inner and outer leaf cover varies with cultivars. Late-maturing cultivars produce more and longer leaves and, therefore, provide good leaf cover. Large leaf size and weight are generally associated with large curd size. Summer- or autumn-type cultivars have less leaf growth and may require leaf tying to protect the curd from discoloration. Tropical cultivars produce relatively few leaves and generally have poor leaf cover.

In general, curd initiation occurs in the postjuvenile plant stage. For early-maturing cultivars, this period is when 15–20 leaves have formed. For late-maturing cultivars, initiation occurs when 25–30 or more leaves have formed.

Important cultivar variables are head size, shape, compactness, surface texture, and color. Pure white curds are preferred, although cultivars producing cream, purple, green, and orange curds are also grown.

Cauliflower cultivars can be broadly grouped into three major maturity types: early (for summer and autumn harvest), intermediate (late autumn and early winter), and late (winter and spring). Late-maturing types required vernalization for curd initiation (Table 19.2). Although all cauliflowers require some amount of vernalization to flower, the required period varies among cultivars. Late-maturing cultivars require longer exposure and lower temperatures than early cultivars.

In northern Europe, late-maturing and overwintering cauliflowers are often referred to as broccoli, or heading broccoli. This distinction should not be confused with sprouting broccoli.

Some of the major cultivar types are as follows:

Italian cultivars of varied curd form and color are grown as annuals and biennials. Examples include Jezi, Romanesco, and Flora Blanca.

Northern European cultivars are grown as annuals during summer and autumn. Examples are Alpha and Snowball.

Northwestern European cultivars are grown as biennials for late winter and spring harvesting. Examples are Roscoff and St. Malo.

Australian cultivars are grown as annuals for late autumn and early winter and sometimes spring harvest. These were developed from European sources. Barrier Reef is an example.

Asian cultivars are grown as early annuals and adapted for high-temperature regions. These often lack uniformity, curd compactness, and color. An example is Panta.

SPROUTING BROCCOLI, *Brassica oleracea* L. var. *italica* Plenck (Group Italica)

Like cauliflower, early-, medium-, and late-maturing types of sprouting broccoli are grown. Unlike cauliflower, the edible plant portion is the inflorescence consisting of immature fully differentiated flower buds and tender portions of the upper stem. Stems are taller, 50–90 cm, and have longer internodes than either cauliflower or cabbage. Divided and petiolate leaves are grayish to bluish green. The primary inflorescence forms at the terminus of the elongated unbranched stem, although the inflorescence itself is highly branched. Flower buds at the terminus of each branch of the inflorescence collectively produce a compact and somewhat hemispheric shaped head (Fig. 19.4). The green or purple bud clusters may be surrounded, but usually are not covered, by subtended leaves. Broccoli heads are less dense than cauliflower and are fully exposed throughout development. With continued growth, the branches of the inflorescence tend to grow apart, resulting in a loss of head compact-

FIG. 19.4.   Hybrid sprouting broccoli, *Brassica oleracea* gp. Italica, in a Taiwan market.

ness and shape. Following growth of the primary inflorescence, small inflorescences form in the axils of lower leaves. The development of these secondary inflorescences is influenced by apical dominance of the terminal inflorescence, the level of suppression varying with cultivars.

Late-maturing and overwintering cultivars are biennial and require vernalization to initiate the inflorescence and subsequent flowering. Annual cultivars do not require vernalization to produce the inflorescence. Important types of broccoli include purple sprouting (overwintered branching biennial), purple cape (overwintered single-headed biennial), purple Sicilian (pale purple single-headed annual, sometimes known as "purple cauliflower"), white sprouting (overwintered branching biennial), and calabrese (green sprouting, mostly single headed with annual and biennial forms). The calabrese types are the most widely grown, with many excellent hybrids displacing open pollinated cultivars.

Important crop characteristics include head compactness and shape, extent of branching, size of individual flower buds, stem length, number and length of internodes, and axillary floral development.

BRUSSELS SPROUTS, *Brassica oleracea* L. var. *gemmiferae* Zenk. (Group Gemmiferae)

Brussels sprouts are readily identified by the numerous vegetative buds (sprouts) that develop in leaf axils of an elongated unbranched stem. The tall (50–90 cm) stem terminates with a large dominant apical bud that strongly influences axillary bud development.

Axillary buds continue to be produced until growth ceases; the production of more than 100 buds is not uncommon. Early-maturing cultivars tend to be shorter and produce fewer sprouts than tall-growing late cultivars. However, tall cultivars are more prone to lodging. Bud compactness, size, shape, color intensity, harvest period, and productivity are important attributes.

KOHLRABI, *Brassica oleracea* L. var. *gongylodes* L. (Group Gongylodes)

The fleshy swollen tuberlike enlargement of the short unbranched stem characterizes kohlrabi. If the stem were stretched out, kohlrabi plants would have some resemblance to marrow stem kales, from which they may have evolved. Leaves on the bulblike stem develop in a compressed spiral pattern with the youngest leaves terminating at the apex. Stem enlargement occurs at or just above the soil surface. The taproot at the base of the enlarged stem is relatively slender and very fibrous with many small lateral roots. Fiber develops rapidly in the enlarged stem with advancing maturity.

Important crop characteristics are size, shape, and color (whitish green or purple) of the tuberlike stem. Total leaf number, remnant leaf scars, fiber development, and tendency for stem cracking are also important.

KALE/COLLARD, *Brassica oleracea* L. var. *acephala* DC. (Group Acephala)

These plants are recognized by the rosettelike whorl of foliage formed toward the apex of erect unbranched stems. For many cultivars, the stem is usually relatively short, although stem heights of some cultivars may exceed 1 m. Short-stem cultivars are easier to cultivate and less likely to lodge, especially when overwintered.

FIG. 19.5.   Collards, *Brassica oleracea* gp. Acephala.

The variability among different kale types is large. Most vegetable kales have many large upright heavily curled leaves. Curly leaf forms occur because of disproportionate growth along leaf margins, whereas the savoy appearance is due to nonuniform growth within portions of the leaf laminae. Because the thick stems and vigorous foliage produce considerable biomass, some kales, such as marrow stem, are grown exclusively for animal feed.

Collard cultivars are similar to the kales but differ in having larger and smooth leaves and smooth leaf margins. At times, the whorl of leaves of some collards may overlap sufficiently to resemble a loose head (Fig. 19.5). Important kale and collard features are leaf numbers, shape, size, leaf texture, color, and stem length.

CHINESE BROCCOLI, *Brassica oleracea* L. var. *alboglabra* Bailey (Group Alboglabra)

Also known as Chinese kale and white-flowered broccoli, this crop has a resemblance to both sprouting broccoli and kale. Until recently, it was sometimes classified as a separate species, *B. alboglabra.* Plant heights are usually between 40 and 50 cm. Leaves are oblong and semisessile. Although the resemblance to the mustard brassicas is strong, the nine pairs of chromosomes clearly indicate classification as *B. oleracea.* The edible plant portions include the stem, accompanying leaves, and early-developing inflorescence which does not require vernalization (Fig. 19.6).

Propagation

Cole crops are seed propagated; they are directly field sown or transplanted. Not surprisingly, all cole crop seed are similar in appearance, being dark brown, round, and small. Size variability ranges from 220 to 350 seed per gram. Seed are usually planted between 1 and 2 cm deep, with germination most rapid at soil temperatures of 15–20°C.

Crops with relatively short growing periods, such as kohlrabi, kales, and Chinese broccoli are commonly directly seeded. Brussels sprouts, cauliflower, cabbage, and broccoli are frequently transplanted as well as being field sown. Expanded usage of relatively expensive hybrid

FIG. 19.6.   Chinese broccoli, *Brassica oleracea* gp. Alboglabra.

seed for many cole crop cultivars has resulted in large increases in transplanting.

Transplants are field grown and also nursery produced. They are used as bare rooted seedlings or as soil block or module plug raised plants. Plants are 4–6 weeks old and are usually hardened before field planting. The practice of trimming tops and sometimes roots to facilitate handling is harmful because of delayed establishment and the potential for spread of black rot and other diseases.

Growth and development are affected by field spacings. Plants of sprouting broccoli or kale tend to accommodate each other at close spacings without significantly affecting production, whereas, with other cole crops, spacing is more critical. Representative spacing are shown in Table 19.3. A recent trend is for higher plant densities for most fresh market production. Plant spacings are often used to regulate desired product size. For example, small sizes of Brussels sprouts, 30 mm or less in diameter, are preferred by freezer processors, and they are produced from plants grown at high densities relative to those for fresh market. Similarly, to meet changing market preferences, fresh market cabbages are also grown at high populations in order to produce small heads that many markets prefer, whereas cabbages for sauerkraut are widely spaced to achieve large head sizes.

Soils, Nutrition, and Moisture

Many soil types are satisfactory for cole crop production, although fertile sandy or silty loams are better suited for early and heavy soils for late and overwintering crops. Soil pH should be within a range of 6–8. Soils with a pH less than 6 should be limed, especially if *clubroot* disease is present. Cabbages and kales are fairly tolerant of soil salinity; brussels sprouts and sprouting broccoli have a moderate tolerance, and kohlrabi and cauliflower are fairly sensitive. Chinese broccoli is moderate to fairly sensitive to salinity.

In general, cole crops utilize large quantities of nutrients during

TABLE 19.3. RANGE OF COLE CROP PLANT SPACINGS AND FIELD POPULATIONS

Crop	Distances (cm)		Population × 1000 per hectare
	In row	Between row	
Broccoli	20–40	40–90	27–125
Brussels sprouts	40–60	60–90	18–41
Cabbage	25–50	50–90	22–80
Cauliflower	30–60	60–90	18–55
Chinese broccoli	15–25	40–75	53–166
Kale/collards	30–40	50–75	33–66
Kohlrabi	15–20	40–75	66–166

growth, which is optimized with a uniformly available supply of fertilizer. Cauliflower, broccoli, cabbage, and brussels sprouts are usually supplied more fertilizer than kohlrabi or Chinese broccoli. The latter two have short growing periods and therefore utilize less total nutrients. Excessive nutrition, especially nitrogen, can have detrimental effects if excessive vegetative growth occurs which can delay cabbage heading and cauliflower curd formation. Lodging of brussels sprouts and *hollow stem* development in broccoli and cauliflower are other examples. Hollow stem of broccoli and cauliflower can be caused by boron deficiency, which differs from the hollow stem caused by excessive nutrition and rapid growth rate. Molybdenum deficiency causes a cauliflower disorder called *whip tail,* symptoms of which are bleached or whitish leaves having narrow leaf lamina.

Cole crops require a uniform level of moisture for steady stress-free growth, especially at the stage when the edible product is developing and enlarging. To meet crop moisture requirements, soils are frequently maintained near field capacity throughout growth. Moisture requirement varies with prevailing temperatures in addition to stage of plant development. The evapotranspiration of some cole crops can be high. A rate of more than 4 mm per day has been reported for a cabbage crop.

Temperature and Growth Response

For most cole crops, optimum growing temperatures are between 15°C and 20°C, and the best quality occurs when plants mature during uniformly cool to moderate temperatures. Temperatures in excess of 30°C generally suppress growth, and for some cole crops, 25°C may limit growth. Growth is slow at 10°C, but even at 5°C, some growth does occur. Cole crops vary greatly with regard to cold tolerance, and the stage of plant development greatly influences the level of tolerance. Young plants are more tolerant of low and high temperatures than mature plants. When appropriately acclimated, most plants tolerate frost and some even freezing temperatures. Kales are the most cold hardy and also tolerate high temperatures better than other cole crops. Brussels sprouts also are low-temperature tolerant; broccoli and cauliflower are less tolerant of frosts, especially near harvest.

The various cole crops differ in their temperature response with regard to flowering. Young plants with relatively few developed leaves and/or those with stem diameters of less than 5 or 6 mm are considered to be at a juvenile growth stage and do not vernalized. However, plants developed beyond the juvenile stage and exposed to temperatures less than 10°C for several weeks are sensitive and become vernalized. Sensitivity varies with crop type and cultivars, as well as temperature and duration (Table 19.2). Low temperatures accelerate the initiation of cauliflower curds and the broccoli heading of both annual and biennial types. Such premature development is a disadvantage if plants have not achieved adequate vegetative growth and thereby fail to produce a useful curd or head. In some situations a period of high temperature following cold induction can reverse or lessen the vernalization effect.

Cabbage growth with regard to temperature exhibits a pattern common to most leafy crops whereby foliage growth follows a sigmoid curve with most of the development occurring in the latter stages of vegetative growth. Stem and root growth follow a linear pattern, (Fig. 19.7). Cabbages, in general, have a broad temperature adaptation, with satisfactory growth at low and also relatively high temperatures. Vegetative growth is optimum between 15 and 20°C. Temperatures greater than 25°C adversely affect head density and shape.

Cauliflower vegetative growth is promoted by warm temperatures. Warm temperatures also tend to delay curding and flowering, whereas cool temperatures favors curd initiation. It is undesirable for curd formation to occur until plants have achieved sufficient leaf number or size, because the resulting curd will be small. Ideally, growers attempt

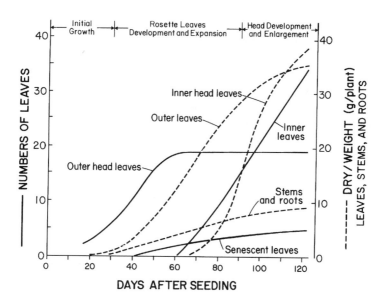

FIG. 19.7.   Growth stages of cabbage, *Brassica oleracea* gp. Capitata. Source: Redrawn from data of Hara and Sonoda (1979).

to produce a large plant before curding occurs. Premature curd formation is a common problem for some annual and tropical-type cultivars. Winter-type cultivars will not produce a curd unless low-temperature exposure is adequate. When these cultivars are grown in the tropics they remain vegetative. Knowledge of how temperatures affect cultivar growth is used in determining planting schedules.

Once initiated, the reproductive phase is not easily reversed. Curd initiation diminishes leaf development and causes lateral buds to elongate into shoot apices that comprise the convex surface of the curd. For many cultivars, the best curd quality develops at 17–18°C, and quality declines above mean temperatures of 20°C. However, active curd growth of some winter cultivars occurs even at 10°C, whereas curd development of tropical cultivars occurs at temperatures as high as 30°C. Once initiated, high-temperature accelerates the rate of curd development but also tends to reduce compactness.

Curd disorders attributable to temperature effects are buttoning, riceyness, bracting, and misshapen curds. Buttoning is caused by premature generative growth before adequate vegetative development, resulting in the formation of very small curds. The development of a velvetlike curd surface is a precocious formation of flower buds and is called riceyness. It is attributed to cold inductive temperatures followed by warm temperatures, so that rapid growth of stem apices is promoted. Inconsistency in the appearance of riceyness is probably due to cultivars maturing at different times and differing exposure to cold. Bracting is the result of rapid vegetative growth of small leaves normally suppressed. However, in response to high temperature, leaves grow between and above the lateral branches of the curd. Rapid and/or extended growth will similarly elongate lateral branches of the curd which results in misshaped heads in addition to bracting.

A condition known as blindness was believed to be caused by low-temperature damage of the apical meristem. Because of the absence of the apical meristem, affected plants develop few leaves and growth is severely restricted. However, because the condition also occurs when plants are not exposed to low temperatures, there apparently are other causes.

Marketing requires that curds be white and not discolored by sun exposure. For many cultivars, foliage development is usually adequate to overlap and cover the curd when small, but with enlargement, reflexing of the subtending leaves exposes the curd. Blanching is achieved by breaking or tying outer leaves over the curd. Cultivars such as Self-Blanche and Stovepipe were developed specifically for long upright leaves that provide shade without tying. High-density plantings and high-nitrogen fertilization also provide large erect leaf growth that can shield curds. Winter types generally produce vigorous leaf growth and often need not be tied. Because these require a long growing period, their production has diminished in most areas, except in northwestern France, some regions of England, and The Netherlands. The effects of sun discoloration is of less concern for tropical-type cultivars. Yellow colored and low-density curds are acceptable because they are difficult to avoid because of the relatively open growth characteristics of these cultivars.

Sprouting broccoli is less sensitive to temperature extremes than cauliflower. Best quality heads are produced in a range from 13°C to 20°C. Above 25°C, compact heading is difficult to maintain because of excessive stem elongation and rapid floral development. Low temperatures during early plant growth have a tendency to initiate premature heading. Growth is very slow at temperatures below 5°C. The inflorescence has some frost tolerance but is usually damaged by freezing.

Brussels sprouts foliage and axillary bud growth are very responsive to temperature. The crop is well adapted to regions of cool temperatures and moderately severe winters such as those of northwestern Europe, where this crop is sometimes overwintered for spring harvesting. Vegetative growth can continue, although slowly, at temperatures as low as 5°C. Plants have frost tolerance and, when acclimated, survive freezing temperatures as low as −5°C to −10°C. High temperatures can suppress stem elongation, and although axillary bud growth may be advanced, bud compactness decreases.

Kohlrabi plants are sensitive to low temperature and are readily vernalized; an exposure of a week at 10°C is sometimes sufficient to initiate bolting. Preferred growing temperatures range between 18°C and 25°C and is optimum at 22°C. At low temperatures, enlargement of the tuberous stem tends to be globular, whereas at high temperatures, the shape is elongated. High temperatures also tend to increase the rate of fiber development and possible cracking of the tuberlike stem.

Kales and *collards* are the hardiest cole crops and, when properly acclimated, can tolerate temperatures of −10°C or lower, and therefore often can be overwintered. Kales and collards have good tolerance to high, 25–30°C, temperatures, although growth is arrested above 30°C.

Chinese broccoli plants clearly have a preference for cool, 15–20°C, temperatures, especially at the approach of harvest maturity. Temperatures greater than 25°C cause excessive stem elongation, fiber development, and early flowering. Although low temperatures are not necessary for flowering, such conditions accelerate bolting.

Growth Periods

Growth periods for the different cole crops vary greatly. From seed sowing to harvest may require less than 50 days for kohlrabi or as many as 180 or 200 days for late-maturing Brussels sprouts or winter cauliflowers. The developmental period for different cole crops are shown in Table 19.4.

Harvesting

Obtaining high yield and quality rely on crop uniformity and harvest at an optimum period. Some cole crops are able to maintain high quality over a long harvest period; others must be harvested within a relatively short time because peak quality quickly declines. The size of harvest portions is determined by cultivar selection and is also influenced by plant density.

Cabbage is usually harvested when the desired head firmness is obtained. Delayed harvest results in excessive stem elongation, loss of textural quality, and possibly head bursting. When harvested, the stem is cut close to the ground at the base of the head. The coarse outer leaves are trimmed. Although most fresh market cabbage is manually harvested, cabbage for processing (sauerkraut) is frequently mechanically harvested. The trend for many European and U.S. fresh markets is a preference of head weights of 1–2 kg. Reduced head sizes are achieved by cultivar choice and use of high-density plantings. Cabbages for processing are larger than those for fresh market, with weights ranging from 5 up to 10 kg and head diameters ranging from 20 to more than 40 cm.

Cauliflower curds are harvested before a further increase in size will result in loss of compactness. Accompanying the loss of compactness, the usually convex-shaped curd becomes flattened or even concave. Prediction of cauliflower harvest is difficult because growth is very

TABLE 19.4.   GROWTH PERIODS FOR COLE CROPS FROM SEED EMERGENCE TO HARVEST MATURITY

Broccoli	70–95 days; late-maturing cultivars require as many as 130 days
Brussels sprouts	120–180 days, longer when over wintered.
Cabbage	60–90 days, to 130 days for some sauerkraut and over wintering cultivars
Cauliflower	75–100 days, to 200 days for some late-maturing and overwintering cultivars; Tropical (Indian)-types mature as early as 60 days after planting
Chinese broccoli	40–60 days
Kale and collard	50–90 days, longer if overwintered
Kohlrabi	50–70 days

responsive to temperature. Several harvests are commonly required because of plant-to-plant variation and nonuniform curd development.

When harvested, the stem is cut well below the base of the curd. Some basal leaves surrounding the curd are retained. The attached whorl of leaves are usually trimmed to allow the curd to be viewed while providing protection from physical damage; the extent of trimming is varied according to market preferences. Head weights range from 0.5 to 2 kg with diameters from 15 to 30 cm. Field packaging of individually plastic-film-wrapped trimmed curds is practiced in many countries in order to limit desiccation. Product marketed close to production areas may not be wrapped. Selective harvest machinery has been tested, but the usage is not presently feasible.

Sprouting broccoli harvests are made when the terminal inflorescence has attained maximum size while still compact and before floral buds have opened. Bud size varies among cultivars. The terminal head is cut or snapped from the upper 15–25 cm of the stem. The basal end is cleanly trimmed and large leaves are stripped. Continued plant growth allows for the production and harvest of axillary inflorescence, which are small and short compared to the terminal inflorescence.

Terminal head weights range widely, from 100 to 1000 g, with diameters from 10 to more than 25 cm. For fresh market, two to four heads are often tied together as a bunch or marketed as individual heads. A recent practice is packaging broccoli as "crowns," which are individual intact heads that have the stem trimmed back to the base of the head. In another procedure, the lateral branches collectively comprising the head are segmented and packaged as "florets."

Like cauliflower, harvest mechanization has not proceeded beyond experimental efforts. The introduction of hybrid cultivars has greatly increased crop uniformity and yields. Improvement in crop uniformity has reduced the number of harvests. Yields from commercial production vary considerably in a range from 7 to 20 t/ha.

Brussels sprouts are usually hand harvested. Individual axillary buds (sprouts) are removed by breaking them away from the stem. The most advanced buds are at the base of the stem and are harvested first. Successive harvests are made as axillary buds continue to develop and enlarge on upper portions of the stem. Hand harvest can occur over a period of several weeks and as long as 4 months. For multiple hand harvesting, the apical bud is not removed and the stem can continue to elongate and produce additional axillary buds. However, as height increases, plants may lodge, which interferes with cultural and harvest operations.

Apical bud removal (topping) is commonly performed to improve machine or single harvests and is made several weeks before intended harvest. Topping enhances the determinatelike development of the axillary buds. Although further stem elongation is prevented, total growth is not hindered, and axillary buds near the stem apex enlarge at a faster rate than those near the base. Most of the bud growth is achieved by the redistribution of nitrogen and carbohydrates from the leaves. Accordingly, apical bud removal should not occur before plants have achieved appropriate growth. The growth retardant SADH (succinic acid 2,2-dimethyl hydrazine) has been used to substitute for apical bud removal to arrest apical dominance.

The practice of a single destructive harvest using labor aids or partial mechanization has greatly increased, especially for freezer processing. In this procedure, plants are stripped of leaves, the stems cut, and then sprouts are mechanically sheared from the stem. The fresh market crop is similarly harvested in situations where the cost for selective hand harvesting is prohibitive.

Individual sprouts weigh from 20 to more than 50 g, with a diameter range from 20 to 60 mm, the preferred diameter for freezer processing being between 15 and 25 mm. Yields range from 6 to 20 t/ha; yields for multiple harvested fresh market crops are generally higher than those obtained for processing.

Kohlrabi plants generally develop uniformly and usually can be harvested in one or two operations. Plants are easily hand pulled from the soil. The taproot is cut, and varying with market practices, several plants can be tied together and marketed as bunches, or trimmed for bulk handling. Preferred weights of individual kohlrabi tubers weigh between 150 and 250 g, with diameters ranging from 6 to 9 cm. Large kohlrabi tend to become very fibrous and often exhibit surface cracking. The foliage can be used as greens.

Kales and *collards* are harvested for their leaves, and generally do not have a precise maturity stage, although adequate leaf size is the important criteria. Plants may be harvested over a long period by removal of outermost foliage and allowing young leaves to enlarge for later harvests. Because sequential harvesting is labor intensive, single harvests of the entire rosette of leaves are made by hand or machine. Harvested plants and/or foliage are bulk handled, for processing, or tied into bundles, for fresh market. For some markets and processors, midribs may be stripped from leaf blades.

Chinese broccoli tends to grow very rapidly, and the ideal harvest stage occurs when the first flower buds of the inflorescence are about a day before opening. Some markets prefer seeing a trace of white petal color. The harvested portion, consisting of the upper tender stem, attached leaves, and inflorescence, is hand cut from the plant. Typical market presentation consists of tying several stems together in bunches.

Postharvest and Storage

Cabbage and kohlrabi have relatively good postharvest features in terms of how long product quality can be maintained. In those terms, brussels sprouts are only fair. Because of their high rates of respiration, cauliflower, sprouting broccoli, kales, and collards are prone to rapid desiccation. Therefore, it is important to provide rapid cooling, low temperature, and high humid conditions until the commodity is consumed. Cabbage can be stored for long periods; some cultivars are grown specifically to be stored for off-season marketing. These cultivars can be maintained in acceptable condition for 2–3 months, and up to 6 months with the use of modified atmosphere storage (1% O_2 and 5% CO_2 at 0°C to –1°C and 98–100% RH.). Other cole crops, such as sprouting broccoli or kale, are not usually stored beyond 1 or 2 weeks, and cauliflower not more than 2–3 weeks. Brussels sprouts can be stored for 3–4 weeks, but Chinese broccoli is seldom stored more than a week. Broccoli, brussels sprouts, kales, and collards are often packaged with ice to retain freshness. All cole crops should be stored free from ethylene to avoid premature senescence and tissue injury.

Processing

Cole crops are well adapted for processing. Cabbage is fermented to produce sauerkraut and also pickled, frozen and dehydrated. Brussels sprouts and cauliflower are also pickled. Substantial amounts of cauliflower, sprouting broccoli, brussels sprouts, kale, and collards are processed by freezing. Kale and collards are also processed as canned products.

Diseases and Pests

Important bacterial diseases affecting cole crops and several other crucifers include black rot (*Xanthomonas campestris* pv. *campestris*), bacterial leaf spot or peppery leaf spot (*Pseudomonas syringae* pv. *maculicola*), and a leaf spot caused by *Xanthomonas campestris* pv. *armoraciae*. The soft rot that follows is caused by *Erwinia carotovora*, which is usually a secondary invader; *Pseudomonas* species are also involved in bacterial soft rots.

Fungal diseases injurious to cole crops and other crucifers include the following:

Alternaria leaf spot and decay	*Alternaria brassicae, A. brassicicola, A. raphani*
Black leg	*Leptosphaeria maculans (Phoma lingam)*
Bottom rot/wire stem	*Rhizoctonia solani*
Cercospora leaf spot	*Cercospora brassicicola*
Cercosporella leaf spot	*Cercosporella brassicae*
Clubroot	*Plasmodiophora brassicae*
Damping off	*Pythium* spp., *Fusarium* spp., *Rhizoctonia solani*
Downy mildew	*Peronospora parasitica*
Fusarium rot	*Fusarium oxysporum* f. sp. *raphani*
Gray mold	*Botrytis cinerea*
Horseradish Ramularia spot	*Ramularia armoraciae*
Olpidium	*Olpidium brassicae*
Phytophthora root rot	*Phytophthora megasperma*
Powdery mildew	*Erysiphe cruciferarum*
Radish black root rot	*Aphanomyces raphani*
Ring spot	*Mycosphaerella brassiciola*
Scab	*Streptomyces scabies*
Sclerotinia rot	*Sclerotinia sclerotiorum, S. minor*
Turnip anthracnose	*Colletotrichum nigginsianum*
Verticillium wilt	*Verticillium albo-atrum, V. dahliae*
White rust/blister	*Albugo candida*
White spot/stem	*Pseudocercosporella capsellae*
Yellows/Fusarium wilt	*Fusarium oxysporum* f. sp. *conglutinans*

Viruses affecting cole crops and other crucifers include cauliflower mosaic virus (CaMV), turnip yellow mosaic virus (TYMV), cabbage black ring spot virus, also know as turnip mosaic virus (TuMV), and radish mosaic virus (RaMV).

Sugar beet cyst nematode *Heterodera schachtii* and cabbage cyst nematode, *H. cruciferae,* as well as species of rootknot *(Meloidogyne)* and stubby root *(Paratrichodorus)* are frequent nematode pests.

Lepidopterous insect pests are especially injurious to crucifer crops, some of which include the following:

Armyworm	*Pseudaletia unipuncta*
Beet armyworm	*Spodoptera exigua*

Cabbage looper	*Trichoplusia ni*
Cabbage moth	*Barathra brassicae*
Cabbage webworm	*Hellula undalis*
Corn ear worm	*Heliothis obsoleta*
Cutworms:	*Spodoptora littoralis*
Black cutworm	*Agrotis ipsilon*
Granulate cutworm	*Feltia subterranea*
Variegated cutworm	*Peridroma saucia*
Diamond back moth	*Plutella xylostella*
Gramma moth	*Phytometra gamma*
Imported cabbage worm	*Pieris rapae*
Leaf webber	*Crocidolomia binotalis*
Oblique-banded caterpillar	*Cacoecia costana*
Pyralid caterpillar	*Crocidolomia binotalis*
Salt march caterpillar	*Estigmene acrea*

Additional insect pests are as follows:

Bugs	*Eurydema* spp.
Cabbage aphid	*Brevicoryne brassicae*
Cabbage flies	*Chortophila brassicae*
Cabbage maggot/fly	*Hylemya brassicae*
Cabbage seed pod weevil	*Ceutorrhynchus assimilis*
Cabbage stem weevil	*Ceutorrhynchus napi*
Cabbage whitefly	*Aleurodes proletella*
Crickets	*Gryllus* spp.
Darkling beetles	*Blapstinus* spp.
False wireworms	*Eleodes* spp.
Flea beetles	*Phyllotreta* spp.
Green peach aphid	*Myzus persicae*
Harlequin bug	*Murgantia histrionica*
Leafminers	*Liriomyza* spp.
Mustard beetles	*Phaedon* spp.
Seed corn maggot	*Hylemya platura*
Seed pod midge	*Dasyneura brassicae*
Silverleaf whitefly	*Bemisia argentifolii*
Springtails	*Papirius maculosus*
Sweet potato whitefly	*Bemisia tabaci*
Thrips	*Thrips tabaci*
Turnip sawfly	*Athalia rosae*
Turnip aphid	*Hydaphis erysimi*
Wireworms	*Agriotes* spp., *Limonius* spp.

TABLE 19.5.   WORLD PRODUCTION OF CABBAGE, 1994

	Area (ha × 10³)	Yield (t/ha)	Production (t × 10³)
World	1,713	23.5	40,250
Africa	33	25.4	841
North and Central America	106	20.1	2,139
South America	25	23.2	569
Asia	983	24.0	23,566
Europe	563	23.1	13,007
Oceania	4	29.0	127
Leading countries			
China	419[a]	23.5	9,850[a]
Russian Federation	180[a]	26.0	4,680[a]
India	200[a]	16.5	3,300[a]
Japan	68[a]	40.0	2,700[a]
Korea, Rep.	47[a]	55.0	2,600[a]
Poland	57	29.3	1,672
United States	76[a]	21.6	1,650[a]
Indonesia	61[a]	21.8	1,332[a]
Ukraine	69	12.9	893
Korea, D.P. Rep.	44[a]	19.7	865[a]

[a]estimated
Note: The reported cabbage statistics may also include some Chinese cabbage production, especially for the production indicated for many Asian countries.
Source: 1994 FAO Production Yearbook, Vol. 48, FAO, Rome, 1995

Some common physiological disorders are *tipburn* of cabbage and brussels sprouts, *hollow stem* of broccoli and cauliflower, *riceyness* and *blindness* of cauliflower, and *suckering* in cabbage. *Brown bead* of broccoli, *black speck* of Chinese cabbage, and *oedema* in many crucifer species are additional disorders.

Production

Asia clearly leads in cabbage production having 58% of world production; see Table 19.5. China's production leads the world with over 24% of the total cabbage produced. The former USSR republics collectively accounted for more than 18% of the total, with the combined production of the Russian Federation and Ukraine contributing more than 13% of world production.

European cauliflower production many years ago was greater than that of Asia. Presently, Europe accounts for about 22% of world production, whereas Asia produces about 70% of the total. The leading individual country is India, with 44% of world production having recently replaced China as the world's largest producer. See Table 19.6.

TABLE 19.6. WORLD PRODUCTION OF CAULIFLOWER, 1994

	Area (ha × 10³)	Yield (t/ha)	Production (t × 10³)
World	606	18.0	10,888
Africa	9	17.6	160
North and Central America	28	13.7	383
South America	5	13.7	67
Asia	407	18.9	7,708
Europe	152	16.2	2,463
Oceania	5	21.0	107
Leading countries			
India	270[a]	17.8	4,800[a]
China	88[a]	25.7	2,265[a]
France	44	12.6	553
Italy	23	19.7	443
United Kingdom	26[a]	15.7	414
United States	22	13.6	294
Spain	14	20.1	271
Poland	14	15.7	220
Germany	6	25.2	156
Pakistan	9[a]	16.7	145[a]

[a]Estimated.
Source: 1994 FAO Production Yearbook, Vol. 48, FAO, Rome, 1995

OTHER BRASSICAS

In addition to *B. oleracea,* other *Brassica* species which include Chinese cabbage, various mustards, and turnip also have worldwide vegetable importance. Mustard is often used in a generic sense to identify somewhat morphologically similar brassicas even though they are different species. Examples are black mustard *(B. nigra),* white mustard *Sinapis alba (B. hirta),* and Ethiopian mustard *(B. carinata* and *B. juncea,* known as Indian, brown, and yellow mustard. Another very important vegetable *Brassica* species is *B. napus,* which includes rutabaga, Siberian kale, as well as oil seed rape *(Brassica napus* var. *oleifera*), a crop of worldwide importance.

The many different vegetable commodities of *B. rapa* occupy considerable importance in the food supply and nutrition of millions of people, particularly in Asian countries. The origin of *B. rapa* is not definitely known. Some evidence suggests that the eastern Mediterranean region and regions of Afghanistan, Iran, and west Pakistan are probable sites. However, it is believed that some members of the species are indigenous to China and eastern Asia. The earliest forms were likely domesticated for the edible seed oil. The variability of the many forms of *B. rapa* suggests multiple areas of origin.

CHINESE MUSTARDS, *Brassica rapa* L. subsp. *chinensis* (Rupr.) Olsson (Group Chinensis)

Many of the mustards in this group are commonly known as *pak choi* (meaning white vegetable in the Cantonese language) or *bok choy,* and many kinds resemble Swiss chard in appearance (Fig. 19.8). Pak choi production parallels that of Chinese cabbage and has been cultivated since the 5th century. It continues to be an important Asian vegetable, especially in China. The nonheading relatively erect or semiprostrate tight spiral of shiny, dark green, nearly oval or petiolate long leaves are attached to the compressed stem. White or light green petioles are thick and fleshy; plants are from 15 to 30 cm in height. Morphological variations and maturity periods are quite extensive for cultivars of this group. Leaf forms with different shades of green and purple occur, and dwarf cultivars are also known.

Pak choi is less temperature sensitive than Chinese cabbage, and therefore has a broader adaptation. Minimal vernalization is usually

FIG. 19.8. Pak choi, *Brassica rapa* gp. Chinensis.

required for bolting. Flowers are pale yellow. Directly seeded or transplanted, field populations are high and commonly grown at about 20–25 plants/m^2, and dwarf cultivars at twice that density.

Early cultivars mature within 40 days; and some require as long as 80 days following planting. The fully expanded tender leaves and petioles are usually cooked, stir fry being a favorite method. It is important that plants are not permitted to over mature since the quality quickly declines. Pak choi has a relatively short postharvest life, but product quality can be maintained at 0 C and 95% RH for about 10 days.

Choi sum (meaning vegetable heart in Cantonese), also mistakenly called flowering Chinese cabbage, is usually classified as *B. rapa* subsp. *parachinensis* (Bailey) Tsen & Lee (Group Parachinensis). This nonheading plant resembles pak choi and is sometimes called mock pak choi. It is also widely cultivated in Southeast Asia. Because of a short growth period and early harvest of young relatively small plants, high populations are grown. After about 5–6 weeks of growth, the choi sum plants are from 20 to 60 cm tall. The bright dark green or purple foliage and inflorescence is harvested at or slightly after flowering. Plants readily bolt when grown during long days; and the small flowers are yellow. The harvested plant is used like pak choi. The postharvest life of choi sum is short even with good storage conditions.

MIZUNA/MIBUNA, *Brassica rapa* L. subsp. *japonica* (Group Japonica)

Mizuna and mibuna are often called Japanese greens. Mizuna cultivars have many leaves, either entire, lyrated, or finely dissected on long petioles which, with enlargement by tillering, present a bushy growth. (Fig. 19.9) Mibuna cultivars have a similar appearance, but leaf blades are entire and have an elongated oval shape. These plants possess characteristics of *B. rapa* and *B. juncea* and are thought to have resulted from introgression of the two species.

CHARLOCK/KABER, *Brassica kaber* (DC.) Wheeler
(Brassica arvensis, Sinapis arvensis)

Charlock is a related, seldom cultivated, weedy annual mustard that possibly originated in the Mediterranean region. The open rosette of ovate lobed leaves grows to a height of 50–70 cm and are eaten as a potherb. Flowers are yellow and produce small 2-cm-long siliques. The seed can have condiment use, although pungency is low.

FIG. 19.9.   Mizuna, a type of *Brassica rapa.*

TURNIP RAPE, *Brassica rapa* L. subsp. *oleifera* DC. (Group Oleifera)

The primary uses of turnip rape is for its foliage as a potherb, and for seed sprouts and garnish. The use of edible oil extracted from seed is minor. This is in strong contrast to plants of *B. napus* var. *oleifera,* often mistaken for turnip rape. Both species have summer and winter types and seed with high oil and protein content. However, it is *B. napus* var. *oleifera,* widely known as oil seed rape or *canola,* that is primarily grown for seed oil. Oil seed rape cultivars with low erucic acid and low glucosinolate content have greatly improved oil quality. About 10% of the world's edible vegetable oil is obtained from rape seed; and *turnip* rape is becoming increasingly important because of the high content of unsaturated oils. Oil seed rape foliage has relatively little vegetable use.

CHINESE FLAT CABBAGE, *Brassica rapa* L. subsp. *narinosa* (Bailey) Olsson (Group Narinosa)

A low, compact plant, Chinese flat cabbage produces a stout cluster of thick, wrinkled leaves with broad, white petioles. It is known for having very good cold tolerance and is a popular potherb.

CHINESE CABBAGE, *Brassica rapa* L. subsp. *pekinensis* (Lour.) Olsson (Group Pekinensis)

Chinese cabbage is the most important vegetable of this species. It is widely known as pe-tsai (meaning white vegetable in the Mandarin language) and in the United States as napa or nappa cabbage. This vegetable is of major importance in China and Korea and, recently, only less important than radish and cabbage in Japan. Production areas of China, Korea, and Japan are about 300,000, 50,000, and 35,000 ha, respectively. Other Asian countries also have significant production, and its recent popularity has resulted in a considerable increase in Europe and the United States. Chinese cabbage is believed to be native to China and possibly evolved from natural crossings with nonheading pak choi and/or turnip both of which have been grown for more than 1600 years. Distinctive cultivar development of Chinese cabbage has occurred only in the past 600 years.

Chinese cabbage is grown as an annual crop. Most cultivars are biennial, although some exhibit annual flowering behavior. The height of most cultivars ranges from 20 to 60 cm. Initially slender, the taproot becomes prominent, less so if transplanted, and produces a finely branched extensive root system; most root development occurs within 30 cm of the soil surface.

The stem of Chinese cabbage is compressed and unbranched, until bolting occurs. Leaf blades of the sessile, mostly ovate, and slightly wrinkled leaves extend to the bottom of the broad, flat, and lightly colored midrib. Chinese cabbage resembles cos lettuce in appearance. Numerous cultivars are grown: Some produce tightly compact, drum-like heads, others are loose or nonheading (Fig. 19.10). Those with erect cylindrical head forms identified as *B. rapa* subsp. *pekinensis* var. *cylindrica* are called chichili; round and compact heading types of *B. rapa* subsp. *pekinensis* var. *cephalata* are called *chefoo;* another form, *B. rapa* subsp. *pekinensis* var. *laxa* is open headed. Having been cultivated in China since the 5th century, it is not surprising that natural crossing occurred, and by selection, distinctive types resulted. There are many local distinctions for the various forms. In general, maturity characteristics are used to select cultivars and to schedule planting and harvest periods.

A productive, relatively short growing period and broad adaptation provided by the many cultivar choices make this a popular crop for growers. Moderate day and cool night temperatures result in high productivity and quality. The preferred growth temperature ranges from 12°C to 22°C; head formation is optimum between 16°C and 20°C. Temperatures

FIG. 19.10.   Medium longitudinal section of wong bok type of Chinese cabbage, *Brassica rapa* gp. Pekinensis.

greater than 25°C tend to delay heading and reduce quality, except for heat-tolerant and early-maturing cultivars. These cultivars develop narrow leaves having a high ratio of midrib to leaf blade tissues. Additionally, high temperatures during head formation cause a loss of head compactness and can increase the incidence of tipburn. Temperatures greater than 32°C are injurious to flower development.

Light intensity and temperature interact to influence growth and development. High light intensity promotes broad leaf development and early heading. Conditions of low light result in narrow leaves and yield loss. Day length does not affect head formation but may influence growth rate. Prolonged low temperatures are conducive to bolting; for some cultivars, less than 13°C is sufficient. Vernalization is necessary to induce flower stalk development; the longer the chilling period, the more rapid the bolting response. To avoid premature bolting, temperatures should be above 18°C. Whereas low temperatures stimulate bolting, long days enhance flowering. For some cultivars, long days are a greater stimulus to bolting than low temperatures. Under continuous long days, some sensitive cultivars can bolt and flower without experiencing temperatures less than 15°C.

Most of the Chinese cabbage crop is grown in the cool dry subtropics and in temperate regions having moderate temperatures. Nevertheless, the crop is also grown in the tropics at high elevations where temperatures are moderate. Breeding objectives for the lowland tropics include heat tolerance and tipburn resistance. Many cultivars bred for production of heading types during high temperatures have been introduced.

Tipburn, to which Chinese cabbage is more susceptible than other crucifers, is a frequent disorder responsible for large losses. Excessively rapid growth rate during heading tends to increase the disorder. Some control is possible if nitrate rather than ammonium nitrogen is used during growth. Foliar applications of $CaCl_2$ have also helped to reduce the incidence.

Seed and/or transplanted seedlings are used to establish the crop. Field populations are usually between 60,000 and 80,000 plants per hectare. In general, most cultural practices are similar to those applicable to cole crops. Plants have a relatively high requirement for nitrogen and calcium, as well as soil moisture. Soil moisture should not be less than 70–85% of field capacity. Good drainage is important because tolerance to flooding is poor; tolerance to salinity is also poor. Raised plant beds are frequently used to improve drainage.

Early growth is rapid and head formation follows soon after a rosette of 12–15 leaves have formed. Sometimes, growers tie the outer plant leaves together in order to advance and improve head formation. Some cultivars mature in as few as 50 or 55 days; others require as many as 100 days. Plant weight also varies greatly, ranging from 0.5 to 5 kg; 10 kg is possible. Leaf numbers range from as few as 20 to 150, and yields as high as 60 t/ha are achieved.

Chinese cabbage is extensively used as a cooked vegetable. In many Asian countries much of the Chinese cabbage is also processed by pickling; the product is known as kimchi in Korea. Leaves are also dehydrated for later use in soups and other food preparations. In Western cuisine, the foliage is often used raw in salads. Because of a high vitamin C content, Chinese cabbage is a highly valued vegetable. Provitamin A values may also be high, although levels vary greatly with cultivars. Crosses of turnip and Chinese cabbage have been made to improve interior leaf color and thus increased carotene content of heading-type Chinese cabbages. Given proper temperature and humidity management, Chinese cabbages have good storage characteristics and may be stored for several months. After storage, it often is necessary to remove the outer leaves as they become senescent or diseased.

TENDERGREEN MUSTARD, *Brassica rapa* L. subsp. *perviridis* Bailey (Group Perviridis)

This mustard probably originated in areas east of the Mediterranean region. Plants produce large, dark green foliage that has a mild flavor compared to most mustards. The rapidly growing, early-maturing plants are hardy; most are annual, and some are biennial. In addition to the edible foliage, the enlarged hypocotyl crown is also used fresh and pickled.

BROCCOLI RAAB, *Brassica rapa* L. *rapa* (DC.) Metzg. (Group Ruvo Bailey)

Broccoli raab is a highly regarded mustard potherb in Italy and other Mediterranean countries. It is frequently called Italian turnip, cima de rapa, or rapini. Broccoli raab is an annual resembling sprouting broccoli, but it develops a much smaller, less compact inflorescence. Leaves are petiolate, deeply lobed, dark green, and glossy. Tender young shoots, foliage, and the young inflorescence are edible and used as a potherb (Fig. 19.11). Plants usually attain harvest maturity in 40–60 days. Harvests are made before flower buds open. Two major

FIG. 19.11.   Broccoli raab, *Brassica rapa* gp. Ruvo.

cultivar types are grown: one for fall or early winter harvests, the other for spring harvests.

TURNIP, *Brassica rapa* L. subsp. *rapifera* Bailey (Group Rapifera)

The turnip possibly originated in eastern Afghanistan and western Pakistan; the Mediterranean region may be another primary center. Turnips are widely distributed in Asia Minor, a secondary center, and were known to the Greeks and Romans in the pre-Christian era. They remain more popular to Europeans than Asians. Wild forms resemble weedy rape or field mustard. Plants are biennial and largely self-incompatible.

During domestication, selections were for the enlarged fleshy storage organ, consisting of hypocotyl and upper root tissue. Turnips vary in form: They may be round, cylindrical, or flattened globes. The taproot is distinctive, with few secondary roots developing from lower portions, and the upper portion, because of the close attachment of leaves, is often smooth without an evident neck (Fig. 19.12). Exteriors may be white, green, bronze, purple, or combinations of these colors; the interior is white or light yellow. Cultivars are grown that mature within 50 days; some require as long as 100 days. Others are overwintered for spring harvest and are marketed as bunched plants or as trimmed

FIG. 19.12. Turnip, *Brassica rapa*.

roots. The best quality roots are those less than 10 cm in diameter; larger roots are prone to cracking and pithiness.

Turnip leaves are closely attached near the apex of the highly compressed stem. They are bright green, thin, and slightly pubescent and are an important and highly nutritious potherb. Some cultivars are produced specifically for the foliage, and others have fairly wide use as a fodder crop.

Oriental turnips differ from the European forms in having entire and glabrous leaves. Oil seed forms, called turnip rape, are hardy plants that have good climatic adaptation but yield less seed oil than *B. napus* rapes.

SEVEN TOP TURNIP, *Brassica rapa* L. subsp. *septiceps* Bailey (Group Septiceps)

This biennial, also grown for its mustard like foliage, produces several very leafy, tall, edible stems that develop form from the thickened crown. The young shoots resemble broccoli raab.

BLACK MUSTARD, *Brassica nigra* Koch.

This mustard apparently originated in areas of Turkey and Iran. A short-season, semierect plant, it was a primary source for culinary mustard. Although still preferred for its high pungency, its use has been largely replaced by *B. juncea,* because seeds of *B. juncea* do not readily shatter and, thus, are adaptable to mechanical harvest. Major production of *B. juncea* and *B. nigra* as seed crops for condiment use occurs in North America, United Kingdom, Denmark, and eastern Europe.

MUSTARDS of *Brassica juncea* (L.) Czernj. & Coss.

The likely origin of *B. juncea* is central Asia near the Himalaya foothills. Migration occurred to secondary centers of domestication in India, central and western China, and the Caucasus mountains region. Sanskrit records indicated that these crops were grown as early as 3000 B.C. Plants of this annual self-pollinating, generally hardy species are also widely known as Indian, brown, or yellow mustard. The classification for members of *B. juncea* is confusing because of many different

FIG. 19.13.   Chinese mustard, *Brassica juncea;* one of many leafy forms.

forms and because some are occasionally also referred to as Chinese or Oriental mustards (Fig. 19.13).

Examples of the various kinds of *B. juncea* mustards are heading, large, small, curled leaf, large petiole, green petiole, root, big-stem, multishoot, and an almost endless list other names. These have been identified by subdivision as botanical varieties. Being known by numerous common names adds to the confusion. Nevertheless, they are widely grown and are produced in great volume.

These mustards vary considerable in pungency. Most forms of these mustards are used as cooked potherbs. Inner leaves are milder and preferred for salads; older outer leaves are strongly flavored and therefore usually cooked. The nutritious foliage is high in pro-vitamin A and ascorbic acid.

ETHIOPIAN MUSTARD, *Brassica carinata* A. Braun

Also known as Abyssinian mustard because of its East African origin, Ethiopian mustards are rapid-growing, mild-flavored plants, tolerant of low temperatures. The foliage can be harvested in as few as 35–40 days, and plants can be harvested more than once if regrowth is permitted. Production of Ethiopian mustards is considerably less than that of Indian mustards; Ethiopian mustard is reported to be less palatable.

RUTABAGA, *Brassica napus* L. var. *napobrassica* (L.) Reichb. (Group Napobrassica)

Rutabaga is sometimes also identified as *B. napobrassica.* Of relatively recent domestication, its likely origin was in European Mediterranean regions as an amphidiploid of *B. rapa* and *B. oleracea;* wild forms are unknown. *B. napus* plants are generally self-fertile; in contrast, those of *B. rapa* and *B. oleracea* are not.

Sometimes called Swedish turnips, swedes, or turnip rooted cabbages, they are broadly adapted high-yielding crops, preferring cool to moderate growing temperatures of less than 25°C. Most are produced in high-latitude temperate regions.

Rutabagas are biennial plants grown as annuals for the globe or subglobose swollen hypocotyl/root axis (Fig. 19.14). The leaves are smooth and bluish white with thick petioles and have a waxy bloom. The roots are yellow, green, bronze, purple, or a combination of these colors. The high dry matter flesh is either pale yellow or white. Second-

FIG. 19.14. Rutabaga, *Brassica napus* gp. Napobrassica.

ary roots arise from the lower underside of the enlarged organ and from the taproot. Rutabagas exhibit a prominent thickened neck.

Flavor improves during low temperatures or following light frosts. Harvest maturity is usually achieved in about 90 days. Winter hardiness and long-term storage correlates with high dry matter content. Long storage with minimum quality loss is a major attribute of the crop.

SIBERIAN KALE, *Brassica napus* var. *pabularia* (DC.) Reichb. (Group Pabularia)

This leafy vegetable resembles collards or salad rape and is used fresh in salads and as a cooked potherb. Leaves are curled, but less than the kales of *B. oleracea* group Acephala. Another name for this commodity is Hanover salad.

OTHER CRUCIFER VEGETABLES

RADISH, *Raphanus sativus* L.

Of the six *Raphanus* species, only *R. sativus* is cultivated. Although some taxonomists consider China to be the center of origin, others do not. The greatest diversity is found in the eastern Mediterranean and eastward to regions near the Caspian Sea. Wild radish such as *R. raphanistrum* are found in eastern Europe. *R. maritimus* is found in the same region and also along the European northwest and western coasts, and *R. rostratus* is found near Greece and eastward to the Caspian Sea. Radish was used for food in ancient Egypt as early as 2700 B.C. and in China and Korea about 400 B.C.

Botanical varieties of R. sativus are as follows:

var. *radicula* (garden or European radish)
var. *longipinnatus* (Daikon, Chinese winter radish, varied types)
var. *raphanistroides* (Japanese winter radish)
var. *niger* (black or Spanish radish)
var. *caudatus* (rat-tail radish, grown for its long edible tender seed pods)
var. *moughi* (fodder radish, no enlarged storage root)
var. *oleifera* (oil seed radish)

Most cultivated forms are annuals; long photoperiods tend to advance flowering. The thickened fleshy hypocotyl and upper portion of the root is the primary edible portion; secondary roots branch from the lower tap-root. Storage root length and width range from short to very long and slender to thick, and shapes may be spherical, cylindrical, tapering, or combinations of these. Skin colors may be red, white, yellow, purple, black or green; flesh color is usually white. A red fleshed form is popular in China.

Leaves are alternate, smooth, or slightly hairy, of variable shape but most often oblong linear; the growth habit is mostly rosette. Under short days, storage roots are well shaped and top growth small. With long days, root shape is altered and top growth increases and flowering occurs early. After low temperature induction followed by long days, bolting is expressed. The white to lilac flowers are self-incompatible. Silique size differs among cultivar types. Radish seeds are larger than those of Brassica species; about 100 seed weigh 1 g.

Radishes are seed propagated and directly sown. Plant spacing vary greatly because of cultivar differences. Garden radishes are grown at high densities such as 2–5 cm between plants in rows and from 10 to 20 cm between rows; sometimes, broadcast plantings are made, often when the crop is to be mechanically harvested. On the other hand, the daikon and shogoin radish cultivars require more space because of their large storage roots (Fig. 19.15a).

With many garden-type cultivars, the relatively small enlarged fleshy hypocotyl/root axis is subtended by a rosette of short leaves arising from the compressed stem (Fig. 19.15b). Preferred storage root sizes range from 2 to 5 cm in diameter. The most common root shape is globe; other cultivars produce elongated or tapered roots, some with lengths of more than 15 cm. Skin colors vary from red to white, or a combination of both. Another cultivar group, popular in Asian countries, produces long (10–25 cm) and narrow (1.5–3.0 cm), tapered white

a   b

FIG. 19.15.   Radish roots, *Raphanus sativus,* daikon (a) and garden radish (b).

storage roots but has a slightly longer growing period. Excessive growth rate can result in early pithiness. Overmaturity and/or bolting also lead to pithiness; the extent varies among cultivars.

Garden radish cultivars are generally a temperate region crop and mostly grown in Europe. They are often classified according to seasonal use and are commonly harvested within a short period that may range between 20 and 40 days. Even with good temperature management, 0°C and 95% RH, most garden radish cultivars have a relatively short, 1–2 week, postharvest life. If stored with tops removed, shelf life can be extended to 3–4 weeks.

Compared to garden radish, Chinese winter radish is by far the most important. The volume of consumption is huge, especially in Japan, China, Korea, and other Asian countries. Winter radishes require a longer growth period to produce the large storage root and foliage biomass. Leaf blades are often strongly notched and may be erect, spreading, or prostrate.

Daikons are one form of winter radish; and they are a major vegetable in many Asian countries and are highly appreciated in Japanese diets. Another that differs in size and shape is the shogoin type. Daikon root size can range from 10 to more than 50 cm in length, with diameters from 4 to 10 cm, and can weigh between 1 and 4 kg, but larger roots, as heavy as 20 kg, are obtained. However, edible quality tends to diminish with excessive size.

Raised-bed cultivation of daikon cultivars is beneficial for the long growth of the storage root of these cultivars and also facilitates harvest. A common spacing is 15–25 cm apart in rows and about 40–60 cm between rows. Some cultivars exhibit considerable above-soil surface development of the storage organ, which can result in green coloration. Varying with cultivars, harvests are made between 50 and 90 days for optimum quality. Harvest is performed by hand and carefully dug so as not to break the storage roots. When overmature, the incidence and severity of pithiness increases. Stress-affected growth results in bitter taste, tissue toughness, and pithiness.

Winter radishes, under proper conditions, have fair to relatively good storage characteristics. When tops are removed, many daikon cultivars can be maintained in good condition for 3–4 weeks with cool temperatures and high relative humidity.

Daikon is eaten fresh, cooked, pickled, and as a dried product. Shogoin types are generally cooked. Pungency is due to the presence of 4-methylthio-3-butenyl isothiocyanate, and the content varies with cultivars but can be modified by environmental conditions.

Spanish radish is also a annual winter radish type that produces round, usually black-skinned roots, with preferred sizes ranging from

FIG. 19.16.   Radish sprouts, *Raphanus sativus,* used for garnish and in salads.

10 to 15 cm in diameter. Roots mature in 50–90 days after planting. Spanish radishes have a long, 2–3-month, postharvest storage life when properly refrigerated and held at high relative humidity.

Although best recognized for the crisp texture and pungency of the storage root, other radish plant parts are also consumed. Young leaves are eaten as salad or as potherbs. Young seed pods, especially those of rat-tail radish (*R. sativus* var. *caudatus*) can be 30 cm long and are eaten as fresh, cooked or pickled, and as snack foods, particularly in India, and also in East Asian countries. Sprouted seedlings at the cotyledon stage are used as garnish in many countries (Fig. 19.16).

HORSERADISH, *Armoracia rusticana* (Lam.) P. Gaertn., B. Mey., et Scherb.

A native of southeastern Europe, horseradish is a hardy perennial, well adapted to temperate areas and grown for the piquant flavor of

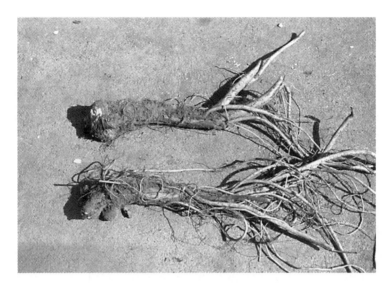

FIG. 19.17.   Horseradish, *Armoracia rusticana,* side roots are trimmed before market presentation.

the fleshy rhizomes (Fig. 19.17). Allyl isothiocynates are responsible for the flavor. The lower leaves are large (30 cm and larger) and half as broad, with jagged margins. Upper leaves are lanceolate with smoother margins. Flowers are small, white, and form on a terminal panicle.

Some growers follow a cultural practice that carefully removes attached side roots (actually rhizomes) from the primary rhizome during growth order to improve its size and smoothness. Harvests are made at the end of the growing season or when the primary rhizomes are sufficiently large. Small and/or secondary rhizomes are commonly used for propagation; these exhibit a distinct polarity.

UPLAND CRESS, *Barbara verna* (Mill.) Asch.

Also known as Belle Isle cress, the foliage of this biennial plant is harvested from cultivated and noncultivated plants in western and southwestern Europe, and is most popular in France and Belgium. Its prostrate growth resembles watercress, but plants do not grow in water. However, a cool temperature and moist soil environment is preferred. The rosette of leaves are pinnately lobed and are used like watercress for salads, garnishes and as a potherb. Seed are also spouted for salad use.

WINTER CRESS, *Barbara vulgaris*

Winter cress is similar to upland cress and used for the same purpose, but is usually a less preferred alternative. It is also known as yellow rocket or rocket cress. Plants frequently become weedy pests. "Cress" is a term that unfortunately has been applied to many plants with resulting confusion as to their specific identity.

SHEPHERDSPURSE, *Capsella bursa-pastoris* (L.) Medikus

Although best known as a weed, the young rosette foliage of some forms can serve as a spinach substitute. Leaves have high nutritional value.

SEA KALE/CRAMBE, *Crambe maritima* L.

Sometimes known as Abyssinian kale, sea kale should not be confused with *Crambe abyssinia,* an oil seed plant. Sea kale is native to the coastal areas of western Europe and is also found in the Black Sea region. The name and reputation was acquired from its use as a vitamin C source for sailors to avoid scurvy.

Sea kale is a hardy perennial growing 60–90 cm tall with large bright green leaves which are curled at the edges. Leaf shoots are the main edible portions. Emerging leafy shoots, about 10–15 cm length are harvested while still tightly folded and before leafing out, although expanded leaves are also eaten. Previously harvested and vernalized roots can be forced in the dark and shoots harvested. Plants are propagated from seeds or cuttings.

ROCKET SALAD, *Eruca vesicaria* (L.) Cav. subsp. *sativa* (Mill.) Thell.

This is a plant of many names, some being roquette or arrugula. Originating in southern Europe and western Asia, rocket salad is an old crop known to have been used by the ancient Romans. An annual, the plant produces dull green, deeply cut, compound foliage that resembles radish leaves. The leaves have a sharp pungent taste resembling horseradish. The cool season plant matures in 60–90 days, but young tender leaves or small shoots can be harvested at any time. Mature foliage and/or foliage grown during periods of high temperature have an especially hot burning flavor. Mean temperatures greater than 30°C are conducive to bolting. Rocket salad foliage is used fresh in salads, or cooked as a potherb, usually being mixed with other vegetables. Seeds are also used for the highly pungent oil.

GARDEN CRESS/LAND CRESS, *Lepidium sativum* L.

This cress is also called pepper cress and is an annual of European origin. Its leaves are used for their pungent taste in salads, and as a garnish. Seed are also sprouted for salad use.

WHITE MUSTARD, *Sinapis alba* L. *(Brassica alba)*

White mustard originated in eastern Mediterranean and Aegean Sea regions. This species is photoperiod sensitive and requires long (16 h) days to flower. Relatively little vegetable use is made of either black or white mustard as potherbs.

Future Improvements

Future increases in productivity and consumption of many crucifers is promising because of recognition of their excellent nutritional values and in some measure to the proclaimed health benefits, namely in cancer prevention. The cole crops, broccoli in particular, has shown exceptional increases in consumption and other brassicas such as Chinese cabbage remain as a major component in Asian diets. The relatively broad adaptation of *Brassica* species suggests that many subtropical and tropical regions will increase production. Breeding of these crops is extremely intense as evident by introductions of many hybrids. Utilization of new technologies will further improve adaptation as well as pest resistances and nutritional content.

SELECTED REFERENCES

Baggett, J.R., and Kean, D. 1985. Inheritance of annual flowering in *Brassica oleracea*. HortScience *24*, 662–664.

Bose, T.K., and Som, M.G., eds. 1986. Vegetable Crops in India. Naya Prokah Publ., Calcutta.

Dickson, M.H., and Wallace, D.H. 1986. Cabbage breeding. In Breeding Vegetable Crops. M.J. Bassett, ed. Chapman & Hall, New York, pp. 395–432.

Elers, B., and Wiebe, H.J. 1984. Flower formation of Chinese cabbage. I. Response to vernalization and photoperiods. Sci. Hort. *22,* 219–231.

Fisher, N.M. 1974. The effect of plant density, date of apical bud removal and leaf removal on the growth and yield of single harvest Brussels sprouts. J. Agric. Sci. *83,* 489–496.

Fujime, Y., and Hirose, T. 1979. Studies on thermal conditions of curd formation and development in cauliflower and broccoli. I. Effects of low temperature treatment of seeds. J. Japan. Soc. Hort. Sci. *48,* 82–90.

Fujime, Y., and Hirose, T. 1980. Studies on thermal conditions of curd formation and development in cauliflower and broccoli. II. Effects of diurnal variation of temperature on curd formation. J. Japan. Soc. Hort. Sci. *49;* 217–227.

George, R.A.T., and Evans, D.R. 1981. A classification of winter radish cultivars. Euphytica *30,* 483–492.

Gray, A.R. 1982. Taxonomy and evolution of broccoli (*Brassica oleracea* var. *italica*). Econ. Bot. *36,* 397–410.

Hand, D.J., and Antherton, J.G. 1987. Curd initiation in the cauliflower. I. Juvenility. J. Exp. Bot. *38,* 2050–2058.

Hara, T., and Y. Sonada, 1979. The role of micronutrients for cabbage head formation-growth performance of a cabbage plant and potassium nutrition with the plant, Soil Sci. Plant Nutr. 25, 103–111.

Hill, C.B., Williams, P.H., Carson, D.C., and Tookey, H.L. 1987. Variation in glucosinolates in oriental *Brassica* vegetables. J. ASHS *112,* 309–313.

Howard, H.W. 1976. Watercress. In Evolution of Crop Plants. N.W. Simmonds, ed. Longmans, London, pp. 62–64.

King, G.J. 1990. Molecular genetics and breeding of vegetable brassicas. Euphytica *50,* 97–112.

Kowalenko, C.C. 1983. Broccoli response to nitrogen application, Res. Review May–Aug: pp. 8–9, Res. Sta. Agassiz, B.C. Canada

Kuo, C.G., and Tsay, J.S. 1981. Physiological responses of Chinese cabbage under high temperature. In Chinese Cabbage, Proc. First International Symposium, N.S. Talekar and T.D. Griggs, eds. AVRDC, Taiwan, pp. 217–224.

Kuo, C.G., Tsay, J.S., Tsai, C.L., and Chen, R.J. 1981. Tipburn of Chinese cabbage in relation to calcium nutrition and distribution. Sci. Hortic. *14,* 131–138.

Magnifico, V., Lattanzio, V., and Sarli, G. 1979. Growth and nutrient removal by broccoli. J. ASHS *104,* 201–203.

Miller, C.H., Konsler, T.R., and Lamont, W.J. 1985. Cold stress influence on premature flowering of broccoli. HortScience *20,* 193–195.

Moe, R., and Guttormsen, G. 1985. Effect of photoperiod and temperature on bolting in Chinese cabbage. Sci. Hortic. *27,* 49–54.

Monteiro, A.A., and Williams, P.H. 1989. The exploration of genetic resources of Portuguese cabbage and kale for resistance to several *Brassica* diseases. Euphytica *41,* 215–225.

Sherf, A.F., and McNab, A.A. 1986. Vegetable Diseases and Their Control. John Wiley & Sons, New York.

Silva-Dias, J.C., Barnard, J., Phippem, W.B., and Kresovich, S. 1990. Genetic diversity of landraces of Portuguese cabbages and kales, *Brassica oleracea* L. In Proc. Sixth Crucifer Genetics Workshop, J.R. McFerson, S. Kresovich, and S.G. Dwyer, eds. Cornell University, Ithaca, NY.

Stevens, C.P. 1983. Watercress. In ADAS/MAFF Ref. Book 136. Grower Books, London.

Swarup, V., and Chatterjee, S.S. 1972. Origin and genetic improvement of Indian cauliflower. Econ. Bot. *26,* 381–393.

Talekar, N.S., and Griggs, T.D., eds. 1981. Chinese Cabbage, Proc. First International Symposium. AVRDC, Taiwan.

Tsunoda, S., Hinata, K., and Gomez-Campo, C., eds. 1980. *Brassica* Crops and Wild Allies. Japanese Science Society Press, Tokyo.

U, N. 1935. Genome analysis in *Brassica.* Japan. J. Bot. *7,* 389–452.

Vaughan, J.G. 1977. A multi-disciplinary study of the taxonomy and origin of *Brassica* crops. BioScience *27,* 35–40.

Weatheritt, N. 1992. Cauliflower maturity prediction: A commercial service in the United Kingdom. Horticulturae *32* (1), 4.

Weibe, H.J. 1975. The morphological development of cauliflower and broccoli cultivars depending on temperature. Sci. Hort. *3,* 95–101.

Wen, F., Sun, D., Ju, P., Sun, Y., and An., Z. 1991. Effect of NAA on calcium absorption and translocation and prevention of tipburn in Chinese cabbage. Acta Hortic. Sinica *18,* 148–152.

20

Carrot, Celery, and Other Vegetable Umbels

Family: Apiaceae (Umbelliferae)

There are about 250 genera and more than 2500 species of the widely distributed Apiaceae. Most are cool season plants, although some species and cultivars are adaptive to subtropical environments and are grown at high elevations in tropical areas. Vegetable Apiaceae are recognized for their aromatic flavor characteristics. Usually it is the seed that contain the essential oils responsible for the flavor; for some species, all plant parts are aromatic. The composition of the essential oils are fairly specific for each species. Several species, in addition to vegetable use, also are widely used for flavoring. Nearly all of the vegetable Apiaceae have had some medicinal attributes assigned to them; some are also used as ornamentals. On the other hand, some species contain furanocoumarins, compounds that can caused dermatitis. Although not all individuals are sensitive, the dermatitis response is intensified with exposure to ultraviolet light and can result in skin discoloration.

The characteristic of this family is the umbrellalike inflorescence called an umbel (Fig. 20.1). Pedicels of each flower radiate from a common point at the apex of the inflorescence stalk that resemble the ribs of an umbrella. Umbels are compounded, each primary ray is terminated by a secondary umbel (umbellet). Another common Apiaceae characteristic is the production of schizocarp fruit, which consists of two attached carpels, each containing a seed (mericarp).

Vegetable Apiaceae (Umbelliferae)

Species	Common names
Anthriscus cerefolium	Salad chervil
Apium graveolens var. *dulce*	Celery

FIG. 20.1. Carrot, *Daucus carota,* umbels; the inflorescence is characteristic of other Apiaceae species.

Apium graveolens var. *rapaceum*	Celeriac
Apium graveolens var. *secalinum*	Smallage, leaf celery
*Arracacia xanthorrhiza	Arraccacha
Centella asiatica	Asiatic pennywort
Chaerophyllum bulbosum	Turnip rooted chervil
Coriandrum sativum	Coriander, Cilantro
Cryptotaenia japonica	Japanese hornwort
Daucus carota var. *sativa*	Carrot
Foeniculum vulgare	Florence fennel
†Oenanthe javanica (Sium javanicum)	Water dropwort
Pastinaca sativa	Parsnip
Petroselinum crispum	Parsley
Petroselinum crispum var. *tuberosum*	Turnip rooted parsley
*Perideridia gairdneri	Epos
Sium sisarum	Skirret

*See Chapter 14.
†See Chapter 26.

Condiment Apiaceae

Anethum graveolens	Dill
Angelica archangelica	Angelica
Carum carvi	Caraway
Coriandrum sativium	Coriander
Cuminum cyminum	Cumin
Levisticum officinale	Lovage
Myrrhis odorata	Sweet cicely, myrrh
Pimpinella anisum	Anise

Celery, fennel, coriander, parsley, and chervil also have condiment usage.

CARROT: *Daucus carota* L. var. *sativa* Hoffm.

Of the Apiaceae, carrot is the most widely grown and important vegetable. Progressing from its initial medicinal usage, carrot has become a major vegetable and is largely recognized for the high alpha- and beta-carotene content of the taproot. Both carotenes are important in human nutrition as precursors of vitamin A.

Origin

Afghanistan is considered a center of origin, because the greatest diversity of wild types is found there. Wild types are also found in southwest Asia and eastern Mediterranean regions, which are considered secondary centers of diversity and domestication. Carrot cultivation has been traceable to the 10th century in Asia Minor. Purple and yellow carrot variants were introduced into Europe possibly as early as the 11th century. Introduction into India and China occurred during the 13th or 14th century and to Japan about the 17th century.

Taxonomy and Domestication

Wild carrot, *Daucus carota* var. *carota,* also known as Queen Anne's Lace, is thought to be the ancestor of the present-day carrot, although that opinion has been challenged. It is an annual and readily crosses with the cultivated carrot and often contaminates carrot seed production. Other wild carrot species are *D. maritimus, D. commutatis, D. hispanicus, D. gummifer, D. fontanesii, D. bocconei,* and *D. major.*

Carrot roots occur in different colors, sizes, and shapes. Primitive carrots contain anthocyanin and have purple root tissues. Yellow root mutants were preferred to purple and increasingly became more important. White- and orange-fleshed carrots probably were obtained after repeated selections from the yellow types. Plant breeding during the 17th century in The Netherlands resulted in improving root smoothness and orange color which became landrace cultivars known as Long Orange and Horn types. These served as the basis for much of the germplasm of modern cultivars. It is speculated that mutation and selection was more responsible for the development of the cultivated carrot than by hybridization with wild germplasm.

A major separation of root types are the European and Asian cultivars. In general, European cultivars are firm textured, sweet, highly flavored, yellowish orange to strongly orange in color, slow bolting, and acclimated to cool temperatures. Cultivars grown in Asia have a slightly softer texture, are less sweet, low flavored, bolt easily, are adapted to warm temperatures, and often have scarlet or reddish orange colored roots.

Banga (1976) suggested that along with increased taproot tissue color development, temperate region selections resulted in changing photoperiod response from short to long day and annual to biennial habit. However, that viewpoint and photoperiod response is not recognized by all carrot researchers. Bolting clearly appears to be only a cool-temperature response because photoperiod effects have not been reported, and thus temperature rather than photoperiod is generally considered the major influence. Low temperatures, less than 5°C, tend to accelerate floral induction. The period of low-temperature exposure varies from several weeks to as much as 12 weeks for bolt-resistant cultivars. Some tropical cultivars can be induced to bolt at temperatures less than 15°C. Roguing of early bolters in the seed crop when rigorously performed reduces the population of premature bolters in subsequent crops. Generally, temperate-zone cultivars are biennials; tropical cultivars exhibit an annual habit and are grown during short-day conditions. Being grown at low latitudes, the tropical types thus appear to have a short-day preference.

Botany

Carrot plants form a rosette of leaves and a large fleshy storage taproot during the first year. The stem is very compressed, almost platelike in the first year of growth with foliage height between 25 and 60 cm, occasionally taller. Leaves that arise from the stem have long petioles that flare and are sheathlike at their basal attachment. Leaf blades are repeatedly divided with small, narrow, highly lobed segments. The foliage to root ratio varies considerably among cultivars. Those developing large tops generally produce large roots but require more growing time, whereas small top cultivars produce small roots, but in a shorter growth period.

The taproot, initially a long, thin, vertically growing organ, begins to rapidly elongate and achieves its potential length in 12–24 days after emergence. Yield potential increases with root length. Those longer than 30 cm are difficult to harvest and handle. The taproot is comprised of hypocotyl and primary root tissues. Fibrous roots are absent in the upper hypocotyl area, but numerous, very fine, and highly branched fibrous roots grow from the lower portion of the taproot, some to depths of more than 75 cm. Anatomically, the root is comprised of primary xylem and phloem tissues with cambium sections joining together in a ring. The cambium produces a secondary xylem toward the interior and a secondary phloem toward the outside. For high edible quality, the root ideally should have a minimum of core (xylem) relative to cortex (phloem) and with minimal color differences between these tissues. Xylem color is typically less intense than that of phloem tissue.

In a longitudinal section (Fig. 20.2), the periderm is the outermost

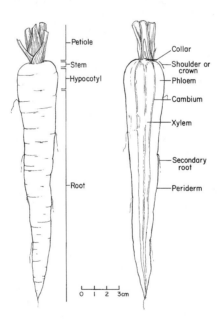

FIG. 20.2.   Carrot, *Daucus carota,* root and longitudinal section.

tissue. Moving inward are phloem, cambium, and xylem tissues. Oil ducts in the intercellular spaces of the pericycle contain essential oils that are responsible for the characteristic carrot odor as well as flavor. The taproot stores a fair quantity of sucrose and other sugars. Usually conical, roots may also be cylindrical, round, or of intermediate shape. At the widest portion, root diameters vary from 1 to more than 10 cm. Root lengths range from 5 to more than 50 cm; between 10 and 20 cm is most common. Flesh colors can be white, yellow, orange, red, or purple. Anthocyanin is responsible for the reddish purple color. Alpha- and beta-carotene, responsible for yellow and orange color, respectively are the major carotenoid pigments. Beta-carotene usually is 50% or more of the total carotenoid content; the ratio of alpha to beta is usually about 1 : 2. Red coloration in the flesh of some cultivars is from lycopene. The carotenoids are not uniformly distributed in the root. Carotene synthesis proceeds from the proximal to the distal tissues of the taproot. Phloem tissue usually contains about 30% more pigment than the xylem.

Differences in carotene content are also influenced by temperature and plant maturity as well as cultivar. The carotene content of the most widely grown carrot cultivars ranges from 60 to more than 120 µg/g fresh weight. Breeders have greatly increased the carotene content in new cultivars (Table 20.1). The amount of lycopene is relatively low in most carrots except in some red-fleshed cultivars such as those of Kintoki-type cultivars that are popular in Japan.

When bolting occurs, the stem elongates and produces many rough, bristly hispid branches. Usually several floral stalks develop, ranging from 1 to 2 m in height. The inflorescence is a terminal compound umbel consisting of many umbellets bearing small white flowers. The umbels are encircled by long-lobed bracts, and umbellets are similarly surrounded by bracts. A large primary umbel of a flower stalk may

TABLE 20.1. CAROTENE CONTENT OF SEVERAL CARROT CULTIVARS AND GERMPLASM

Cultivar/germplasm	Total carotene content (µg/g fresh weight)
Chantenay	41
Nantes	59
Hicolor 9	65
Danvers 126	71
Imperator 58	78
A Plus	168
Beta III	270
High color mass selection	475

Source: Simon (1987). Permission Prentice-Hall.

contain 50 umbellets, each with as many as 50 flowers. Secondary, tertiary, and quaternary umbels are progressively smaller and develop later. The quaternary umbels are less productive, and their seed frequently fail to adequately mature, a major reason for poor seed quality. The flowering period can continue for more than a month. Flowers are usually bisexual with protandrous and centripal floral development, and are mostly insect pollinated. The conspicuous umbels and flowers having nectaries are features that encourage insect visits. The bilocular fruit is a schizocarp.

At maturity, peduncles of the outer umbellets curve inward and the umbel appear concave and looks somewhat like a bird nest. Seeds are flat, ribbed, spiny, and vary greatly in size, with a range from 500 to 1000 seed per gram. With biennial cultivars, flower stalks and seed are produced in the second year, but with appropriate scheduling of growth period and vernalization, seed can be obtained within 12–13 months.

Climate, Soils, Moisture, and Nutrition

Both root and foliage growth are optimum between 16°C and 21°C. Growth slows at temperatures below 10°C; acclimated plants have some frost tolerance. Temperatures greater than 21°C tend to make roots short and stubby, whereas temperatures less than 16°C tend to produce long slender roots. Large diurnal temperature fluctuations are conducive to rapid growth, and if night temperatures are sufficiently cool, carrots can be grown in the tropics. Foliage growth is less affected by temperature and more tolerant of high temperatures than root growth. At temperatures greater than 30°C, foliage growth diminishes and root quality is also adversely affected because of the development of strong flavors.

Carotene synthesis is affected by temperature and is optimum from 16°C to 25°C, and poor below 16°C and above 25°C. Pigment synthesis lags behind root growth, a reason why young roots are pale. With continuing growth, carotene accumulates, attaining a maximum after about 90–120 days of growth, and then usually remains constant or may slowly decline.

An ideal soil for carrot production is a deep, friable, fertile, well-drained sandy loam or peat. Carrots, especially long-rooted cultivars, are adversely affected by shallow or compacted soils. Root length is markedly shortened by compacted soils; root form is also affected. Plants have a fair tolerance to acidity; a favorable pH is between 5.5 and 7.0. Fertilization is usually adequate when 75–150 kg/ha of N,

50–100 kg/ha or P, and 50–200 kg/ha of K is supplied; carrots generally have a relatively high uptake of K. Most recommendations suggest avoiding excessive nitrogen because of the tendency to promote foliage growth preferentially to root enlargement.

Most carrot crops require about 30–50 mm of water per week or from 450 to 600 mm for a cropping season. Uniformity of available water is important; low moisture causes roots to develop strong flavors. High soil moisture can cause root splitting or cracking and tends to inhibit color development. Best yields and water efficiencies are obtained from water applications whenever about 40% of available moisture has been depleted from the root zone.

Culture

Uniform emergence and stand establishment are constant concerns because of variable seed quality and normally slow germination. Because seed are formed on different umbels, they are of different physiological maturity, and yet, with few exceptions, are harvested together. The mixture of seed maturities often leads to variability in germination and emergence. Seed thermodormancy is another factor that can affect germination.

Specialized planting equipment and uniformly sized coated seed are increasingly used for precision sowing carrot seed. In some situations, pregerminated seed are planted using water or a gellike material as a carrier medium. Seed are sown between 5 and 20 mm deep.

Sowing rates (1–3 million seed/ha) are determined depending on germination percentage, seed vigor, and the anticipated effect of field and environmental conditions on emergence. This determination is important because thinning of carrot seedlings is seldom feasible; thus, plant density is predetermined by intended usage. For the fresh market, the density generally ranges from 80 to 100 roots per m². Field populations for small rooted cultivars range from 100 to 200 per m², and for the production of minimally processed "cut and peel" carrots, similar plant densities are used; these are often incorrectly called "baby" carrots. However, for very small carrots that are harvested very early, 5 million seed/ha are planted. These roots are not processed and are produced for a limited market. For large-rooted cultivars, which are usually processed, field densities range from 40 to 70 roots per m². In general, for a particular cultivar, roots will be large when populations are low, and small at high densities.

Nonuniform stands occur because seedlings often fail to become established or grow poorly due to soil crusting, adverse temperature, dry conditions, and weed competition. To improve emergence uniformity, seed priming, precision sowing, and sprinkler irrigation practices are employed to assure that germination is initiated at the same time. A widening use of hybrid cultivars has improved uniformity of germination, seedling growth, and vigor. Adequate moisture for uniform emergence and subsequent early seedling growth is critical. Late-emerging seedlings compete poorly with earlier seedlings and often do not produce marketable roots.

Stands are established in single or multilined rows or as a narrow band of randomly placed seeds. Ideally, spacings should provide uniform distances between roots, but this is difficult to achieved, even with precision planters. To accommodate most mechanical harvesting equipment, the width of seed lines or bands are usually less than 10 or 12 cm.

Early growth is very slow because emergence can vary from 7 to more than 20 days and the first true leaves are not developed until 3–4 weeks after planting. Such slow growth makes carrot a poor competitor with weeds, which, when uncontrolled, can readily overwhelm carrot seedlings. Several selective herbicides are effectively used for weed management.

Cultivar Types

Cultivars are commonly grouped into types that reflect similar morphology. Although all cultivars can have fresh market use, some cultivars are better suited for processing, and some have dual-purpose features. Figure 20.3 illustrates the shape and relative size of some better known cultivar types.

Different countries have preference for certain carrot types and root colors. In Japan, where carrots are seldom eaten raw, long reddish orange colored roots and thick cylindrical carrots are preferred. In Europe, a strong preference exists for relatively short, slender, yellow–orange Nantes and Nantes-like cultivars, whereas the preference in North America is for long, deep orange Imperator and Imperator-type cultivars.

The commercial use of hybrid cultivars has markedly increased, especially for fresh market types. Their major advantage is uniformity of size, shape, and color. Cytoplasmic male sterility (CMS) is employed for producing hybrid cultivars. Two sources of CMS, brown anther and petaloid, are used. With brown anther, sterility results because of anther dysfunction. Sterile cytoplasm and at least two recessive genes with complementary action are involved. Petaloid sterility results from

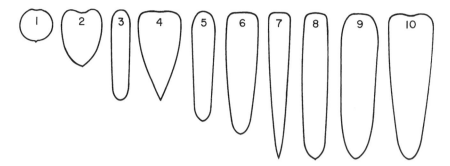

FIG. 20.3. Carrot, *Daucus carota,* types: 1–Parisian market; 2–Oxheart; 3–Amsterdam forcing; 4–Chantenay; 5–Nantes; 6–Danvers; 7–Imperator; 8–Flakkee; 9–Berlikum; 10–Kuroda.

the formation of petallike structures in place of anthers, and again sterile cytoplasm and at least two dominant genes with complementary action are involved.

In addition to uniformity, other major objectives for the improvement of carrot cultivars include increased growth rate, yield, root surface smoothness, and cracking resistance. Additional goals are improved flavor, texture, bolting and pest resistance, and greater adaptation for high-temperature growth, especially in subtropical or tropical regions.

Bolting

Except for seed production, floral seed stalk formation is undesirable because of fibrous development of the central core (xylem). Bolting susceptibility is conditioned by temperature, cultivar, and root size. Some cultivars have strong biennial characteristics and are more tolerant of low temperatures that induce bolting.

Floral induction is promoted by a 6–10-week period of exposure to 10°C or less. Plant susceptibility to vernalization varies with root size. Seedlings with root diameters of about 6 mm or larger are responsive to cold-temperature induction; small or juvenile seedlings are not. The juvenile stage lasts longer for large-rooted cultivars. Young seedlings also tolerate low temperatures and frosts better than older plants. Cultivars having an annual character readily bolt under cold temperatures. Following vernalization, 4–5 months are required to produce mature seed. The incidence of bolting can be avoided or limited by planting schedules that minimize plant exposure to long periods of low temperature. Slow-bolting cultivars have greatly alleviated this problem.

Harvest and Postharvest

Carrot harvesting is not determined by a clearly defined maturity stage. For various reasons, crops are often harvested before the potential full root size or maximum yield is obtained. Depending on cultivar and prevailing growing conditions, the period from planting to harvest can be less than 70 to more than 150 days. Earliness is achieved by small size and/or rapid growth rate. Occasionally, roots are field stored and harvested as needed. Processing carrots are grown for a longer time to increase weight, color, sweetness, or dry matter. High dry matter content is associated with better storage and handling. However, extensive harvest delay is often accompanied by increases in fiber and strong flavor.

Hand harvesting is a difficult task and is infrequently performed except for home garden or small-scale production. With one type of machine harvesting, as the roots are undercut, gripper belts simultaneously grasp the foliage and lift plants from the soil. This action is followed by removal of the foliage which is discarded while the roots are collected into large containers or wagons. Strong and healthy foliage is an important feature for efficient harvesting. However, excessive foliage growth can interfere with harvest efficiency. With other harvesters, the foliage is first cut off, then the roots are dug and lifted out, similar to harvesting and handling of white potatoes.

Because of nonuniform root development, a large proportion of roots are undersized and fail to contribute to marketable yield. Conical shaped carrots tolerate unfavorable growing and harvesting conditions better than cylindrical shaped roots because conical shapes are physically more resistant to spitting during growth and breakage during and after harvest. Smooth root surfaces are important for appearance and preparation. Roots without tops are usually bulk handled until graded for processing, or packaged for the fresh market.

Postharvest preparation of topped carrots can take several routes. For fresh marketing, carrots can be handled and presented in bulk or prepackaged. Fresh bulk handled carrots can be marketed after washing, a practice not followed in all countries. Washed carrots are marketed prepackaged in small plastic bags and have become very popular because the package helps in maintaining quality and contains clean and relatively uniform carrots.

Although much less frequently, bunched carrots with intact tops are

FIG. 20.4. Bunched carrots, *Daucus carota,* for the fresh market.

still marketed (Fig. 20.4). The primary purpose of having healthy and attractive intact foliage that usually is not consumed is to indicate freshness. During harvest, roots are undercut and hand lifted from the soil. Several plants with roots of uniform size are tied together around the base of the foliage to form a bunch; each bunch contains about the same number and weight of carrots. Following bunching, carrots are washed to remove soil from the roots. Several bunches are then placed into containers and hydrocooled or iced to reduced product temperature and respiration. The carrots should be cooled to 1°C or 2°C as soon as possible in order to retain quality and reduced wilting which is especially important with bunched carrots.

Storage

Carrots store best at 0°C and 95% RH; sugars increase during cold storage. Root respiration rates are relatively low compared with other vegetables, and several months of storage are possible with good holding conditions. With such conditions, film-packaged carrots readily maintain good quality for 6–7 weeks. However, bunched carrots store poorly and roots lose firmness because moisture is extracted by the tops and thereby significantly reducing shelf life to as little as 7 days, as well as a reduction of root quality. Postharvest storage of "minimally processed," film-packaged, small carrot sections is usually restricted to less than 20 days.

Exposure to ethylene in storage results in the formation of bitter-tasting isocoumarin compounds. Hence, carrots should not be stored together with ethylene producing commodities, such as apples, bananas, melons, and so forth.

Diseases and Other Pests

Carrot Diseases

Alternaria black rot	*Alternaria radicina*
Alternaria leaf blights	*Alternaria dauci, A. radicina*
Bacterial blight	*Xanthomonas campestris* pv. carotae
Bacterial soft rot	*Erwinia carotovora*
Cavity spot	*Pythium violae,* other *P.* spp.
Cercospora leaf blight	*Cercospora carotae*
Cottony soft rot	*Sclerotinia sclerotiorum*
Crater rot	*Rhizoctonia carotae*
Crown rot	*Rhizoctonia solani*
Downy mildew	*Erysiphe heraclei*
Fusarium dry rot	*Fusarium roseum*
Gray mold	*Botrytis cinerea*
Licorice rot	*Centrospora acerina*
Powdery mildew	*Erysiphe polygoni*
Purple root rot	*Helicobasidium brebisonii*
Pythium root dieback	*Pythium ultimum.* other *P.* spp.
Rhizopus wooly soft rot	*Rhizopus* spp.
Rusty root	Complex of several fungi, and possibly tobacco necrosis virus
Violet root rot	*Rhizoctonia crocorum*
White rust	*Albugo candida*

To eradicate *Xanthomonas carotae,* infected seed can be hot-water treated at 51°C for 30 min; however, the treatment sometimes reduces seed vigor and germination.

Symptoms of motley dwarf virus are expressed when both carrot mottle virus (CaMoV) and carrot red leaf virus (RLV) are present; neither alone is adequate to cause disease. The willow aphid *Cavariella aegopidii* is the insect vector. Other viruslike disorders are aster yellows, a phytoplasm agent, and a insect-transmitted agent known as beet leafhopper transmitted virescent agent (BLTVA).

Nematodes are serious pests because they directly affect the roots, causing galls and distortion. Soil fumigation is commonly practiced when damaging populations are present. Important root knot species of *Meloidogyne* are *M. hapla, M. incognita, M. javanica, M. arenaria,* and *M. chitwoodi; Heterodera carotae* is another serious pest.

Carrots are usually less troubled by insect pests than many other vegetables; nevertheless, the following insects are known to cause substantial damage.

Armyworm	*Spodoptera* spp.
Black cutworm	*Agrotis ipsilon*
Blister beetle	*Epicauta* spp.
Carrot caterpillar	*Papilio polyxenes*
Carrot beetle	*Bothynus gibbosus*
Carrot weevil	*Listronotus oregoneusis*
Green peach aphid	*Myzus persicae*
Leafhopper	*Macrosteles divisus, M. fascifrons*
Lygus bug	*Lygus hesperus, L. elisus*
Rust fly maggot	*Psila rosae*
Willow aphid	*Cavariella aegopidii*
Wireworms	*Limonius, Melanotus,* and *Conoderus* spp.

The two-spotted spider mite, *Tetranychus urticae,* and crown mite, *Tyrophagus dimidiatus,* can also cause crop damage.

Production

Most of the world's carrot production occurs in the temperate zone countries. Europe and the Asian continent together account for more than 74% of world volume. The leading countries, China, the United States, and the Russian Federation produced more than 17%, 9%, and 8%, respectively, of the world carrot supply. See Table 20.2.

Uses and Composition

Carrots are highly regarded for their nutritional value because they are an important source of pro-vitamin A, and are consumed either raw or cooked. Although infrequently used, young and tender foliage

TABLE 20.2. WORLD PRODUCTION OF CARROTS, 1994

	Area (ha × 10³)	Yield (t/ha)	Production (t × 10³)
World	666	21.3	14,176
Africa	66	11.1	739
North and Central America	64	30.8	1,958
South America	40	17.2	685
Asia	199	21.5	4,279
Europe	294	21.4	6,285
Oceania	6	35.3	229
Leading countries			
China	108[a]	22.9	2,475[a]
United States	43	32.3	1,384
Russian Federation	100[a]	12.5	1,250[a]
Poland	36	22.1	786
United Kingdom	17	46.3	769
Japan	24[a]	30.0	720[a]
France	17	35.3	640
Italy	11	44.3	470
Netherlands	8[a]	56.2	461[a]
Canada	8[a]	40.7	305[a]

[a]Estimated.
Source: 1994 FAO Production Yearbook, Vol. 48, FAO, Rome, 1995

is also edible. Roots are processed by canning, freezing, pickling, and dehydration. A new "minimally processed" value-added product is known as cut and peel, reformed, or baby carrots. These are prepared by cutting roots into 4–7-cm sections, followed by machine abrasive peeling that results in small slender carrot pieces, suitable for a snack food and other fresh uses. Normally small and/or immature carrots are also marketed as baby carrots and are an important specialty product. An additional use of carrot is the carotene that is extracted for coloring margarine and as a natural carotene source. Carotene is also added to poultry feed to intensify skin and egg yolk color, and carrot seed oil is extracted for flavoring. Carrot flavor is complex and is strongly influenced by the presence of volatile terpenoids, which may be mild to harsh, depending on cultivar and environmental conditions. Adequate levels of sugars are also necessary for sweetness to meet consumer expectations.

CELERY, *Apium graveolens* L. var. *dulce* (Mill.) Pers.

Other names: Stalk celery, apio

CELERIAC, *Apium graveolens* L. var. *rapaceum* (Mill.) Gaud.

Other names: Knob or root celery

SMALLAGE, *Apium graveolens* L. var. *secalinum* Alef.

Other name: Leaf celery

Stalk celery, *Apium graveolens* var. *dulce,* the most important member of about 15 *Apium* species, is widely grown in North America and temperate Europe. However, smallage, *A. graveolens* var. *secalinum* is most popular in Asian and Mediterranean regions and exceeds the volume of stalk celery. Celeriac, *A. graveolens* var. *rapaceum,* is most often grown in northern and eastern Europe in areas where stalk celery is not adapted.

Origin

The origin of celery and its allied varieties is not clear. Wild forms can be found in marshy areas throughout temperate Europe and western Asia. Although the eastern Mediterranean region appears to be the most logical center of domestication, the distribution of wild types raises some doubt.

Ancient literature documents that celery or a similar plant form was cultivated before 850 B.C. The earliest plant form probably most resembled leaf celery (smallage) and the first usage, common to some other Apiaceae, was medicinal. The early forms of stalk celery had a tendency to produce hollow stems and petioles, an adaptation to its marshy origins. During domestication, selection altered this heritable characteristic and reduced associated bitter and strong flavors. Blanching, a common cultural practice in past years, was intended to reduce bitterness and increase tenderness.

Botany

Although cultivated as annuals, these plants are normally biennial. However, if subjected to early vernalization, the plants can complete the life cycle within a year. Stalk celery is characterized by enlarged, long, thick, fleshy, and solid edible petioles; celeriac, by the development of a fleshy globe-shaped enlargement of hypocotyl and root tissues; and smallage, by the rosette of many, long, thin, petioled leaves. Celery usually grows to a height of 60–90 cm, celeriac about 50–60 cm, and smallage shorter than either (Fig. 20.5).

The root systems of each *A. graveolens* variety develop a considerable mass of fine, well-branched roots that form within a relatively short radius from the strong fleshy almost tuberous taproot. Except for the taproot, roots are relatively shallow, most occur within 30 cm below

FIG. 20.5.   Celery (a), celeraic (b), and smallage (c), variations of *Apium grave-olens.*

the surface. During transplanting, the taproot is inadvertently destroyed, and as a result, adventitious lateral roots grow profusely from the taproot remnant at the base of the plant. The stem of the vegetative celery plant is short, highly compressed, and fleshy, and when the plant becomes reproductive, the stem elongates and becomes fibrous.

Leaves grow in a rosette, are pinnately compound with five or seven leaflets, and are attached to the stem by long fleshy petioles. Leaflets are ternately compound and further divided many times. Leaf petioles are erect and broad with sheathing or shingling bases; succeeding inner petioles are increasingly tender.

A mature celery plant has 7–15 clearly distinguishable petioles; additional petioles developing at the apical meristem are hidden by the outer larger ones. Leaves become progressively smaller and self-blanched because of light exclusion. This grouping of small, tender, inner leaves is referred to as "the heart."

Petioles are glabrous, crescentic in trans-section, with prominent ribbing on the abaxial surface and a smooth adaxial surface. Ribbing is a result of separate collenchyma bundles on the abaxial side. Collenchyma tissue is strong, four times stronger than vascular tissue. Vascular bundles provide strength to the petiole as well as its fibrous texture. Vascular bundles and collenchyma strands collectively make celery petioles stringy. Parenchyma tissue constitutes the bulk of the petiole. Celery tends to sucker, axillary stems producing tillers, which because of late development contribute little to the stalk, and are usually trimmed during harvest. Breeders have significantly reduced the tendency for tillering.

Varying with cultivars, celery leaves are a yellowish green to dark green; celeriac foliage is generally a darker green than most celery cultivars. Smallage leaves are also dark green, although cultivars containing anthocyanin are also grown. Celeriac and smallage petioles are shorter and not as fleshy; also many more petioles are formed compared to celery. Frequently, celeriac petioles are partially hollow. The mostly hypocotyl and upper root portion of celeriac plants enlarges into a white, fleshy, globe-shaped storage structure about 10–15 cm in diameter; the upper portion is exposed above the soil.

Flowers are small, greenish white and borne in compound umbels. Protandrous flowering behavior limits natural selfing, and although self-fertile, flowers are mostly insect pollinated. The fruit is a very small (1 mm) schizocarp that splits when mature in two single-seeded mericarps. The oval seed are very small; about 2500 weigh 1 g. Celery plants are prolific seed producers.

Cultural Requirements

Celery is an environmentally demanding crop. For high quality and yield it must be grown under exacting climatic conditions. Optimum production occurs when mean temperatures range between 16°C and 21°C. Breeders have introduced cultivars that extend the upper temperature range. With these types, celery can be grown in some subtropical regions. Temperatures of 10°C or less are conducive to premature seed stalk formation. Celery is sensitive to freezing temperatures; celeriac and smallage are slightly less sensitive. When acclimated, celery can tolerated a light frost (–1°C or –2°C) for a short time with minimal damage. Leaf celery has more heat tolerance than either celeriac or stalk celery.

Celery and celeriac require soils having good water holding as well as good drainage capabilities; peat and clay loam soils are usually well suited for production. Celery is moderately sensitive to salinity, and a soil pH of about 5.8 is satisfactory for peat soils and about 6.7 is optimum for mineral soils.

Being shallow rooted, celery and celeriac require that water be available and replenished regularly to avoid stress, especially during the latter period of development when growth is very rapid. Accordingly, in most celery-producing regions, if rainfall is insufficient, irrigation must be made available. About 50 mm per week or 750–900 mm of water per crop are utilized for producing most celery crops. Celeriac and smallage moisture requirements are less than celery.

Excessive drainage below the root zone should be avoided in celery growing because of NO_3^- leaching and runoff of other nutrients into groundwater. This is a environmental concern because of potential nitrate toxicity in humans. Drip irrigation is one means for efficient and lower water use and can improve nitrogen use efficiency.

Celery nutrient needs are high, especially for nitrogen, which is required for the large biomass produced. Accordingly, soils of high fertility are usually used for production. Supplemental nutrient applications averaging about 300 kg N, 75 kg P, and 250 kg K per hectare are used on mineral soils; on muck soils, less nitrogen may be used. Rapid plant growth often magnifies minor element deficiencies and the addition of minor nutrients with the applied fertilizer is often practiced. A deficiency of boron results in cracked stem and brown checking symptoms; calcium deficiency causes blackheart, and inadequate magnesium results in leaf chlorosis. These minor nutrients can be applied by topical sprays for rapid response, although soil applications made in advance of planting are preferred and more effective.

Propagation

Seed germination and emergence of these crops are slow, even when conditions are favorable. Germination at 10°C requires more than 15 days, and 30 days or more at 5°C. At optimum temperatures between 15°C and 20°C, germination requires 7–12 days. It is much slower at 25°C, and because of thermodormancy, no germination occurs at temperatures greater than 30°C. Some cultivars exhibit thermodormancy even at 25°C. A seed-soaking treatment at 10°C using growth regulators, $GA_{4/7}$, and ethephon at 1000 ppm can overcome this dormancy.

There are significant cultivar differences in these responses. Seed are more heat sensitive when germinated in the dark. As temperatures increase, more light is required. The presence of light, periods of low-temperature alternation, and growth regulators help alleviate thermo-dormancy. Far-red light or sunlight exposure improves germination percentages. GA treatment also overcomes the light requirement. Seeds are sown shallow to enhance light exposure.

In addition to direct temperature affects, other reasons for poor ger-mination include variable seed quality and postharvest seed dormancy. As with carrots, celery seed are of variable maturity because they are a product of sequentially formed umbels that develop over a fairly long period. Newly harvested seed exhibit dormancy due to a water-soluble germination inhibitor that is removable by seed soaking to leach the inhibitor, aging, or the use of growth regulators. This dormancy is usually not a concern unless freshly harvested seed is planted.

Because of its small size, celery seed is often pelleted to facilitate precision seed placement for either transplant production or direct field sowing. Sowing of pregerminated seed in gel or liquid carriers are sometimes made to accelerate seedling emergence. Directly seeded crops mature in 160–180 days; transplanted fields are usually har-vested in 90–125 days.

Seed priming permits partial imbibition and the initial germination processes to occur, but radicle emergence is inhibited by the osmotic solution. In this procedure, seeds are brought to a uniform pregermina-tion stage, removed from the solution, dried, and then planted. Primed seeds germinate and emerge more uniformly than those not primed. Because of the difficulty in establishing stands, only limited amounts of celery are produced from directly sown plantings, the majority being made with transplants.

A major disadvantage of celery transplanting is its high cost, part of which is due to slow seedling development. A growth period of about 2 months is required to produce seedlings suitable for transplanting. However, this is partially compensated because transplanting reduces the growth period in the field. Transplanting provides full stands and uniform spacings and also reduces some cultural inputs, such as fertil-ization, irrigation, pest, and weed management.

Good quality transplants are critical for a successful crop. Transplant production has become a highly specialized procedure. Because con-tainer-grown or modular-tray-grown transplants rapidly establish in the field and produce uniform growth, the use of bare-rooted transplants has declined. Although more costly to produce, container-grown trans-plants are of high quality and better adaptable to mechanical trans-planting equipment, some of which are fully automated. Mechanical transplanting results in more uniform planting depth and spacing than hand transplanting. Spacing arrangements generally range from 12 to

20 cm in rows, and from 50 to 75 cm between rows. Typical celery field populations are between 50 thousand and 80 thousand per hectare.

Celeriac is most often directly sown because transplanting results in excessive lateral rooting and roughness of the enlarged globelike root. Celeriac spacings range from 30 × 50 to 40 × 60 cm, with field populations between 40 thousand and 50 thousand per hectare. Smallage, being a smaller plant, is planted closer and field populations are about 100 thousand per hectare.

Growth and Development

Early seedling growth of celery is slow. The growth rate steadily increases so that most of the biomass accumulation occurs during the 3–4 weeks before harvest. Gibberellic acid has been used to accelerate plant growth and earliness with some success. However, applications occasionally result in long, thin, and light-colored petioles.

Crispness and tenderness are major quality attributes of celery petioles which have little value when wilted, pithy, or excessively fibrous. When growth is very rapid, the petioles become susceptible to pithiness. However, petiole age is the major cause. Pithiness is a softening and degradation of parenchyma tissues resulting in a loss of density. If severe, petioles develop hollow areas and the stem plate can also be affected. Moisture stress or freezing can cause similar symptoms. Hollowness in celeriac root is similarly caused.

When grown continuously at 15°C or higher, celery plants remain vegetative. Seedlings with less than four or five true leaves are considered to be in the juvenile stage and insensitive to vernalization, but once beyond the juvenile stage, plants are susceptible. Bolting is initiated when sensitive plants are exposed to low temperatures, usually less than 10°C. Cultivars vary as to their sensitivity threshold with regard to temperature and duration. Some tolerate low temperatures and long exposure better; others are sensitive even when temperatures are 13°C or 14°C. The inductive response is more acute with old plants and lower temperatures. With warmer temperatures following the cold induction, seed stalk elongation is accelerated. Plant stress can also increase the predisposition to vernalization. The presence of the seedstalk reduces quality and marketability (Fig. 20.6).

Cultivars with slow-bolting characteristics are seldom grown because of their less desirable market and quality features. To avoid bolting, field plantings should be scheduled so that low-temperature exposure is minimized. Similarly, exposure to temperatures of less than 15°C in transplant nurseries should be avoided. An acclimation treatment

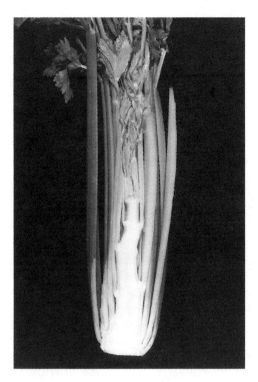

FIG. 20.6.   Medium longitudinal section of bolting celery plant, *Apium graveolens.*

between 25°C and 30°C for 10–20 days to nursery seedlings has been shown to reduce transplant sensitivity to low field temperatures.

In some regions, blanching is a cultural practice in the production of celery to reduce the harsh flavor and petiole fiber, and to increase succulence. The application of ethylene gas to celery following harvest was an occasional alternative to field blanching. The blanching is performed a few weeks before harvest by placing soil, boards, paper, or other opaque materials against petioles but not covering the leaf blades. Peat soils are preferred because of the ease of moving soil against the petioles. To facilitate blanching with soil, seedlings are initially transplanted into shallow furrows which are filled-in during the growth period. Soil blanching also provides a small measure of frost protection.

Major reasons for the decline of blanching were labor and material costs, and the presence of soil within the harvested plant, which required careful washing. An interim development between blanched (white) and nonblanched (green) celery production was the introduction of "self-blanching" cultivars that did not require soil or other coverings. These cultivars resembled blanched celery in having light yellowish green leaves and light petiole color. Presently, the majority of celery cultivars grown is not blanched and the foliage and petioles are dark green.

Harvest and Postharvest

The celery crop is normally harvested when the majority of plants are determined to be of marketable stage, but some size variation is inevitable. A delay can result in some plants becoming pithy, but early harvest results in fewer large stalks.

Harvesting is primarily by hand labor. Rather than selectively harvesting plants at the same stage of development, all plants are harvested and then size-graded according to diameter. Important celery harvest characteristics are plant weight, width, and height, and the thickness and number of petioles. Leaf celery has a shorter growing season and multiple harvests can be made when the leaves are cut sufficiently above ground to allow regrowth of new foliage.

Celery plants are undercut to sever roots, lifted and trimmed of side roots and suckers. Further trimming removes a portion of the leaf blades and upper petioles to a length that meets packaging requirements. Untrimmed leaf blades are prone to wilting. Trimming and packing are performed directly in the field or in mobile or stationary packing sheds.

A limited amount of machine harvesting of celery occurs, mostly for processing. Machinery is commonly used for digging celeriac, and for trimming and washing the globelike product. Celery crops frequently yield about 60–70 t/ha; celeriac yields are about 40 t/ha.

Harvested celery is often hydrocooled following packing, although vacuum (hydrovac) and forced air cooling are also used. Hydrovac equipment introduces water during the vacuum procedure to limit moisture lost from the product, as even a slight amount of water loss results in visible wilting.

Optimum storage conditions for celery and celeriac are 0°C and a high RH (95%). With these parameters. both products store well for a month. Although now rare, some field storage in plastic-covered, straw-insulated trenches or clamps are still used for winter storage. Controlled atmospheric storage can be used to maintain marketable quality for relatively long periods. Such storage requires 0°C and high RH in an atmosphere of 1–2% O_2, 4% CO_2, and with ethylene removal.

Diseases and Other Pests of Celery and Other Apiaceae Vegetables

Diseases
  Bacteria

Bacterial leaf spot	*Pseudomonas syringae* pv *apii*
Bacterial soft rot	*Erwinia carotovora* var. *carotovora*
Brown/Southern bacterial blight	*Pseudomonas cichorii*
Northern bacterial blight	*Pseudomonas apii*

  Fungi

Cephalosporium brown spot	*Cephalosporium apii*
Centrospora storage rot	*Centrospora acerina*
Cercospora blight	*Cercospora apii*
Damping off	*Rhizoctonia, Sclerotinia, Pythium,* and *Fusarium* spp.
Fusarium yellows	*Fusarium oxysporum* f. sp. *apii*
Parsley leaf spot	*Didymaria petroselini*
Phoma root rot/crown rot	*Phoma apiicola*
Rhizoctonia crater and Petiole rot	*Rhizoctonia solani*
Sclerotinia pink rot	*Sclerotinia sclerotiorum*
Septoria blight	*Septoria apiicola*
Septoria leaf spot, parsley	*Septoria petroselini*

In the United States, the largest celery-producing country, cultivars used for blanching were susceptible to a race of *Fusarium oxysporum*. Plant resistance was found in dark green colored, nonblanched (Tall-Utah type) cultivars which soon replaced the blanched types. These cultivars are now susceptible to another race of *F. oxysporum* and are being replaced by new resistant cultivars.

Viruses and phytoplasm (mycoplasm)
  Celery latent virus
  Celery mosaic virus
  Celery spotted wilt virus
  Celery yellow net virus
  Celery yellow spot virus
  Celery yellow vein virus
  Cucumber mosaic virus (CMV)
  Poison hemlock ringspot virus

Strap leaf of celery
Western celery mosaic virus
Celery aster yellow, a phytoplasm (mycoplasm)

Physiological disorders

Blackheart	Calcium deficiency
Brown checking, cracked stem	Boron deficiency
Foliage chlorosis	Magnesium deficiency

Insect and other pests

Armyworm	*Spodoptera* spp.
Black cutworm	*Agrostis ipsilon*
Cabbage looper	*Trichoplusia ni*
Celery leaf tier	*Phlyctaenia rubigalis, Oeobia rubigalis*
Celery looper	*Autograph falcifera*
Corn wireworm	*Melanotus communis*
Green celery worm	*Platysentor sutor*
Leafminers	*Liriomyza* spp.
Melon/cotton aphid	*Aphis gossypii*
Parsley caterpillar	*Papilo zelicaon*
Potato wireworm	*Conoderus falli*
Tarnish bug	*Lygus pratensis*
Willow aphid	*Cavariella aegopodii*
Celery rust mite	*Aculus eurynotus*
Red spider mite	*Tetranychus* spp.

Nematodes

Celery root knot	*Meloidogyne hapla, M. incognita*
Pin and Root lesion	*Pratylenchus* spp.
Sting	*Belonolaimus gracilis*
Stubby root	*Trichodorus* spp.

Prompt and effective weed management is mandatory for directly sown celery because of the very slow seedling growth and early inability to compete with weeds. Chemical herbicides are widely used and often supplemented with hand hoeing.

Uses and Composition

Stalk celery leaf petioles are consumed raw or cooked. Celery is also processed by pickling, canning, and dehydration. The major consumption of celery occurs in North American and temperate European countries. Root celery or celeriac is usually cooked, but is also used raw or pickled in salads and is canned. The major use of celeriac occurs in northern and eastern Europe and areas where stalk celery is not adapted. Smallage plants, of which there are many variations, produce many slender leafy petioles, these are generally used for flavoring and garnish. Smallage is widely grown and used in Asia and Mediterranean regions. Celery seed are used for condiment and flavoring purposes, and some medicinal use is also made of extracted seed oil. Apiin, (apigenin 7-apiosylglucoside) is the main flavoring glycoside of celery leaves and celeriac roots.

OTHER UMBELLIFEROUS VEGETABLES

PARSLEY, *Petroselinum crispum,* Mill. Nym. var. *crispum* *(P. hortense, P. sativum)*

Parsley, an ancient crop of the Mediterranean region, where it is believed to have originated, has been used for centuries for its medicinal and food-flavoring properties. The volatile oil responsible for its unique flavor is apoil. Parsley is a biennial or short-lived perennial often cropped as an annual. When grown for processing, the crop is grown as a perennial, which allows for multiple harvests.

Plants have a stout and deep taproot; the foliage growth tends to be semiprostrate and usually not more than 30 cm in height. Foliage is dark green, shiny, with three pinnate leaflets, but cultivars exhibit a diversity of leaf forms (Fig. 20.7). The two main foliage types are plain (Italian) and curled leaf. Each having dual usage for flavoring and garnish. Another type is turnip root parsley (Hamburg), sometimes also known as petrouska; both roots and foliage are frequent ingredients in soups and stews. The different leaf types are sometimes identified as botanical varieties whereby *P. hortense* var. *crispum* or *filicinum* designates curly leaf, var. *neapolitanum* or *latifolium* for plain leaf, and var. *tuberosum* or *fusiformis* represents turnip root parsley.

Although a cool season crop, parsley has a broad climate and soil adaptation, and grows best at a mean temperature range from 7°C to 16°C. Most cultivars have relatively good frost tolerance; Hamburg parsley is winter hardy. In the first year, plants produce a rosette of leaves. When vernalized, seed stalks develop with compound umbels bearing many small greenish yellow flowers. Although flowers are self-fertile, cross-pollination usually occurs. The small seed are ribbed; about 650 weigh 1 g.

FIG. 20.7.   Parsley, *Petroselinum crispum,* curled leaf form.

Propagation is almost always by direct sowing because parsley does not transplant satisfactory. Seed are very slow to sprout, requiring from 7 to 20 or more days. Poor germination may be due to immature embryos and inhibitors. Usual plant spacings are about 15 × 60 or 75 cm. For fresh market, plant populations are often twice those for processing, and plants are harvested at a young stage. Fertilizer applications typically are 100 kg/ha each of N, P, and K. Excessive N results in very long petioles, and that is undesirable.

Parsley is subject to many of the same diseases and pests affecting carrot and celery. Leaf spot, *Didymaria petroselini (Cercospora petroselini),* and stemphylium (*Stemphylium* spp.) are the more common diseases.

Parsley can be harvested about 90 days after sowing. The fresh market crop is often hand harvested, a labor-intensive process in which several small plants are tied together as a bunch. Another procedure is to mechanically cut the foliage which is handled as loose leaves. The processed crop is usually mechanically harvested by cutting above the apical growing point to allow regrowth for repeat harvests. Some fields are harvested several times each year for several years. In addition to its fresh, garnish, and cooked vegetable uses; considerable amounts of parsley are dehydrated. Parsley is a rich source of pro-vitamin A and vitamin C. Parsley stores well at 0°C and at a high relative humidity for as long as 1 or 2 months.

PARSNIP, *Pastinaca sativa* L.

Parsnips have had a long history of cultivation and are believed to have originated somewhere between the Mediterranean basin and west of the Caucasus region. The slow-growing, deep-rooted biennial plants often require 120–150 days of growth before harvest. The edible portion is the long, tapered, fleshy axis formed by hypocotyl and taproot tissues. Parsnips are aromatic and have a sweet taste and a slight mucilaginous texture. Important quality aspects are size and surface smoothness, preferable without lateral branching. Roots are usually harvested when about 25 cm long and the diameter at the shoulder is 5–8 cm.

During the first year of growth, the plant develops the storage root. If vernalized, seed stalks elongate in the second year; these are grooved, highly branched, and hollow, and can reach 2 m in height. Yellowish green flowers occur in compound umbels. Flowering is protandrous and embryo development is often incomplete and frequently results in seed poor and slow in germination. Seed viability is normally relatively short. Germination is best between 19°C and 24°C. Parsnip seed differ from other Apiaceae in having broad, flat, winglike appendages or margins. Seed are 5–8 mm long; about 275 weigh 1 g.

Propagation is with seed because transplants tends to produce undesired taproot branching. Parsnips are commonly spaced about 10×60 to 75 cm for a plant density of 13 to 16 plants per m². A deep, light to medium textured soil with good drainage is preferred; fine textured soils tend to encourage root branching. Plant growth is optimum at mean temperatures between 15°C and 18°C.

Parsnips are harvested by plowing or hand digging. The foliage is usually removed before roots are lifted. Harvests are often delayed until low temperatures or frosts have occurred, because root quality is improved under such conditions. With advancing maturity, starch stored in the root is converted to sugars; the conversion is accelerated by cold exposure. Parsnip can be field stored unless freezing is severe; also roots store well in soil clamps or under refrigeration at 0°C and greater than 90% RH. Because roots easily desiccate, high humidity is important for extended storage. As with carrots, parsnips exposed to ethylene during storage become bitter.

Parsnips are subject to many of the diseases affecting other Apiaceae crops. Some diseases unique to parsnip are as follows:

Blight	*Pseudomonas marginalis*
Canker and leaf spot	*Itersonilia perplexans*
Downy mildew	*Plasmopara nivea*
Cercospora leaf spot	*Cercospora pastinacae*
Cylindrosporium leaf spot	*Cylindrosporium pastinacae*
Powdery mildew	*Erysiphe umbelliferarum*
Root rot	*Phoma* spp.

Because of slow initial growth, parsnips are poor weed competitors; therefore, early effective weed management is important.

CORIANDER/CILANTRO/CHINESE PARSLEY,
Coriandrum sativium L.

Coriandrum sativium when cultivated for its foliage is also known as cilantro or Chinese parsley, but when grown to produce a condiment seed crop, it is known as coriander. This versatile plant has been cultivated for more than 3000 years for both its foliage and dry seed used in flavoring. Leaves are frequently used fresh or cooked in salads, soups, and other dishes, and even the roots are eaten, especially in China and Thailand. The foliage is indispensable in many Indian, Chinese, and Mexican meals for the unique flavor provided by its foliage. The dried seed contain a flavorful volatile oil that makes them useful for pickling and curry preparations. Immature seed have a sagelike, tangy citrus taste, and roots have a nutty taste. Like many other Apiaceae, coriander is sometimes used medicinally.

The plant is believed to have originated in the region between the eastern Mediterranean and the Caucasus mountains. It is a low-growing annual, reaching 50–60 cm in height when flowering. Leaves are a yellowish light green color with slender petioles; some cultivars have red petioles (Fig. 20.8). The lower leaves are broad with shallow lobbing; the upper leaves are finely cut with narrow linear lobes. The taproot is slender. Small white or pink flowers are borne on compound terminal umbels. The small, 3-mm-wide, aromatic two-seeded globular fruit is hard, yellow brown, and ribbed; about 80 weigh 1 g.

Growth is favored by moderate temperatures and high light intensities. The plants have a fairly broad climatic adaptation and have some tolerance to light frosts. High temperatures promote early seed stalk

FIG. 20.8.   Coriander foliage, *Coriandrum sativium.*

development. However, European cultivars require long days for initiation, whereas those in India and Southeast Asia require shorter days. Because transplants do not establish readily, propagation is with seed even though they are slow to germinate. Plant spacings commonly used are 20 cm in rows and 50–75 cm between rows. Crops require 3–4 months to mature.

When adequate foliage has developed, leaves are cut in such a manner as to allow for regrowth and repeated harvests. Coriander foliage has a relatively short shelf life and requires low temperatures and high relative humidity to maintain quality. However, once flowering begins, foliage growth becomes negligible; in this case, seed becomes the harvested product. Seed shatter is a harvest problem.

FLORENCE FENNEL, *Foeniculum vulgare* Mill. var. *dulce* Fiori

Other names: Sweet anise, finocchio

Fennel, a native of the Mediterranean region, is grown for the bulblike enlargement of fleshy overlapping petiole leaf bases. The strongly flavored leaf bases are used fresh for salads, as a cooked potherb or pickled. The bulbous enlargement usually ranges from 10 to 15 cm in diameter (Fig. 20.9). The bulb shape varies from strongly ovoid to round; a round and compact form is preferred.

All parts of the plant are aromatic and edible. The anise-flavored, finely divided leaves are used as potherbs, for seasoning and garnishes. The principal aromatic compounds are anethole and carvone. Seed are used for flavoring but should not be confused with the pungent herb anise, *Pimpinella anisum.*

Fennel is a perennial, generally cultivated as an annual. Average plant height of the vegetative plant is 75–90 cm. The taproot is large, long, fleshy, and well branched. Stems are highly compressed, as in celery. Leaves are alternately branched, and the bright green smooth petioles terminate in feathery fernlike or filigree leaves. From multiple growing points, suckers can occur; these are undesirable for marketing purposes.

Seed germination tend to be variable, and transplants are often used in place of direct seed sowing. Common field spacings are 20 to 25 ×

FIG. 20.9. Florence fennel plants, *Foeniculum vulgare.*

75 to 90 cm. Depending on prevailing temperatures, growth periods from sowing to maturity range from 100 to 160 days. Fennel grows well in temperate regions, and temperatures most favorable for growth are between 15°C and 20°C. Warm temperatures reduce petiole size and succulence.

Long days as well as vernalization are a stimulus for bolting, which results in the development of tall seed stalks that bear numerous small yellow flowers on large umbels. Seed are greenish brown, noticeably ridged, and 5–6 mm long; about 200 weigh 1 g.

Some off-season glasshouse production occurs; plant densities are higher than in the field, with spacings often at 25 × 25 cm. Plants are harvested while small and are not intended to achieve the usual size of field-produced plants. Spacing does influence the shape of the bulb; very close spacings result in oval shapes. However, when widely spaced, suckering and basal splitting of outer leaf petiole bases increase. For some markets, fennel is blanched a week or more before harvesting.

Harvests are made by cutting plants free just above the taproot. Upper petiole and leaf blades are trimmed relatively close to the bulb. Fennel has good postharvest storage qualities if maintained at 0°C and high relatively humidity.

Additional types of fennel cultivated are *F. vulgare* var. *azoricum* and *F. vulgare* var. *piperitum,* known as Italian fennel. Wild fennel, *F. vulgare (F. officinale),* although not cultivated, is also eaten; in some locations, it has become a persistent weedy pest.

TURNIP ROOTED CHERVIL, *Chaerophyllum bulbosum* L.

The area from central Europe to western Asia is the probable center of origin for turnip rooted chervil. The hardy biennial plant is grown as an annual for its edible leaves and roots. The plant grows to a height of 90 cm. Leaves are harvested 6–10 weeks after planting. Also harvested are the spindle-shaped, 5–10-cm-long roots which are blackish gray with yellowish white flesh. Plants readily bolt, especially when subjected to low temperatures and grown during long days. The crop can be overwintered if the climate is mild. The flavor of the root improves with maturity; roots have good storage characteristics.

SALAD CHERVIL, *Anthriscus cerefolium* L. (Hoffm.)

The origin of salad chervil is in the region of southeastern Europe and western Asia. A hardy annual, salad chervil grows to a height of about 60 cm. Propagation is directly from seed, which require exposure to light for germination. Plants are harvested 50–60 days after planting but can be overwintered. The foliage is used for salads and garnishes. Roots of salad chervil are white and thin and usually not eaten. Foliage types, plain or curled, differ with cultivars. Repeated hand harvesting of the aromatic leaves are made. High temperature promotes early seed stalk elongation; the inflorescence bears tiny white flowers. Seed have a noticeable beak; about 450 weigh 1 g.

SKIRRET, *Sium sisarum* L.

Skirret probably is of East Asian origin. Other *Sium* species are found in Northern Hemisphere locations, although one species is native to South Africa. Skirret is a perennial grown as an annual for the cluster of tuberous roots. Plant height ranges up to 90 cm. Roots are grayish white, firm, and white fleshed and have a taste resembling salsify. Leaves are lanceloate and toothed. Plants are propagated by seed or root divisions. The crop is frequently overwintered.

JAPANESE HONEWORT/MITSUBA, *Cryptotaenia japonica* Hassk.

Mitsuba (meaning three leaflets in Japanese) is a minor but important crop in Japan and is also cultivated in other Asian countries. Its origin is presumed to be Japan because wild forms are found there. The distinctive aromatic flavor makes it popular for use much like celery, either raw or cooked in salads, soups, and flavoring.

Mitsuba is a cool season perennial usually grown outdoors and harvested during the spring. Erect plant height ranges from 20 cm to slightly more than 1 m. Leaves are trifoliate; the lowest are the largest and have long petioles (Fig. 20.10). Petioles with attached leaflets are usually harvested when 15–20 cm long. When blanched, which is fairly common, petioles elongate to lengths as much as 30 cm.

In Japan, during periods or in locations unfavorable for outdoor cultivation, production is obtained from plants forced using hydroponic culture in glass or plastic houses. Propagation is with seed and also by rhizome sections. *C. canadensis,* of North American origin, is a wild species known as white chervil; it has a strong resemblance to mitsuba.

FIG. 20.10.   Japanese honewort, *Crytotaenia japonica,* bunched for marketing, Hamamatsu, Japan.

ANGELICA, *Angelica archangelica* L. *(A. officinalis)*

Angelica is believed to have originated near Syria. The perennial plant is used as flavoring for its unusual musky odor and sweet taste, and also for some medicinal attributes. The purple-colored roots are the preferred edible portion, although stems are used for garnish; oil is extracted from seed. Angelica oil is frequently confused with oil extracted from seed of star anise, *Illicium verum.*

The tall, 1.5-m, perennial plants resemble flowering wild carrot and poisonous water hemlock *(Cicuta maculata).* The pinnate leaves have large petioles, are coarse toothed, and oval; the terminal leaflet is three-lobed. The stout stem is hollow and purplish. Tiny greenish white flowers grow in a globe-shaped compound umbel. Mature fruit are oblong and contain two yellow winged seeds. About 120 days from anthesis are required for seed to mature; they easily shatter.

Angelica plants have a preference for moist locations near streams and grow best in cool climates. Plants are easily propagated vegeta-tively. Seed, also used for propagation, require light for germination. Usual plant spacings are 10 or 15 × 50 or 60 cm. Roots are commonly harvested during the first year; if not, then production is for the stems and/or seed.

ANISE, *Pimpinella anisum* L.

Anise, which probably originated near Mediterranean Egypt, is an ancient aromatic plant used primarily for flavoring; both leaves and seed are used for that purpose. The extracted seed oil (anethole) is used for medicinal purposes. Anise is often confused with Florence fennel because they have a similar aromatic odor and flavor, and also with star anise *(Illicium verum)* which is sometimes used as an anise-flavoring substitute.

Anise is a pubescent annual, but is considered perennial in warm climates. Plants have a long taproot and the main stem is hollow. The lower seedling leaves are simple, broad, and pinnate, resembling those of flat leaf parsley. The upper leaves with long petioles are compound, highly divided, and lacelike. Plants in the vegetative stage grow to a height of about 60 cm.

Small yellowish white flowers develop in a large loose umbel. Seed are flat-ovate, ribbed, and usually slow to germinate; germination is best at 20°C. Uniform soil moisture is required for high yields. Most plant spacings are about 25 cm and rows are spaced 50–75 cm apart.

CARAWAY, *Carum carvi* L.

Caraway is of central European and western Asia origin, and has a long history as an important herb, in addition to some medicinal uses. The crop is widely distributed in temperate regions. Leaves, stems, and roots are eaten, although seed are the major product. Seed are rich in the oil carvone which combines with limonene to provide the distinctive caraway flavor. Caraway, in addition to flavoring rye bread, is also used for flavoring liqueurs and cheese.

Caraway is a biennial plant, usually cultivated as a seed-propagated annual. Plants grow to a height of about 70 cm and produce edible, thick, long tuberous roots. The slender stems are glabrous, grooved, and hollow. Leaves are compound and finely cut. Excessive foliage growth is discouraged because it reduces seed yields. Plants are intolerant of moist soils.

Small, yellowish white flowers are formed in compound terminal umbels. Seed which readily shatter are curved, ribbed, about 5 mm long, and have a pentagonal cross section; about 300–350 weigh 1 g.

CUMIN, *Cuminum cyminum* L. *(C. odorum)*

Cumin is an old crop of Mediterranean origin. Seed are used in curry powder, soups, pickles, and even in perfumes. Cumin is a short (25 cm), highly branched plant with bluish green leaves. The small white or pink flowers are borne on a compact terminal umbel; the grayish elongated oval seed are aromatic and pungent.

DILL, *Anethum graveolens* L.

Dill is a native of Eurasia with a long history of medicinal and food-flavoring uses. Both seed and foliage are used for flavoring; foliage is frequently processed by drying. Plants grow to a height of 1 m. The taproot is singular and slender. The long hollow stem is smooth (glabrous) and grayish green. Leaves are compound, finely dissected, and a feathery blue green color (Fig. 20.11).

Dill does not transplant well and therefore is usually directly sown. Common plant spacings are 45 × 90 cm. Dill is a long-day plant having annual and some biennial forms; small greenish yellow bractless flowers develop on large open umbels. Seed, which easily shatter, are light brown, flattened, ribbed, and strongly fragrant due to the essential oil carvone. About 400–500 seed weigh 1 g.

LOVAGE, *Levisticum officinale* W.D.J. Koch

Lovage is a perennial of Mediterranean origin. The leaf stalks and blanched lower stem portions, which taste like celery, are eaten. Seeds, leaves, and roots are used for flavoring.

The plants which grow up to 1.5 m have compound, shiny, wedge-shaped dark green leaves. Stems are stout and hollow. Tiny greenish yellow flowers form in compound umbels. Plants are propagated by seed or root divisions. Related species are *L. canadense* and *L. scoticum* (Canadian and Scotch lovage, respectively).

SWEET CICELY, *Myrrhis odorata* (L.) Scop.
Other names: Myrrh, sweet chervil

Sweet cicely, a native of Europe, has a taste similar to celery and sweet anise. The root is eaten after steaming, and seeds are used in

FIG. 20.11. Dill plant with foliage and inflorescence, *Anethum graveolens*.

flavoring. The perennial plant produces a tall, 60–90 cm, pubescent branched stem; roots are fleshy. Leaves are fernlike, with the underside having a whitish fuzzy appearance. Leaf stalks tend to wrap about the stem. Plants produce many small white flowers in compound umbels. Inner flowers are male; others are bisexual. Shiny dark brown ridged seed are about 2 cm long. Because seed germination is slow and often poor, plants are propagated from divisions or seedling transplants.

ASIATIC PENNYWORT, *Centella asiatica* (L.) Urban

Other name: Indian pennywort

Although indigenous to the southeastern United States, Asiatic pennywort is especially popular for the aromatic pungent flavor of its leaves in Sri Lanka and which are of some importance in the region from Bangladesh to Indonesia. Leaves are used fresh for salad greens, as a potherb and cooked with other foods. Plants are also appreciated for their purported medicinal benefits, some are claimed to enhance longevity and memory, and as a cure for leprosy and some skin wounds.

A perennial, the plant produces slender, runnerlike, long stolons having wide internodes. Clusters of petiolate leaves form at the nodes. The inflorescence has few umbels and few flowers per umbel; petals are white. Seed are formed but propagation relies on vegetative cuttings.

The preferred growth habitat is for wet or moist soil and partial shade rather than full sun. Shade favors leaf growth with long petioles. Two plant types are cultivated: a small-leaf creeping form, and a large-leaf erect bushlike form; the latter is faster growing and more productive.

Harvest occurs when leaves have achieved full size; edible quality declines with further maturity. With good horticultural practices, harvests can be made at about 2-month intervals for several years. The small-leaf type is harvested and marketed as intact plants bundled together. For the bushy type, the leaves are cut and also bundled; the long petioles facilitate handling and leaf tying as bundles. Without refrigeration, postharvest life is seldom more than 2–3 days.

SELECTED REFERENCES

Banga, O. 1976. Carrot. In Evolution of Crop Plants, N.W. Simmonds, ed. Longmans, London, pp. 291–293.

Buishand, J.G., and Gabelman, W.H. 1979. Investigations of color and carotenoid content in phloem and xylem of carrot roots (Daucus carota L.). Euphytica 28, 611–632.

Cserni, I., and Prohaszka, K. 1988. The effect of N supply on the nitrate, sugar and carotene content of carrots. Acta Hortic. 220, 303–307.

Erickson, E.H., Garment, M.B., and Peterson, C.E. 1982. Structure of cytoplasmic male sterile and fertile carrot flowers. J. ASHS 107, 698–706.

Gray, D., Brocklehurst, P.A., Steckel, J.R.A., and Dearman, J. 1984. Priming and pregermination of parsnip (Pastinaca sativa L.) seed. J. Hort. Sci. 59, 101–108.

Gross, J. 1991. Pigments in Vegetables—Chlorophylls and Carotenoids. Van Nostrand Reinhold, New York.

Ivie, G., Wayne, D.L., Holt, L. and Ivey, C.C. 1981. Natural toxicants in human foods: Psoralens in raw and cooked parsnip root (Pastinaca sativa). Science 213, 909–910.

Lee, C.Y., Simpson, K.L., and Gerber, L. 1989. Vegetables as a major vitamin A source in our diet. Food Life Sci. Bull. 126, 1–11.

Mathews-Roth, M.M. 1985. Carotenoids and cancer prevention—experimental and epidemiological studies. Pure Appl. Chem. 57, 717–722.

Nakamura, S., Teranishi, T., and Aoki, M. 1982. Promoting Effect of polyethylene glycol on germination of celery and spinach seed. J. Japan. Soc. Hort. Sci. 50, 461–467.

Peterson, C.E., and Simon, P.W. 1986. Carrot breeding. In Breeding Vegetable Crops, M.J. Bassett, ed. Chapman & Hall, New York, pp. 321–356.

Pressman, E., Negbi, M., Sachs, M., and Jacobsen, J.U. 1977. Varietal differences in light requirements for germination of celery (*Apium graveolens* L.) seeds and the effects of thermal and solute stress. Aust. J. Plant Physiol. *4*, 821–831.

Sachs, M., and Rylski, I. 1980. The effects of temperature and daylength during the seedling stage on flower stalk formation in field grown celery. Sci. Hort. *12*, 231–242.

Shattuck, V.I., Yada, R., and Lougheed, E.C. 1988. Ethylene induced bitterness in stored parsnips. HortScience *23*, 912.

Simon, P.W. 1987. Genetic improvement of carrots for meeting human nutritional needs. In Horticulture and Human Health, ASHS Symposium Series No. 1, B. Quebedeaux and F.A. Bliss, eds. New Jersey: Prentice-Hall, Englewood Cliffs, NJ, pp. 208–214.

Simon, P.W., and Wolff, X.Y. 1987. Carotenes in typical and dark orange carrots. J. Agric. Food Chem. *35*, 1017–1022.

Simon, P.W., Wolff, X.Y., and Peterson, C.E. 1985. Selection for high carotene content in carrots. HortScience *20*, 586.

Temple, N.J., and Basu, T.K. 1988. Does beta-carotene prevent cancer? A critical appraisal. Nutr. Res. *8*, 689–701.

Thomas, T.H., Palevitch, D., Biddington, N.L., and Austin, R.B. 1975. Growth regulators and the phytochrome-mediated dormancy of celery seeds. Physiol. Plant *35*, 101–106.

21

Spinach, Table Beets, and Other Vegetable Chenopods

Family: Chenopodiaceae

The Chenopodiaceae (goosefoot) family of about 75 genera has only a few vegetable members and several important field crops; most species are weedy. Some family characteristics are succulent tissues, tolerance to salinity, and small wind-pollinated inconspicuous flowers.

Principle Vegetable Species	*Common Name*
Spinacia oleracea	Spinach
Beta vulgaris var. *crassa*	Table beet
Beta vulgaris var. *cicla*	Swiss chard
Beta vulgaris var. *orientalis*	Spinach beet/palak
Atriplex hortensis	Orach

Sugar beets, an agronomic crop of worldwide importance, and mangel-wurzels, a livestock feed crop, are members of the same species as table beets. Quinoa *(Chenopodium quinoa),* endemic to Peru, is a valuable, high-protein, high-carbohydrate, small-grain staple. Huauzontle *(Chenopodium nuttalliae)* is closely related to *C. quinoa;* the young plants are cooked as potherbs; unripe fruit clusters are eaten boiled or fried and have a taste like broccoli. Lambs quarter *(Chenopodium album),* a weed of wide distribution, is often gathered and eaten as a leafy potherb. The young fleshy leaves of epazote *(Chenopodium ambrosioides)* are often eaten with bean dishes. The plant is also known as "Mexican tea" and is used to make a tea having an aromatic flavor pleasant to some and not to others. Other *Chenopodium* species having limited vegetable use of their foliage are *C. capitatum,* known as Strawberry blite, *C. bonus-henrious,* known as Good King Henry or Mercury, *C. berlandieri, C. bushianum,* and *C. pallidicaule.*

SPINACH, *Spinacia oleracea* L.

Origin

The origin of spinach has been placed near Iran where it has been cultivated for at least 2000 years. Its cultivation in North Africa and Europe began about 1000 A.D. Related wild types are *S. tetranda,* a possible ancestor, and *S. turkestanica.*

Botany

Spinach is an annual plant grown in temperate regions exclusively for its leaves. The spinach root system consists of many shallow fibrous laterals developing from a slightly thickened taproot, with few large laterals. Soon after the seedling stage, plants assume a rosette growth habit with many fleshy leaves attached to a short stem. Spacing and environmental conditions influence leaf number and size. Leaf blades range from ovate or nearly triangular to long and narrow arrowhead shapes, the latter is a characteristic of primitive types. Leaf margins are smooth or wavy and surfaces are smooth, semisavoy to heavily savoyed (Fig. 21.1). The blistered appearance of the savoy tissue results from differential growth of parenchyma tissues between leaf veins. Petioles are usually as long as the leaf blade and often become hollow when leaves are fully expanded. Leaf growth habit varies from prostrate to upright, partly affected by plant spacings.

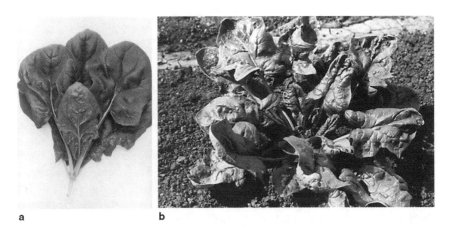

a   b

FIG. 21.1.   Spinach, *Spinacia oleracea,* plants of different foliage type: (a) smooth leaf, (b) savoy leaf.

Spinach is classified as a dioecious plant which is not strictly true because varying sexual types occur. Plant types are either male, female, or both male and female; the degree of monoeciousness is genetically and environmentally influenced. Occasionally, hermaphoditic flowers appear.

Plant types are classified by flowering or sex expression characteristics as follows:

Extreme males produce only staminate flowers, minimal foliage, flower early, and die after flowering.

Vegetative males produce only staminate flowers, more foliage, and flower slightly later than extreme males.

Monoecious plants produce staminate and pistillate flowers, well-developed foliage, and are slow to flower.

Female plants produce only pistillate flowers, have well-developed foliage, and are very late to flower.

The predominate sex expression of spinach plants is difficult to determine prior to flowering, although small plants usually are male. Female and vegetative males are preferred because they are larger, slower bolting, and higher yielding.

Apetalous staminate flowers are clustered on spikes. Pistillate flowers, also without petals, are attached to the base of the calyx that encloses the ovary. After pollination, mostly by wind, the fertilized ovary develops into a one-seeded fruit *(utricle)*. The calyx persists and together with the pericarp provides the hard covering that encloses the seed. About 100–110 seed weigh 1 g. A seed inhibitor is present in the outer seed layers. Parthenocarpic fruit often occur.

Two seed types exist having either a smooth round shape or an irregularly prickly shape. Prickly seeded cultivars are considered winter types, and the round seeded cultivars are considered summer types. Prickly seeded cultivars are infrequently grown. Prior to Linnaeus, taxonomists identified these round and prickly types as different species, *S. spinosa* and *S. inermis,* respectively. It is believed the prickly seed form preceded the round seeded types.

Physiology

Bolting is the response to lengthening days; high temperatures interact to accelerate this effect. Cultivar choice relies on critical day length characteristics relative to the growing latitude and seasonal temperatures. The critical photoperiod ranges between 12.5 and 15 h. High temperatures following initiation of flowering causes rapid stem elongation which is often attributed to temperature alone, although the prime stimulus is day length. High plant densities can also contribute to bolting. Slow-bolting cultivars are selected when plantings are scheduled to mature during long days and warm temperatures.

At the initiation of flowering, vegetative growth almost stops, new leaves formed are small, narrow, and pointed. The stem elongates and forms lateral branches from which clusters with as many as 20 small, inconspicuous, greenish flowers appear.

Climatic, Soil, Moisture, and Nutritional Requirements

Spinach growth is best at temperatures averaging 18–20°C; growth is slow at 10°C. When acclimated, plants can tolerate freezing to as low as –10°C. Temperature also affects leaf qualities; low temperature tends to increase leaf thickness but decreases size and smoothness.

Earliness is related to growth rate; early cultivars grow faster. Growers balance cultivar choice with growing conditions to obtain fast growth and high yield while avoiding bolting. Spinach is grown on a wide range of soil types; those with high moisture holding capacity and with good drainage are preferred. Plants have some tolerance to salinity but are sensitive to acidity; the favorable pH range is 6.5–8.0. The moisture requirement usually is not high because transpiration is low during the cool seasons when spinach is usually grown; about 250 mm is often sufficient for a crop. Nevertheless, because of the shallow root system, plants can easily be stressed from inadequate moisture. Waterlogged soil is also detrimental.

Nitrogen fertilization generally increases production of winter-grown spinach because little nitrification occurs at low soil temperatures. Spinach is well fertilized to increase leafiness and to meet the demands of very rapid growth that occurs in the short period before harvest. About two-thirds of total crop biomass is produced during the last third of the growth period. To meet this need, appropriate scheduling of fertilizer applications is important.

Culture

Spinach seed germination is optimum at 20°C and germination is better at lower (5–10°C) temperatures than at 25°C. However, emergence is slower at low temperatures; seed will germinate at 0°C, but emergence is extremely slow. Thermoquiescence occurs above 30°C (Table 21.1). Germination can be improved with seed presoaking or acid treatment to remove water-soluble inhibitors and to soften the pericarp.

Seed are frequently sown in multirows or in narrow (10 cm) bands

TABLE 21.1.   EFFECT OF SOIL TEMPERATURE ON
SPINACH SEED GERMINATION

Temperature (°C)	% Germination	Days to emergence
0	83	63
5	96	23
10	91	12
15	82	7
20	52	6
25	30	5
30	30	6
35	0	—

Source: Harrington and Minges (1954).

into level or raised beds at depths ranging from 1 to 3 cm. Sowing rates vary with intended crop use. Plant densities for fresh market production average about 60 plants per m², whereas for processing, densities are doubled. High plant populations result in more upright leaf growth which is desirable for machine harvesting. Low plant densities facilitate hand cutting for bunching of leaves or intact plants.

The fresh market crop is seldom thinned; thinning is labor prohibitive for the processing crop. To obtain optimal populations, seeding rates are determined according to germination percentage, seed vigor, seed size, and field conditions. Improved yields have been achieved through high-performance hybrids having high levels of femaleness.

Weed management is a critical factor, especially for the processing crop because weeds are a contaminant and some are similar in appearance to spinach and, therefore, difficult to separate. Frequent cultivation and effective selective herbicides can minimize this problem.

Harvest, Postharvest, and Storage

Harvests are made when plants have achieved marketable size, which, depending on seasonal period and temperature, can occur as soon as 30 days or as long as 80 days after planting, and as much as 150 days in some overwintered production. Most plants suitable for harvest will have five to eight fully developed leaves. A total of about 25 leaves may be produced from seedling to harvest; older leaves having died or are yellowed, others being at various stages of development. Harvest delays can increase plant weight; however, leaf quality may be affected.

For fresh market, plants are hand pulled or undercut below the stem. Intact plants are trimmed and several are tied together in bunches and packaged, but not all hand-cut spinach is bunched. To minimize leaf damage, plants are sometimes harvested later in the day when they are less turgid. Mechanical harvesting is also used for the fresh market crop, the leaves being collected and bulk handled rather than bunched (Fig. 21.2).

Processing crops are usually mechanically harvested. Harvest equipment is adjusted to cut 10–15 cm above the growing point to reduce

a

b

FIG. 21.2.   Hand (a) and machine harvest (b) of spinach, *Spinacia oleracea*.

the amount of petiole tissue removed with leaf blades, and also to allow regrowth for a possible second harvest 3–4 weeks later. A high ratio of leaf blade to petiole tissue is preferred which increases the value of the processed product. Processing crop yields range from 5 to as much as 20 t/ha, the latter from multiple harvests; fresh market yields are usually much less.

Smooth or semisavoyed cultivars are usually used for processing because they grow faster, yield more, and are more easily washed, whereas leaf surfaces of savoy cultivars are not easily washed free of soil. Surface texture of savoy cultivars is an advantage for fresh market because such leaves resist compression during packing and thus allow for better aeration, cooling, and postharvest life.

A limited amount of glasshouse spinach is produced in northern Europe during winter and early spring when adverse temperatures prevent outdoor production. Cultivars used are bred expressly for rapid foliage growth during the short days and low temperatures.

Harvested spinach is highly perishable because of the large amount of delicate leaf surface area and high rate of respiration. Overheating quickly destroys quality, and rapid cooling is essential to prevent wilting and weight loss. For spinach, hydrocooling is very effective for lowering product temperature. Spinach is also vacuum cooled. Hydrovac cooling, where moisture is added during vacuum cooling is useful in limiting wilting. For short-term holding, spinach should be iced or held at 0°C and 95% RH. Under these conditions, a 10-day storage life is possible.

Spinach Diseases and Pests

Diseases

Alternaria leaf spot	*Alternaria spinaciae*
Anthracnose	*Colletotrichum dematium,* f. sp. *spinaciae*
Ascochyta leaf spot	*Ascochyta spinaciae*
Cercospora	*Cercospora beticola, Cladosporium macrocarpum, C. variable*
Damping off	*Pythium, Aphanomyces, Fusarium,* and *Rhizoctonia* spp.
Downy mildew	*Peronospora farinoas* f. sp. *spinaciae (P. effusa)*
Fusarium wilt	*Fusarium oxysporum* f. spi-

	naciae
Heterosporium leaf spot	*Heterosporium variabile*
Leafspot	*Phyllosticta chenopodii*
Phytophthora root rot	*Phytophthora megasperma,*
	P. crystogea
Ramularia leaf spot	*Ramularia spinaciae*
Red rust	*Puccinia aristidae*
Stemphyllium leaf spot	*Stemphyllium* spp.
White rust	*Albugo occidentalis*

Viruses
- Beet Curly Top Virus (BCTV)
- Beet Western Yellows Virus (BWYV)
- Broad Bean Wilt Virus (BBWV)
- Cucumber Mosaic Virus (CMV)
- Lettuce Mosaic Virus (LMV)
- Yellow Dwarf Virus (YDV)

Insects

Alfalfa looper	*Autographa californica*
Armyworms	*Pseudaletia unipuncta*
Bean aphid	*Aphis favae*
Beet leafhopper	*Circulifer tenellus*
Cabbage looper	*Trichoplusia ni*
Corn ear worm	*Helicoverpa zea*
Cucumber beetle	*Diabrotica undecimpunctata*
Cutworms	*Peridroma, Agrotis,* other spp.
Green peach aphid	*Myzus persicae*
Lygus	*Lygus* spp.
Melon/cotton aphid	*Aphis gossypii*
Seed corn maggot	*Hylemyia cilicrura*
Spinach leaf miner	*Pegomyia hyoscyami*
Web worm	*Hymenid* spp.
Wireworms	*Limonius* spp.

Nematodes

Root knot	*Meloidogyne* spp.
Cyst	*Heterodera* spp.

Uses and Composition

Spinach is a traditional potherb, but it is widely used uncooked in salads. It is also processed by canning and freezing. Leading spinach-producing countries are Italy, France, Germany, Netherlands, and United States.

Excessive nitrate accumulation in spinach tissue has been associated with the disease methemoglobinemia. Nitrate accumulation varies with cultivars; smooth leaf types accumulate less than the savoyed. High nitrogen fertilization, low light, and low temperatures also favor accumulation.

Spinach tissues produce oxalic acid which combines with calcium to form insoluble calcium oxalate as well as reducing dietary magnesium and iron availability. Oxalic acid content also varies among cultivar types; savoy types contain less than smooth leaf types. Oxalic acid formation is greater at low than at high temperatures.

TABLE BEET, *Beta vulgaris* L. var. *crassa* (Alef.) J. Helm

Table beet is a relatively popular cool season, temperate-zone crop grown for its edible fleshy taproot and leafy tops. The related sugar beet is grown for the high sucrose content of the roots. Sugar levels of present-day sugar beet cultivars approach 20% of fresh weight, those of table beet being about 6% or less.

Origin

Wild species of beets are believed to have originated in parts of the Mediterranean region and North Africa with geographic distribution eastward to western India and westward to the Canary Islands and the western coast of Europe that included the British Isles and Denmark. A prevailing theory suggests table beet possibly originated from hybridization of *B. vulgaris* var. *maritima* (sea beet) with *B. patula*. Related wild species are *B. atriplicifolia* and *B. macrocarpa*. Initially, red beets may have been used primarily as a leafy vegetable, and interest in the storage root occurred later, possibly after A.D. 1500. Fodder beets probably began to be cultivated about 1800, and sugar beets apparently originated from the fodder beet population.

Botany

Beets are herbaceous biennials producing a closely spiraled leafy rosette and a fleshy storage taproot during the first year (Fig. 21.3). Leaf blades are oblong or triangular, cultivars may have wavy or straight margins, and smooth or crinkled surfaces. Leaf color varies from light green to dark red. Petioles are thin and variable in length. Soon after

FIG. 21.3. Various forms of table beets, *Beta vulgaris* var. crassa. Source: Courtesy of National Garden Bureau, Downers Grove, Illinois.

emergence, the taproot grows rapidly and deeply into the soil, sometimes to a depth of 2 m. Secondary rooting is extensive, most of which is within 50–60 cm of the surface. The beet root system is very efficient and provides fair drought tolerance.

Anatomically, the beet storage root consists of an enlarged hypocotyl-root axis formed near the soil surface and the narrowing tapered true root portion. Secondary roots are produced at the lower portion of the taproot. Before domestication, wild-type beet roots were white, long, thin, and well branched. Current beet characteristics are the product of domestication.

The enlarged hypocotyl-root axis consists of alternating zones of conducting and storage tissues appearing as rings. Increase in thickness results from cambial growth accompanied by division and cell enlargement of parenchyma tissues.

Color differences (zoning) between xylem and phloem tissues is due to different levels of pigmentation. High temperatures tend to reduce pigmentation. The red color of table beets is due to the pigment betacyanin, a nitrogen-containing compound with chemical properties similar to anthocyanin; 70–90% of betacyanin is betanin. Beets also contain betaxanthin, a yellow pigment. The ratio of these pigments varies with cultivars and can change because of environmental conditions. The intense red color indicates that little betaxanthin is present, and yellow color indicates that betacyanin is absent; white indicates that neither is present.

Storage root sizes range from as little as 2 cm to over 15 cm in diameter. Shapes are variable and may be globe, cylindrical, toplike, or flattened.

Bolting is initiated when mature plants experience temperatures below 10°C for a period of 15 or more days. The developed seed stem is tall (1.5 m) and well branched into a large compound open inflorescence bearing small greenish, almost sessile flowers in clusters of two or three or singularly in the axils of bracts on open spikes. The flowers are perfect. Flowering is protandrous, stigmas having about a 2-week period of receptivity, whereas anthers dehisce pollen and shrivel within 2–3 days. Pollen is wind transported, although some insect pollination occurs. Male sterility is used in hybrid seed production.

Multiple-seeded fruit are formed when two or more flowers grow together and are encircled by the fused dry corklike bracts forming a hard irregularly shaped seed ball. When planted, such seed balls (fruits) can produce several seedlings, thus giving the appearance of high or sometimes more than 100% germination. Individual seed are reddish brown, measuring about 1.5 × 3.0 mm. About 55–60 seed weigh 1 g. Monogerm cultivars produce single rather than multiple flowers in bract axils; therefore, the fruit will contain only a single seed.

Climate, Soil, Moisture, and Nutritional Requirements

A hardy, cool season crop, beet edible quality is best when grown at 16–20°C. Low temperatures tend to increase leaf thickness, a quality beneficial for edible top use. In contrast to spinach, beets are tolerant of warm growing temperatures.

Well-drained sandy loams are most suitable for root growth. Although beets grow satisfactorily in muck and heavy textured soils, soil compaction can interfere with root shape. Beets are sensitive to soil acidity; pH should be within 6–8. Beets tolerate salinity fairly well; an old cultural practice was the use of salt for controlling weeds. Small amounts of boron, zinc, and sodium are often applied to satisfy minor element needs. Unlike many vegetables, beets are tolerant of high soil boron levels. Crop moisture demands are not excessive; 300 mm of water uniformly available is often adequate.

Culture

Beets are seed propagated; germination occurs over a range from 8°C to 30°C, with an optimum between 18°C and 20°C. The use of monogerm seed has made plantings more precise and improved stand uniformity compared to use of multigerm seed, although germination of monogerm cultivars has been lower than that of multigerm cultivars.

Seed are planted 1.5–2.5 cm deep in single or multiple rows or scattered within a narrow band. Spacings vary according to intended crop use. For fresh market, in-row spacings of 4–5 cm are made with rows 40–60 cm apart. For processing, seed are placed closer because processed whole small beets usually have a greater value. Large roots, with diameters greater than 15 cm, are susceptible to being cracked, fibrous, and poorly colored. Beets are seldom thinned, because the work is tedious and because slight overcrowding can be tolerated.

Harvest and Postharvest

Beets are easily hand pulled, bunched together with leaves and roots attached, and often packed directly in the field. Beets for processing are usually mechanically harvested, but some fresh market beets are also machine harvested. In most mechanical harvesting procedures, foliage is topped and roots are bulk handled much like other root crops. Beet foliage (beet greens) are a common potherb and can be harvested separately from storage roots and handled as bundled tops.

Beet greens and bunched beets with tops can be maintained in good condition for 10–14 days if held at 0°C and at 95% RH. Topped beets under the same conditions can be stored for 4–6 months. Leaf tissue respiration is much higher than that of the beet root and requires rapid postharvest cooling. Hydrocooling and icing procedures are commonly used.

Beet Diseases and Other Pests

Bacteria

Bacterial black ring	*Pseudomonas wieringae*
Bacterial blight	*Pseudomonas aptata*
Bacterial rot	*Erwinia carotovora* subsp. *carotovora*
Beet crown gall	*Xanthomonas beticola*
Leaf silvering	*Corynebacterium betae*

Fungi

Alternaria leaf spot	*Alternaria tenuis, Urophlyctis leproides*

Aphanomyces black rot	*Aphanomyces cochlioides*
Cercospora leaf spot	*Cercospora beticola*
Damping off/root rot complex	*Pythium ultimum, Rhizoctonia solani*
Downy mildew	*Peronospora schachtii, P. farubisa* f. sp. *betae*
Phoma heart rot	*Phoma beta*
Powdery mildew	*Erysiphe polygoni*
Ramularia leaf spot	*Ramularia beticola*
Rhizoctonia	*Rhizoctonia solani*
Rust	*Uromyces betae, Puccinia aristidae*
Scab	*Streptomyces scabies*
Septoria leaf spot	*Septoria betae*

Viruses
Beet Curly Top Virus (BCTV)
Beet Mosaic Virus (BMV)
Western Beet Yellows Virus (WBYV)
Yellow Net Virus (YNV)

Physiological disorders
Boron-deficiency internal black spot

Insects

Aphids	*Aphis* spp.
Beet armyworm	*Spodoptera exigua*
Beet leafhopper	*Circulifer tenellus*
Beet leafminer	*Pegomya hyoscyami, P.* spp.
Beet webworm	*Hymenid* and *Loxostege* spp.
Blister beetle	*Epicauta* spp.
Click beetle	*Melanotus* spp.
Cucumber beetle	*Diabrotica undecimpunctata*
Cutworm	*Agrotis ipsilon*
Flea beetle	*Epitrix* spp.
Garden symphylan	*Scutigerella immaculata*
Green peach aphid	*Myzus persicae*
Variegated cutworm	*Peridroma saucia*
Wireworms	*Ctenicera* and *Limonius* spp.

Nematodes

Cyst	*Heterodera schachtii,* other *Heterodera* spp.

| Root knot | *Meliodogyne* spp. |
| Stubby root | *Paratrichodorus* spp. |

Other pests
| Spider mite | *Tetranychus* spp. |
| Gray garden slug | *Agriolimax reticulatus* |

Uses and Composition

Beet roots are eaten cooked and a large proportion of the crop is processed by canning and pickling; some are dehydrated. Small-diameter whole beets are a premium item. Those of intermediate size are usually processed sliced and the larger sizes are better suited to dicing and strips. Intense uniform color with minimum zoning is an important quality. The red pigment can be extracted and used as a natural dye product. Beet greens are cooked as potherbs and compete with spinach, Swiss chard, and some mustards. Nutritionally, the pro-vitamin A and ascorbic acid content of beet greens are similar to that of spinach and Swiss chard (Appendix Table C).

SWISS CHARD, *Beta vulgaris* L. var. *cicla* L.

Although closely related and cultivated like table beets, Swiss chard has several important differences. Plants are grown exclusively for the large, fleshy, semisavoyed crispy and glossy leaves rather than the root (Fig. 21.4). The taproot exhibits relatively little expansion but does branch and produces many deep secondary roots.

Other differences are the wide and differently colored leaf midrib. Midrib colors are white, red, or green. Leaf blade colors range from light green to dark green. Although a cool season crop, Swiss chard has fairly good heat tolerance. Foliage growth is usually vigorous within a broad temperature range, as well as tolerant to light frosts.

Like table beets, the crop is directly sown, but plant stands are thinned to spacings of 20 cm within rows and between-row spacings of 60–75 cm to facilitate harvesting.

Normally, harvesting is begun about 50–60 days after seeding. The larger outer petioles are pulled away from the base of the rosette growth. Inner leaves are left to enlarge for subsequent harvests, which often are made weekly. Rather than frequent serial leaf harvesting, mechanized harvesters are frequently used for crops intended for canning or freezing. A repeated harvest is sometimes possible when adequate regrowth occurs.

FIG. 21.4.   Swiss chard plants, *Beta vulgaris* var. *cicla*.

SPINACH BEET/PALAK, *Beta vulgaris* L. var. *orientalis* Hort. (var. *bengalensis*)

Palak, Palang sag, or Palanki, are regional names for this important northern India crop. It is also important in parts of Indochina where it apparently originated and where it has a long history as a medicinal and food plant. Palak, grown during the cool part of the year, is widely used in salad and stew preparations.

This annual plant has a prominent taproot and long petiolate leaves that grow from the base of the compressed stem. Plants resemble Swiss chard having reddish and green midribs. Harvests are made about 3–4 weeks after sowing and every 15–20 days following regrowth. An average of five harvests are made and yields from 9 to 12 t/ha are obtained. Bundled harvested leaves are marketed as soon as possible, because postharvest life is only about 1 week, even with cold storage.

ORACH, *Atriplex hortensis* L.

Other names: Mountain or French spinach

Orach is considered to have originated in northern India and has been used as a medicinal and food plant for more than 2000 years.

Orach is a hardy branching monoecious annual with tender leaves and shoots used as a potherb. First to develop is a rosette of leaves which may be followed by a seed stalk that can grow to a height of 2–3 m. Plants have a tolerance to drought, salinity, and are adapted to a broad temperature range. Although stems quickly elongate, flowering is slow and plants tolerate growing temperatures too high for spinach.

Flowers, yellow in color, are wind pollinated. They are sometimes pinched off to encourage greater vegetative growth. Seed require 4–5 months to mature and are surrounded by persistent disklike bracts which should be removed before sowing.

Seed are sown about 1 cm deep in rows 50–75 cm apart and initially with in-row spacings of 3 cm. Harvest of young seedling, 10–20 cm in height, and thinning occur together. This results in the remaining plants being spaced about 25 cm for continuing growth and repeated harvests. Shoot harvest stimulates new upper leaf growth. Leaves are smooth and have a triangular or shieldlike shape on moderately long petioles. Three main cultivar types are white (pale green leaves), red (dark reddish stems and leaves), and green (dark green stems and foliage). The white type is the most tender and best flavored. Orach has a mild flavor, much like spinach but contains less oxalic acid. Even when the plant has bolted, young leaves are usable. However, old leaves are not palatable and are not harvested.

SELECTED REFERENCES

Atherton, J.G., and Farooque, A.M. 1983. High temperature and germination in spinach. II. Effects of osmotic priming. Sci. Hortic. *19*, 221–227.

Babb, M.F., and Kraus, J.E. 1939. Orach, its culture and use as a greens crop in the great plains region. USDA Cir. No. 526, USDA, Washington, DC.

Benjamin, L.R. 1987. The relative importance of cluster size, sowing depth, time of seedling emergence and between-plant spacing on variation in plant size in red beet (*Beta vulgaris* L.) crops. J. Agric. Sci. Cambr. *108*, 221–230.

Cantliffe, D.J. 1972. Nitrate accumulation in spinach grown under different light intensities. J. ASHS *97*, 152–154.

Cantliffe, D.J. 1972. Nitrate accumulation in spinach grown at different temperatures. J. ASHS *97*, 674–676.

Carlsson, R., and Wendy Clarke, E.M. 1983. *Atriplex hortensis* L. as a leafy vegetable, and as a leaf protein concentrate plant. Qual. Plant Foods Human Nutr. *33*, 127–133.

Correll, J.C., Morelock, T.E., Black, M.C., Koike, S.T., Brandenberger, L.P., and Dainello, F.D. 1994. Economically important diseases of spinach. Plant Dis. *78*, 653–660.

Fennell, A., and Li, P.H. 1987. Freezing tolerance and rapid cold acclimation of spinach. J. ASHS *112*, 306–309.

Harrington, J.F., and Minges, P.A. 1954. Vegetable Seed Germination. University of California, Berkeley.

Huyskes, J.A. 1971. The importance of photoperiodic response for the breeding of glasshouse spinach. Euphytica *20,* 371–379.

Maynard, D.N., and Barker, A.V. 1974. Nitrate accumulation in spinach as influenced by leaf type. J. ASHS *99,* 135–138.

Pandita, M.L., and Lal, S. 1986. Leafy vegetables: Spinach beet (Palak) and spinach. In Vegetable Crops in India. T.K. Bose and M.G. Som, eds. Naya Rrokash, Calcutta, pp. 645–669.

Rosa, J.T. 1925. Sex expression in spinach. Hilgardia *1,* 258–274.

Wolyn, D.J., and Gabelman, W.H. 1990. Selection for betalain pigment concentration and total dissolved solids in red table beets. J. ASHS *115,* 165–169.

22

Peas, Beans, and Other Vegetable Legumes

Family: Fabaceae (Leguminosae)

LEGUME CHARACTERISTICS

The Fabaceae constitute a broad and very large botanical family, consisting of more than 450 genera and over 12 thousand species. Many species are important as food sources for humans and animals. Legumes are dicot annuals and perennials; most cultivated vegetable and grain legumes are annuals. Grain legumes, often identified as pulse crops, rank second to cereal grains as a primary world food source. The term pulse refers to edible seed of pod-bearing plants and the word gram, also refers to some legume seed-bearing plants. In India, the term dal or dahl refers to a food preparation usually made with pigeon pea, *Cajanus cajan.* In French horticultural writings, legume refers to a garden vegetable or pot herb.

In terms of volume and calories, the importance of grain legumes outweighs that of vegetable legumes. Nevertheless, the food value of the latter is very significant. Vegetable legumes in addition to fresh use of pods, foliage, tender shoots, and storage roots also produce a wide array of dried seeds. Some species are used extensively as forage, cover, and green manure crops, and for edible oils as well as timber, dyes, gums, and many other industrial products. A listing of vegetable legumes is given in Table 22.1.

Several vegetable legumes contain toxic substances such as saponins, lathrogens, cyanogenic and other glycosides, protease and amylase inhibitors, and hemagglutinins (Chapter 5). Certain leguminous plants grown in high selenium or molybdenum soils can absorb excess quantities of these elements, which can be toxic when consumed.

Legume crops have been cultivated in various parts of the world for more than 6000 years. Wild forms of many present-day legumes cannot

TABLE 22.1.   VEGETABLE LEGUMES

Species	Common name	Edible vegetable portion(s)[a]
Apois americana[b]	Apois	2, 3, 7a
Arachis hypogaea	Peanut	3, 5, 6, 9
Canavalia ensiformis	Jack bean	3, 4
Canavalia gladiata	Sword bean	3, 4
Cajanus cajan	Pigeon pea	2–4
Cicer arietinum	Chick pea	2–4
Cyamopsis tetragonolobus	Cluster bean	4
Dolichos lignosa	Australian pea	3
Glycine max	Soybean	1–4, 8, 9
Lablab purpureus	Hyacinth bean	2–4, 8
Lathyrus sativus	Grass pea	2, 3, 6
Lens culinaris	Lentil	3, 4
Lupinus spp.	Lupines	3
Pachyrhizus ahipa[b]	Ahipa	7
Pachyrhizus erosus	Yam bean	4, 7, 9
Pachyrhizus tuberosus[b]	Potato bean	7
Phaseolus acutifolius	Tapary bean	3
Phaseolus coccineus	Scarlet runner bean	2–4, 7, 10
Phaseolus lunatus	Lima bean	1–6
Phaseolus vulgaris	Snap and common bean	2–6
Pisum sativum	Garden and field pea	1–6, 8
Psophocarpus tetragonolobus	Winged bean	1–7, 8
Pueraria lobata[b]	Kudzu	9
Sphenostylis stenocarpa[b]	African yam bean	7
Trigonella foenum-graecum	Fenugreek	3, 5, 6
Tylosema esculentum	Marama bean	3, 7
Vicia faba	Broad bean	2, 3
Vigna aconitifolia	Mat bean	3, 4
Vigna angularis	Adzuki bean	3, 4, 8
Vigna mungo	Urd bean	3, 4
Vigna radiata	Mung bean	3, 4, 8
Vigna subterranea	Bambara groundnut	1, 3
Vigna umbellata	Rice bean	3, 4
Vigna unguiculata cultigroup cylindrica	Catjang cowpea	2–4, 6
Vigna unguiculata cultigroup sesquipedalis	Yardlong bean	4
Vigna unguiculata cultigroup unguiculata	Cowpea	2–6

[a] 1 = Immature seed; 2 = fresh mature seed; 3 = mature dry seed; 4 = immature pod;
5 = tender shoot tips; 6 = tender foliage; 7 = tuberous roots; 7a = tubers, rhizomes;
8 = sprouted seed; 9 = starch or oil, or other food forms; 10 = flowers.
[b] See Chapter 14.

be found, although, according to Vavilov, some legume species appeared to be associated with certain identified centers of origin. For example, he assigned certain species of *Phaseolus* to a New World (Americas) origin, and several others identified as *Phaseolus* to the Old World. However, those ascribed to the Old World origin previous identified as *Phaseolus* are now correctly classified in the large and widely distributed genus *Vigna*.

Several cultivated legumes are perennial, but most are annuals.

Generally, leaves are alternate and mostly compound, and either pinnate, trifoliate, or digitate. Flowers are perfect and characteristically papilionaceous (butterfly shaped), consisting of an upright dorsal petal (standard), two lateral petals (wings), and two lower petals often more or less united into a keellike structure. Enclosed within the petals are the stamens and pistil. Self-pollination is common; when cross-pollination occurs, bees are often the main agent. Seed develop in a pericarp (pod), which is usually dehiscent.

Another legume characteristic is the cotyledonary position of germinated seedlings. Peas, *Pisum sativum,* have hypogeal germination in which cotyledons and the seed coat remain beneath the soil because of limited hypocotyl elongation. The epicotyl is fully differentiated prior to germination and pushes through the soil. Epigeal germination is typical for the common bean, *Phaseolus vulgaris,* where the cotyledons appear above the surface because of rapid hypocotyl elongation. Seed orientation in the soil can affect hypocotyl growth and, thereby, the time and success of emergence. The seed coat is sometimes left in the soil or may be carried above and shed as the cotyledons unfold. The peanut, *Arachis hypogaea,* exhibits a different behavior, in which the cotyledons remain at the surface rather than below or above the soil. Further growth is due to aboveground epicotyl elongation; the length of the hypocotyl depends on the depth of sowing.

Generally, the cultivated legume species have a broad adaptation to temperature. Some species prefer cool and humid conditions; others thrive in hot and dry environments; and many are frost sensitive. Photoperiod requirements for flowering also vary from short day to neutral day lengths. Most cultivated legumes perform best when grown in well-drained, light textured, well-aerated soils, and are injured by waterlogging. A slightly acid to nearly neutral soil pH is usually preferred.

Nitrogen fixation is a widely recognized legume characteristic. *Rhizobium* bacteria species are able to fix atmospheric nitrogen while living symbiotically in root nodules. The bacterial enzyme responsible for nitrogen fixation are nitrogenase; adequate iron and molybdenum is required for enzyme activity. The nitrogen fixed in the nodules is used in growth and also can enrich the soil after plant death. Under some conditions, the bacteria are able to excrete soluble nitrogenous compounds to the soil.

Rhizobium organisms are free-living and mobile in soil, and, as such, are unable to fix atmospheric nitrogen. However, after the bacteria infect plant root hairs, nodules are formed, the bacteria undergo change, and are then able to fix nitrogen. Host–organism relationships are specific, and when not complementary, the symbiosis will be ineffective or will not develop. The presence of bacteria and soil environment interactions affect the level of symbiosis. When the proper *Rhizobium* species in the soil is not present or present in low amounts, it is sometimes recommended that seed be inoculated with the proper strain for the particular crop.

In general, *Rhizobium* fixed nitrogen is not supplied in sufficient quantity to provide for vigorous growth during the early stages of plant growth. This can have a negative effect on yield because there is a metabolic cost for N_2 fixation in that carbohydrates are utilized as a source of energy for the process. It is questionable to inoculate seed of crops with short growth periods with *Rhizobium* bacteria, because of the time required for colonization and nitrogen fixation to occur. In a low-input production system, the nitrogen supplied by *Rhizobium* may be adequate, but not for intensive production. For the latter system, nitrogen fertilizers can be applied.

Increasing concerns about environmental and nutritional impacts of nitrogenous fertilizer usage have increased research directed to the unique nitrogen-fixation capability of legumes. Transformation of this capability to nonlegume species would be a great benefit and accomplishment. Presently, because of genetic complexities, it is doubtful that such a transformation will be achieved.

GARDEN AND FIELD PEAS, *Pisum sativum* L.

There are three major pea types and their harvested products include (1) plants producing well-developed, succulent but immature seed, (2) plants producing immature, succulent, edible pods and seed (Figs. 22.1), and (3) plants producing fully developed, mature, dry seed. In parts of the Orient, the tender shoots of pea plants are used as greens. English, garden, vining, and green peas are English language terms that refer to plants of the first type. The early involvement of English plant breeders with the pea crop is believe to be the origin of the term "English" peas.

Origin and Taxonomy

The wild progenitor of peas is unknown and views differ about the possible origins. One view suggests central Asia, Abyssinia, and the Mediterranean basin as primary centers, with the Near East a secondary center. Another considers the Mediterranean basin as the primary

FIG. 22.1. Pea, *Pisum sativum,* (a) plant, (b) pods, and (c) edible podded peas. Source for (b) Courtesy of Burpee Seed Co.

center with the Near East and the central plateau of Ethiopia as secondary centers.

Peas are among the oldest cultivated plants; carbonized pea seed remains were identified as being between 7000 and 9000 years old. Domestication resulted in a segregation for seed, fodder, and vegetable types; the edible podded pea forms are thought to be recent developments (Fig. 22.1c).

Present garden pea cultivars are the opposite of primitive forms which had rough, tough, and slightly bitter seed coats. Modern cultivars also differ from wild types in having large seed and a short compact growth habit. Although less frequently grown, some tall-growing, small-seeded cultivars are still cultivated.

Edible pod pea types are commonly known as snow, sugar, or China pea; they have been assigned botanical variety status as *P. sativum* var. *saccharatum*. Another edible pod type, designated as *P. sativum* var. *macrocarpon,* is known as sugar snap pea. The major characteristics of each type is the succulent pod wall and very slow seed development. The sugar snap pea is a relatively new introduction, having fleshy and succulent edible pods that closely resemble those of a snap bean in contrast to the flat, wide, and thin-walled sugar pea. Pods of each type tend to remain relatively succulent, even when some seed enlargement occurs, because the fibrous parchment layer of the interior carpel wall does not form.

Field or dried peas are grown for the production of mature dry seed and are usually considered an agronomic crop. A previous classification of field pea as *Pisum arvense* is not appropriate, and this crop is not different from *P. sativum.*

Botany and Morphology

Peas are short-lived, herbaceous annuals, having alternate leaves with the tip of the compound leaf modified as a tendril. Growth habits range from indeterminate, vine, and climbing types to determinate bush or dwarf forms. Foliage types vary, ranging from cultivars having extensive leaflets to others having almost all leaflets converted to tendrils; the latter forms are known as leafless.

The genetic control of many pea leaf traits is well identified. When homozygous recessive, the *af* (afila) gene results in a plant without leaflets and many tendrils. The *tl* gene results in plants with extra leaflets and no tendrils, and the *st* gene results in reduced straplike stipules and small leaflets. This array of genes and several others can be genetically manipulated so that a wide variety of foliage forms can be produced. The *af* and *afst* foliage forms have improved standing ability because tendrils intertwine and provide mutual support. Leafless cultivars also appear to be less affected by insects and diseases because of improved drying conditions in the leaf canopy. However, the altered reduced leaf or leafless forms are generally less productive for fresh pea production, except at very high plant densities that tend to compensate for the low leaf area index. Afila cultivars produce fewer light colored berries than leafy cultivars because of greater light penetration into the canopy.

Foliage surfaces have a noticeable waxy cuticle layer, and leaf colors range widely from yellowish green to deep blue green. Stems are slender, angular, glaucous, usually singular, and hollow, except near the base. Some are upright, but generally not self-supporting, and branching usually is limited.

Taproots can reach 80 cm in depth. Nevertheless, the root system, even with many lateral roots, is not extensively developed. The symbiotic association with *Rhizobium* species is usually beneficial.

Hermaphoditic flowers, borne at leaf axils, are usually white, but may be pink, purple, or a blended color; field pea flowers are usually purple. Self-pollination usually occurs before flowers fully open so that the frequency of cross-pollination is very low. Very early-maturing cultivars produce flowers after as few as five or six nodes have formed. Some late cultivars flower after 15 or more nodes have formed. The number of nonbearing nodes produced is a fairly consistent cultivar characteristic. Flowers are initiated sequentially, beginning at the lowest flowering node and advancing with node development.

Pod number per node is genetically determined but is influenced by environment; thus, plant stress can reduce node numbers. Early-flowering cultivars average between one and two pods per node; late-flowering cultivars develop more pods, averaging more than two per node. There are cultivars which develop four or more pods per node. Some determinate cultivars have a fasciated character, where flowers are grouped in a terminal inflorescence instead of among flowering nodes. Development of the terminal inflorescence stops further vegetative growth.

The pod is a dehiscent fruit with its two sides formed from the carpels of the flower. Seed are attached alternately to the side of the fused carpels. Pod walls of garden and field peas have a hard parchment layer consisting of several layers of lignified cells. The pod walls of sugar and sugar snap pea cultivars do not form the lignified layer. Pod size as well as seed numbers are a stable cultivar characteristic, but pod size and seed number vary among cultivars.

The pea seed is a berry consisting of two large cotyledons surrounding the embryo and enclosed by the seed coat (testa), which may be pigmented or colorless. Pea seed exhibit hypogeal germination, and the cotyledons tend not to shed the seed coat.

Mature smooth seed types are associated with rapid and higher starch accumulation and a lower sugar content than wrinkled seed. The mealy texture of smooth pea seed is due to the high starch content. Smooth-seeded cultivars, especially those of field peas, have better cold tolerance than wrinkled seed types.

Climate, Soil, and Moisture

Optimum mean temperatures for vegetative growth are between 13°C and 18°C; growth essentially stops above 29°C. Plants are very temperature responsive, especially during vegetative development, and although a cool season vegetable, they are frost sensitive. Blossoms and pods are more susceptible to injury from frost than leaves and stems; young plants are more tolerant to low temperatures.

Correlations between number of nodes to first flower, growing temperatures, and time to harvest maturity are very high; they are useful for scheduling plantings and harvest periods. Heat units (degree days) are also commonly used to predict harvest dates for processing peas. For peas, a base temperature of 4°C with an upper limit of 29°C is usually the range for calculating accumulated heat units. Plant growth is negligible at temperatures below or above this range and, thus, has little influence toward accumulated heat units (see heat units, Chapter 6).

From growth and performance tests, cultivars are identified by their heat unit requirements. Early-producing cultivars require as few as 1000 heat units to achieve harvest maturity, whereas late cultivars may require more than 1600. Generally early-flowering cultivars, are day neutral; flowering is accelerated by long days for late cultivars. Moderately high temperatures also decrease time to flowering, but temperatures above 30°C can cause flower or ovule abortion. Moderate diurnal temperature fluctuations usually improve plant growth.

Culture

Peas are grown in many soil types, from light sandy loams to heavy clays. An acceptable pH range is from 6.0 to 7.5. Manganese deficiency is occasionally observed at high soil pH. Poor soil drainage and compaction strongly impair productivity and increase susceptibility to root diseases. Crop rotations help reduce some root rot diseases, and pea crops should not be grown in the same field more than once every 4 years.

In temperate climates, peas are usually planted in the spring, or in late fall and early winter if heavy frosts do not occur. Peas can tolerate light frosts before flowering. In the tropics and subtropics, plantings are made at high elevations and during seasonal periods when temperatures are relatively cool.

Partly because of its bulk, seed is a significant cost in pea production. High-quality seed is very important, especially if plantings are made when soils are cold. Seed quality can be evaluated by measuring the electrical conductivity of a water solution in which the seed have been soaked. Increased conductivity of the solution is due to leakage from seed. The higher the conductivity, the greater the loss of quality.

Germination can occur over a wide range of soil temperatures and is best at 20°C. At temperatures greater than 25°C, germination percentage decreases. Seed can be treated with fungicides and insecticides to minimize disease and insect injury during emergence and seedling growth.

For machine harvest, seed are planted into moist soil of good tilth at uniform depths of about 3–5 cm, and at spacings of 3–5 cm within rows and 20–30 cm between rows. Plant densities of 80–90 plants per m^2 are common. At these populations, tillering is minimized. With garden peas, pods that develop on tiller stems mature too late to contribute to market yield, especially when machine harvested; therefore, single stem development is preferred. Lodging reduces yield by making mechanical harvest difficult. Because of the mutual support of intertwining plants, high populations can reduce the tendency for lodging. To limit lodging, determinate and short-stature cultivars are used. Except for home gardens or for edible podded pea production, indeterminate cultivars and trellising are seldom used.

Grain drills are commonly used for sowing. Seeding rates are extremely variable because of the large differences in seed size among cultivars, so that the amount of seed sown can vary from 70 to more than 200 kg/ha. Total yield may increase with higher densities, but seed costs may exceed the value of the yield increase. Hand-harvested, fresh market peas are grown in rows spaced 75–90 cm apart.

Nutrition and Irrigation

Nitrogen must be carefully managed in order to avoid excessive foliage growth. Lush vegetative growth competes with pod and seed development. Phosphorus is important for lateral root development and potassium for plant growth and seed enlargement. Fertilizer should be well incorporated before planting with placement close to but not in direct contact with the seed. With high-density plantings having many parallel narrow rows, preplant applications can be made across plant rows in order to reduce direct contact with seed.

Peas have a relatively low moisture requirement, and depending on seasonal conditions, 75–150 mm of water often can be adequate; many pea-producing regions rely entirely on rainfall. Peas are sensitive to excessive moisture, so waterlogging should be avoided, especially at the flowering stage. The most sensitive moisture periods are just before flowering and also during pod enlargement. Leafless pea cultivars have higher soil moisture tolerance than conventional leafy forms. Rain or irrigation during flowering may increase the incidence of disease.

Harvest

The optimum harvest time is when pods are well filled and the seed are still soft and immature. As seed mature, they increase in firmness, the seed coat thickens, becomes tough, and sugars are converted to starch. Peas mature rapidly during high temperatures and are at optimum quality for only a day or two. Therefore, it is important to properly select cultivars and planting dates in order to predict harvest periods as well as to achieve high yield and quality. To extend production, cultivars with different maturity periods are grown.

For most fresh market production, multiple harvests are made by hand labor as pods develop. Although very labor intensive, hand harvesting minimizes physical damage and helps to retain quality longer. During marketing, peas are retained in the pods and not removed until ready for preparation and consumption. When removed from pods, peas have a greatly reduced postharvest life.

Machines, many being self-propelled combines, are used for harvesting peas for processing. These machines separate pods from vines and then peas from pods. The peas are collected while the pods and vines are discarded. In some situations, the vines are harvested with machinery or by hand and taken to a stationary viner where peas are separated from pods. In order that harvest equipment function effectively, plants should have a determinate habit, short stiff stems, and the pods concentrated close to the top of the plant. Harvest combines function more effectively when plants have sparse foliage; this is an advantage of the leafless-type cultivars.

The use of single mechanical harvest can result in some peas being overmature and others immature. This is more of a concern for processing rather than fresh use. Peas of different maturity of the same cultivar are commonly separated by size. Because shelled peas of different maturity have different specific gravities, they can also be separated by floatation using different concentrations of brine.

Because pea seed development increases rapidly over a brief period, it is important to equate yield to an identifiable state of seed development; maturity in this sense meaning edible quality. Edible quality is usually objectively determined using a tenderometer. This instrument measures resistance of a sample of peas to crushing pressure; the greater the resistance, the less tender are the peas. Tenderometer readings, when adjusted for temperature, correlate well with edible quality, as judged by sweetness, tenderness, and starchiness. The amount of alcohol-insoluble solids in seed is another method for quality determination.

Edible podded pea and snap pea cultivars are not machine harvested. Hand harvesting is sometimes carried out daily in order to avoid pod overmaturity. The proper harvest stage is when the pods are about full size but before seed development is apparent. With snap peas, the seed are slightly more developed but still immature. Harvests of indeterminate, edible podded pea plants can continue for many weeks, and frequent harvesting tends to favor additional flowering.

Field pea harvests are delayed as long as possible to allow late-set pods to mature. Harvest delay can result in seed shatter. Foliage desiccants are sometimes used to facilitate harvesting. Early harvesting can compromise yield and also require more time and energy for artificial drying. Seed moisture content of about 40% is often a guide to harvest. In the fully dried state, moisture content is less than 15%.

Postharvest and Storage

Temperature is a fundamental factor in the conversion of sugar to starch; therefore, it is important that harvested garden peas are not exposed to high temperature. A period as little as 3 h at ambient temperature can result in significant quality loss. Rapid, preferably wet cooling, is essential for quality maintenance. Harvested pods or shelled peas should be cooled as quickly as possible to 0°C to limit sugar conversion as well as fiber development. Peas retain higher quality when unshelled than those shelled, because of reduced desiccation. The high CO_2 content within the pod provides a minicontrolled atmosphere. Shelled peas are processed the same day, often within hours after harvest. The quality of fresh garden pea pods and edible podded peas can be maintained in relatively good condition for 1–2 weeks when stored at 0°C and 90% RH. Dry field peas, because of their low moisture content, present few postharvest or storage concerns.

Uses and Composition

In addition to fresh use, peas are frozen, canned, and dehydrated. It is more common for wrinkled-seed cultivars to be processed by freezing, and smooth-seeded ones to be canned. However, cultivars having wrinkled seed are also increasingly canned. In the past, processors preferred to use dark-colored peas for freezing and light-colored peas for canning; presently, this distinction is observed less. In processing, preference is given to small sizes. In general, small peas are those 3.5–5 mm in diameter; medium are 5–7 mm in diameter, and large have a diameter greater than 7 mm. Another measure uses seed weight per 1000 seed. One thousand fresh peas weighing less than 150 g are small, those 150–250 g are medium, and those more than 250 g are large. There is a perception that small peas (petite pois) are sweeter and more tender than larger ones. Within a cultivar, large peas are usually less tender than small ones, but between cultivars, size is not an indication of tenderness.

Dried peas are reconstituted by soaking and are used directly or are further processed by canning. When processed as a dry pulse, the seed coats are removed. A popular use of the intact or split cotyledons are in soups, and they are also ground into a powdered product. Because of the high protein content, good amino acid balance, and high digestibility, considerable use of dried field peas is for animal feed.

Diseases and Other Pests

Diseases

Aphanomyces root rot	*Aphanomyces euteiches*
Anthracnose	*Colletotrichum pisi*
Ascochyta blight complex	*Ascochyta pisi, A. pinodella, Mycosphaerella pinodes*
Bacterial blight	*Pseudomonas syringae* pv. *pisi*
Black root rot	*Thielaviopsis basicola*
Botrytis	*Botrytis cinerea*
Downy mildew	*Peronospora pisi, P. viciae*
Fusarium wilt and near wilts	*Fusarium oxysporum* f. *pisi,* other races of *F. oxysporum*
Fusarium root rot	*Fusarium solani* f. sp. *pisi*
Leaf and pod spot	*Phoma medicaginis* var. *pinodella*
Powdery mildew	*Erysiphe polygoni,* other *Erysiphe* spp.

Pythium root rot	*Pythium ultimum,* other *Pythium* spp.
Rusts	*Uromyces pisi, U. fabae, U. trifolii*
Sclerotinia	*Sclerotinia sclerotiorum*
Septoria blight	*Septoria pisi*

Viruses (most are aphid vectored)
Bean Leaf Roll Virus
Bean Yellow Mosaic Virus (BYMV), pea strain
Pea Enation Mosaic Virus
Pea Streak Virus
Pea Stunt Virus
Pea Seed Borne Virus
Pea Early Browning Virus
Pea Top Yellow Virus
Common Pea Mosaic Virus
Cucumber Mosaic Virus (CMV)

Insects

Cutworms	*Agrotis segetum, A. exclamationis, Noctua pronuba*
Pea aphids	*Acyrthosiphon pisum, Macrosiphum pisi*
Pea borer	*Heliothes* spp.
Pea leafminers	*Liriomyza huidobrensis,* other *Liriomyza* spp., *Phytomyza* spp., and *Agromyza* spp.
Pea midge	*Contarinia pisi*
Pea moth	*Cydia nigricana*
Pea weevil	*Sitona lineatus, Bruchus pisorum*
Thrips	*Thrips angusticeps, Kakothrips pisivorus, K. robustus*

Nematodes

Lesion	*Pratylendrus* spp.
Pea cyst	*Heterodera* spp.
Root knot	*Meloidogyne* spp.

Marsh spot (hollow heart) is a physiological disorder that occurs in crops deficient in manganese; high soil pH can limit manganese availability.

TABLE 22.2.   WORLD GREEN PEA PRODUCTION, 1994

	Area (ha × 10³)	Yield (t/ha)	Production (t × 10³)
World	742	5.9	4346
Africa	50	5.0	248
North and Central America	145	8.5	1233
South America	60	2.5	148
Asia	199	4.7	938
Europe	270	6.1	1650
Oceania	18	7.3	129
Leading countries			
United States	117	9.5	1111
China	63[a]	7.8	490[a]
France	29	14.3	413
India	99[a]	2.7	268[a]
United Kingdom	49[a]	5.4	262[a]
Belgium–Luxembourg	12[a]	17.9	210[a]
Russian Federation	70[a]	2.0	140[a]
Hungary	15[a]	7.3	110[a]
Egypt	9[a]	11.7	105[a]
Italy	16[a]	6.1	100[a]

[a]Estimated
Source: 1994 FAO Production Yearbook, Vol. 48, FAO, Rome 1995.

Weeds

The pea plant is a poor competitor with many weed species unless given a headstart; therefore early weed elimination or control is needed to obtain good yields. In order to have narrow row spacings without a dependency on mechanical or hand cultivation, several selective herbicides are widely used. It is important in selecting a herbicide to consider crop use following peas. Crop rotations are also an important component in weed management.

Production

Europe with almost 38% and North America with 28% of the world's total production are clearly the leading green pea producers. The United States individually produces about 25%; the majority of that production is for processing. See Table 22.2.

Dry field pea production is more than three times that of fresh peas. Countries leading in dry pea production in 1994 were France, the Russian Federation, Ukraine, China, and Canada with approximately 3.4 million, 2.5 million, 2.5 million, 1.5 million and 1.4 million tons, respectively. India's production of 579,000 tons is also significant. The production from all the former USSR republics was almost 36% of the 1994 world production of 14.5 million tons of dried peas, and France individually supplied about 23% of the total. These production levels do not necessarily imply domestic consumption because some is for export and for animal feed.

Phaseolus SPECIES

The genus *Phaseolus* includes several very important vegetable species grown for the fleshy pods, immature, and mature seed. The important cultivated New World *Phaseolus* are

Snap and common dry bean	*P. vulgaris*
Lima bean	*P. lunatus*
Runner bean	*P. coccineus*
Tepary bean	*P. acutifolius* var. *latifolius*

In previous literature, several legume species were identified as members of the genus *Phaseolus,* whereas now they are classified as members of *Vigna.* Among those reclassified are the moth, adzuki, mung, rice, and urd beans (Table 22.1).

SNAP BEAN, *Phaseolus vulgaris* L.
Other names: French bean, garden bean, string bean, stick bean, haricot bean, bush bean, pole bean, and green bean

Snap bean is the best known vegetable member of this genus. Although snap beans do not provide the quantity of protein and calories supplied by dry beans, they, nevertheless, are an important protein, vitamin, and mineral source. In addition to the consumption of cooked snap bean pods, the potherb use of shoots and leaves is common in many areas of Africa and Latin America. Firm, large, but immature seed (shelled beans) and, to a lesser extent, dried seed of some snap bean cultivars are also eaten.

Origin
Vavilov proposed southern Mexico and warm regions of Guatemala as a primary center of origin, with Peru, Ecuador, and Bolivia as secondary centers. Other evidence suggests that the common bean and its many biotypes evolved from wild *P. aborigineus* in the Andean regions. Another theory suggested the common bean was domesticated in South America and was transported to Central America, where maximum diversity is found. However, Singh et al. (1991) indicated two distinct

gene pools of common bean, one of Andean origin and the other in Mesoamerica (Central America and Mexico). Most snap beans appear to have had an Andean origin with introgressions from the Mesoamerica group.

Through radiocarbon dating, the common bean was identified as existing more than 7000 years ago. European explorers were responsible for the early export of the New World *Phaseolus* species, especially *P. vulgaris,* to other parts of the world, where they were well adapted and rapidly accepted. In the wild state, the common bean is found from low to high elevations as well as dry to humid environments. However, the fleshy pod (snap) types appear to have less climatic adaptation than dry bean types.

Domestication resulted in diminished branching with an increase in flower number, pod, and seed size. Although seed size increased, seed number per pod decreased. Overall, pod dehiscence and pod fiber development were reduced, and pod fleshiness increased with snap bean types. Water permeability of the dry seed increased, and seed hardness decreased. A shift from short-day photoperiod response to day length neutrality occurred with many biotypes. Although wild-type *Phaseolus* have perennial and annual forms, most current vegetable *Phaseolus* are annuals.

According to use, the following grouping of *Phaseolus vulgaris* beans are recognized:

French bean: Green, yellow or purple fleshy pods containing underdeveloped seed are eaten. Pods do not have string or parchment layer.

Haricot filet bean: Pods contain string, but the fleshy immature pods are eaten.

Haricot: Fresh seed are eaten, pods contain string and fiber and generally are not consumed.

Dry (field) bean: Dry trashed seed is eaten, pods have string, fiber, strong parchment layer and are not eaten.

Botany

Generally, the root system of many snap beans is not large or extensive, and lateral branching is shallow. The prominent taproot is usually short, but in deep friable soils, can extend to about 1 m. In the presence of *Rhizobium* bacteria, nodules develop on lateral roots. A root system that strongly anchors the plant is an important feature for mechanical harvesting.

Many *P. vulgaris* cultivars are warm season twining vinelike plants.

Besides indeterminate climbing and nonclimbing forms, there are dwarf (bush) determinate and intermediate forms. The present determinate bush cultivars differ from earlier indeterminate climbing types in having less apical dominance and little or no short-day photoperiod response. Indeterminate vine and upright cultivars branch more and, by providing more flowering nodes, have a greater yield potential. Stem length in climbing types can be as long as 3 m with more than 25 flowering nodes. These forms lodge severely, and thus are generally supported on poles or a trellis. Determinate bush forms are short, some not more than 60 cm in height, nodes are few, and inflorescence are terminal.

Foliage is pinnately trifoliate. Present-day cultivars have small leaves which improve light penetration into the canopy, especially for high-density plantings. Although this characteristic tends to increase total yield, small leaf size is linked to small pod size.

Flowers are large and showy and may be white, pink, or purple. They are perfect and, like the pea, have 10 stamens, 9 of which are united into a tube enclosing the long ovary; one upper stamen is free of the others. Flowers are self-pollinating and generally little out-crossing occurs.

Pods are almost always considerably longer than wide; lengths range from 8 to 20 cm or more, with widths of less than 1 to several cm (Fig. 22.2). Depending on cultivar, pod ends may have a pointed or blunt tip; cross-sectional shapes vary from round to elongated oval, and some are heart shaped. Pods of most present-day cultivars are relatively straight, although some are normally curved. Most cultivars have light to dark bluish green pods; others are yellow (wax), purple, or multicolored.

The amount of pod fiber and rate of development also varies. Through selective breeding, fiber has been greatly reduced. The stringless character was introduced more than 100 years ago. Presently, only heirloom and other old bean cultivars possess the strong stringlike suture fiber. Calvin Keeney, a seed grower from LeRoy, New York, is credited with introducing the first stringless cultivar in about 1800. Stringless is a recessive character and is incorporated in most presently grown cultivars. Stringless-type cultivars also contain less wall fiber. Nevertheless, within the United States, "string bean" tends to remain as a generic term to identify snap beans. The word "string" was used because of the strong stringlike fibers at dorsal and ventral sutures of the pod, the dorsal being the strongest. As seed fully matured, the pod would split open. The term "snap" probably resulted from the sound resulting when fresh pods are broken. Most snap bean pods are glabrous; a few

FIG. 22.2. Snap beans, *Phaseolus vulgaris, cv.* Derby. Source: Courtesy of All America Selections, Downers Grove, Illinois.

exhibit some pubescence. Pods do not have the persistent calyx that peas do.

Seed number is another cultivar characteristic; most snap bean cultivars contain three to five seed; dry or common bean types tend to have several more. Mature seed sizes exhibit enormous variation in size and weight, ranging between 5 and 20 mm in length and individual seed weight of some cultivars varying from 0.15 to more than 0.80 g. Seed shapes are round, orbicular, ovoid, oblong, and kidney. Seed coat colors are cultivar specific and can occur in numerous colors and combinations, and are of some importance. It is interesting that different countries of Latin America prefer a certain seed coat color: black seed coat—Brazil, El Salvador, Mexico and Venezuela; red—Colombia and Honduras; yellow—Peru; white—Chile. Cultivars produced for snap bean processing usually have white or light-colored seeds. Other important features of mature seed are thickness and adherence of the testa and cotyledon resistance to cracking; resistance to cracking is genetically linked with seed coat color.

Climate, Moisture, and Soil

A mean air temperature of 20–25°C is optimal for growth and high yield. Pole snap beans tend to grow better at slightly cooler temperatures and are more sensitive to high temperatures at flowering than bush types. Heat stress negatively affects pod set, and some cultivars are more tolerant than others. A favorable soil temperature range is 18–30°C.

Snap beans are sensitive to drought and flooding. Ideally, moisture should be evenly distributed throughout growth; 250–450 mm is usually sufficient. For high yields, good moisture management is essential. Soil moisture should be near field capacity, particularly during flowering; flooding causes anoxia, to which beans are sensitive, and leads to increases in the incidence of root rot. Moisture stress affects pod yield, seed number, and size, in addition to color, fiber, and firmness. Besides germination, flowering and pod development are most sensitive to water deficit. Dry winds can cause blossom drop. Well-drained, friable, medium textured, clay loams are well suited for snap bean production. Growth is strongly reduced by soil compaction. Slightly acid soils are preferable; the optimum pH range is between 6.0 and 6.5.

Most present-day snap bean cultivars are photoperiod insensitive. Nevertheless, some cultivars are still used that only develop flower buds during short days. An interesting behavior is the characteristic of bean leaves to face and follow the sun, a feature that enhances photosynthetic efficiency. Conversely, during periods of excessive heat and low soil moisture, leaves will turn parallel to the sun's rays, thus decreasing leaf temperature. Yields are reduced when plants are shaded.

Nutrition

Most cultivars have relatively small root systems, and because these often have limited absorptive capabilities, it is common to supply supplemental fertilizer. Determinate cultivars, in particular, do not have early or adequate access to *Rhizobium* fixed nitrogen; hence, fertilization is necessary for vigorous crop development. The nitrate form of nitrogen is preferred to the ammonium form. Phosphorus is especially important during early plant growth. Application rates should consider plant populations because high-density plantings generally require high levels of supplemental fertilizer. Snap beans are sensitive to salinity, and during planting, seed should not come into direct contact with fertilizer. Beans are highly sensitive to excessive soil boron.

Planting and Spacing

The large cotyledons of snap bean seed are susceptible to cracking from physical damage, which can occur during seed harvest, cleaning, and planting. Bean seed germination is epigeal, and loss or partial loss of a cotyledon can affect seedling growth and subsequent yield. An adverse result of reducing seed hardness was an increased susceptibility to damage. Improvements in seed harvesting equipment, such as using rubber belts for seed separation instead of metal cylinders, greatly reduced damage. Seed germination is optimum between 25°C and 30°C; temperatures less than 10°C and above 35°C do not permit germination (Table 22.3). Under good conditions, emergence occurs within 7–12 days.

When planted into cold soils, germination is slow and seed decay often occurs. Protective fungicide seed treatment is beneficial for minimizing decay. Seed are often given cold-soil germination tests to determine their vigor potential under adverse conditions. Poor emergence was frequently noted with white seed coat cultivars in comparison to otherwise comparable dark-colored seed. Plant breeders have corrected this characteristic in new white seed coat cultivars.

Sowing depths range from 3 to 8 cm. Sometimes, seed are planted in a shallow furrow that is later filled in during cultivation. For high yields, many bush beans are grown at densities of about 40 plants/m². High densities are achieved with closely spaced rows or by broadcast plantings. Such cultivation is not suitable for hand harvesting but is well suited for mechanical harvesting.

Wide spacings are used to permit hand harvesting in producing pole snap beans, as these indeterminate plants are harvested many times. Trellised pole snap bean are usually spaced about 10 cm apart in the row with between-row spacings ranging from 120 to 150 cm. Pole-supported hill plantings are spaced equidistant, from 90 to 120 cm. Hill plantings are usually sown with five to six seed per hill, later thinned to about three plants. In some parts of the tropics and subtrop-

TABLE 22.3. DAYS FOR SNAP AND LIMA BEAN SEED GERMINATION AT DIFFERENT TEMPERATURES

Temperature (°C)	Days	
	Snap bean	Lima bean
10	a	a
15	16.1	30.5
20	11.4	17.4
25	8.1	6.5
30	6.4	6.7
35	6.2	a
40	a	a

[a]No germination.
Source: National Garden Bureau, Inc. Downers Grove, IL.

ics, pole beans are grown with or following maize or okra crops, often using the stems of these crops for support.

Plant densities for hand harvesting of bush beans are between 45 thousand and 60 thousand plants per hectare, whereas from 250 thousand to 450 thousand plants per hectare are grown in mechanically harvested, high-density plantings. The use of front-end multirow harvesters has facilitated narrow-row, high-density planting practices. Although very close plant spacings tend to reduce pod color, this is often an acceptable sacrifice for high yield. However, high-density plantings increase the potential for disease. The wide use of precision-placement seed planters provides accurate spacings and has decreased the number of seed planted; high seed costs also tend to reduce seeding rates.

Growth and Development

Under favorable growing conditions, bush-type snap beans are capable of producing a crop in 60–70 days; pole beans generally require about 10–20 days longer. Depending on growing temperatures, pods are usually suitable for harvest about 7–15 days after anthesis, although the best pod quality is obtained when harvested before full elongation occurs.

Initiation and flower development is greatly delayed under suboptimum temperatures. During temperatures less than 10°C, fertilization may not occur, or if partial seed development does occur, small and misshapen pods result. Blossom drop and ovule abortion can be caused by temperatures greater than 35°C.

Lateral branching provides more flowering nodes and extends the flowering period and yield potential. Branching habit is also beneficial because it permits differential flower and pod development. This characteristic is useful, especially following flower or pod abortion, because resumption of flowering is possible. Because pole bean cultivars generally have more lateral branching, they usually have a broader adaptation to stress. Determinate cultivars are more susceptible to stress that interfere with pod set, which can result in a low single-harvest yield. However, when growing conditions are favorable, determinate plants have the advantage of highly uniform pod development. To extend edible quality, plant breeders have produced snap bean cultivars with slow seed maturation.

Production methods for the common dry bean cultivars are similar to those for snap beans. Dry beans are commonly grown on a larger scale than snap beans, being less perishable and usually requiring less management and labor.

Harvesting

Harvest decisions for snap beans are based on the stage of pod development. For high yields, snap bean pods should achieve maximum length before significant seed enlargement and while still succulent. The ideal situation is to have all pods at the same stage of development. Degree-day measurements are frequently used for predicting and scheduling harvests of snap beans in developed countries.

Hand harvest of bush and pole snap beans is common in most of the world because machinery use is not feasible in many countries. Pole beans having indeterminate flowering can be harvested over an longer period relative to bush types; consequently, yields usually are higher. Besides a greater yield potential, pole bean production advantages include better adaptation to high-rainfall conditions, with reduced humidity within the foliage canopy and a lower incidence of disease. Additionally, because pods are less likely to contact the soil, they are clean and grow straight. However, bush bean production continues to increase relative to pole beans because of lower production costs.

For successful machine harvesting, both equipment and plants must be compatible. Present cultivars facilitate mechanization because they have concentrated flowering and pod set, an upright growth habit with pod set midway or high on the plant, reduced foliage, strong root attachment to the soil, and resistance to some diseases. All these factors improve the effectiveness of harvest equipment, which relies on the pod stripping action of many metal tines combing through the foliage. The detached pods and leaves are subsequently separated. Machine harvesting was initially used only for the processing crop because pod damage usually was not a serious product defect if the pods are processed within a short time. However, damage becomes readily visible and is unacceptable for harvest of fresh market snap beans. Many current mechanical harvesters have been modified to greatly reduce pod damage and are capable of harvesting pods for the fresh market.

Shelled bean cultivars are harvested when seed have achieved full size and are relatively firm. Seed moisture is very high compared to dry bean seed. Hand labor and/or machinery are used to harvest the pods. The seed are separated from the pods, the latter discarded because they are fibrous and not succulent. Shelled snap beans have the characteristic of remaining firm following cooking, much like that of most cooked common dry beans and unlike the sluffing and softening of cooked snap beans.

Postharvest and Storage

Snap bean pods have a high respiration rate and should be rapidly cooled to about 5°C and maintained at 95% RH. Hydrocooling is a preferred method to achieve rapid cooling and maintaining pod turgor. Temperatures less than 3°C for more than several days are avoided because they are conducive to chilling injury. Pod shelf life of acceptable quality for 2–3 weeks is achievable with holding at 5–10°C and 95% RH.

Disease and Other Pests

Bacteria

Bacteria wilt	*Corynebacterium flaccumfaciens* pv. *flaccumfaciens*
Bacterial brown spot	*Pseudomonas syringae* pv. *syringae*
Common blight	*Xanthomonas phaseoli, X. campestris* pv. *phaseoli*
Fuscous blight	*Xanthomonas phaseoli* pv. *fuscans*
Halo blight	*Pseudomonas phaseolicola, P. syringae* pv. *phaseolicola*

Fungi

Alternaria leaf and pod spot	*Alternaria* spp.
Angular leaf spot	*Isriopsis griseola*
Anthracnose	*Colletotrichum lindemuthianum,* other *Colletotrichum* spp.
Ascochyta leaf spot	*Ascochyta* spp.
Cercospora leaf spot	*Cercospora cruenta*
Downy mildew	*Phytophthora phaseoli*
Fava downy mildew	*Peronspora viciae*
Fusarium root rot	*Fusarium solani* f. sp. *phaseoli*
Fusarium yellows	*Fusarium oxysporum* f. sp. *phaseoli*
Gray mold	*Botrytis cinerea*
Phyllosticta leaf spot	*Phyllosticta phaseolina*
Pod blight	*Diaporthe phaseolorum*
Powdery mildew	*Erysiphe polygoni*
Pythium root rots and damping-off	*Pythium* spp., also *Aphanomyces,* and *Thielaviopsis* spp.

Rhizoctonia	*Rhizoctonia solani*
Rust	*Uromyces phaseoli,* other *Uromyces* spp.
Scab, lima bean	*Elsinoe phaseoli*
Stem anthracnose	*Colletotrichum truncatum*
Verticillium wilt	*Verticillium albo-atrum*
White mold	*Sclerotinia sclerotiorum*

Viruses
  Bean Common Mosaic Virus (BCMV)
  Bean Golden Mosaic Virus (BGMV)
  Bean Pod Mottle Virus (BPMV)
  Bean Southern Mosaic Virus (BSMV)
  Bean Yellow Mosaic Virus (BYMV), bean strain
  Cucumber Mosaic Virus (CMV), bean strain
  Sugar Beet Curly Top Virus (SBCTV)
  Peanut Stunt Virus

Insects and other pests

Bean aphid	*Aphis craccivora*
Bean leaf beetle	*Cerotoma trifurcata*
Bean weevils	*Acanthoscelides obtectus, Zabrotes subfasciatus*
Beet leafhopper	*Circulifor tenellus*
Black bean aphid	*Aphis fabae*
Diabrotica	*Diabrotica decimpunctata*
Leafhoppers	*Empoasca fabae, E. kraemeri*
Lygus bug	*Lygus hespersus, Chauliops fallax*
Mexican bean beetle	*Epilachna varivestis*
Pod borer	*Apion pisi, A. godmani*
Red spider mite	*Tetranychus urticae*
Root weevil	*Sitona lineatus*
Seed corn maggot	*Hylemya cilicrura*
Stem fly	*Ophiomyia phaseoli*
Thrips	*Thrips tabaci, Heliothrips* spp.
White fly	*Bemisia tabaci*
Lepidoptera pest members of the following genera:	*Ascotis, Spilosoma, Amsacta* and *Euproctis*

The most common nematode pests of beans are species of root knot, *Meloidogyne*. Lesion nematodes, *Pratylenchus* spp., are also frequent pests. An occasional physiological disorder is hypocotyl necrosis which becomes evident during germination and is the consequence of low calcium in the seed.

For weed management, considerable use is made of selective herbicides, especially for high-density plantings where narrow row widths make hand cultivation impractical. Early control is especially needed until plant growth can effectively compete or suppress weeds. Large weeds can interfere with harvest machinery.

Uses and Composition

Snap bean pods, shoots, leaves, and immature (shelled) seed are consumed after cooking. Processing by canning or freezing of pods and seed of shelled-type cultivars represent a significant volume of the total production.

Interestingly, consumers have strong preferences for snap bean pod shape and color and seed color. Seed color is of little significance for fresh use of pods, but white-seeded cultivars are preferred for canning, because dark-colored seed coats tend to discolor the canning liquid. Although white-seeded cultivars are preferred, dark-seeded cultivars are also used for freezing. Processed whole pods are a high-value product. Snap bean protein quality is good, although the level of the sulfur-containing amino acids is low.

Production

Asia and Europe with more than 50% and 33% of world production, respectively, are the dominant snap bean producers; see Table 22.4. China and Turkey are the leading individual countries, each with more than 17% and 13% of world production, respectively. Production from many small holdings and home garden plantings are not reported; thus, accurate collection of production statistics is less than ideal. Also confounding data collection is the widely used practice of intercropping snap beans with other commodities.

LIMA BEAN, *Phaseolus lunatus* L.

Origin and Taxonomy

Remnants of large-seeded lima beans found in Peru were identified as more than 7000 years old. Small-seeded lima remnants found in Central America were about 2500 years old. Wild types continue to be found in Mexico, Central America, and throughout the Andean regions. Cultivation is widespread, with crops being grown by subsistence farm-

TABLE 22.4.   WORLD PRODUCTION OF SNAP BEANS, 1994

	Area (ha × 10³)	Yield (t/ha)	Production (t × 10³)
World	462	6.9	3176
Africa	29	6.5	189
North and Central America	33	5.6	185
South America	29	3.2	94
Asia	233	6.8	1595
Europe	130	8.2	1070
Oceania	8	5.3	44
Leading countries			
China	50[a]	11.2	560[a]
Turkey	53[a]	8.4	440
Spain	25	9.8	240
Italy	26	8.1	212
Indonesia	38[a]	4.1	157[a]
France	10[a]	10.7	109
Egypt	13[a]	8.1	105[a]
United States	17[a]	5.1	87[a]
Thailand	21[a]	4.0	84[a]
Netherlands	5[a]	18.1	83[a]

[a] Estimated.
Source: 1994 FAO Production Yearbook, Vol. 48, FAO, Rome, 1995.

ers in northern Brazil as well as having become an important pulse staple crop in some African and southeast Asian regions.

Common names identifying small-seeded lima beans are sieva and sometimes navy bean. Butterbean and Madagascar bean refer to large-seeded types. "Fordhook" is a well-recognized name for the large-seed, "potato lima" group.

Disagreement about the taxonomic classification of lima beans exists relative to the species designation of *P. lunatus* and *P. limensis.* An earlier classification identified thick-podded, large-seeded limas as *P. limensis,* and thin-podded, small-seeded forms as *P. lunatus;* different forms were assigned botanical variety designations which often are found to be spelled differently. Separate species status for the different types of limas is questionable and probably not justified because all types are interfertile. Currently, all limas, wild and cultivated, are identified as *P. lunatus.*

Botany

Limas are considered perennials or long-lived annuals, but are cultivated as annuals. Both climbing and bush cultivars are grown. Climbing cultivars can reach lengths of 3–4 m, whereas bush forms are shorter (50–90 cm). Plants have a highly branched root system of moderate depth, often more than 1 m. Roots are able to develop *Rhizobium* containing nodules.

Flowering is indeterminate, and flowers are self-fertile, although some out-crossing occurs. The seed is somewhat moon shaped, hence its species name. The length of the slightly curved oblong pods range from 5 to 15 cm and in width from 2 to 3 cm (Fig. 22.3). Most cultivars usually contain two to four seed, although with others, pods may contain as many as six seed. Pods of some cultivars are thick; others are relatively thin. The large, flat, oblong seed of some plant types are as long as 3 cm. Other seed types are also flat but are more rounded and about 1 cm in length; seed of each type are smooth.

Commonly grown cultivars have a light green or white seed coat; others may be red, purple, brown, or black. The two large-seed cotyledons comprise most of the seed volume. Seed of wild types have a high content of cyanogenic glucosides and must be leached before or during cooking. The glucoside content of present-day cultivars, especially those with light-colored seeds, contain little or none. Nevertheless, consumption of raw lima beans is not recommended.

Culture

The crop is more sensitive to the environment than snap beans and require slightly warmer weather for growth. Favorable mean tempera-

FIG. 22.3. Lima beans, *Phaseolus lunatus*. Source: Courtesy National Garden Bureau, Downers Grove, Illinois.

tures range between 15°C and 25°C. Large-seeded limas are generally adapted to lower temperatures and higher humidities than small-seeded ones, especially with regard to pollination. For large-seeded limas, temperatures above 30°C and relative humidity less than 60% during flowering may result in blossom drop. This sensitivity tends to restrict production to rather specific environments. The growth regulator NAA has been used to improve pod set.

Seed germinate well between 15°C and 30°C; above or below this range, germination is poor. Planting depths differ with seed size and may range from 3 to 6 cm, with large-seeded types planted deeper. Emergence is most rapid at about 25°C. Seed are very susceptible to mechanical injury.

Light textured, warm, and well-drained soils result in higher yields than the use of fine textured soils. A slightly acid soil between pH 6 and 7 is also preferable. Other than temperature restrictions, lima bean cultural procedures are similar to those for snap beans, but the longer growth period requires more moisture and nutrients. Moisture stress is most critical during flowering. Supplemental nutrition is commonly provided with fertilizers containing nitrogen, phosphorus, and potassium. The response to nitrogen fertilization is greater than that shown by snap beans. Many of the same diseases and pests affecting snap and other bean species also affect limas. For bush-type plantings, a usual spacings is 60–90 cm between rows with in-row spacings of 10–15 cm. Pole-type cultivars are spaced wider to accommodate hand harvesting.

Harvest and Postharvest

Depending on cultivar and growing temperatures, the period from seeding to harvest can vary from 70 to 110 days, longer if temperatures are low. Compared to snap beans generally grown for the fleshy pods, limas require more time before harvesting in order to produce the enlarged but immature seed. As seed mature, the seed coat color changes from green to cream or white. During seed maturation, the pod bulges as a result of seed enlargement. At the proper harvest stage for fresh use, pods are green and seed moisture is at 60–70%. Because of different flowering times, pods do not mature uniformly, which is less critical for fresh than processing use.

Because of the high labor requirement for hand harvesting, a substantial portion of the fresh market crop is machine harvested. The difficult hand removal of immature seed from pods is a further encouragement for mechanization. Efficient harvest mechanization requires dwarf-type plants with a determinate flowering habit. Yields of shelled fresh green limas range between 2 and 3 t/ha.

Machine-harvested, shelled lima beans have a short postharvest life, and are therefore processed as soon as possible, usually by freezing or canning. Lima beans of different maturity are separated based on specific gravity with the use of brine solutions. Freshly harvested lima beans in pods can be held in good condition for several weeks at 5–7°C and 90% RH. At lower temperatures, chilling injury may occur.

A considerable amount of limas are grown for the dry, fully mature seed. The dry seed crop is also machine harvested and handled much like common field beans. Lima beans, being at a more advance maturity than snap beans, have a higher seed dry matter, protein, and carbohydrate content. However, if corrected for moisture content, they are fairly comparable.

SCARLET RUNNER BEAN, *Phaseolus coccineus* L. *(P. multiflorus)*

Scarlet runner beans, native to the tropical highlands of Mexico and Central America, are perennials, but are usually grown as annuals. They have been identified as having been cultivated for more than 2000 years. As a long-day plant, the runner bean is well suited for growth during summers in many temperate regions or at high elevations in the tropics. Although having some cool weather adaptation, runner beans, like other *Phaseolus,* are frost sensitive.

Plants exhibit twiny vine growth; vines of some cultivars grow as long as 4 m. The root system has a thick taproot and is fibrous. Flowers are large, abundant, and attractive with a scarlet, white, or variegated-color corolla. In the United States, the plant is often regarded as an ornamental, but it is appreciated much more in Europe for the excellent taste of the immature pods that resemble snap beans. The pods are larger than those of snap beans, and are somewhat flattened with prominent wrinkled sutures. Immature as well as dry seed are a popular food in Central America. The oblong seed are about 2.5 cm long and multicolored, usually dark purple or red. Protein content of mature seed is about 17%, with carbohydrate content about 65%. The thickened, starchy, tuberous roots are also consumed, although some reports suggest they may be poisonous. In some regions, flowers are also consumed.

Propagation is usually with seed, although cuttings may be used to start plants. Bush types exist, dwarfness being an inherited trait. With some dwarf plants, pods can range between 10 and 30 cm in length. These long pods may contact the soil and become crooked, dirty, and/or diseased. Plant spacings are about 15 to 30 × 90 cm for bush-type cultivars, and about 15 to 30 × 150 cm for pole- or trellis-supported production. The cost of support materials influences plant density choices. To reduce vine length and increase determinatelike bearing, growers pinch out the growing point to encourage branching and thus limit plant height. However, the labor required and limited benefits tend to discourage this practice. About 4 months are required to complete the annual growth cycle. In Central America, the crop may be interplanted, usually with maize.

TEPARY BEAN, *Phaseolus acutifolius* A. Gray var. *latifolius* G. Freem.

The tepary bean originated in northern Mexico and southwestern United States, where it probably was first cultivated about 5000 years ago. The attributes of this plant are its tolerance to high temperatures and low relative humidities. Pods can set at high temperatures (35°C), whereas those of other *Phaseolus* species cannot. The cultivated plant is a short, semierect annual with short-day responsiveness; wild forms are climbers. Flowers are white or pale violet. Pods are small, 6–7 cm long, and hairy. Seed are flat, with much variation in seed colors. Seed mature in about 60–90 days; dried seed contain about 60% carbohydrate and 22% protein.

Tepary bean cultivation has been limited to its specific climatic niche and is mainly grown by indigenous peoples in the southwestern United States. The crop was introduced to Africa, where a limited amount of production occurs, and it is almost exclusively grown in areas of high temperature and limited rainfall. Tepary bean is resistant to common bacterial blight, and plant breeders have transferred this resistance into common bean.

PEANUT/GROUNDNUT, *Arachis hypogaea* L.

Other names: Goober, pindar, earth nut

The peanut is known to have been cultivated for more than 5000 years, with early domestication occurring east of the Andes mountains in southwestern Brazil, Bolivia, Paraguay, or northern Argentina, the possible center of origin. Presently, the peanut is grown throughout the tropics, much of the subtropics, and in the absence of frost, even in temperate zones as far as 40° latitude from the equator.

Most cultivated plants are tetraploids and annuals; there are also perennial forms. Of two major plant forms, the runner type is prostrate and spreading, and the bunch type is more upright and less spreading. Both are about 60 cm in height. Peanut performs best on friable and well-drained soils with a preference for sandy soils. Light textured soils facilitate penetration and pod development, which normally occurs below the soil surface. Soil calcium availability is important for good seed development. In solid stands, bunch-type plant spacings commonly are about 30×60 cm or 15×75 cm; more space is allowed for runner types. Peanuts are frequently intercropped with other species.

For some cultivars, short days are required for flowering. Yellow flowers are borne in leaf axils, and after self-pollination, the stalk (carpophone) bearing the fertilized ovary becomes geotropic and penetrates into the soil. The pod (fruit) develops at the end of this peg-like structure.

Bunch-type cultivars, also known as Spanish or Valencia peanuts, are early and are harvested in 3–5 months. The runner or Virginia types require 6 months or more before seed mature. Mature seed have little or no dormancy and delayed harvest may result in seed germinating within the pod. Planted seed exhibit neither hypogeal nor epigeal germination; rather, the cotyledons are pushed to the surface by the hypocotyl and remain at the surface. Further growth is due to aboveground epicotyl elongation. Propagation commonly is with seed; cuttings can be, but are seldom used.

Being indeterminate bearers, especially the runner types, plants are harvested by uprooting the plants when sufficient pods have set and matured. Plants are usually windrowed and allowed to dry before pods are removed. Hot and dry weather at the time of pod maturity is desirable for development and seed quality.

Tender shoots and leaves are used as greens and immature pods as a cooked vegetable. Worldwide use is made of roasted seed and extracted seed oil. Dry seed carbohydrate composition ranges between 10% and 25%, the protein content is about 30%, and for some cultivars, seed oil content is as high as 40–50%.

JACKBEAN, *Canavalia ensiformis* (L.) DC.

Other name: horsebean

Jackbean is native to Central America and the West Indies, where a large diversity of plant types are found. Vegetable use is made of tender young pods and immature seed; plants are also used for fodder or green manure.

The hardy and bushy erect annual plant is more than 1 m tall, deep rooted, and drought resistant. Self-pollinated flowers are rose to violet in color. The pendent pods are large, 20–30 cm in length and 2–2.5 cm wide, containing 8–20 slightly flattened white seed, which require detoxification by boiling and rinsing before consumption. Dry seed carbohydrate content is about 55%, and protein is about 24%. Immature fresh pods have about 13% carbohydrate and about 7% protein.

SWORDBEAN, *Canavalia gladiata* (Jacq.) DC.

Unlike the jackbean, *C. gladiata* is of Old World origin and may have been derived from *C. virosa,* which is found wild in tropical Asia and Africa. Young swordbean pods and seeds are widely used vegetables in the tropics. especially in Asia. Although widely cultivated, production is largely limited to home gardens and for local markets.

Although grown as an annual, *C. gladiata* differs from *C. ensiformis* in being a perennial and a climber. Flowers, mostly self-pollinated, are pinkish white to white. The ratio of pod length to width is used to identify swordbean from jackbean; the ratio of swordbean pod length to width is less than that of jackbean. The name swordbean is probably from the appearance of the pendant pods, 15–40 cm long and 4–5 cm wide, commonly containing 5–10 dark red seed with thick tough seed coats. Although rare, white seed are preferred because of their flavor. Seed are relatively large but variable, and individual seed weight can range from 1 to 4 g. Another identifying feature is the seed hilum scar, which for swordbean is more than half the length of the seed, whereas the scar is less than half of the seed length for jackbean.

For optimum growth and yield, swordbean plants require a tropical climate with mean temperatures between 20°C and 30°C. Because roots penetrate deep into the soil, the plants are tolerant to drought. Plants can be grown in nutrient-poor soils, can withstand some salinity, and can even tolerate some shade.

Seed are used for propagation with spacing commonly 50–60 cm between plants in rows that are 75–90 cm apart. Germination and early growth is usually rapid; ideally, plants should be provided some support. After 3–4 months, green pods are harvested when about 10–15 cm in length and before observed seed swelling. Yields of 4 t/ha can be achieved. Five to 10 months of growth are required for harvest of fully matured seed; yields of mature seed range between 700 and 900 kg/ha.

Young green pods are used as boiled vegetables; while still succulent, the seed are also consumed. Sometimes, flowers and young leaves are used as potherbs. Like jackbean, mature seed should be treated before eating to remove toxicants that can interfere with the human body's nutrient absorption. This inconvenience tends to limit the use and popularity of swordbeans.

PIGEON PEA, *Cajanus cajan* (L.) Huth., *(C. indicus)*

Other names: Red gram, Congo pea

Based on findings of wild types, the origin of pigeon pea is attributed to Africa. However, this is questionable because the variability found in India clearly makes that a center of diversity and perhaps origin. Pigeon pea is a crop of great importance in India, where its production occupies a huge area of crop land. Its importance is also significant in other Asian countries. Pigeon pea ranks sixth in world production of dry legume seed. The wide climatic and soil adaptation of the pigeon pea is a reason for its extensive use. The deep root system enhances drought resistance. Plants are intolerant of waterlogging or shading.

Plants grow to heights from 1 to 4 m, are somewhat woody, and, although short-lived perennials, are usually cultivated as annuals. The crop can be continued for 3–4 years when used for fodder. Lanceolate-shaped leaves are from 5–10 cm long. Pigeon pea is usually seed propagated, but stem cuttings can be used.

Flowers of most cultivars are yellow orange; others are red or purple. Most cultivars are short day, but some dwarf forms appear to have little or no photoperiod sensitivity. Flowering is indeterminate, and flowers are self-fertile, but insect visits can result in considerable cross-pollination.

Pods form within 3–4 months from sowing and are typically flat and broad, being 4–10 cm long and 1–3 cm wide and contain round to oval-shaped seed. Seed colors vary among cultivars. Six months or more are required to mature seed with early cultivars, and 9–12 months for late cultivars.

Two botanical varieties recognized are *C. cajan* var. *indicus,* known as "tur," which is a short, early-maturing, plant type with green pods usually containing three seed. The other, *C. cajan* var. *bicolor,* known as "arhar," is a large, bushy, late-maturing plant type with dark-colored pods containing 4–5 seed.

For vegetable purposes, immature seed are used fresh; however, considerable quantities are processed by canning. Fresh green pods are also consumed in large quantities. Overall, the major importance of pigeon pea is as a pulse crop mainly for the preparation of dahl. Dry seeds contain about 57% carbohydrate and 19% protein, whereas for succulent seed, the values are 20% and 7%, respectively.

CHICK PEA, *Cicer arietinum* L.

Other names: Garbanzo, Bengal gram, cece, homos, and chana

Progenitors of chick pea have not been found. The possible center of origin may be western Asia, specifically southeast Turkey. Southwest Asia, the Mediterranean, and Ethiopia are identified as centers of diversity. Chick peas are the second most important pulse crop in the world. Of the 7.9 million tons of world production in 1994, 70% was in India. Most of the production occurred in countries with developing economies.

Chick peas are well adapted to cool (8–22°C), dry climates and less suited to regions of frequent precipitation and high temperatures. Plants are bushy, erect annuals about 50–60 cm tall and covered with glandular hairs that exude an acrid fluid comprised of malic and small amount of oxalic acid. Worker outer garments, especially shoes, can disintegrate from long contact with the acids. Plantings are closely spaced and often sown by broadcasting. Row plantings are usually 75 cm between rows with in-row spacing of 30 cm. Long days induce early flowering, but the response differs among cultivars.

Self-pollinated flowers are usually white and, depending on the cultivar, can range from pink to blue. Pods are short, oblong, and somewhat puffy (Fig. 22.4). They are 2–3 cm long and 1–2 cm wide, each usually containing one or two seed, sometimes three, and sometimes none. Considerable variation exists in seed size and shape among cultivars. Seed maturation requires 4 months after anthesis.

Two seed forms are recognized. The macrocarpa or kabuli form, with one to two large beige-colored seed per pod that are characterized by a "ram's head" shape and also a beak. The microcarpa or desi form produce two to three small dark-colored, irregularly shaped seed per pod.

Immature green pods and tender shoots are used as vegetables. Mature seed is consumed fresh and also processed by canning or freezing. Nevertheless, on a worldwide basis, the greatest usage is as a pulse in preparing dahl and for flour. Dry seed carbohydrate and protein content are about 60% and 23%, respectively; the seed have very good digestibility. Dry seed yields usually average about 700 kg/ha.

FIG. 22.4.   Chick pea foliage and pods, *Cicer arietinum.* Note: foliage is not trifoliate.

CLUSTER BEAN, *Cyamopsis tetragonolobus* (L.) Taub. *(C. psoralioides)*

Other name: Guar

The origin of cluster bean is not completely clear because wild types have not been found. The species probably originated in Africa, and domesticated in dry regions of western Asia, following introduction by Arab traders. The crop is cultivated extensively in India, Pakistan, and Myanmar for vegetable and industrial uses.

Plants are erect and bushy, often as tall as 3 m, but dwarf forms also exist. Being drought tolerant the crop is well suited for dry land culture, but is sensitive to flooding. Tolerance to salinity is better than many other legumes. Self-pollinated flowers of this short-day annual are white or pinkish white initially, then change to blue.

Pods, compressed and clustered like many stiff and erect fingers, are from 4 to 10 cm long and contain from 2 to as many as 10 seed, each about 5 mm in diameter. Cultivars grown for pod production tend to have fewer seed.

Propagation is with seed, usually broadcasted. When grown in rows, plant spacings are about 15 cm within rows and 60 cm between rows. Pods for vegetable use are usually harvested 3–4 months after planting; mature dry seed are harvested after 5–7 months. Dry seed are infrequently used for human food but are widely used for animal feeds and industrial products. The mucilaginous galactomannan gums in the seed endosperm is used in textile and paper manufacturing and other uses. The carbohydrate content of dried seed ranges from 40% to 45%, with a protein content between 30% and 33%.

SOYBEAN, *Glycine max* (L.) Merr.

Other name: Soya, edamame, daizu (Japan), mao duo (China), poot kong (Korea)

As a major source of plant protein and edible oil, soybeans are without doubt the world's most important food legume. The United States, Brazil, and China are the largest producers. In 1994, the United States produced about 50% of the world production of 136 million tons, Brazil produced about 25 million tons, and China about 16 million tons. Although mainly produced for its dry seed, the extensive use of the immature seed, especially in eastern Asia, makes soybeans an important vegetable.

Wild forms of *Glycine max* have not been found. *G. soya* is an annual found in eastern China that will hybridize with *G. max* and perhaps is the wild ancestor. Another thought is that *G. max* may be a cultigen resulting from the hybridization of *G. ussuriensis* with *G. tomentosa;* both are found wild in eastern Asia and south China. Undoubtedly a plant of eastern Asian origin and domestication, soybean cultivation has an ancient history, known to be grown since 2800 B.C. in China. Initially a short-day and subtropical plant, soybean domestication has resulted in numerous landraces, and plant breeders have introduced cultivars adapted to different latitudes. The ability to be widely grown is a major asset for this crop.

Plants are erect (70–150 cm tall), bushy, pubescent annuals with extensive root systems. Soybeans generally have a wide soil adaptation with a preference for well-drained, light to medium texture soils; plants are sensitive to saline conditions. Leaves are alternate and trifoliate. Growth is optimized at temperatures between 20°C and 25°C. Temperatures from 12°C to 20°C are sufficient for most growth processes but will delay seed germination and emergence and also flowering and seed development. At temperatures greater than 30°C, photorespiration tends to negate photosynthetic gain.

Flowering habit varies from strongly indeterminate to strongly determinate. The start of flowering is cultivar dependent and can vary from 80 to as many as 150 days. White, lilac, or purple flowers are self-pollinated. Produced in clusters, pods usually contain two to three seed which are globular or flattened and especially rich in protein and oil. Seed color can differ among cultivars.

Indeterminate-type cultivars are used for vegetable production and are grown at close spacings, usually 5–15 cm within rows and 50–60 cm between rows. Seed are sown between 2 and 4 cm deep. *Rhizobium japonicum* is specific for soybean nodulation, and seed are often inoculated before planting. In contrast to many legumes, soybeans are responsive to fertilizers, therefore, applications are made to increase yields.

Pods are harvested when fully expanded and prior to senescence, and when seed are fairly soft. The immature shelled seed are boiled in salty water, or cooked while in the pod (Fig. 22.5). Those cooked in the pod are a favorite snack food in China and Japan; the pods are too fibrous to be eaten. Production of mature seed generally requires an additional month of growth, but can vary from 4 to 6 months after planting. Pod dehiscence and seed shatter are limitations to dry seed production in arid environments.

Food use is generally correlated with seed color. Green and yellow-green seed are produced mostly for vegetable (edible seed) use. Large yellow, seeded cultivars are used to make tofu (bean cake). Large black seed are used in meals for celebration or other special occasions, and small, flat, black seed are made into a spicy fermented garnish. Generally, yellow-seeded cultivars are rich in oil and have a relatively low protein content, whereas black-seeded cultivars are high in protein and low in oil. Depending on seed type, carbohydrate content can range between 15% and 25%, the protein content as high as 50%, and some cultivars contain as much as 25% oil. Pods of seed oil cultivars commonly contain one or two seed; those of vegetable cultivars usually contain two to three seed.

Certain cultivars have specific vegetable use for which the seed is used directly, and for others, significant use is also made of sprouts produced from germinated seed. Seed of certain cultivars are used specifically to produce oil; soybeans are a major source of edible vegetable oil for frying and margarine production. Additional uses of soybeans range from animal fodder to a wide array of industrial products.

FIG. 22.5.   Vegetable soybean plant and pods, *Glycine max.*

HYACINTH BEAN, *Lablab purpureus* (L.), *(Dolichos lablab, D. nigar, Lablab vulgaris)*

Other names: Indian bean, Egyptian bean

The likely origin of hyacinth bean is India, where wild forms are still found and where this crop has been cultivated since ancient times. Nevertheless, others, including Vavilov, suggested the species was introduced into Asia from Africa. The immature pods and tender seeds are a popular vegetable in India and many tropical regions. Dry mature seed also are an important food, as are sprouted seed. The large starchy root can be eaten, and plants are occasionally used as ornamentals.

The plant is a short lived perennial, but is mostly grown as an annual to produce long edible pods. Plants grow well from sea level to high (2200 m) elevations and in low-rainfall, high-temperatures regions, and are intolerant of waterlogging. Although dwarf cultivars are grown, the typical hyacinth bean has a climbing growth habit, with vines 6–10 m long when supported. The trifoliate leaves are large (15 cm), nearly rhomboid, and contribute to the large biomass produced.

Cultivars vary in photoperiod response; both long- and short-day forms exist. Cultivars also show wide variations of stem, flower, and seed colors. Flowers are white, pink, or purple and mostly self-pollinated. Green or purple pods are thin, flat, oblong, and often curved. Harvest occurs when pods are between 5 and 10 cm long and before seed are mature. Pods contain three to six small, round seed which fully mature in 3–5 months.

Seed colors are usually white or black, but reddish brown and speckled ones occur; all have a prominent, long, white hilum, (Figure 22.6). White-seed cultivars contain low and nontoxic amounts of a cyanogenic glucoside and a trypsin inhibitor, whereas dark-seed cultivars have high levels of both.

Hyacinth bean is frequently intercropped with sorghum or maize. This association is favorable because the main growth of the bean crop occurs after the companion crop is harvested, and the sorghum or maize stems help support the vines. For fresh pod production, plants are often hill planted to include two or three plants grouped together at equidistant spacings of about 100 cm. Plants are supported in order to

FIG. 22.6.   Hyacinth bean, seed and dry pods, *Lablab purpureus.*

facilitate multiple harvests. The crop is usually grown for the pods, but when grown for the mature seed, broadcast plantings are often made. With row plantings, spacings commonly are 10–15 cm within rows and 50 cm between rows.

Fresh pods are cooked and consumed like snap beans. They contain about 4–5% protein. Dry seed have a carbohydrate content between 50% and 60%, and contain 20–25% protein.

AUSTRALIAN PEA, *Dolichos lignosus*

The Australian pea, *D. lignosus,* has a bushy (1 m tall) growth habit. Plants are a longer lived perennial than *D. lablab,* but are also cultivated as an annual. *D. lignosus* further differs in producing leaves and nonpalatable pods that are much smaller than those of *D. lablab.* Other than these phenotypic differences, the crop shares many cultural and growth similarities. The related *Dolichos uniflorus,* known as horse gram, is a hardy, semierect, annual dry land plant also grown as a pulse crop.

GRASS PEA, *Lathyrus sativus* L.
Other name: Chickling pea

Southern Europe and western Asia is considered the origin of grass pea. This bushy, nearly erect, weedlike, hardy, cool-weather pulse crop is widely cultivated in India and to some extent in Africa. It is often on poor soils and frequently used as a "catch" crop. The plant has the ability for a reasonable yield even when moisture is limited, as well as having some waterlogging tolerance. Furthermore, it does not have a high nutrient demand.

Grass pea is a branched annual with a well-developed taproot, angled stems, and blue or purple flowers. Short, flat, 4-cm-long, winged pods contain three to five wedge-shaped white, brown, gray, or mottled seed. Harvest of dry mature seed is achieved in 4–5 months. Dry seed contain about 58% carbohydrate, 28% protein, and 9% oil. Young leaves are used as a potherb. Excessive or prolonged consumption of seed can result in lathyrism, a muscle paralysis disorder in humans and animals.

LENTIL, *Lens culinaris* Medikus, *(L. esculenta)*

Lentils are another of the world's important pulse crops. The western Mediterranean basin and Southwest Asia are considered centers of origin and where lentils have been cultivated for thousands of years. Lentils have cool season and broad soil adaptation, tolerate some salinity, but do not perform well in the hot, humid tropics.

Plants are much-branched, small (25–40 cm) semierect, short-season annuals. Bluish white to pink flowers are mostly self-fertilized. Immature pods are used as vegetables, although the primary use of the seed is for preparing dahl and other pulse dishes. Lentils are grown and used in many European countries, although India appears to be the largest producer and consumer. Lentils are frequently grown in mixed cultivation.

Large-seed (macrospermae) and small-seed (microspermae) cultivar groups are recognized. Pods are 2–3 cm long and produce one or two yellow or brown, convex, lens-shaped seed. Large-seeded cultivars produce seed about 6–9 mm in diameter, and those of small-seeded cultivars range from 3 to 6 mm. Seed maturation occurs in 5 months, and yields average about 750 kg/ha. The carbohydrate content of dry seed is about 55%, the protein content is about 25%.

LUPINES, *Lupinus* spp.

The important cultivated species of lupine originated in the area of the Mediterranean basin and include white lupine *(L. albus),* yellow lupine *(L. luteus),* and European blue lupine *(L. angustifolius).* Egyptian lupine is sometimes identified as *L. termis* but is now considered a form of *L. albus* and is also widely grown. Lupines are an ancient crop and were cultivated for centuries in many European and African countries as a grain legume and also as a cover and green manure crop. The genus Lupinus also has centers of origin in South and North America. *L. mutabilis* of South American Andean origin is cultivated and known as sweet lupine. *L. pubescens* is a wild form found in South America and possibly the progenitor of sweet lupine.

Varying with species, plants range in height from 30 to more than 150 cm, with foliage palmately divided into many leaflets; Old World species are long-day plants. Flower color, also species-specific, are white, yellow, or violet, and sometimes variegated. Flowers are usually self-pollinated.

Pods of the large-seeded white lupine and sweet lupine are hairy and compressed and are 7–15 cm long. They contain three or four seeds, which are oval and flat, resembling lima beans. Pods of small-seeded lupines are between 3 and 6 cm long; seed vary in weight from 150 to 330 mg.

Being drought, high-temperature, and moderately frost tolerant gives the crop a broad adaptation. It is frequently relegated to grow in marginal and nutrient-poor soils, where it generally performs better than other pulses. Nevertheless, well-drained, coarse textured soils are preferred because lupines are susceptible to root rots often associated with cold, wet, and fine textured soils. Plants are erect, cool season annuals, grown during the summer in temperature latitudes and during the winter in the subtropics. Crops are frequently sown by broadcasting seed, to provide about 30 plants/m^2; crops mature in 6–7 months. Yields range from less than 1 to as many as 5 t/ha.

Lupine seed contain lupinine, a bitter, poisonous alkaloid, which is removed by soaking the seed in repeated changes of water for several days. However, many cultivated species of *Lupinus* are practically free of this alkaloid as result of selection and plant breeding. Lupines are relatively free of antimetabolic and flatulence factors other food legumes possess. The cooked mature beans generally are a low-cost protein source, ranging in protein content from 14% to 20%. Seed oil content varies greatly between cultivars. Mature seed contain about 50% carbohydrate. Lupine flour can be mixed with wheat flour to produce an acceptable pasta product.

YAM BEAN/JICAMA, *Pachyrhizus erosus* (L.) Urban

Other names: Sincamas or sinkamas (Philippines), Fan-ko and
   sar-gott (China), Dolique bulbeus (France)

The yam bean is a perennial indigenous to a broad region of tropical America, from Mexico to northern South America. It is widely grown for the tuberous roots throughout these regions and in areas of the Philippines and south China with similar growing environments. Related cultivated species are *P. ahipa* (domesticated in Bolivia and northern Argentina) and *P. tuberosus* (domesticated in the Amazon River headwaters). Wild forms of yam beans are found in Mexico and northern Central America; two recognized wild species are *P. panamensis* and *P. ferrugineus*. The widely used name, "jicama," is the Spanish form of the Nahuatl Indian word "xicamatl."

Jicama plants grow well in a hot humid environment and require a long, warm, frost-free growing season. A moist, light textured, well-drained soil and short-day conditions are preferred for optimum production of the fleshy tuberous roots.

Jimaca is a climbing plant having long vines of 3 or 4 m length, sometimes longer. Leaves are entire or of lobed rhomboidal shape and about 15 cm long. With the occurrence of short days, flowering commences, new vegetative growth diminishes, and storage root enlargement accelerates. During long days, although storage root initiation may occur, vine growth strongly competes with root enlargement. Growers commonly remove flowers because pod fill is a strong sink and competes with root bulking.

Violet or white flowers develop in erect racemes, producing pods 7–14 cm long and 1–2 cm wide. Although immature pods are edible as cooked vegetables, mature pods, foliage, and seed are poisonous. Seed are somewhat flattened, mostly rounded, and 5–10 mm wide, and in contrast to other *Pachyrhizus* species, they are never kidney shaped. Normally, 10 months are required to produce mature seed. Cultivars with greenish brown-colored seed are preferred because they are more productive than those with either green or brown seed.

The preferred tuberous root shape is a flattened sphere, much like a turnip, although elongated, spindlelike, and occasionally lobed roots also occur (Fig. 22.7). The exterior skin is colored tan to light brown. The interior flesh is white, does not discolor upon exposure, and has a watery crisp texture and sweet taste, comparable to that of Chinese water chestnut. The best quality roots usually average between 10 and 15 cm in diameter and weigh about 2 kg, although some are as wide as 30 cm and weigh more than 3 kg. Overly enlarged roots tend to be fibrous and starchy at the expense of crispness and sweetness. In Mexico, two major types are grown. The "jicama de leche" type has a dark skin, produces a slender spindle-shaped root that is not very succulent, and has a milky taste. The "jicama de agua" type has a light skin, produces a turnip-shaped root that is very succulent, and has a sweet watery taste.

Propagation is almost exclusively with seed (about five seed weigh 1 g) which are planted 2–4 cm deep. Germination usually occurs within 6–12 days. Plants often develop a strong symbiosis with *Rhizobium* bacteria. Basal sprouts are occasionally used for clonal propagation. Plants are row planted and are frequently cultivated as hill plantings. Spacings range from 15 to 30 cm in rows, with rows 100 cm apart; lower densities are used in hill plantings or when intercropped. Staked or trellised plants are more productive than those not supported.

Roots are harvested by hand digging or after being plowed out. Although yields can range from 4 to more than 45 t/ha, typical yields are about 15 t/ha. Depending on the growing conditions, from 4 to 8 months are required to produce market-sized roots. They are usually cleaned of soil and washed before marketing.

Storage roots can be field stored and harvested as needed. An interesting procedure used by some growers, especially for field storage, is to withhold moisture for several weeks before harvest, which causes some shrinkage. Several days before harvest, the field is irrigated and the

FIG. 22.7.   Yam bean/jicama, *Pachyrhizus erosus.*

storage roots readily absorb the moisture and regain turgor and weight. After harvest, the skin of the tuberlike root tends to thicken and limits moisture loss. Storage roots can be stored for more than a month at 13–15°C; but left for a long time at temperatures less than 12.5°C, chilling damage can occur. Temperatures greater than 15°C help to reduce the incidence of mold.

Yam bean is usually eaten raw in vegetable and fruit salads and appreciated for its mild flavor and succulent crispy texture; often it is consumed as a snack food. Tissue texture also remains crisp after cooking or pickling. Yam beans have a low caloric content; carbohydrates comprise less than 10% and proteins slightly more than 1% of fresh weight. Rotenone contained in mature pods and seed has insecticidal properties. The irritant hairs of the foliage contain pachyrhizid, a poisonous glycoside.

WINGED BEAN, *Psophocarpus tetragonolobus* (L.) DC. *(Tetragonolobus purpureus)*

Other names: Goa bean, four-angled bean, Manila bean, kok-tau.

Winged beans were once considered to be a miracle plant because the pods, seeds, flowers, stems, tuberous roots, and leaves are edible and nutritionally valuable; even the seed oil is of high nutritional value (Fig. 22.8). It is reported to have nutritive qualities similar to soy beans. Nevertheless, extensive cultivation in large holdings has not

FIG. 22.8.   Winged bean, *Psophocarpus tetragonolobus;* (a) plant with pods, (b) storage roots, and (c) seed. Source: Courtesy of Anson E. Thompson.

materialized. However, winged bean is an excellent home garden vegetable and is widely used as a semidomesticated crop in subsistence situations in areas of Southeast Asia, and especially in Papua New Guinea.

The east African coastal region is suggested to be a center of origin, but it is possible that winged bean has a tropical Asian origin. It is interesting that the crop is of little importance in Africa. The species is well distributed in south Asia and Southeast Asia and many Pacific Islands. Large diversity is found in Papua New Guinea.

Although winged bean is a fast-growing perennial, with vines achieving lengths between 2 and 4 m, the crop is usually grown as an annual. The trifoliate leaves are broadly ovate and the many shallow roots have long laterals. As a tropical plant with good subtropical adaptation, winged bean is well suited to humid conditions. Days of 30°C and nights of 22°C are most favorable for vegetative growth. Day temperatures of 24°C and 13°C nights are favorable for storage root enlargement. The plant is exceptional in having numerous root nodules and, thus, can utilize bacterial fixed nitrogen. Plants can be relatively productive when grown in low fertility soils, but yields are improved when supplemental nutrition is supplied. Winged beans are highly sensitive to water-logging.

Propagation is usually with seed, but stem cuttings can be used. Seedling emergence occurs in 5–7 days at 25°C, which is an optimum mean temperature for growth. Spacings vary with cultivar type. For winged bean pod production, staking is required with most spacings averaging 20 cm between plants and 90 cm between rows. Winged beans are frequently interplanted with other crops. When grown for tuberous root production, plants usually are not staked and populations may exceed 200,000/ha. Staking does not affect root yield but advances the peak period of pod production by as much as 1 month. Although the tuberous roots require about 8 months to reach a diameter of 3–4 cm. Pods can develop in as little as 3 months, which is about 1 month after flowering; seed mature in 5 months.

Most cultivars are short-day plants with indeterminate flowering; cultivars are available that are almost day neutral. Breeding efforts are directed for greater photoperiod insensitivity, determinate flowering, and plant dwarfness. Self-pollinated flowers are white, pale blue, or purple. Pods often are the primary production objective, but in order to enhance root enlargement, growers commonly prune flower buds and young shoots. Pods occur in all shades of green, and some have purple coloration. All are four sided, with ragged, thin, winged edges the full length of the pod. Pod length ranges from 5 to 35 cm, and widths from 2 to 5 cm. Fiber development quickly occurs if pods are allowed to overdevelop; dried pods commonly shatter. Pod seed numbers can range from 5 to 20. Seed are almost round in shape, the largest about 1 cm in diameter; an average seed weighs about 250 mg. Seed colors are usually white or black, but yellow and brown seed also occur.

Fresh pod yields of 10–15 t/ha are reported, with as much as 30 t/ha in experimental plantings. Tuberous root yields range between 5 and 10 t/ha, and seed yields range from 1 to 1.5 t/ha. Immature fresh pods have a protein content between 1% and 3%, and fresh leaves range between 5% and 7% protein as well as containing a high level of pro-vitamin A and vitamin C. Fresh tuberous roots, which are the preferred product in Papua New Guinea, contain 8–10% protein. The protein content of dry seed is about 33%, and carbohydrate and oil about 32% and 16%, respectively. Fresh pods have a short shelf life and usually are not stored, whereas storage roots, if necessary, can be stored as long as 2 months.

FENUGREEK, *Trigonella foenum-graceum* L.

Other names: Fenugrec, metha

Endemic to the Mediterranean region, fenugreek is an annual herb grown for centuries in the Middle East and India as a food, and also as a fodder and green manure crop. In India, young leaves are used as fresh greens and also are sun dried for later use.

Plants grow to heights of 40–90 cm. White flowers bloom 50–80 days after planting and produce long-beaked, slender pods 8–15 cm long. Pods contain 10–20 seed that require 180–210 days to mature. The mature seed are pungent and are used in preparing curry powder and other seasonings. Seed contain about 25% protein and 50% carbohydrates and are reported to have some medicinal qualities.

MARAMA BEAN, *Tylosema esculentum, (Bauhinia esculentum)*

Marama bean is a perennial, drought-tolerant plant native to southern Africa. Plants are cultivated and also harvested from the wild for their seed that have a protein content and quality comparable to soybeans and an oil content similar to peanut. Edible tuberous roots are also produced.

Prostrate vines grow to lengths of 4–6 m and need trellising. Leaves are bilobed and 8–15 cm in width. Yellow flowers are produced in clusters of three to nine. Pods are flat and oblong, about 6 cm long, and dark brown when mature. Each pod usually contains two flat, oblong, seed, each about 2 cm long.

After a few years of growth, tuberous storage roots can weigh more than 10 kg; plants having produced roots larger than 100 kg have been reported. However, it is usually the young, small, tender, 1–2 kg roots that are baked, boiled, or roasted. Seed are usually roasted but also boiled. Because they contain a strong trypsin inhibitor, they should not be consumed without cooking.

BROAD BEAN, *Vicia faba* L.

Other names: Fava bean, horsebean, field bean, Windsor bean, tick bean

Wild ancestors of broad bean are not known. The probable origin of *Vicia faba* is the Near East and western Mediterranean regions, with secondary centers of diversity in Afghanistan and Ethiopia. The species has an old history of cultivation in the Mediterranean and subtropical regions, and also in many temperate regions because of its cool-climate adaptation.

Broad beans, grown for the enlarged, succulent, immature seed, are harvested as shelled beans. They are also produced for the mature seed. *Vicia faba* is the best adapted grain legume to European climates, and very competitive with field peas. In some areas, the crop is overwintered for early spring production, but in the tropics, production is usually restricted to high elevations. Major production occurs in China, Egypt, Ethiopia, Italy, and Morocco.

Plants are erect annuals, from 50 to 180 cm in height, with pinnate rather than trifoliate leaves. Plants produce a strong, deep taproot. Stems are square shaped and partially hollow. Self-fertile flowers are white with a purple tinge.

Broad bean pods are fleshy, highly pubescent with a white velvetlike inner lining, and usually contain four to six seed (Fig. 22.9). Pod size varies greatly; those of vegetable forms usually are about 15 cm long but can be as long as 30 cm. Seed are 2–3 cm long and have a flatten oval or nearly rectangular shape. The small-seeded forms have pods between 5 and 10 cm in length, and the seed are nearly globe shaped with an average diameter about 1 cm. The color of the mature seed varies from light brown to black.

Variations in pods and seed allow for a further classification within the species. The large-seeded form, *V. faba* var. *major,* are known as broad beans and are the main vegetable type. The fresh immature seed of broad beans are an important summer vegetable in Great Britain and in other European countries. Mature, dried broad bean seed are also used as a vegetable. The small-seeded types, *Vicia faba* var. *equina* and *V. faba* var. *minor,* known as field beans are more widely grown than broad beans, but are more often used for animal feed. Another type of broad bean sparingly grown is *V. faba* var. *paucijuga.*

Adaptable to many soils and because of rapid biomass accumulation, they are frequently used as a green manure crop. Plants are moderately tolerant to salinity. Typical plant field densities of small-seeded cultivars is about 40 plants/m^2, and 15/m^2 for large-seeded broad beans.

Seeds are high in tannins and contain vicine or convicine, which can cause "favism" in certain populations, particularly those of Mediterranean ancestry (see Chapter 5). Favism is less common if the beans are well cooked. Carbohydrate content of dry seed is between 50% and 60%, and protein content is between 22% and 36%; oil content is low at 1–2%.

FIG. 22.9.   Broad bean pods and seeds, *Vicia faba*.

MOTH/MAT BEAN, *Vigna aconitifolia* (Jacq.) Marechal, *(Phaseolus aconitifolius)*

Indigenous to India, Pakistan, and Myanmar (Burma), moth beans are found in arid and semiarid environments from sea level to above 1000 m, but are unsuited to the wet tropics. The short-day, hot-weather, drought-resistant plant is a well-branched annual with a short ground-hugging trailing growth habit. Small, yellow, self-fertile flowers produce pods, 5–6 cm long and 5 mm thick, that contain six to nine small (5 mm) yellow to dark brown, rounded, ricelike seed. The crop matures in 2–3 months and often is interplanted with cereals.

Moth bean is a popular crop in India where the green immature pods are used as vegetables; the mature seeds which contain about 60% total carbohydrates and 23% protein have many food uses.

ADZUKI BEAN, *Vigna angularis* (Willd.) Ohwi & Ohashi (*Phaseolus angularis*)

The origin of adzuki bean is uncertain; some researchers think the Indo-Burma-China region to be a center of origin with diversity associated with domestication in regions of southern and central China, Japan, China, Korea, and India. The use of adzuki was identified to have occurred as early as 1000 B.C. in Korea; it is assumed that production occurred even earlier in China. One hypothesis suggest adzuki was domesticated from wild forms, another that it evolved from the rice bean, *Vigna umbellata*. Adzuki means little bean, and when translated from Japanese to English, adzuki is spelled azuki. Adzuki remains as important pulse crop in these Asian countries. China is the leading producer, followed by Japan, the Korean peninsula, and Taiwan with production from 670 thousand, 120 thousand, 30 thousand, and 20 thousand hectares, respectively.

Adzuki is a short-day annual best adapted to production in areas between 35 to 49° latitude. Humid and warm to high growing temperatures between 25°C and 30°C are most favorable for high seed yields; dry air at harvest can increase the incidence of seed shatter. Many cultivars are landraces that vary in the degree of determinate growth habit. Early-maturing cultivars are strongly determinate, and those later maturing are less determinate. Most cultivars are bushy and about 70–75 cm in height with erect foliage that resembles cowpeas; some cultivars are vinelike and have a prostrate growth habit. Plants have good drought tolerance, but being intolerant of waterlogged soils, they require good drainage.

Adzuki plants are seed propagated and planting densities vary considerably between growers with a range from 150 thousand to over 300 thousand plants per hectare. Plantings are made by broadcasting as well as in rows. Seed are sown 2–5 cm deep, about 8–10 cm apart in rows that range from 20 to 60 cm apart.

The bright yellow flowers are self-fertile, although out-crossings frequently occur. Mature cylindrical pods, resembling those of mung bean, are 6–12 cm long, about 0.5 cm wide, and usually contain up to 10 oblong red seed; other seed colors are black, green, gray, yellow, white, and mottled combinations of these colors; red or maroon is preferred. Seed shapes is generally subcylindrical with subtruncated ends, with a protruding ridge on the side of the hilum, and range in length from 5 to 10 mm. Cultivars producing large-size seed are known as "dainagon." The seed mature about 40–50 days after anthesis.

Harvest for seed usually occurs after 90–140 days of growth, but depending on cultivar and growing season, that may extend to as much as 160 days. Immature pods are used like fresh snap beans, but are relatively fibrous. Vegetable use is also made of sprouted seed. Dry seed contain about 60% carbohydrate, 20% protein, and a low 1% of lipids. A major use of dry adzuki beans in Japan and other Asian countries is as "an" (or ahn) a red, sometimes white, pastelike product consisting of cooked mashed adzuki beans and sugar. This has many uses in baked and confectionery preparations and as topping for desserts and ice cream. The cooked bean are also a common ingredient in rice dishes (seki-han) and soups. A mixture of adzuki and wheat flour is used for noodles.

URD BEAN, *Vigna mungo* (L.) Hepper *(Phaseolus mungo)*

Other name: Black gram

Although urd cultivation in India is ancient, the wild form is unknown. The plant is a semierect annual, well branched, but less than 1 m tall. Stems are covered with long, dense, brown hairs. Flowers are self-pollinating and a light yellow color. The hairy pods, which mature about 20 days after anthesis, are relatively short (4–7 cm) and slender (0.5 cm). They contain 6–10 small, usually black, oblong seed. Of the two main cultivar types, one is early maturing with large black seed, the other later maturing with small olive green seed. Dry seed carbohydrate content is about 57% and protein about 23%.

Urd is an important grain staple crop in India, but differs from mung bean in having cultivars with a black seed coat. Plants have a wide adaptation to drought, soils, and temperature. Green pods are utilized as a vegetable, but the major use is the dried seed in dahl and flour preparation. Millions of hectares are grown and the crop is an important staple for its dietary protein contribution.

MUNG BEAN, *Vigna radiata* (L.) Wilcz., *(V. aureus, Phaseolus aureus)*

Other names: Green and golden gram, chop suey bean, moong

Vigna radiata is believed to have its origin in the India-Burma region of Southeast Asia from where it was introduced to many other areas of the world. Wild mung bean, *Vigna vexillata,* is a viney plant that grows wild in the Himalayan foothills and parts of northern India, but is sometimes cultivated. However, wild forms of V. *radiata* have never been found, although wild progenitor species have been identified in India, which is a major production area.

The annual semierect plants are between 0.5 and 1 m tall, with many branches that are covered with short brownish hairs and trifoliate foliage that resembles that of cowpeas. Short- and long-day cultivars are grown. The self-fertile flowers produce pods 5–10 cm long and 0.5 cm thick that mature in 20 days after flowering. Pods commonly contain 10 or more small oblong to round seed of a dark olive green or yellow color; some plants produce brown or black seed.

Seed is usually sown broadcasted; but when row planted, a common row spacing is from 5 to 10 cm, with a distance of 70–90 cm between rows. Row plantings are often used because they facilitate cultivation for weed control, which broadcast plantings do not. Annual production is estimated to be between 2.5 million and 3 million tons from about 5 million ha; this is about 5% of all pulses produced. The major production area ranges from southern Asia to Southeast Asia.

Mung beans are a very important crop in India, where green immature pods are used as vegetables, although a major use is as a pulse for preparing dahl, a porridgelike food. Cultivars producing yellow seed (golden gram) are principally used for this purpose. One reason proposed for the popularity of mung bean is its tendency to cause less flatulence.

In other countries, especially China, green-seeded cultivars (green gram) are grown to produce seed that are used after sprouting. Seed are soaked, sprouted, and allowed to grow in the dark for several days before being harvested for consumption. One gram of seed results in the production of 6–8 grams of fresh sprouts. The etiolated hypocotyl and young cotyledonary leaves and young root radicle are eaten, either cooked or uncooked, with another vegetable dish. The sprouts are a good source of vitamin C. The dry seed carbohydrate content is between 55% and 60% with a protein content of about 23%. The tuberous roots of mung bean plants are of some interest because of their protein content of almost 15%.

BAMBARA GROUNDNUT, *Vigna subterranea* (L.) Verdn. *(Voandzeia subterranea)*

Vigna subterranea is indigenous to west central Africa, and most cultivated bambara groundnut production occurs in dry regions of tropical western Africa. Fresh vegetable use is made of the semimature seed after cooking. Mature seed are used as a pulse crop, mostly to prepare porridge and similar foods. Dry seed are very hard and require a long cooking time to be palatable. The trypsin inhibitor is also inactivated during cooking.

The branched, relatively low, lateral growing plants produce adventitious roots at stem internodes as they spread across the ground. The taproot is well developed and forms many lateral roots. Because of internode spacing, plant growth habit varies from bushy to spreading; bushy types generally mature earlier. Leaves are pinnately trifoliate on erect, grooved petioles. Plants are annuals with indeterminate flowering, and some cultivars require short days for flowering.

The crop is appreciated because of its drought tolerance and performance in poor soils, which assures some yield. A well drained, slightly acidic (pH 5.0–6.5) and friable soil is preferred, especially to facilitate pod entry into and development in the soil. High fertility is avoided as it tends to increase foliage growth in preference to pod development.

Plantings are established with seed usually sown about 5 cm deep. Seed viability is relatively short and because of the hard seed coat, germination is slow and often poor. Crops are grown in mixed cultivation about as often as in pure stands. Plant densities in pure stand cultivation range from 6 to 12 plants/m². The higher population is used when adequate moisture is available. A uniform moisture supply of about 900 mm is optimum.

Favorable growing temperatures are between 20°C and 28°C, and a minimum period of 3–4 months is required to produce a mature crop, late cultivars require 5–7 months. Plants are intolerant of frost, but tolerate high temperatures.

Flowering can begin 40–50 days after seedling emergence and is continuous. The pale yellow flowers are self-pollinated. After pollination occurs, peduncles lengthen and grow to or slightly below the soil surface. The nearly round pods are 2–3 cm in diameter and usually resemble groundnuts (peanuts) in their underground development. Pods typically contain one smooth, nearly round seed; seed colors vary with cultivars. Seed mature as early as 50 days after fertilization, but some late cultivars require more than 100 days. Mature seed are 8–10 mm in diameter, and individual seed weigh between 0.5 and 0.7 g.

Plants are harvested by uprooting and removing the pods. Gleaning of detached pods from soil is often necessary to recover the full yield. Bush-type cultivars with their concentrated production are easier to harvest than the spreading types. Seed removal from the pods is difficult but is usually done manually. Crop yields greater than 3500 kg/ha have been obtained. In the rain-fed dry regions of Africa, yields average about 750 kg/ha. Worldwide annual production is about 330,000 tons, most of which is produced in western Africa.

RICE BEAN, *Vigna umbellata* (Thumb.) Ohwi & Ohashi
(Phaseolus calcaratus)

Other name: Red bean

The origin of rice bean is unknown, although wild forms are found over a wide geographic area from the Himalayan foothills through mid-China and south to Malaysia. Plants have high-temperature tolerance and fair drought resistance. They are short-day annuals with semierect to climbing growth habits, with stems reaching lengths between 1 and 3 m. Yellow self-fertile flowers produce long (6–12 cm) and slender (0.5 cm) pods containing 6–12 small oblong seed.

Immature pods and young leaves are consumed as vegetables. Dried beans are frequently prepared with or as an alternative to rice dishes, hence the common name. Dry seeds contain about 55–60% carbohydrate and 21% protein and store well. Plants are often selected to be grown in rotation with rice plantings either for vegetable use and occasionally as a green manure crop.

COMMON COWPEA, *Vigna unguiculata* L. Walp.
cultigroup *unguiculata*

Other names: Blackeye pea, southern pea, crowder pea, frijole, coupe, lubia, niebe, kaffir bean

CATJANG COWPEA, *Vigna unguiculata* L. Walp.
cultigroup *cylindrica*

Other names: Bombay cowpea, Jerusalem pea, marble pea

YARDLONG BEAN, *Vigna unguiculata* (L.) Walp.
cultigroup *sesquipedalis*

Other names: Snake bean, asparagus bean, sitao, bodi bean

Common cowpea, catjang, and yardlong bean are collectively considered cowpeas and these three *V. unguiculata* cultigroups can readily intercross. The term "cultigroups" is more appropriate than subspecies for these cowpeas because genetic differences are small. Cowpea is an important vegetable and is also widely recognized as a major pulse crop. The domestication of the common cowpea most likely was in the tropical west African savanna, but the diversity of wild relatives of cowpeas is in southeastern Africa. The common cowpea is grown extensively in Africa, whereas the related catjang and yardlong bean are not, but are well represented in India and southeast Asia, respectively. The yardlong bean is extensively grown in China.

These cultigroups of *V. unguiculata* have been cultivated for centuries. Edible tender shoots, leaves, immature pods, fresh green, and dried seed are edible products of each. Their combined annual production from more than 5 million hectares contributes significantly to the dietary protein needs for millions of people. The value of this crop is especially obvious in tropical and subtropical regions of Africa, where common cowpea is the second most important pulse crop; the crop is also very important in Brazil.

The different cowpea species share some similarities. They are annuals that develop strong taproots with many laterals. Growth is optimum at 27–30°C days and 17–22°C nights. Cowpeas tolerate heat and dry conditions better than common field or lima beans, but they are very sensitive to air and soil temperatures of less than 20°C. Sandy loams are the preferred soil textures. Plants are sensitive to waterlogging.

Cultivated common cowpea cultivar types range widely from procumbent indeterminate short-day to erect determinate and day-neutral plants. Flowers are usually self-pollinated, with colors ranging from yellowish white to purple. Plants produce many fingerlike pods that are pendant. Pods are 10–30 cm in length and contain seed typically having a dark outline or eye around the hilum, and therefore some cultivar types are frequently called blackeye beans or peas. Other major types are called purple hull and cream, because of the pod color of the former and seed color of the latter. The "crowder" type generally refers to cowpeas having seed-crowded pods. Many cultivars exist within each type, but only a few produce edible fresh pods. Early-maturing cultivars produce immature pods in as few as 40 days and fresh mature seed in as few as 60 days. In some situations, common cowpeas are grown in mixed plantings rather than in pure stands. If not broadcasted, plant spacings are commonly 5–10 cm apart in the row and 70–90 cm between rows.

Fresh immature seed and pods are eaten cooked; fresh or dried seeds are processed by canning and freezing in the United States. The carbohydrate and protein content of dry common cowpea seed is more than 50% and 20%, respectively. Tender shoots and leaves are eaten as potherbs. Cultivars that tolerate frequent shoot clipping and leaf removal are grown for potherb use. Some cultivars are used this way with relatively little loss of either fresh pod or seed yield.

The yardlong bean is the most vegetable-like of the cowpeas. The stems of the trailing or climbing plants are several meters long. During growth, yardlong bean vines are supported on a trellis in order to prevent pods from contacting the ground and to assure straight pod development. Plants are easily stressed by limited moisture but tolerate rainfall and high humidity better than other cowpeas.

Flowers start to appear as early as 4–6 weeks after seedling emergence, and edible pods are formed about 2 weeks after anthesis. However, harvest most often begins about 70 days after planting and may continue for 25–30 days. Pods are 30–80 cm in length, and sometimes longer (Fig. 22.10). The slender, pendent pods are used like snap beans. The short postharvest life of yardlong bean pods is due to high respiration and wilting. Although low-temperature storage will extend shelf life of harvested pods, yardlong beans are chilling sensitive and are injured even after a few days at temperatures below 10°C.

Pods have an appearance as if inflated, and as seed mature, the pods tend to become constricted. Fresh mature seed, although less preferred than those of common cowpea, are also eaten as shell beans. Mature seed, which are seldom eaten, are kidney shaped, vary in length from 6 to 12 mm, and usually are reddish brown or black. Yardlong bean pods are a popular vegetable in Southeast Asia, China, the Philippines, and the Caribbean. Production in China for edible pods is in excess of 250,000 ha; yields range from 4 to 10 t/ha. In Indonesia, young leaves and shoots are eaten as a potherb. Bush types (bush sitao) have been developed by crossing yardlong and common cowpea. In contrast to yardlong bean, this plant produces shorter edible pods and does not

FIG. 22.10. Yardlong bean, *Vigna unguiculata* cultigroup *sesquipedalis,* pods bundled for marketing in China.

require trellis support. Its popularity is strongest in the Philippines and is likely to expand throughout Southeast Asia. In the United States, bush-type cultivars have produced 25–28 t/ha of edible pods.

Catjang cowpeas are semierect; pods grow upright, are 7–12 cm long, and produce many small seeds that are used as vegetables when immature. The crop is popular in India for use as a pulse crop. Additional use is made of the plant for animal forage.

SELECTED REFERENCES

Alvarenga, A.A., and Valio, I.F.M. 1989. Influence of temperature and photoperiod on flowering and tuberous root formation of *Pachyrrhizus tuberosus*. Ann. Bot. *64,* 411–414.

Basterrechea, M., and Hicks, J.R. 1991. Effect of maturity on carbohydrate changes in sugar snap pea pods during storage. Sci. Hortic. *48,* 1–8.

Bean Improvement Cooperative Annual Reports, Department of Plant Pathology, New York State Agric. Expt. Station, Geneva, NY.

Ben-ze'ev, N., and Zohary, D. 1973. Species relationships in the genus *Pisum O. Israel.* J. Bot. *22,* 73–91.

Biddle, A.J., Knott, C.M., and Gent, G.P. 1988. The PGRO Pea Growing Handbook. Processors and Growers Research Organization, Peterborough, England.

Cantwell, M., Orozco, W., Hernandez, L., and Rubatzky, V. 1992. Postharvest handling and storage of jicama roots. Acta Hortic. *318,* 333–343.

Carter, T.E., Jr., and Shanmugasundaram, S. 1993. Vegetable soybean (Glycine). In Pulses and Vegetables. J.T. Williams, ed. Chapman & Hall, London, pp. 219–239.

CIAT Annual Reports. Centro Intenacional de Agricultura Tropical, Cali, Colombia.

Data, E.S., and Pratt, H.K. 1980. Patterns of pod growth, development, and respiration in the winged bean *(Psophocarpus tetragonolobus).* Trop. Agric. (Trinidad) *57,* 309–317.

Dickson, M.H., and Petzoldt, R. 1989. Heat tolerance and pod set in green beans. J. ASHS *114,* 833–836.

Erickson, H.T. 1992. Inheritance of growth habit and qualitative flowering response in lima beans *(Phaseolus lunatus* L.) HortScience *27,* 156–158.

Faris, D.G. 1965. The origin and evolution of the cultivated forms of Vigna sinensis. Can. J. Genet. Cytol. *7,* 433–452.

Frey, R.L. 1987. Genetics of Vigna. Hort. Rev. *9,* 311–394.

Gepts, P. 1990. Biochemical evidence bearing on the domestication of *Phaseolus* (Fabaceae) beans. Econ. Bot. *44*(3 Supplement), 28–38.

Gritton, E.T. 1986. Pea breeding. In Breeding Vegetable Crops. M.J. Bassett, ed. Chapman & Hall, New York, pp. 283–319.

Gutierrez Salgado, A., Gepts, P. and Debouck, D.G. 1995. Evidence for two gene pools of the Lima bean, *Phaseolus lunatus* L., in the Americas. Genet. Res. Crop Evol. *42,* 15–28.

Hagedorn, D.J., ed. 1984. Compendium of Pea Diseases. American Photopathological Society, St. Paul, MN.

Hamdi, A., Erskine, W., and Gates, P. 1991. Relationships among economic characters in lentil. Euphytica *57,* 109–116.

Haq, N. 1993. Lupins *(Lupinus species).* In Pulses and Vegetables. J.T. Williams, ed. Chapman & Hall, London, pp. 103–130.

Hebblethwaite, P.D., ed. 1983. The Faba Bean. Butterworths, London.

Hebblethwaite, P.D., Heath, M.C., and Dawkins, T.C.K., eds. 1985. The Pea Crop. Butterworths, London.

Herklots, G.A.C. 1972. Vegetables in South-East Asia. George Allen & Unwin, London.

ICARDA Annual Reports. International Center for Agricultural Research in the Dry Areas, Aleppo, Syria.

ICRISAT Annual Reports. International Crops Research Institute for the Semi-Arid Tropics, Patancheru, India.

Kadam, S.S., and Salunkhe, D.K. 1984. Winged bean in human nutrition. CRC Crit. Rev. Food Sci. Nutr. *21,* 1–40.

Kaplan, L. 1981. What is the origin of the common bean? Econ. Bot. *35,* 240–254.

Konsens, I., Ofir, M., and Kigel, J. 1991. The effect of temperature on the production and abscission of flowers and pods in snap beans (*Phaseolus vulgaris* L.). Annu. Bot. *67,* 391–400.

Ladizinski, G. 1986. Pulse domestication before cultivation. Econ. Bot. *41,* 60–65.

Linnemann, A.R., and Azam-Ali, S. 1993. Bamboo groundnut (*Vigna subterranea*). In Pulses and Vegetables. J.T. Williams, ed. Chapman & Hall, London, pp. 13–58.

Lumpkin, T.A., and McClary, D.C. 1994. Azuki Bean Botany, Production and Uses. CAB International, Wallingford, United Kingdom.

Munger, H.M. 1995. Personal communication.

Noda, H., and Kerr, W.E. 1983. The effects of staking and inflorescence pruning on the root production of yam bean (*Pachyrrhizus erosus* Urban). Tropical Grain Legume Bull. *27,* 35–37.

Pisum Newsletter, Annual newsletter for International audiences concerning breeding and genetics of *Pisum* spp. Dept. Hort. Sci., New York State Agric. Expt. Station, Geneva, NY.

Poehlman, J.M. 1991. The Mungbean. Westview Press, Boulder, CO.

Powell, A.M. 1987. Marama bean (*Tylosema esculentum,* Fabaceae) seed crop in Texas. Econ. Bot. *41,* 216–220.

Purseglove, J.W. 1968. Tropical Crops—Dicotyledons. Longman, London.

Saxena, M.C., and Singh, K.B., eds. 1987. The Chick Pea. CAB International, Wallingford, United Kingdom.

Schoohoven van, A., and Voysest, O., eds. 1991. Common Beans—Research for Crop Improvement. CIAT, Cali, Colombia/CAB International, Wallingford, United Kingdom.

Sherf, A.F., and Macnab, A.A. 1986. Vegetable Diseases and Their Control. 2nd ed. John Wiley & Sons, New York.

Siemonsma, J.S., and Piluek, K., eds. 1993. Plant Resources of South-East Asia, No. 8, Vegetables. Pudoc Scientific Publ., Wageningen, Netherlands.

Singh, S.P., Gepts, P., and Debouck, D.G. 1991. Races of common bean (*Phaseolus vulgaris* L., Fabaceae). Econ. Bot. *45,* 379–396.

Snoad, B., and Davies, D.R. 1972. Breeding peas without leaves. Span *15,* 87–89.

Sorensen, M., Grum, M., Paull, R.E., Vaillant, V., Venthou-Dumaine, A., and Zinsou, C. 1993. Yam bean (*Pachyrhizus* species). In Pulses and Vegetables, J.T. Williams, ed. Chapman & Hall, London, pp. 59–102.

Wehner, T.C., and Gritton, E.T. 1981. Horticultural evaluation of eight foliage types of peas near-isogenic for the genes *af, tl,* and *st.* J. ASHS *106,* 272–278.

Woodroff, J.G. 1983. Peanuts, Production, Processing, and Products, 3rd ed. Chapman & Hall, New York.

23

Tomatoes, Peppers, Eggplants, and Other Solanaceous Vegetables

INTRODUCTION

Solanaceae is mainly a tropical family of about 75 genera and 2000 species. The more important vegetable genera are *Solanum* (potato and eggplant), *Lycopersicon* (tomato), and *Capsicum* (pepper). The Solanaceae, widely known as the nightshade family, also includes some poisonous alkaloid-containing species such as belladonna *(Atropa belladonna)*, mandrake *(Mandragora officinarum)*, henbane *(Hyoscyamus niger)*, Jimson weed *(Datura stramonium)*, climbing nightshade *(Solanum dulcamara)*, and widely used tobacco *(Nicotiana tabacum)*. Some genera, particularity *Solanum*, can be extremely poisonous and caution is advised before consumption. The potato, *Solanum tuberosum*, is discussed in Chapter 9. Following are some of the cultivated solanaceous vegetables, most, with the exception of Chinese box thorn, are grown for their edible fruit.

Cultivated Solanaceous Vegetables

Genus and species	Common name
Capsicum annuum and *C. frutescens*	Pepper
Cyphomandra betacea	Tree tomato
Lycium chinense	Chinese box thorn
Lycopersicon lycopersicum	Tomato
Lycopersicon pimpinellifolium	Currant tomato
Physalis alkekengi	Chinese lantern
Physalis ixocarpa	Tomatillo
Physalis peruviana	Cape gooseberry
Physalis pruinosa	Husk tomato
Physalis pubescens	Ground cherry

Solanum aethiopicum	Mock tomato
Solanum americanum	Glossy nightshade
Solanum gilo	Jilo
Solanum hirsutissimum	Lulita
Solanum hygrothermicum	Peruvian potato*
Solanum incanum	Sodom apple
Solanum indicum	Indian nightshade
Solanum macrocarpon	African eggplant
Solanum melongena	Eggplant
Solanum muricatum	Pepino
Solanum melanocerasum	Garden huckleberry
Solanum quitoense	Naranjilla
Solanum sessiliflorum	Cubiu/cocona
Solanum torvum	Turkeyberry
Solanum tuberosum	Potato†

TOMATO, *Lycopersicon lycopersicum* (L.) Karsten (*L. esculentum* (L.) Mill.)

Second to potato, tomato is the most widely grown solanaceous vegetable. The acid sweet taste and unique flavors account for its popularity and diverse usage. Because of its high per-capita consumption, tomatoes are nutritionally valuable for their high pro-vitamin A and vitamin C content. Interestingly, although indigenous to western South America, its dietary and economic importance in the regions of origin has lagged in comparison to other parts of the world.

Origin and Domestication

The Vera Cruz and Puebla areas of Mexico are generally regarded as centers of domestication. However, as judged by the distribution of wild species, the progenitor of the tomato is considered to have originated in the narrow, dry, tropical, coastal areas of Ecuador and Peru and portions of northern Chile. Of the wild forms, *L. lycopersicum* var. *cerasiforme,* is considered the probable immediate ancestor of the cultivated tomato. This wild form spread from Ecuador and Peru throughout tropical America.

The initial introduction of tomato to Europe appears to have been from Mexico rather than the Andean regions. It is suggested that the

*See Chapter 14.
†See Chapter 9.

name "tomato" comes from the Nahuatl language of Mexico. The early introductions into Europe were associated with a reputation of being a dangerous food because of the relationship to poisonous Solanaceous species such as belladonna and mandrake. Except in Italy, this belief restricted initial acceptance and the first use was as an ornamental. In France, the fruit were called "pomme d'amour" or love apple. In Italy, they were called "pomi di oro" or golden apple, suggesting those first introduced were yellow fruited. Following its European introduction and acceptance, tomato cultivation quickly spread throughout the world, especially during the 20th century.

Taxonomy

The two cultivated *Lycopersicon* species have red and smooth fruit, are self-pollinated, and are included in the subgenus *Eulycopersicon*. Wild species belong to the subgenus *Eriopersicon* and have green and pubescent fruit, and some rely completely on cross-pollination. Plant breeders have used wild species to transfer disease and nematode resistance, high-fruit solids, and other useful characteristics to improve the cultivated species.

Cultivated Species and Botanical Varieties

Lycopersicon lycopersicum cultivars are perennials, which in temperate regions are grown as annual crops. Among the cultivars, flowering habit ranges from highly indeterminate to strongly determinate; flowers are usually self-pollinated. They have erect to prostrate stems, pubescent foliage, and glandular trichomes. Some characteristics of the major *L. lycopersicum* botanical varieties are as follows:

var. *cerasiforme* (Cherry tomato)
Leaves and fruit are smaller than cultivated *L. lycopersicum*.
Flowers form in long clusters; fruit are usually bilocular.
var. *commune*
Leaflets are small; fruit are globular.
var. *grandifolium*
Potatolike foliage.
var. *pyriforme*
Fruit are either pear shaped or oval and usually small.
var. *validum*
Plants are dwarf, upright, and sturdy with curled and crowded leaf growth.

The currant tomato, *L. pimpinellifolium,* is an annual. It has slender, extensively branched, finely pubescent stems with many flowers that occur in clusters. Flowering is typically indeterminate. Flowers are self-compatible and easily cross with *L. lycopersicum. L. pimpinellifolium* is used as a source of disease resistance. Fully developed fruit are red, currantlike, and about 1 cm in diameter.

All wild *Lycopersicon* species are perennial in their native habitat; some of their individual characteristics are as follows:

L. chmielewskii—Resembles *L. pimpinellifolium.* Cross- and self-pollinated; a source of high-fruit-soluble solids.

L. cheesmanii—Self-pollinated, fruit globose; one accession found to be highly salt tolerant.

L. chilense—Erect plants with narrow leaves and heavy but brittle stems. Self-incompatible; source of drought, cold, and alkali tolerances. Resistant to some gemini viruses.

L. hirsutum—Self-incompatible in the center of its distribution; self-compatible in periphery. Fruit are small and green when mature. Foliage and fruit exhibit extreme pubescence and have many glandular trichomes. Best source of insect resistance, also resistance for late blight, black leaf mold, and potato virus Y, in addition to being a source of cold tolerance.

L. parviflorum—Also similar to *L. pimpinellifolium.* Flowers are self-pollinated. Fruit are small (1–1.5 cm), whitish green, and slightly hairy.

L. pennelli—Mostly self-incompatible; fruit are yellow at maturity. Also a source of drought, salinity, and insect (tomato fruitworm) resistance.

L. peruvianum—Highly self-incompatible; plants produce small greenish white to purple highly acid fruit. Source of cold tolerance. Resistant to TMV, nematodes, and blight.

Barriers to interspecies crosses have been circumvented with varying degrees of success. Hybridization is usually facilitated where *L. lycopersicum* serves as the female parent; embryo culture can also facilitate introgression. Intergeneric exchanges with some *Solanum* species are possible.

Botany

Tomatoes are usually annuals in temperate regions or short-lived perennials in the tropics. Plants grow from 0.5 to 2.0 m tall, with solid and thick stems. Some dwarf cultivars, grown as novelties, are less than 30 cm tall. Growth habit can vary from erect to semiprostrate and some also exhibit substantial vining. Taproots usually are strong and deep; some occasionally reach depths of 3 m. Small glandular hairs that appear on stems, leaves, and peduncles have a noticeable odor. Leaves are compound pinnate, coarsely toothed, and often curled, but also can be smooth.

Plant growth characteristics range from indeterminate to highly determinate (Fig. 23.1). Inflorescence are borne opposite and between leaves. Although some cultivars have 30 or more flowers per cluster, usually 4–12 flowers develop on a broad, flat raceme. Flowers are perfect, about 2 cm in diameter, and often pendent with a yellow star-shaped corolla; yellow anthers are united to form a tube. Self-pollination is commonly observed. Flowers do not produce nectar, although cross-pollination, usually by bees, occurs with varying frequencies.

Pedicels typically have an abscission zone about midlength. Many recent cultivar introductions have a "jointless" characteristic where the abscission layer does not develop. Therefore, fruit can easily separate without the pedicel attached. When the pedicel portion remains attached to the fruit, puncture of other fruit can occur during handling, leading to undesirable postharvest losses.

The tomato fruit is a fleshy berry, the surface being slightly hairy when very young, but smooth when mature. Fruit of most cultivars are globose; other shapes are elongated, plum, and pearlike. Noticeable lobes are present with some cultivars, an indication of fruit that have multiple ovaries. Ripe fruit colors, usually solid, are red, pink, tangerine, orange, yellow, or colorless. Red color is due to lycopene pigmentation; yellow because of other carotenoid pigments. Intermediate colors are due to differing ratios of these pigments in combination with skin color. Red tomatoes have a yellow skin and red flesh (pericarp); pink

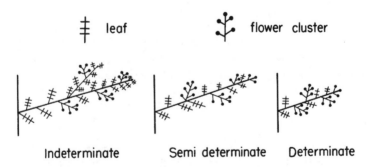

FIG. 23.1. Growth characteristics of the tomato plant.

cultivars also have red flesh, but because of a recessive gene, the skin is colorless. Yellow flesh, the result of another recessive gene, when overlaid by yellow skin produces bright yellow fruit; if combined with colorless skin, the fruit is pale yellow.

At maturity, seed are surrounded by a gelatinous material that normally fills the locules (Fig. 23.2). Fruit usually contain many seed, which are flat and a light cream to brown color. Seed are typically 2–3 mm long; about 300–350 seed weigh 1 g.

Culture

Tomato cultivation is adaptable to many environments, so that production is found well distributed from high-elevation regions near the equator to temperate areas far from the equator. Exceptions are the humid tropics because of high disease incidence and temperate regions where low temperatures and short growing seasons limit growth.

Soils and Moisture

Tomatoes are grown successfully on a wide range of soil types, from sandy to fine textured clays, as well as in soils of high organic content. A soil pH range from 5.5 to 7 is usually satisfactory for most cultivation. Plants grow best when provided with uniform moisture and well-drained soils. They are intolerant of waterlogging, especially shortly after germination and at the period of fruit maturation. Excessive

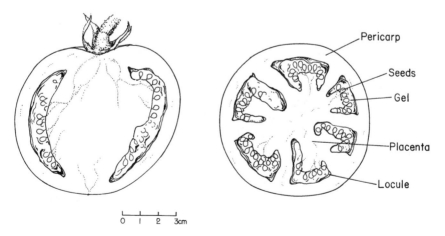

FIG. 23.2. Longitudinal and cross-sectional structure of a tomato fruit.

moisture is often conducive to damping off and root rot diseases. Where drainage is a problem, raised-bed culture is recommended. Soils with disease histories such as *Fusarium* or *Verticillium* wilt should be avoided or resistant cultivars utilized. Crop rotation is strongly recommended to minimize disease problems.

Tomato plants generally have an extensive root system, most within the upper 60 cm; taproots can grow to a considerable depth when not restricted by hard pans or high water tables. The deep-rooted system provides the plant some drought tolerance. However, when rainfall is insufficient, irrigation should be provided to avoid yield reduction. Water application methods include surface and furrow flooding, subbing (raising subsurface water level), and, to a lesser extent, overhead sprinkling. Trickle or drip irrigation, which has achieved rapid acceptance in some areas, also is an effective method. The latter method is initially costly, but once installed, it often requires less labor, utilizes less water, reduces the occurrence of soil diseases, such as *Phytophthora* root rot, and also restricts between-row weed growth. Other advantages, in addition to water conservation, include the ability to use drip systems to effectively apply fertilizer and pesticides.

Water usage commonly is about 25–30 mm weekly, and on a hot, dry day, evapotranspiration can exceed 10 mm. Although irrigation frequency and amounts varies, processing tomato crops grown in California are usually supplied with 600–900 mm of water. In order to increase the soluble solids content, irrigation is sometimes curtailed during the latter stage of fruit development for the processing crop. Moisture stress during fruit enlargement can contribute to the incidence of blossom end rot, a physiological disorder.

Temperature
The extensive array of cultivars enable producers to grow tomatoes over a wide range of temperatures. Tomatoes can be grown in most open-field locations where there is a minimum of 3–4 months of warm, frost-free weather, with an average temperatures above 16°C. Vegetative and reproductive growth at lower temperatures is very limited, and an extended period of plant growth at 12°C or less can result in chilling injury. Although frost sensitive, tomatoes are hardier than peppers or eggplants. Day temperatures of 25–30°C with night temperatures between 16°C and 20°C are optimal for growth and flowering. A wide diurnal difference between day and night temperatures tends to improve flowering, growth, and fruit quality. Fruit set is best between 18°C and 24°C; it is poor below 15°C or above 30°C. Night temperatures are more critical than day temperatures for fruit setting.

Crop Nutrition

Nitrogen is very important for vegetative growth. A mixture of NO_3^- and NH_4^+ nitrogen with NO_3^- in a higher proportion than NH_4^+ generally gives best results. It is important to obtain sufficient plant size prior to flowering. However, excessive vegetative growth can reduce early and subsequent fruit set. Adequate phosphorus is also important for early plant development and flowering. Achievement of high-fruit-soluble solids relies on adequate potassium, and calcium is important for cell wall development.

Usually a starter fertilizer is applied prior to or at planting. Both phosphorus and potassium are commonly applied preplant with a portion of the total nitrogen. Additional nitrogen applications are frequently made at the start of flowering and again when fruit are enlarging.

Propagation

Plants are propagated by direct field seeding or with transplants. Although infrequently done, indeterminate cultivars can be vegetatively propagated from stem cuttings. The minimum soil temperature for seed germination is 10°C; the maximum is about 35°C. Between 25°C and 30°C, seedling emergence occurs within 6–9 days. Table 23.1 shows the effect of temperature on emergence. Seed priming procedures can also be used to improve low-temperature germination rate and uniformity. Seed are sometimes coated to produce uniform size and shaped pellets that improve seed handling and placement when used with precision planters.

Seed is often treated with fungicides to reduce damping off diseases. With direct seeding, especially for processing tomatoes, it is common to sow a slight excess of seed that provide a complete and uniform population; surplus plants are removed. Precise final spacing of individual plants is less critical for good results, because closely spaced plants, within reason, often are able to adapt without significantly affecting total yield. Thus, precise thinning is not necessary. For fresh market tomatoes, plant-to-plant competition is more critical because it will affect uniformity of fruit set and size.

To obtain full stands, the processing tomato grower often make

TABLE 23.1. INFLUENCE OF TEMPERATURE ON TOMATO SEEDLING EMERGENCE

Temperature (°C)	5	10	15	20	25	30	35	40
Days to emergence	a	43	15	9	6	6	9	a

[a] Little or no germination.

"clump" plantings, which involves sowing several seed at each location. In this procedure, emerging plants are not thinned. Crowded plants produce less fruit per plant, but the additional plants produce more total fruit, and total yields are higher or comparable to other planting practices.

Besides direct seed drilling, another method, called plug planting, uses dry or imbibed seed mixed into a medium containing vermiculite, peat moss, or other material. A portion of this mixture containing three to seven seed is placed at a precise spacing into the seedbed. The mulchlike medium is helpful in supporting early seedling growth. In some situations, with high quality seed and favorable seedbed conditions, a precise sowing rate is used, and thinning is not performed.

A significant change from traditional transplanting practices to those of direct seeding has been achieved through numerous research efforts. The principal intent was to reduce production costs and labor. Direct sowing is used extensively by many growers of processing tomatoes in the United States and elsewhere. However, in regions where earliness is sought and where short growing periods restrict direct seeding, growers continue to establish the crop with transplants. Many fresh market and processing cultivars are now hybrids and the increased commercial use of expensive hybrid seed has resulted in a return to transplant practices. Considerable advances achieved in transplant production make transplant use economically feasible. Efficient use of land, early weed control, and a general reduction of production inputs are some benefits of transplanting.

Transplant procedures range from producing open-field plants sown at high densities and grown with or without protection. After achieving appropriate size, plants are removed and planted as bare root seedlings. Transplants are also grown in individual peat blocks or using synthetic media in multiple cell trays in glass or plastic houses or tunnels. When transplanted, these plants usually have a small root ball.

Transplants are established in the field manually or with machines, some of which are fully mechanized. To advance early production, hot caps, small plastic tunnels, row covers, windbreaks, and mulches can also be used.

Spacing

Cultivar growth habit greatly influences plant spacing. Determinate plants are grown at higher densities than those that are indeterminate. The end use and harvest method influences spacing. Wide, often raised beds are used for mechanically harvested processing crops, where total yield and not fruit size is the major objective. Bed widths range from 150 to 180 cm, with in-row spacing from 30 to 60 cm to provide populations of 10,000 to 20,000 plants per hectare. On the other hand, plant density of field-grown fresh market tomatoes range from 8000 to 14,000, and about 6000 to 8000 per hectare if staked. Typical spacing range from 60 to 75 cm within rows and from 120 to 150 cm between rows. Indeterminate cultivars are usually staked. Staking or trellis support of plants avoids fruit contact with soil, makes harvest easier, and usually incurs fewer diseases by improving airflow and pesticide coverage. Because of increased light interception and a long harvest period, high yields are achieved. Staking is a preferred method for producing ripened fruit. However, because staking practices require significantly higher labor costs, there is considerable interest to grow plants for fresh market without supports. Determinate cultivars are used, although their yield potential is usually less than those of indeterminate cultivars.

Flowering and Fruit Set

Fertilization of flowers for most presently grown cultivars is generally favorable at day temperatures between 21°C and 30°C, and night temperatures between 15°C and 21°C. Depending on temperature, fertilization occurs within 48 h after pollination. For many cultivars, day temperatures above 32°C reduce fruit set, and at 40°C fruit set is negligible.

Flowers open during the day, and stigmas are receptive to pollination for 4–7 days. Style elongation occurs within the anther cone and usually coincides with pollen release from dehiscing anthers. High temperatures interfere with viable pollen production and its dispersion, and also affect ovule viability. Hot, drying winds can have similar effects. Cool temperature and high humidity and/or low light intensity limit pollen shedding. At or below 15°C pollen formation and function is greatly inhibited. Cold temperature can also affect ovule viability. The most temperature-sensitive periods regarding fruit set occur about 5–10 days before anthesis, and 2–3 days following pollination.

High light intensity tends to accelerate flowering in many cultivars, whereas low light intensity limits vegetative growth and may also delay flowering. Tomatoes grown in protective structures often are provided with supplemental light when intensity is low and day lengths are short. To facilitate self-pollination, plants or flower clusters are shaken to disburse pollen. The practice is often accomplished using small vibrators.

Once flowering begins, the fruit becomes the major photosynthetic sink with proportionally less directed to vegetative growth. The appropriate level of vegetative growth should be achieved before flowering begins so that the plant is able to support subsequent fruit development. Underdeveloped plants typically yield poorly. Low light intensity and high night temperatures are conducive to excessive vegetative growth which competes with the fruit for photosynthates. Also low light intensities and night temperatures less than 10°C or greater than 27°C can cause early fruit abscission. Some cultivars are better adapted to temperature extremes and, thus, are able to set fruit during adverse conditions.

During cool periods, hormones such as naphthalene acetic acid (NAA), indoleacetic acid (IAA), or para-chlorophenoxyacetic acid (4-CPA) can be used to increase fruit set. However, the resulting fruit, whether fully or partly parthenocarpic, often appear puffy because of poorly filled carpels. To reduce this defect, gibberellins are applied in combination with fruit-setting hormones.

Fruit Ripening

Varying with cultivars, most tomato fruit mature 35–60 days after anthesis. Prevailing temperatures influence the rate of ripening. The degree-day procedure is used by some growers to schedule planting and harvest periods and is most frequently used for scheduling processing tomato production. Optimum temperatures for fruit maturation and color development are between 20°C and 24°C. At favorable temperatures, red color (lycopene) develops during ripening even without light; however, light accelerates development and intensity of color. Below 13°C, fruit ripening is poor and slow; at 10°C ripening stops and chilling injury can occur. Lycopene synthesis is inhibited to a greater extent than other carotenoids at temperatures greater than 32°C or less than 10°C. Under these conditions, mature fruit exhibit a yellowish to an orange red color instead of deep red. At temperatures greater than 40°C, fruit tend to remain green because chlorophyll degradation is inhibited.

For most cultivars, in addition to color change during maturation which is accompanied by fruit softening, there are changes in fruit sugars and organic acids (Fig. 23.3). The accumulation of sugars and aromatic compounds in the presence of acids give the fruit its characteristic flavor and aroma.

Diseases and Pests of Tomatoes and Other Solanaceous Crops

Bacteria

Bacterial canker	*Corynebacterium michiganense* pv. *michiganense*
Bacterial speck	*Pseudomonas syringae* pv. *tomato*

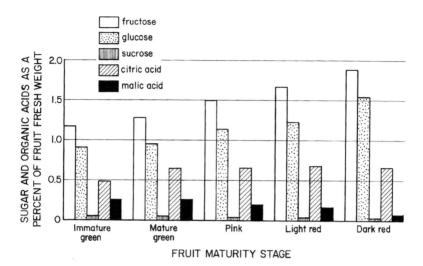

FIG. 23.3. Changes in sugar and acid composition during tomato fruit ripening.
Source: Redrawn from data of Picha (1987).

Bacterial spot   *Xanthomonas campestris* pv.
                                              vesicatoria
Southern bacterial wilt   *Pseudomonas solanacearum*

Fungi
  Anthracnose rot   *Colletotrichum coccodes*, other
                                              Colletotrichum spp.
  Blackmold/alternaria stem   *Alternaria alternata* f. sp. *ly-*
    canker   *copersici*
  Cercospora leaf mold   *Cercospora fuligena*
  Corky root   *Pyrenochaeta lycopersici*
  Damping off   *Pythium, Phytophthora,* and
                                              Rhizoctonia spp.
  Didymella stem rot   *Didymella lycopersici*
  Early blight   *Alternaria solani*
  Fusarium crown rot   *Fusarium oxysporum* f. sp. *rad-*
                                              icis-lycopersici
  Fusarium wilt   *Fusarium oxysporum* f. sp. *ly-*
                                              copersici
  Gray leaf spot   *Stemphylium* spp.
  Gray mold   *Botrytis cinerea*

Helminthosporium blight	*Helminthosporium carpo-saprum*
Late blight	*Phytophthora infestans*
Leaf mold	*Cladosporium fulvum (Fulvia fulva)*
Nail-head spot	*Alternaria tomato*
Phoma rot	*Phoma destructiva*
Phytophthora root rot/buckeye rot	*Phytophthora parasitica*
Powdery mildew	*Leveillula taurica (Oidiopsis taurica)*
Pythium fruit rot	*Pythium ultimum*
Rhizoctonia fruit rot	*Rhizoctonia solani*
Sclerotinia/timber rot	*Sclerotinia sclerotiorum, S. minor*
Septoria leaf spot	*Septoria lycopersici*
Southern blight	*Sclerotium rolfsii*
Stemphylium gray leaf spot	*Stemphylium solani*
Verticillium wilt	*Verticillium dahliae, V. albo-atrum*

Viruses and virus like agents
Alfalfa Mosaic Virus (AMV)
Beet Curly Top Virus (BCTY)
Bushy Stunt Virus (BSV)
Cucumber Mosaic Virus (CMV)
Double Streak Virus, combination of TMV and PVX
Potato Virus Y (PVY)
Tobacco Mosaic Virus (TMV)
Tomato Mottle Virus TMoV)
Tomato Ringspot Virus (TRSV)
Tomato Spotted Wilt Virus (TSWV)
Tomato Yellow Leaf Curl Virus (TYLCV)
Tomato Yellow Top Virus (TYTV)
Big Bud Microplasma

Insect pests

Alfalfa looper	*Autographa californica*
Armyworm	*Pseudaletia unipuncta*
Beet and other armyworms	*Spodoptera exigua*, other S. spp.
Beet leafhopper	*Circulifer tenellus*
Black cutworm	*Agrotis ipsilon*
Cabbage looper	*Trichoplusia ni*

Colorado potato beetle	*Leptinotarsa decimlineata*
Cotton aphid	*Aphis gossypii*
Darkling beetles	*Blapstinus* spp.
Epilachna beetle	*Epilachna vigintioctopunctata*
Flea beetles	*Epitrix hirtipennis,* other *Epitrix* spp.
Fruit fly	*Drosophila melanogaster*
Garden symphylan	*Scutigerella immaculata*
Green peach aphid	*Myzus persicae*
Leafminers	*Liriomyza* spp.
Mealybug	*Centrococcus insolitus*
Potato aphid	*Macrosiphum euphorbiae*
Potato tuberworm	*Phthorimaea operculella*
Saltmarsh caterpillar	*Estigmene acrea*
Seed corn maggot	*Hylemya platura*
Silverleaf whitefly	*Bemisia argentifolia*
Spider mites	*Tetranychus* spp.
Stink bug	*Euschistus conspersus*
Sweet potato whitefly	*Bemisia tabaci*
Tobacco budworm	*Heliothis virescens*
Tobacco horn worm	*Manduca sexta*
Tomato fruitworm	*Heliothis zea*
Tomato horn worm	*Manduca quinquemaculata*
Tomato pinworm	*Keiferia lycopersicella*
Tomato russet mite	*Aculops lycopersici*
Thrips	*Thrips tabaci, T. palmi, T.* spp.
Variegated cutworm	*Peridroma saucia*
Western flower thrips	*Frankliniella occidentalis*
Whitefly, greenhouse	*Trialeurodes vaporariorum*

Nematodes

Nematode pests includes various root knot nematodes (*Meloidogyne* spp.) and species of cyst, lesion, stubby root, and stem and bulb nematodes.

Resistances to some of the diseases and pests exist in many cultivars and it can be anticipated that additional resistances will continue to be transferred from wild to cultivated species.

Physiological Disorders

The incidence of nonpathogenic fruit disorders is largely determined by cultivar characteristics and usually incited by features of the environment such as moisture stress and temperature extremes. Disorders include blossom end rot, which is exhibited by a necrotic dark brown sunken area at the fruit blossom end; if severe, it can affect the entire fruit. Confusion exists as to the cause or causes of blotchy ripening which appears as a failure of areas of the fruit wall to color uniformly. The tissue tends to remain firm and may be white or yellow and occasionally becomes brown; internal browning is seen as a browning of vascular tissues in the fruit wall; gray wall is observed as areas of discoloration within the fruit wall.

Additional disorders are zipper, which occurs when floral parts adhere to the surface of the fruit as it expands, fruit puffiness because of inadequate pollination resulting in partially filled locules, and catface, also caused by poor pollination, a result of low temperature and low light conditions that is seen as nonuniform locule formation, an asymmetric extension of the locules and the corklike blossom scar. Radial and concentric fruit cracking and sunscald are additional defects. One form of irregular fruit coloring is ascribed as a reaction to a toxin produced by whitefly feeding.

Weed Management and Parasitic Plants

Weed competition is most damaging early in the growing season. In additional to mechanical and hand cultivation, selective herbicides are used extensively to control most nonsolanaceous weeds. Weed management is easier with transplanted crops, as clean cultivation can be practiced before transplants are established. In addition to numerous weed pests, two parasitic plants of concern to tomato producers are dodder (*Cuscuta* spp.) and broomrape *(Orobanche ramosa)*.

Harvest and Postharvest

The time from planting until first fruit harvest is dependent on cultivar and growing conditions and may range from as few as 70 to as long as 125 days.

All tomatoes were hand harvested until about 1965 when mechanical harvest equipment was introduced for processing tomatoes. Presently in the United States, essentially all processing tomatoes are machine harvested, as are a significant volume of mature green fruit for fresh market. The transition to machine harvesting for the processing crop is rapidly occurring in other major tomato-producing countries. Advancements such as electronic color sorting are increasingly incorporated into harvest machinery.

For machine harvest adaptation, tomato plants were specifically bred for a determinate growth habit with small vines that produce a concentrated fruit set of many small, thick-skinned, uniformly ripened fruit. Fruit for processing into paste and sauces are firm and thick walled with few and small locules. Small fruit with these features are less subject to impact and compression damage. Fruit intended for processing as juice tend to have larger locules. An additional important character is that even fully red-colored fruit remain attached to the vine until harvested. Such vine storage is critical for high yields, and although fruit do not abscise, they are easily detached by harvest equipment. Many newly introduced cultivars have a small pedicle attachment, and many others have the jointless character. Machine-harvested fruit are typically handled in bulk, and processed within a day or two; thus, physical injury, unless severe, has little effect on the processed products. When low temperatures slow the rate of ripening, ethephon (2-chloroethylphosphonic acid) at 2000 ppm is sometimes applied to the foliage to accelerate maturation and color development. However, accelerated foliage senescence caused by ethephon can increase fruit susceptibility to sunscald and sunburn.

Fresh market fruit are harvested when physiologically mature, and range from mature green to the full red-colored stage (see Table 23.2). Although of different firmness, both mature green and fully colored fruit are physiologically mature because seed are capable of germinating. Full color and fruit softening are usually highly correlated, and the terms "red ripe" and "full color development" are used interchangeably. Determination of color stage when fruit are harvested depends on how

TABLE 23.2. STAGES OF TOMATO FRUIT RIPENING AND COLOR DEVELOPMENT FOR RED-FRUITED CULTIVARS

Harvest stages	Days from mature green at 20°C	Description
Immature green		Fruit still enlarging, dull green, lacks skin luster. Gel not well formed; seed is easily cut through when fruit is sliced. Immature seed do not germinate, and fruit do not color properly.
Mature green	0	Bright to whitish green; well rounded, skin has waxy gloss. Seed are embedded in gel and are not easily cut when fruit is sliced. Seed are mature and can germinate; fruit will ripen under proper conditions.
Breaker	2	Showing pink color at blossom end; internally the placenta is pinkish.
Turning	4	Pink color extending from blossom end, covering 10–30% of fruit.
Pink	6	Pink to red color covering 30–60% of fruit.
Light red	8	Pink to red color, covering 60–90% of fruit.
Red	10	Red color at least 90% of fruit.

fruit will be handled and used. Fruit to be transported to distant markets or not intended for immediate use are harvested at the mature green or breaker stage, and further color development occurs during holding or transit. Mature green or breaker stage fruit are firm and better able to withstand postharvest handling than fully colored soft fruit. Breaker or pink stage fruit are harvested when extended postharvest life is less essential. Red fruit are harvested when the period for marketing, transport distance, or consumption are relatively short. With regard to edible quality, preference is usually given to fruit harvested at light red to red stages. It is frequently observed that mature green fruit do not have the flavor and aroma of fruit that develop to a red color on the vine.

Fruit ethylene production is associated with color development and this natural hormone is also involved in fruit softening. Ethylene application can be used to accelerate color development and fruit softening when provided at or slightly beyond the mature green stage. Ethylene (100–150 ppm) is usually applied as a gas to the fruit in a ripening room that can be sealed. The best ripening response to ethylene occurs at a temperature of 20–21°C and 85–90% RH for 12–24 h.

Many studies have been made using the ripening inhibitor mutant gene "rin" to limit ethylene production, and the nonripening "nor" gene to prevent ethylene development in order to extend fruit shelf life. In the heterozygous condition, rin behaves as a recessive and fruit are red with only slight softening. Fruit with the nor gene ripen to a light orange red, and their flavor is inferior. Another mutant gene, "alc" (alcobaça) prolongs the storage of harvested fruit, and has less adverse effects on color development. These mutants, have been incorporated into some cultivars that are used in commercial production.

Through research, a gene that initiates ethylene synthesis (ACC synthetase) and tissue softening by activation of the enzyme polygalacturonase (PG) has been identified. The PG gene itself has also been isolated. Based on knowledge of the PG gene and molecular biological research using antisense introgression technology, the PG enzyme can be inhibited. Thus, fruit tissue softening can be retarded even when fruit are fully colored in transgenic plants. This discovery offers the possibility for harvesting fruit at a late stage of color development without incurring an abbreviated shelf life. Cultivars with low PG enzyme and low ethylene synthesis activity have been incorporated into fresh market cultivars.

Fruit deterioration because of excessive softening is a major reason for marketing losses. Rough handling, poorly designed containers, and exposure to hot and dry conditions also contribute to significant losses.

At many fresh market tomato-packaging facilities, fruit are washed in chlorinated water to remove dirt and to limit postharvest diseases. When cold washwater is used, postharvest decay often increases because of contaminated water entering the fruit though the stem scar. However, when the washwater is equal to or warmer than the fruit temperature, contaminated water is not drawn into the fruit.

Fresh market fruit are graded for uniform size and quality before packaging; in some facilities this is accomplished electronically. Occasionally, fruit are waxed to reduce moisture loss and improve appearance.

Storage

Tomatoes can be stored successfully for several weeks, but recommended storage temperatures differ with stage of fruit maturation. When mature green fruit are stored, temperatures should be between 13°C and 18°C and 85–90% RH. At these temperatures, chilling damage does not occur, but color development is slow. The optimum temperature for ripening mature green fruit is between 18°C and 21°C; below 13°C, fruit will not develop a dark red color. Mature green fruit have been stored for 6 weeks at 13°C in a 3% oxygen, 97% nitrogen atmosphere, and upon ripening, there was no noticeable flavor or other quality impairment.

Red fruit have a short shelf life at room temperature but can tolerate storage at lower temperatures than mature green fruit. Firm red fruit can be held at 7–10°C for several days without significant quality losses. For ripe fruit, temperatures less than 7°C will cause chilling damage and the fruit loses firmness, flavor, and shelf life. Chilling injury is cumulative and increases with length and level of low temperature. Red fruit can be stored as long as 3 weeks at 0–1.5°C in acceptable condition. However, fruit should be used within a day or two following removal from storage because flavor and textural quality becomes unacceptable. The usual recommendation for red fruit to maintain quality is to avoid low-temperature exposure.

Glasshouse Production

Tomato production in glasshouses or other protective structures is important in many regions where climate limits field production. Several northern European countries and Japan are especially well known for this kind of production, which often results in higher yield than field production.

Protected production requires cultural procedures that are more intensive than those for field production. The range of growing media used is broad. For example, plants are grown directly in soil, in peat, or sawdust-filled bags or cylinders, on straw bales, or rockwool slabs. It is fairly common for plants to be individually supplied with moisture and nutrients from drip irrigation systems. Gravel, sand, and hydroponic culture are also used, although less frequently than in the past. A relatively new practice is called nutrient film technique (NTF) whereby plants are grown in a shallow trough through which water and nutrients slowly flow. The flow need not be continuous, but is frequent enough to provide for plant nutritional and moisture requirements. An adequate and a balanced nutrition program is very important, especially when roots are confined to the relatively small area. Fertilizers and/or nutrient solutions are frequently and carefully monitored to control content, pH, and possible pathogen contaminants. It is important that adequate aeration of the root zone is also provided.

Propagation is almost exclusively with transplants because growing directly from seed to the transplant stage would require considerably valuable glasshouse growing time and space. Another consideration is the high cost of hybrid seed currently used by many growers. To produce transplants, seed is germinated at 18–20°C. During early plant development, day temperatures are maintained between 15°C and 21°C, and at night between 14°C and 17°C. During flowering and fruit development, day temperatures are elevated and may range from 18°C to 30°C, whereas night temperatures usually are 14–17°C. Night temperatures should not fall below 13°C. Equally important are soil temperatures for which the minimum is 14–15°C.

Energy for heating and lighting are a large part of production costs. Various practices and materials are used to conserve and efficiently use energy. For example, thermal blankets are used to reduce radiation losses at night, and reflective mulches are used to improve light utilization.

Whenever air temperatures are high, it is critical that adequate ventilation be provided to lower temperature and, if necessary, to lower relative humidity. Low light intensity frequently limits production, so light transmission into structures should not be restricted. Inadequate ambient CO_2 concentration can be another limitation to production. If temperature and/or light are not limiting, CO_2 enrichment of the growing environment will often increase photosynthesis. Enrichment is usually two to three times the normal 300 ppm ambient level of CO_2. To meet the many critical needs for indoor production, plant breeders have developed plants capable of growth, fruit set, and development in low temperature and low light conditions.

An enclosed environment such as a glasshouse is not conducive to sufficiently high levels of self-pollination. Therefore, pollination is often assisted by vibration or plant-shaking devices, and occasionally with introduced colonies of solitary (bumble) bees. Midday is the best time for pollination. Fruit set hormones are infrequently used because fruit quality is usually compromised.

Pruning is an important cultural procedure to enhance the ratio of foliage to fruit production and also for greater light penetration, aeration, disease management, and ease of harvesting. Pruning also allows for some regulation of fruit size and flowering. The typical glasshouse crop should have a production period of 5 or more months, during which time, the indeterminate plant types continue to grow and produce fruit. Therefore, vine training is an ongoing process. Various trellis designs maximize plant use of available light. An optimum plant population is 3–3.5 plants/m^2.

Throughout the crop cycle, disease and other pest management is very important because of the possibility of rapid spread within the confined area of the glasshouse or growing structure. Therefore, if ground beds are used, it is prudent to ensure freedom from disease by sterilization. This can be accomplished with steam or chemical fumigants and, in some areas, by solarization. Alternatives to sterilization include the use of sterile soil, or peat or soilless substitutes, such as rockwool.

Insects can be as troublesome as diseases and may require rapid application of control measures. The restricted environment of the glasshouse presents opportunities for biological control, such as the use of parasitic and predator insects. Glasshouse pests such as whitefly, thrips, mites, and aphids are controlled with pesticides and by management that also integrates cultural and biological practices.

Production

The United States is a leader in yield per hectare and total production and is also the leading producer of processing tomatoes. Countries in Asia produced almost 39% of world supply, China being the primary contributor. See Table 23.3. Generally, the volume of tomatoes produced in the humid tropics is limited because of diseases. Production volume between developed and developing countries is fairly close, but yield per hectare differ greatly. With a continuing strong growth in worldwide demand for fresh and processed tomatoes, many countries have steadily increased production to satisfy domestic and international markets.

TABLE 23.3. WORLD PRODUCTION OF TOMATOES, 1994

	Area (ha × 10³)	Yield (t/ha)	Production (t × 10³)
World	2,852	27.2	77,540
Africa	428	19.1	8,315
North and Central America	326	45.7	14,874
South America	157	34.0	5,335
Asia	1,313	23.0	30,205
Europe	618	29.7	18,378
Oceania	11	40.3	433
Leading countries			
United States	190	63.7	12,085
China	344[a]	26.0	8,935[a]
Turkey	160[a]	39.4	6,300
Italy	109	48.1	5,259
India	321[a]	15.7	5,029[a]
Egypt	148[a]	31.1	4,600[a]
Spain	62	49.8	3,066
Brazil	58	43.6	2,550
Iran	75[a]	25.9	1,940[a]
Greece	41[a]	44.1	1,810

[a]Estimated.
Source: 1994 FAO Production Yearbook, Vol. 48, FAO, Rome 1995.

Uses and Composition

The versatility of tomato usage strongly contributes to its popularity. Tomatoes are eaten raw, cooked, and processed. Processed forms include juice, sauce, puree, paste, and dehydrated. Green tomatoes are pickled, candied, and made into preserves. A useful edible oil is extractable from seed.

Although the tomato fruit is about 90% water, it nevertheless is a good source of pro-vitamin A and vitamin C; the content of both increases as fruit mature and develop color while on the vine. Low light limits ascorbic acid content. Fruit-soluble solids are comprised mainly of sugars and organic acids and are very important quality components. Simple sugars, fructose and glucose, increase and malic acid declines during development to full color; sucrose and citric acid levels are lower and remain relatively constant during ripening.

Fruit flavor relies heavily on the content of sugars and acids and is complimented by many flavor-contributing compounds. The ratio of fruit wall to locular volume is an important feature because the fruit wall contributes more to soluble solids than locular materials. However, a high locular volume is associated with high acidity and flavor.

Fruit pH usually ranges between 4.3 and 4.7. Processed fruit with pH values above 4.5 must be acidified to prevent growth of *Clostridium botulinum,* the bacteria-causing botulism. Small quantities of tomatine and other alkaloids are present in immature fruit. Tomatine is enzymatically changed to a nontoxic form as fruit ripen.

PEPPERS, *Capsicum annuum,* L. C. *frutescens,* L., and other *Capsicum* species

Other names: Chile, chillies, aji, pimiento, paprika, capsicum

The most notable feature of peppers is flavor, and whether sweet and mild or strongly pungent, and no matter what local names they are called, peppers are appreciated worldwide. In many countries they are considered an indispensable food. In addition to their flavor contribution, peppers are an excellent source of pro-vitamin A and vitamin C. Medicinal use is also made of peppers, most notably in Africa and by indigenous peoples of Latin America. Some cultivars are used as ornamental plants.

Origin and Taxonomy

Peppers are endemic to tropical and subtropical America. Evidence of early cultivation was found in Peruvian burial sites, and seed remnants dated older than 5000 B.C. were found in caves at Tehuacan, Mexico. Spanish and Portuguese traders were largely responsible for initiating the worldwide dispersal of peppers.

Capsicum annuum is the most widely cultivated and economically important species and includes sweet and pungent fruit of numerous shapes and sizes. Domesticated forms are classified as *C. annuum* var. *annuum;* wild members as *C. annuum* var. *aviculare.* Most likely, the species was domesticated in the region bordering Mexico and Guatemala.

Capsicum frutescens is a semidomesticated species found in the lowlands of tropical America. Additionally, Southeast Asia is recognized as a secondary area of diversity. These are perennial plants with blue anthers, greenish white corollas, and commonly have two or more fruit developed per node. Wide variations in flavor attributes are observed among accessions. "Tabasco" is probably the best known cultivar; it is widely grown in warm temperate as well as tropical regions.

Capsicum chinense domestication was widespread in tropical America, and the species is frequently cultivated in the Amazon regions. Cultivars of this species produce some of the most pungent fruit known. Except for a ringlike constriction at the base of the calyx, the species resembles *C. frutescens* and *C. annuum.* Fruit are smooth and shapes are as diverse as those of *C. annuum,* but often with puckered fruit walls. *C. chinense* has a unique citruslike aroma.

Capsicum baccatum evolution was largely limited to middle South America (Bolivia). The domesticated form is identified as *C. baccatum* var. *pendulum;* the wild forms as *C. baccatum* var. *baccatum* and var. *microcarpum.* Corollas have yellow, tan, or green spots and a prominent toothed calyx. One to two fruit may be borne at a node; the fruit usually are elongated. Fruit flavor differs from *C. annuum* and *C. chinense.*

Capsicum pubescens is grown in Central America and the Andean highlands. Flowers have purple corolla lobes with purple anthers; seed are wrinkled and black. Leaves are pubescent and rugulose; fruit wall tissues are thick. *C. pubescens* is known as "goat chili" or "rocoto." It is adapted to cool temperature growth at elevations of 2000–3000 m in the tropics. No wild ancestral type is known, but the species has an affinity with other wild South American species such as *C. eximium* and *C. cardenasii.*

Fruit are obtained from wild forms of *C. annuum, C. frutescens,* and *C. chininese.* Additional wild species that are utilized include *C. galapogense, C. chacoense, C. tovarii, C. praetermissum, C. eximium,* and *C. cardenasii.* In Bolivia, *C. cardenasii* is a species very often harvested from the wild. In addition, there are at least 15 other described species.

Although taxonomically questionable, many species are often identified according to the previously mentioned flower and fruit features. Interspecies cross-compatibility is complicated because of diverse sterility. Nevertheless, some exchanges are possible, and intermediate types result. The general characteristics of wild species are high pungency, small fruit size, and fruit abscission. Fruit of domesticated species are variable in pungency and do not abscise.

Botany

Peppers are herbaceous plants, most becoming woody at the stem base, and some become shrublike. Similar to tomato and eggplant, peppers are tropical perennials usually grown as annuals. Generally, plants are erect, highly branched, and from 0.5 to 1.5 m in height. Taproots are strong and deep; rooting is generally well developed. The relatively smooth leaves with few trichomes are simple and thin, but of variable size with lamina portions broadly lanceolate to ovate.

Among the different species, corolla colors vary from white to greenish white, and lavender to purple. Anther colors are blue, purple, and yellow, and seed colors are light yellow, tan, or black. The bell-shaped calyx usually enlarges along with the fruit and covers a part or much of the base of the fruit. All domesticated cultivars are self-pollinated, although out-crossings can occur. The number of flowers per node is a species trait.

Botanically, the fruit is an indehiscent, either pendent or erect, many-seeded berry (Fig. 23.4). Fruit are frequently borne singularly at each node for *C. annuum* cultivars, and with multiple fruit (typically two or three) per node for some other species. As fruit develop, the pericarp grows faster than placental tissues, resulting in a cavity. Carpel walls are fused with the placenta at the base of the fruit and may or may not continue that attachment to the tip. As fruit mature, the texture of exterior surfaces become smooth and glossy. Ripening to mature color is accompanied by the accumulation of simple sugars in the pericarp. In contrast to the tomato, placental tissues and seed are dry, and seed are easily detached. Seed of *C. annuum* cultivars are flat, typically pale yellow, ovoid, 3–5 mm long; about 150–160 seed weigh 1 g.

Pepper fruit colors are highly variable: green, yellow, or even purple when young and later turning to red, orange, yellow, or a mixture of these colors with advancing age. Green color is due to chlorophyll, whereas red and yellow due to the presence of carotenoids, and purple due to anthocyanin. Brown color is due to the persistence of chlorophyll concurrent with lycopene and beta-carotene synthesis. In most commercial cultivars, brown is a transient ripening stage.

As with color, fruit shapes vary greatly, ranging from linear, conical, or globose, and all combinations of these shapes. Fruit may be thick

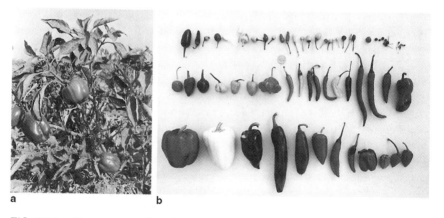

FIG. 23.4. Pepper plant, *Capsicum annuum* (a), and size, color, and shape variations of other *Capsicum annuum* fruit (b). Source: Courtesy Paul Bosland.

or thin walled and range from 1 to more than 30 cm in length, and from 1 to about 15 cm in width. Fruit shapes are erroneously associated with pungency level. Although that association is commonly used to identify sweet and pungent pepper types, it cannot be relied upon.

A frequently used pod type of classification depends on fruit shape. Some examples of *C. annuum* pod types are as follows:

Ancho—Large, heart-shaped, thin-wall fruit with indented stem. Most cultivars are mildly pungent. Fruit are harvested either green, dark brown, or red.

Bell—Fruit are large, blocky, with a blunt end of three or four lobes and thick walled. Most cultivars are sweet and the majority of fruit are harvested as mature green fruit, and some at the red mature color stage. Gold-, yellow-, or orange-colored fruit have been introduced into commercial markets.

Cayenne—Thin wall, tapered, slender and wrinkled, highly pungent, usually harvested when fruit are red.

Cheese—Small- to medium-sized fruit of various shapes (mostly globose) with medium to thick walls. Yellow or green fruit matures to red. Usually not pungent.

Cherry—Small spherical to somewhat flattened fruit, generally thin walled with either sweet or pungent forms, and harvested at either green or red stage.

Chiltepin—Very small, round fruit is egg shaped, thin walled, and highly pungent; often gathered from the wild. Other fruit of this pod type that are oval or somewhat elongated are known as chilipiquin.

Cuban—Fruit have an irregular blunt shape and are mildly pungent. The thin-wall fruit are harvested when yellow green or after becoming red.

Jalapeno—Small, nearly cylindrical with rounded ends. Thick walled, with the outer skin sometimes russeted. These are very pungent, and typically harvested while green.

Long Wax—Most fruit are long and tapered to a point, although some are blunt. Shows tolerance to low and high temperatures. Fruit are harvested at green, yellow, and red stages. Hungarian is a popular and well-known mildly pungent cultivar.

New Mexican—Fruit are long, slender, thin walled with a pointed taper. Moderately pungent, although some cultivars are sweet. Both green and red fruit are harvested. Anaheim is a popular cultivar.

Pimiento (Pimento)—The large, cone- or heart-shaped, thick-walled fruit usually are not pungent and typically harvested when fully red.

Serrano—Elongated, short, nearly blunt, thin walled, and very pungent, usually harvested as green fruit.

Squash—Flattened scallop shape, medium- to thick-walled fruit, usually pungent.

An important example of a *C. chinense* pod type is

Habanero—Small thin, usually puffy-walled fruit, extremely pungent.

An important example of a *C. frutescens* pod type is

Tabasco—The erect, slender, short (3–5 cm), thin-walled, and highly pungent fruit are usually harvested at red stage.

Numerous other pepper pod types are grown; some are recognized as cultivars, many are not, and are identified only by local names.

Culture

Peppers are grown from sea level to elevations of 3000 m; they are frost sensitive and require warm weather and a long growth period to be productive. Mean day temperatures of 20–25°C are ideal; plant growth is improved when night temperatures do not exceed 20°C. Low temperatures tend to limit flavor and color development, and plants and fruit are susceptible to chilling injury.

Peppers are more tolerant to high temperatures than tomatoes. However, flowers are not fertilized at temperatures below 16°C or above 32°C because of poor pollen production. Pollination and fertilization are optimum between 20°C and 25°C. In general, small, fruited cultivars are more tolerant of either high or low temperature extremes. Plants are not photoperiod sensitive. Flowering usually begins between 1 and 2 months after planting, with fruit achieving desired or full size about 1 month after anthesis.

Pepper plants have a moderately extensive root system; taproots can penetrate to a depth greater than 1 m. Crop moisture requirements range from 400 to 1000 mm, which should be uniformly supplied during growth. Although peppers are generally drought resistant, even intermittent periods of moisture and/or nutritional stress can dramatically reduce plant growth and limit fruit size and yield. Lack of moisture during flowering may cause flowers and young fruit to abscise. During fruit development, moisture stress, in association with high temperatures, can increase the incidence of blossom end rot. Peppers should be planted in well-drained soils, because plants are very sensitive to waterlogging. Waterlogged plants tend to defoliate and are subject to root diseases. Peppers are responsive to fertilization and usually the supplied nitrogen is applied preplant, and again prior to first bloom. The most favorable soil pH ranges between 6.5 and 7.0.

Propagation and Plant Spacing

Peppers are seed propagated, seed being sown directly or used to produce transplants. Plants can also be propagated vegetatively, but this practice is rarely used. Seeds germinate in 6–10 days at favorable (30°C) soil temperatures but very slowly at 15°C. Transplants are used when temperatures are too low for rapid seed germination or when early production is desired. Seed priming is a method used to improve germination in cool soil conditions.

Plant density varies with cultivars but frequently provide for 25,000 to 30,000 plants per hectare. Spacings commonly are about 40–50 cm within rows and about 75 cm between rows. Close spacing tend to reduce fruit size, although high densities have the advantage of providing shade that can limit fruit sunburn.

Although less than tomatoes, a significant volume of peppers are grown in glasshouses and other protective structures for the production of off-season as well as high-quality fruit. Night temperatures are not permitted to go below 15°C. Plants are carefully trained and pruned for efficient use of space; usual spacings are 50×90 cm. Well-grown and managed crops can produce yields exceeding 15 kg/m^2.

Harvest and Postharvest

Pepper fruit maturity is not readily determined by appearance alone. Fruit are mature when the seed they contain are capable of germinating. However, fruit size and/or color are often the common determinant as to when harvest occurs. Generally, the surface of older fruit is firmer, shinier, and more waxy than younger fruit. Additionally, the presence of older fruit tends to delay the growth of younger fruit.

Accumulated heat unit measurements are used by some growers for scheduling harvests. Bell-type cultivars are harvested at an early stage while fruit is green, but also at a later stage when fruit color changes. Processed peppers, such as pimientos and paprikas and many chiles, are usually harvested after full red color development.

Fruit are detached by carefully breaking away or by cutting through the pedicel so as to minimize stem breakage. With some cultivars, the pedicel develops an abscission layer. It is important that fruit have an intact pedicel, otherwise the cavity area is subject to desiccation and easily accessible to pathogens.

Mechanical harvesting of peppers is limited because injury to the fresh market product is usually too severe, and when used is limited to the processed crop. Peppers produced for dehydration are allowed to remain attached to the plant until fully colored and desiccated. Field drying reduces the time and energy required for final drying. Total fruit solids increase with fruit age and are usually accompanied by increased carotene and ascorbic acid content. Ethephon is occasionally applied to accelerate fruit coloration, and most often the use is in the production of dry chile and paprika type cultivars.

After harvest, fresh market fruit should be rapidly cooled to about 10°C in high relative humidity. Storage temperatures should be above 10°C because of chilling sensitivity, although a short exposure period to 7°C or 8°C usually is not injurious. High relative humidity is necessary to limit desiccation. Before packing, peppers are handled much like tomatoes and are washed in ambient or warm chlorinated (300 ppm) water to reduce postharvest diseases. Fruit have been waxed to reduce desiccation. However, waxing tends to increase the possible incidence of bacterial soft rot. Shelf life varies among the different pod types. Deterioration is often due to moisture loss; some pod types are very prone to desiccation.

Diseases and Other Pests

Peppers are affected by many of the diseases, viruses, insect, and nematode pests affecting tomatoes. Some specific to peppers include the following:

Anthracnose	*Colletotrichum piperatum, C. capsici*
Bacterial leaf spot	*Xanthamonas campestris* pv. *vesicatoria*
Bacterial soft rot	*Erwinia carotovora*
Cercospora leaf spot	*Cercospora capsici*
Choanephora blight	*Choanephora cucurbitarum*
Downy mildew	*Peronospora tabacina*
Fusarium wilt	*Fusarium oxysporium* pv. *vasinfectum*
Phytophtora blight and root rot	*Phytophthora capsici*
Powdery mildew	*Oidiopsis taurica*

Verticillium wilt   *Verticillium dahliae, V.* spp.

Viruses are often the major disease problem for pepper production. Those affecting peppers include the following:

Alfalfa Mosaic Virus (AMV)
Beet Curly Top Virus (BCTV)
Cucumber Mosaic Virus (CMV)
Pepper Mottle Virus (PeMoV)
Potato Virus X (PVX)
Potato Virus Y (PVY)
Tobacco Etch Virus (TEV)
Tobacco Mosaic Virus (TMV)
Tobacco Ringspot Virus (TRSV)
Tobacco Streak Virus (TSV)
Tomato Mosaic Virus (TomMV)
Tomato Spotted Wilt Virus (TSWV)

Other causes of crop losses are the lesion, stubby root, root knot, and stem and bulb nematodes. Peppers are also susceptible to nonpathogenic disorders such as blossom end rot and sunburn.

Insects attacking pepper plants are commonly those that also attack tomatoes.

Production

Over 48% of the world pepper crop is produced in Asia, with China the leading country. The production in China alone exceeds the entire production of the European countries. Africa's production is also sizable, with more than 15% of the world supply, Nigeria being the major producer. See Table 23.4

Uses and Composition

Peppers are eaten raw in salads, in numerous cooked preparations including salsa, and processed by canning, freezing, pickling, and as dehydrated and powdered condiment products, namely paprika and chili powder. Most European paprikas are mildly pungent; those in Hungary are somewhat more pungent. In the United States, paprika usually is not pungent. Confusion occurs in discussing paprika because paprika is the European name for pepper, and in the United States, paprika is a product of dried powdered nonpungent peppers.

Chili powder is usually comprised of ground, dried, pungent peppers mixed with other spices, such as oregano, cumin, and garlic. Chili

TABLE 23.4. WORLD PRODUCTION OF PEPPERS, 1994

	Area (ha × 10³)	Yield (t/ha)	Production (t × 10³)
World	1,249	9.0	11,192
Africa	213	8.3	1,758
North and Central America	140	11.3	1,586
South America	27	9.2	245
Asia	739	7.3	5,414
Europe	130	16.8	2,189
Oceania	<1	11.0	1
Leading countries			
China	213[a]	14.5	3,077[a]
Turkey	55[a]	18.0	990[a]
Nigeria	90[a]	10.2	920[a]
Mexico	100[a]	8.5	850[a]
Spain	24	31.1	746
United States	26[a]	24.2	630[a]
Indonesia	225[a]	2.0	450[a]
Italy	12[a]	24.2	299
Korea, Rep.	87[a]	3.2	279[a]
Bulgaria	18	11.3	203

[a]Estimated.
Source: 1994 FAO Production Yearbook, Vol. 48, FAO, Rome, 1995.

powders are prepared in different levels of pungency. Various pepper forms, usually chile types, are extensively used in combination with other spices such as turmeric, cumin, and coriander to produce curry powder, the pungency of which depends on the pepper cultivar(s) used. Cayenne powder is a high-pungency condiment produced from dried mature fruit of cayenne-type cultivars.

A single dominant gene controls the presence of fruit pungency. Modifiers of the major gene, and the growth environment influence pungency. Capsaicin ($C_{18}H_{27}NO_3$), the pungent principle in peppers, is contained in the septa and placental tissues, but not in fruit wall or seeds. Capsanthin ($C_{40}H_{58}O_3$) is the most important compound in paprika flavor. Several other compounds, often in minute quantities, that also contribute to flavor have been identified. In general, for a mildly pungent cultivar, the red peppers tend to be milder than green fruit because of a higher sugar content that helps to mask some of the pungency. However, such a difference is not detectable in the fruit of pungent and highly pungent cultivars. Generally, red peppers contain several times more pro-vitamin A than similar green fruit, and about twice as much vitamin C.

Pungency is measured in terms of Scoville units, a subjective organoleptic method that is calculated as the reciprocal of the highest dilution at which pungency can be detected by a taste panel. Capsaicin usually is readily detected at 1 ppm. A Scoville ranking of pepper pungency would place sweet bell peppers and jalapenos, both *C. annuum,* at 0 and 10,000, respectively. Tabasco, *C. frutescens,* often registers 25,000 and habaneros, *C. chinense,* are detected at more than 200,000 Scoville units. Chemical tests using liquid chromatography are more precise than the Scoville scoring.

EGGPLANT, *Solanum melongena* L.

Other names: Brinjal, aubergine, garden egg, Guinea squash

In many parts of the world, eggplant is considered a poor man's food and this may account for is low popularity in some countries. However, in other regions, such as China, India, Japan, and many Mediterranean countries, eggplants are a popular and widely used vegetable. The name "brinjal" is of Indian or Arabic derivation; the English name, "eggplant," was probably derived from plant types bearing white fruit resembling a chicken egg.

Origin and Taxonomy

Eggplant is considered a native to India, where the major domestication of large fruited cultivars occurred, and where wild forms can still be found. A center of diversity is believed to be in the region of Bangladesh and Myanmar, (former India–Burma border). From India, domesticated nonbitter fruit types spread eastward and by the 5th century B.C. was in China, which became a secondary center for domestication of small fruited types. Arabic traders were responsible for subsequent movement to Africa and Spain. Eggplant cultivation in the Mediterranean region is relatively recent.

Eggplant is a member of the *Solanum* genus, of which there are more than 1000 species. Although not widely recognized as botanical varieties of *Solanum melongena,* forms with round, egg-shaped fruit are identified as var. *esculentum.* Those with long slender fruit are identified as var. *serpentinum,* and dwarf plant forms as var. *depressum. Solanum macrocarpon* known as Gboma eggplant is important in West Africa, and Chinese scarlet eggplant, *S. integrifolium,* are grown for food and ornamental uses.

Wild species produce bitter fruit with sharp spines on most plant parts including the fruit calyx. The bitterness is due to glycoalkaloids, but most domesticated cultivars are not bitter and have few or no spines. Fruit tissues contain high levels of phenolics and, when cut or damaged, are quickly oxidized by polyphenol oxidase resulting in a dark brown discoloration of the flesh.

Botany

Eggplants are short-lived perennials in the tropics, and are cultivated as annuals in temperate zones. Plants grow as a bush to a height from 0.5–2.5 m, have an indeterminate growth habit, and depending on cultivar, produce few to many fruit. The bushy growth habit arises from the production of new shoots in leaf axils. Taproots are strong and moderately deep and spreading. Stems are erect and branching, and quickly become woody; some have spines. Leaves are generally large, alternate, and simple and have a dense grayish wooly covering of the under surface, particularly in wild types. Lamina are ovate to ovate-oblong with wavy lobed margins; the base is usually rounded and the apex acute.

Perfect flowers, solitary or multiple in a cyme inflorescence, are usually formed opposite or nearly opposite leaves rather than in leaf axils. Flowers are 2–3 cm in diameter with purplish pubescent corollas and are mainly self-pollinated; out-crossing can occur but is generally low. Flowers may be open for 2 or 3 days and are most receptive in the morning. The calyx is deeply lobed and toothed. The fruit is a large, pendent berry without a cavity. Fruit are round, pear shaped, oblong, or elongated, with lengths ranging from 4 or 5 cm to more than 30 cm (Fig. 23.5). Skin surfaces are smooth and usually shiny. Fruit colors, whether solid or streaked, are varied and may be white, yellow, green, red, purple, black, or mixtures of these colors. Seed are small and light

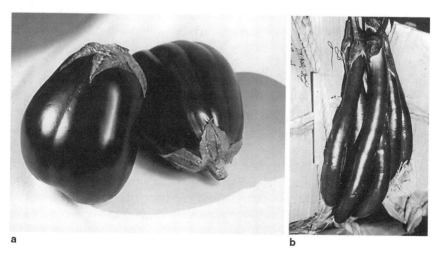

a   b

FIG. 23.5. Eggplants, *Solanum melongena,* globe type (a) and long Japanese type (b).

brown, and imbedded in the placental tissues; about 225 weigh 1 g. Hybrid cultivars are increasingly used because their productivity is generally better than many open-pollinated cultivars. Recently, parthenocarpic cultivars have been introduced.

Culture

Eggplants are well adapted to tropical conditions and midtemperate regions that provide a long period of continuous warm weather throughout growth. Eggplants benefit more from warm temperatures and are more sensitive to low temperatures than either tomatoes or peppers. Favorable daytime temperatures are between 22°C and 30°C and optimal when coupled with night temperatures that are warm, preferably between 18°C and 24°C. At temperatures less than 17°C or greater than 35°C, growth is negligible and pollen dysfunction increases. Cultivars producing elongated fruit tend to be more resistant to high temperature extremes than producing small egg- or oval-shaped fruit. Flowering is considered day neutral. Early cultivars may begin flowering as soon as six leaves have developed, whereas others may not flower until more than 14 or 15 leaves have formed.

Roots are moderately deep and extensive. Most soils are satisfactory unless they impede root development. Because eggplants are sensitive to waterlogging, poorly drained soils should be avoided. Eggplants have better drought tolerance than tomato or pepper, but for high yield, plants require an adequate supply of moisture. Typical crop moisture consumption is about 900–1000 mm. In coarse textured soils, fruit production tends to occur early. A soil pH between 5.5 and 7.5 is preferred. Eggplants have a fairly high nutrient demand and supplemental fertilization is commonly provided.

Propagation

Propagation is either by direct seeding or transplants. Optimum temperatures for seed germination are 24–32°C. Below 15°C and above 35°C, germination is very poor. Bare root or plug seedlings at the two or three true-leaf stage are used for transplanting. Stem layering, in which portions of an attached stem are covered with soil, is a propagating method occasionally used. Adventitious roots form at the nodes, and once rooted, stems are cut free and can be used as transplants. For production in severely disease-infested soils, eggplants can be grafted to resistant rootstocks of *Solanum torvum* or *S. integrifolium*.

Cultivars and cultural practices determine plant spacing requirements. Common field spacings for most nonsupported crops vary from 20 to 30 cm within rows and about 90 cm between rows. Cultivars having large bushy growth are spaced further apart than dwarf types; trellised plantings are also given more space. Trellis plantings are also spaced 20–30 cm within rows, and 100–120 cm between rows.

Harvest and Postharvest

Generally, 3–4 months of growth after seed germination are required before harvest. The period from anthesis to harvest maturity varies among cultivars and temperatures and may be as little as 10 days or as long as 40. Under favorable conditions, flowering and fruit production is continuous.

For best edible quality, fruit are consumed while immature and before seed have enlarged. For example, fruit of elongated-type cultivars can be harvested when they achieve about one-half the size of fully matured fruit. Market or edible maturity of some eggplant types can be determined by pressing a thumb against the side of the fruit. If the indentation retracts, the fruit is immature; if not, the fruit is likely to be physiologically mature. Lack of surface glossiness is another indication of maturity. The flesh of fully mature or overmature fruit begins to dry, becomes bitter and pithy, and the seed darken and become hard. Delayed removal or failure to remove mature fruit reduces flowering and subsequent fruit production.

Eggplants are hand harvested and in the process, the fruit should be cut free rather than torn from the plant, otherwise damage to the plant and/or fruit is likely. Fruit are very susceptible to postharvest physical injury; the damage is often readily apparent. They can be stored in good marketable condition for 7–10 days at 10–15°C and 95% RH. Under low humidity, fruit rapidly desiccate and appear wrinkled and the flesh becomes spongy. Fruit are chilling sensitive and storage below 10°C causes injury, often exhibited as surface pitting and/or decay. Exposure to ethylene or ethylene-producing commodities should be avoided.

Diseases and Other Pests

In addition to some of the diseases and pests affecting other Solanaceae species listed earlier, eggplants are also specifically affected by the following diseases:

Anthracnose fruit rot	*Colletotrichum melongenae*
Cercospora spot	*Cercospora melongenae*
Fusarium wilt	*Fusarium oxysporum* f. sp. *melongenae*
Leaf spot/fruit scab	*Alternaria melongenae,*

	A. tenvis
Phomopsis blight	*Phomopsis vexans*
Phytophthora fruit rot	*Phytophthora parasitica*
Powdery mildew	*Erysiphe polyphaga*

An additional important virus and mycoplasm disease affecting egg-plants are Brinjal mosaic virus (MBV) and little leaf mycoplasm.

Insect pests interfering with eggplant production are eggplant lace bug, aphids, spider mites, leafhoppers, flea beetles, Colorado potato beetle, and fruit borers. Termites and nematodes are additional pests.

Production

Clearly, eggplants are important in Asia where more than 86% of reported world production occurs. Africa and Europe have almost equal shares, each accounting for about 6% of the world supply. Individually, China clearly dominates eggplant production, producing about 60% of world supply. See Table 23.5. However, FAO statistics are not reported for India, which is known to be a major producer of eggplants.

Uses and Composition

Eggplants are prepared for consumption in many ways: boiled, fried, baked, and stuffed; but it is seldom eaten raw. Eggplants provide a

TABLE 23.5.   WORLD PRODUCTION OF EGGPLANT, 1994

	Area (ha × 10³)	Yield (t/ha)	Production (t × 10³)
World	556	16.2	8,979
Africa	27	19.2	527
North and Central America	3	25.0	86
South America	<1	12.4	5
Asia	501	15.5	7,791
Europe	23	24.8	570
Leading countries			
China	326[a]	16.6	5,421[a]
Turkey	35[a]	21.7	760
Japan	14[a]	31.8	430[a]
Egypt	17[a]	20.9	345[a]
Italy	10	26.5	274
Indonesia	55[a]	3.2	175[a]
Iraq	11[a]	14.8	155[a]
Syria	7[a]	21.4	150[a]
Spain	4[a]	32.5	130[a]
Philippines	18[a]	6.4	112[a]

[a]Estimated.
Source: 1994 FAO Production Yearbook, Vol. 48, FAO, Rome, 1995.

relatively low caloric contribution to diets, about 25 calories per 100 g fresh weight, and dietary protein of slightly more than 1%.

OTHER SOLANACEOUS VEGETABLES

CHINESE LANTERN/STRAWBERRY TOMATO, *Physalis alkekengi* L.

This perennial plant is grown mainly as an ornamental. The edible red fruit enclosed by the bright orange or red calyx is called "hozuki" in Japan. Fruit size ranges up to 5 cm (Fig. 23.6a).

a   c

FIG. 23.6. Fruits of different *Physalis* species. (a) Chinese lantern, *Physalis alkekengi,* (b) Husk tomato, *Physalis ixocarpa.* Note: fruit and surrounding husk; (c) Dwarf Cape Gooseberry, *Physalis pruinosa.*

HUSK TOMATO/TOMATILLO, *Physalis ixocarpa* Btoy. ex Hornem.

Tomatillos are of Mexican origin. Plants are annuals and semiprostrate but can reach a height of 1 m. Leaves are glabrous and ovate. Flowers, usually self-incompatible, are bright yellow and 2–3 cm in diameter. The almost spherical fruit is a green- or purple-colored berry with a sticky surface and enclosed in an enlarged modified calyx (husk). Fig. 23.6b The enclosing parchmentlike husk grows with the fruit but its growth slows near maturity, resulting in a tear or bursting of the husk while the fruit continues to enlarge. At maturity, fruit are from 3 to 5 cm in diameter and yellowish green. Immature and small fruit are preferred because seed are undeveloped and fruit are more acid and flavorful. Mature fruit tend to become sweet and less acid. Mashed tomatillo fruit is the major ingredient in many sauces (salsa verde) and are widely used in other ways.

CAPE GOOSEBERRY, *Physalis peruviana* L.

Other name: Uchuba

A native of the Andean region, cape gooseberry is a perennial, widely grown throughout tropical America, and acquired its cape gooseberry name when introduced and grown in South Africa. A bushy perennial about 1 m tall, cape gooseberry is a stronger plant than tomatillo. Plants are tetraploids, whereas other cultivated *Physalis* species are diploids. Leaves are large and hairy; flowers are pale yellow with greenish blue anthers. The smooth, paper-thin calyx that encloses the fruit is five angled and grows at a faster rate than the fruit, and so the greenish yellow spherical fruit, about 1 cm in diameter, is fully enclosed; mature fruit abscise from stems. Fruit are eaten raw or preserved as pickles, and are also used for jams and many kinds of sauces.

DWARF CAPE GOOSEBERRY, *Physalis pruinosa*
GROUND CHERRY, *Physalis pubescens*

Dwarf cape gooseberry and ground cherry plants are annuals; stem growth is short and prostrate, and flowers are small and white. Some authorities indicate these have a North American origin. However, because of some similarities, other authorities consider these plants to be similar to *P. peruviana,* which has a tropical South American origin. Ground cherry differs from dwarf cape gooseberry in having more slender and less pubescent stems. The fruit of each is a smooth, spherical, small, sweet, yellow berry with many seeds. The calyx remains inflated and intact but is easily removed. The fruit are used in ways similar to the other *Physalis* species and are eaten raw or cooked.

TOMATO EGGPLANT, *Solanum integrifolium* Poir.

Tomato eggplant, also called tomato-fruited eggplant and Chinese scarlet eggplant is an annual with spiny and pubescent stems growing to about 1 m in height. Leaves are ovate to oblong-ovate, sinuately toothed at the margins and with a prickly midrib about 25 cm long. The flower has a white corolla about 2 cm in diameter. Edible fruit are scarlet or yellow, ribbed, and globose, and about 10 cm in diameter; fruit are also used as ornamentals.

PEPINO, *Solanum muricatum* Aiton

Pepino is a perennial native to Ecuador and Peru and well adapted to high elevations; wild forms are not known. Domesticated plants are erect to bushy shrubs, 0.5–1.0 m tall, usually without spines, and bear clusters of blue flowers. The fruit, attached to a long stalk, are from 7–20 cm long and either round, ovoid, or elongated. Colors are yellow or light green with purple or red spotting or streaks.

Plants produce fruit within 5–6 months after planting and continue to bear for several years before replanting is required. Successful fertilization of flowers requires cross-pollination, although self-compatibility is common. Fruit often are seedless and plants are commonly propagated with cuttings.

The fruit are aromatic and juicy and slightly acid with either a sweet muskmelon or cucumber flavor, depending on maturity at harvest (Fig. 23.7). Because of a firm, crisp texture, the fruit withstands handling and long-distance shipping. Vitamin C content is generally high. Immature fruit are eaten cooked; mature fruit is eaten as a dessert food.

NARANJILLA, *Solanum quitoense* Lamarck
Other name: Lulo

Native to Ecuador and southern Colombia, naranjilla is grown throughout central and northern South America. Naranjilla is well

FIG. 23.7. Pepino fruit, *Solanum muricatum.*

adapted to tropical high elevations where temperatures are cool, and at other locations growers provide shading to keep temperatures below 27°C or 28°C. Moisture requirements during growth are relatively high.

The spineless semishrub perennials are 2–4 m tall and have large, dark green leaves with purple leaf veins; other plant parts have purple pubescence. Flowers have a white corolla. Fruit are globular, about 5 cm in diameter, and bright orange when mature. They are covered with short, soft, white, wooly hairs which are easily rubbed off. The gelatinous yellowish green pulp is acidic and contains many white seeds. The fresh fruit tastes like a mixture of orange, pineapple, and tomato and is popular for juice, preserves, and flavoring deserts.

GARDEN HUCKLEBERRY, *Solanum melanocerasum,* All.
(*S. nigrum* var. *guineense*)
Other name: Wonderberry

Garden huckleberry is endemic to North America and widely distributed within temperate and tropical regions. Plants have good low-temperature tolerance, resemble peppers, and have clusters of small white flowers. Fruit are small (10–15 mm) green berries which turn black when mature. The interior greenish pulp contains pale yellow seeds. Propagation is usually with seed, although cuttings can be used. Fruit have a hard interior and small seed cavity and are generally cooked; they are also used in preserves and pies. Leaves are sometimes eaten as potherbs; however, caution is suggested because immature berries and leaves may be poisonous.

CUBIU/COCONA, *Solanum sessiliflorum*

A native of northern South America and the upper western Amazon region, cubiu grows best in exposed sunny areas on well-drained soils; its moisture requirements are high. The spineless branched shrub produces large leaves (25 × 25 cm) grows to a height of 1.5 m and is seed propagated. Although smaller, plants resemble those of naranjilla.

Stems are covered with dense, white, wooly hairs. Flowers are white or greenish white. The almost spherical fruit (2–8 cm in diameter) develop in compact clusters; they are covered with short hairs that become yellow to orange-red when mature. The creamlike pulp is acidic and contains many seed. Fruit are used for juice, jam, pie filling, and sauces. Despite a high nutritional value, especially for iron and niacin, cubiu cultivation is limited.

LULITA, *Solanum hirsutissimum*

Lulita is grown in the drier regions of central and northern South America. It is a highly branched, exceptionally spiny plant, producing small (3–4 cm) juicy aromatic fruit. The pulp is acid and seedy. Fruit are covered with a fuzzy covering which is brushed off or peeled before eating. Lulita is eaten either fresh or cooked.

TURKEYBERRY, *Solanum torvum Swartz*

Other names: Devil's fig, takokak

Grown as a shrub or tree in the South American tropics, the immature fruit of turkeyberry are utilized as a vegetable and for curry preparation. Roots are sometimes used as rootstocks for tomato and eggplant cultivation and also have some medicinal use.

FIG. 23.8. Jilo fruit, *Solanum gilo*.

JILO/GARDEN EGG, *Solanum gilo* Raddi

Jilo differs from South American *Solanum* species because it apparently is native to central Africa. In Nigeria, shoots and roots are used in flavoring soup, and the bitter immature fruit are used as cooked vegetables or for seasoning. Jilo was introduced into South America and is most popular in Brazil. The first fruit can be harvested three or four months after planting. Fruit are oval or nearly round and about 5 cm in diameter (Fig. 23.8). Green when immature, fruit become orange-red when ripe and are quite bitter.

AFRICAN EGGPLANT, *Solanum macrocarpon* L.

Solanum macrocarpon is a native of the west African coastal region. It is of some importance, especially in home gardens in the Ivory Coast, the surrounding countries, and Madagascar. The small fruit are similar to eggplant; leaves are also eaten as potherbs.

CUT EGGPLANT/MOCK TOMATO, *Solanum aethiopicum* L.

Grown as a short, 0.3–0.6 m shrub in tropical Africa, the leaves of cut eggplant are used as potherbs and the immature fruit are cooked as vegetables.

SODOM APPLE, *Solanum incanum* L.

Native of tropical India, sodom apple is grown to some extent in west Africa and in portions of the Asian subcontinent. The 1–3-m tall shrub is grown for the immature fruit and is used as a vegetable. The species is related to and will hybridize with eggplant. Ripe fruit are frequently used for medicinal purposes.

INDIAN NIGHTSHADE, *Solanum indicum* L.

Other name: Tiberato

Indian nightshade is a 0.5–1.8-m-tall shrub grown in southeast Asia for its fruit that is used as a vegetable, and for seasoning in curries; roots also have medicinal use.

GLOSSY NIGHTSHADE, *Solanum americanum* Mill.

Glossy nightshade is a noncultivated annual native to central and eastern United States. Occasionally, use is made of its edible leaves and green fruit.

TREE TOMATO/TAMARILLO, *Cyphomandra betacea* (Cav.) Sendtner

A native of Peru, tree tomatoes are 2–6-m-tall shrubs or small trees widely cultivated in Andean regions (Fig. 23.9a). Considerable production also occurs in frost-free areas of New Zealand and to a limited extent in other countries. Plantings are usually replaced after 5 or 6

FIG. 23.9.   Tree tomato, *Cyphomandra betacea,* plants (a) and (b) and fruit (c).

years of production with nursery transplants produced from seed or from rooted cuttings. Plants developing from cutting tend to have a bushy growth habit, those from seed are more treelike.

Tree tomatoes have large heart-shaped hairy leaves (Fig. 23.9b), and the pinkish flowers have long, narrow, corolla lobes. Fruit are usually produced in the second year after planting. About 21–24 weeks after anthesis, fruit are considered sufficiently mature for harvest. The smooth fruit is oval shaped, much like a hen's egg, and attached to a long peduncle (Fig. 23.9c). Ripe fruit are sweet and acidic with a slight tomatolike flavor. Fruit skin and flesh are red, yellow, and purple or combinations of these colors. The hard seed are either light tan or black. When mature, fruit can be eaten raw, but are mostly prepared stewed, juiced, and in preserves. Pro-vitamin A and vitamin C content is high in ripe fruit.

Shelf life at ambient temperatures is about a week. With special treatment, a 10 minute water dip at 50 C and wax covering, fruit can be stored at 3 to 4 C for 2 to 3 months.

CHINESE BOX THORN, *Lycium chinense* Miller

Chinese box thorn, believed to be native to China and Japan, was domesticated in temperate eastern Asia. It is cultivated in much of Southeast Asia and also portions of Europe. The perennial plant is a glabrous, thick-stemmed shrub with a semiprostrate branching growth habit and sometime has thorns. Plants are deciduous and range in height from 1 to 3 m. Young tender ovate to ovate-lancelate shaped leaves and shoot tips, up to 30 cm in length, are eaten as lightly cooked vegetables and for flavoring. Yellow-throated, small, reddish purple flowers produce elongated orange-red berries that are 2–3 cm long, which are not consumed.

Box thorn is adapted to a fairly wide range of temperature and moisture conditions and has a preference for high light intensity. Plants are usually propagated by cuttings although seed can be used. Common in-row spacings are about 30 cm with 50 cm between rows. Plants are often pruned back so as not to exceed 60 cm in height. Harvests commence about 2 months after planting, and sparingly during the first year of growth, but in subsequent years harvest of shoots is often biweekly.

SELECTED REFERENCES

Alcazar-Esquinas, J.T. 1981. Genetic Resources of Tomatoes and Wild Relatives. International Board for Plant Genetic Resources (IPGRI), Rome, Italy.

Andrews, J. 1984. Peppers: The Domesticated Capsicums. University of Texas Press, Austin, Texas.

Atherton, J.G., and Rudich, J., eds. 1986. The Tomato Crop—A Scientific Basis for Improvement. Chapman Hall, London.

Bianco, V.V., and Pimpini, F., eds. 1990. Orticoltura. Patron Editore, Bologna, Italy.

Bieche, B., ed. 1990. Third International Symposium on Processing Tomatoes, Avignon, France. Acta Hortic. No. 277.

Cantwell, M., Flores-Minutti, J., and Trejo-Gonzalez, A. 1992. Developmental changes and postharvest physiology of tomatillo fruits, (*Physalis ixocarpa* Brot.). Sci. Hortic. *50*, 59–70.

Dumas, Y. 1990. Tomatoes for processing in 90's: Nutrition and crop fertilization. Acta Hortic. *277*, 155–166.

El-Beltagy, A.S., and Persson, A.R., eds. 1986. Symposium on Tomato Production in Arid Land. Acta Hortic. No. 190.

Eshbaugh, W.H., Guttman, S.I., and McLeod, M.J., 1983. The origin and evolution of domesticated Capsicum species. Ethnobiology *3*, 49–54.

Heiser, C.B. 1969. Love apples. In Nightshades: The Paradoxical Plants. Freeman, San Francisco.

Huffman, V.L., Schadle, E.R., Villalon, B., and Burns, E.E. 1978. Volatile components and pungency in fresh and processed Jalapeno peppers. J. Food Sci. 43:1809–1811.

Kim, K.Y., Takahashi, K., and Nagaoka, M. 1982. Effect of light intensity, night temperature and CO_2 concentration on the growth, yield and quality of glasshouse tomato. J. Korean Soc. Hort. Sci. 23, 93–108.

Kopeliovitch, E., Rabinowitch, H.D., Mizrabi, Y., and Keder, N. 1979. The potential of ripening mutants for extending storage life of tomato fruit. Euphytica 28, 99–104.

Maissoneuve, B. 1982. Effect d'un traitment a basses temperatures en conditions controlee sur la qualite du pollen de tomate (Lycopersicon esculentum Mill.). Agronomie 2, 755–764.

McLeod, M.J., Guttman, S.I., and Eshbaugh, W.H. 1982. Early evolution of chili peppers (Capsicum). Econ. Bot. 36, 361–368.

Nevins, D.J., and Jones, R.A., eds. 1987. Proc. Symposium on Tomato Biotechnology. Univ. California, Davis, Aug. 20–22, 1986. Alan R. Liss, Inc., New York.

Nothman, J., Rylski, I., and Spigelman, M. 1979. Flowering pattern, fruit growth and color development of eggplant during the cool season in a subtropical climate Sci. Hortic. 11, 217–222.

O'Brien, M. 1980. Tomato harvesting, post-harvest handling and transportation. Acta Hortic. 100, 239–249.

Picha, D.H. 1987. Sugar and organic acid content of cherry tomato fruit at different ripening stages. HortScience 22, 94–96.

Quiros, C.E. 1985. Overview of the genetics and breeding of husk tomato. HortScience 19, 872–874.

Rick, C.M. 1978. The tomato. Sci. Am. 239, 76–87.

Rick, C.M., DeVerna, J.W., Chetelat, R.T., and Stevens, M.A. 1987. Potential contributions of wide crosses to improvement of processing tomatoes. Acta Hortic. 200, 45–55.

Rush, D.W., and Epstein, E. 1981. Breeding and selection for salt tolerance by the incorporation of wild germplasm into a domestic tomato. J. ASHS 106, 699–704.

Smith, P.G., Villalon, B., and Villa, P.L. 1987. Horticultural classification of peppers grown in the United States. HortScience 22, 11–13.

Tanksley, S.D., and Hewitt, J.D. 1988. Use of molecular markers in breeding for soluble solids content in tomato—a re-examination. Theor. Appl. Genet. 75, 811–823.

Tigchelaar, E.C. 1987. Genetic improvement of tomato nutritional quality. In Horticulture and Human Health. B. Quebedeux and F.A. Bliss, eds. Prentice-Hall. Englewood Cliffs, NJ, pp. 185–190.

Tiwari, R.N., and Choudhury, B. 1986. Solanaceous crops. Tomato. In Vegetable Crops in India. T.K. Bose and M.G. Som, eds. Naya Prokash, Calcutta. pp. 248–292.

Villareal, R.L. 1980. Tomatoes in the Tropics. Westview Press, Boulder, CO.

Warnock, S. 1974. Tomato development in California in relation to heat unit accumulation. HortScience 8, 487–488.

Watterson, J. 1985. Tomato Diseases: A Practical Guide for Seedsmen, Growers and Agricultural Advisors. Petoseeds Co., Saticoy, CA.

Wittwer, S.H., and Honma, S. 1979. Greenhouse Tomatoes, Lettuce and Cucumbers. Michigan State University Press, East Lansing, MI.

24

Cucumber, Melons, Watermelons, Squash, and Other Cucurbits

Family: Cucurbitaceae

CUCURBITACEAE

Various species of Cucurbitaceae have served humans as important foods and as many useful products for more than 10,000 years. In particular, species of *Cucurbita* were a significant source of nutrition for pre-Columbian populations in the Americas. The Cucurbitaceae are largely tropical in origin with different genera originating in Africa, tropical America, and Southeast Asia. Besides the tropics, cultivation of cucurbits is extensive in many temperate regions having long periods of warm weather. Species of important vegetable cucurbits are frost sensitive, although some tolerate low temperatures better than others, and some are adapted to xerophytic conditions.

Taxonomy

The Cucurbitaceae family consists of about 120 genera and more than 800 species. Cultivated cucurbits are found in two major tribes: the Cucurbiteae and Sicyoideae. Genera of *Cucurbita, Cyclanthera,* and *Sechium* are of New World origin; others originated in the African or Asian tropics.

Major cultivated species of Cucurbitaceae are as follows:

Genus and Species	Common Name
Benincasa hispida	Chinese winter melon or wax gourd
Citrullus lanatus	Watermelon
Coccinia grandis	Ivy gourd
Cucumeropsis mannii	White-seeded melon
Cucumis sativus	Cucumber

Cucumis anguria	West Indian gherkin
Cucumis melo	Muskmelon, other melons
Cucurbita pepo	Summer and winter squashes, pumpkins and gourds
Cucurbita maxima	Winter squashes, pumpkins
Cucurbita moschata	Winter squashes, pumpkins
Cucurbita argyrosperma (*C. mixta*)	Winter squashes, pumpkins
Cucurbita ficifolia	Figleaf gourd
Cucurbita foetidissima	Buffalo gourd*
Cyclanthera pedata	Caihua or wild cucumber
Hodgsonia macrocarpa	Chinese lard fruit
Lagenaria siceraria	Bottle/white-flowered gourd
Luffa aegyptiaca (*L. cylindrica*)	Smooth loofah
Luffa acutangula	Angular loofah
Momordica charantia	Bitter melon, balsam pear
Momordica cochinchinensis	Sweet gourd
Sechium edule	Chayote
Telfairia occidentalis	Fluted gourd or pumpkin
Telfairia pedata	Oyster nut
Trichosanthes cucumerina var. anguina	Snake gourd
Trichosanthes dioica	Pointed gourd

Botany

Plants of the Cucurbitaceae generally develop a strong, fairly long taproot and a highly branched network of lesser shallow roots that thoroughly explore the soil. Horizontal root extension is often equivalent to that of aboveground growth. For some species, taproots can achieve depths of 1–2 m, and some perennial species such as chayote and buffalo gourd develop large storage roots. In moist environments, adventitious roots develop at the nodes of some *Cucurbita* species.

Trailing vine growth with nodal branching is typical for many species. Branch lengths vary, and for some species can extend to 15 m. Vine growth usually is rapid and may be climbing or prostrate. Usually stems are not woody, although hard stems do occur in some species. Stems often have stiff bristlelike surface hairs, although some are glabrous. Some species have stiff hairs on leaf surfaces. Leaves of most cucurbits are simple, alternate, and palmately lobed. Among species,

*See Chapter 14.

foliage will differ as to size, number, shape, and depth of lobes. Tendrils are borne in leaf axils and may be simple, highly spiraled, or branched. Tendrils are usually absent in bush-type squashes. Bush forms generally exhibit an earlier and more determinate-like fruiting habit than vining forms of the same species. Because bush-type plants can be grown at closer spacings, their harvest index and yield can be greater than vining types.

Flower morphology is often similar within a genus, but among the different genera of Cucurbitaceae, flowers vary considerably with regard to size and color. For example, those of *Citrullus* are smaller and less showy than the large bright yellow flowers of *Cucurbita* or the large showy white flowers of *Lagenaria* or *Trichosanthes* species. Inflorescence types also varies; they can be axillary with flowers either solitary, clustered, or in racemes. Pistillate and staminate flowers may or may not originate at the same node. Commonly, plants are monoecious, although other forms of sex expression occur or can be induced. Various forms of sex expression include the following:

Monoecious: staminate and pistillate flowers on the same plant.
Hermaphrodite: flowers have functional pistillate and staminate organs.
Andromonoecious: some flowers hermaphrodite, some staminate.
Gynomonoecious: some hermaphrodite flowers and some pistillate.
Trimonoecious: plants contain three flower forms: hermaphrodite, pistillate, and staminate.
Gynoecious: all flowers are pistillate.
Androecious: all flowers are staminate.
Dioecious: unisexual flowers formed on separate plants.

A usual flowering sequence for many cultivated cucurbits is first the production of staminate flowers, followed by pistillate and staminate flower formation in a imprecise order. This cycle of flowering tends to be repeated along the lateral branches. Cultivars differ as to when the first female flower node occurs, but within cultivars, this trait is relatively constant. Development of the first female flower influences early yield. The ratio of pistillate to staminate flowers is usually heavily skewed in favor of maleness. Ratios of 25 male flowers to 1 female flower can occur for some species. Although sex expression is inherited, environmental conditions can cause changes in the order of development and ratio of floral gender. Sex expression is associated with the presence and concentration of growth substances in the area of developing flower buds. Environmental conditions can affect plant hormone levels. For example, high temperature and long days usually favor staminate flower production, whereas low temperature and short days tend to favor pistillate flowers. Accordingly, sex expression can also be modified by application of growth regulators.

Ethylene promotes femaleness. Gibberellic acid promotes maleness, GA_{4+7} being most effective. Silver nitrate and silver thiosulfate also induce staminate flowering. Treatments are most effective when made at the two- to four-leaf stage of growth.

Fertilized and developing fruit have an inhibitory influence on the frequency and number of subsequent flowering, particularly with regard to pistillate flowers and also affects vegetative growth. Varying with species and cultivars, plants tend to establish a balance with respect to flowers produced, fruit set, and vegetative growth. Frequent harvest of immature fruit tends to extend flower production. This is a common observation in summer squash and cucumber production. Parthenocarpic fruit appear to have little influence on subsequent plant growth or flowering.

Staminate and pistillate flowers may or may not be of similar size, although both usually contain nectaries that attract insect and other pollinators. Bees, which are very capable of transferring the heavy sticky pollen of many cucurbits, are the most common pollinators. For many species, flowers are receptive to pollination for only a short period, frequently limited to mornings or early afternoons. Flowers of some species remain open during the night or longer than 1 day. When inadequately pollinated, flowers abort or resulting fruit are underdeveloped and misshapen.

Although stem tips and flowers of some species are used for food, the primary edible product of most cucurbits is the fruit. Botanically, most cucurbit fruit are fleshy berries or hard-rinded pepos. Extreme variation occurs in fruit shape, color, and size; the world's largest fruit are cucurbits. However, seed, storage roots, and foliage and flowers of some species are also consumed. Cucurbit seed do not have an endosperm, but the large cotyledons contain considerable amounts of starch and fat and are the food source for the embryo. Seed of several cucurbits are used for human consumption, and seed proteins are often comparable in nutritive value to legume seeds. Storage roots of some species also have a high carbohydrate content.

Many cucurbit species contain a class of bitter terpenoid compounds called cucurbitacins, which are found in varying concentrations during growth in different plant parts. Cucurbitacin content usually is highest near the fruit epidermis and stem end, and gradually diminishes toward the blossom end; it is also present in roots and the foliage contains the least. Bitterness is due to a single dominant gene, although several modifying genes are known. A single recessive gene was discovered that prevents the biosynthesis of cucurbitacin that make cucumber fruit bitter. It is interesting that high foliar cucurbitacin content confers resistance to insects such as aphids, but susceptibility to cucumber beetle.

Culture

Warm temperatures (25–30°C) are optimal for the growth of most cucurbits. Several species grow successfully at temperatures that range between 15°C and 20°C. For most, growth is limited by temperatures less than 10°C, and all plants are frost sensitive. In general, cucurbits require a long warm season to achieve crop or reproductive maturity. Some crops can be considered to have a relatively short growth period if harvested at an immature stage.

Soil moisture requirements vary greatly. The majority require a continuous and uniform supply, but others have excellent drought tolerance; all are intolerant of cold and poorly drained soils. Sandy or silt loam soils with a pH range between slightly acid to slightly alkaline are the most suitable.

Most cucurbits are seed propagated and are usually sown directly in the field. Seed germination is most rapid at 30–35°C. Cold soils delay germination and predispose seed to decay; some species are especially susceptible. Transplants are occasionally used for early production, for disease avoidance with grafted plants, to maximize use of expensive hybrid seed, and for other reasons.

Row plantings are made on flat surfaces, raised beds, or mounds. Usually an excess of seed are sown, and after establishment, seedlings are thinned. Depending on the species and cultivars, vining types may be untrained, trained, or trellis supported. Within-row and between-row spacings are generally wide for vining forms, and much less for bush-type cultivars.

Cucurbits are grown in a wide range of soil types or varying levels of fertility. Generally, roots are effective in foraging for crop nutrients; thus for cucurbit production, less nitrogen is necessary compared to many other vegetable crops. Excess nitrogen often results in the stimulation of vine growth causing delays in flowering.

Cucurbit Diseases and Pests

Bacteria

Angular leaf spot	*Pseudomonas syringae* pv. *lachrymans*
Bacterial leaf spot	*Xanthomonas campestris* pv. *cucurbitae*

Bacterial rind necrosis   *Erwinia carnegieana*
Bacterial soft rot   *Erwinia carotovora*, other *Erwinia* spp., *Pseudomonas* spp.

Bacterial spot, squash   *Xanthomonas cucurbitae*
Bacterial wilt   *Erwinia tracheiphila*
Brown spot   *Erwinia ananas*

Fungi
Alternaria leaf spot/blight   *Alternaria alternata, A. cucumerina*

Anthracnose   *Colletotrichum orbiculare*
Blackrot   *Phoma cucurbitacearum*
Cercospora leaf spot   *Cercospora citrullina*
Charcoal rot   *Macrophomina phaseolina*
Damping off   *Pythium, Phytophthora, Rhizoctonia, Thielaviopsis, Fusarium, Acremonium* spp.

Downy mildew   *Pseudoperonospora cubensis*
Fusarium foot rot   *Fusarium solani* f. sp. *cucurbitae*

Fusarium wilt of
  cucumber   *F. oxysporum* f. sp. *cucumerinum*

  melon   *F. oxysporum* f. sp. *melonis*
  watermelon   *F. oxysporum* f. sp. *niveum*
Gummy stem blight,   *Didymella bryoniae*
  fruit black rot
Monosporascus root rot   *Monosporascus cannonballus*
Phompis black root rot   *Phompis sclerotioides, Phompis cucurbitae*

Phytophthora root rot   *Phytophthora capsici*, other *Phytophthora* spp.

Pink root   *Phoma terrestris*
Powdery mildew   *Erysiphe cichoracearum, Sphaerotheca fuliginea*

Scab   *Cladosporium cucumerinum*
Sclerotinia stem rot   *Sclerotinia sclerotiorum*
Septoria leaf spot   *Septoria cucurbitacearum*
Southern blight   *Sclerotium rolfsii*
Target leaf spot   *Corynespora cassiicola*
Verticillium wilt   *Verticillium dahliae, V. albo-atrum*

Fruit rots

Belly rot	*Rhizoctonia solani*
Blue mold rot	*Penicillium* spp.
Choanephora rot	*Choanephora cucurbitarum*
Cottony leak	*Pythium* spp.
Diplodia stem end rot	*Diplodia natalensis*
Fusarium rot	*Fusarium* spp.
Gray mold	*Botrytis cinerea*
Phytophthora rot	*Phytophthora capsici,* other *Phytophthora* spp.
Pink mold rot	*Trichothecium roseum*
Rhizopus soft rot	*Rhizopus stolonifer*

Viruses are serious threats to production. Because of the difficulty of controlling vectors, the main defense is genetic resistance. Important viruses damaging to cucurbits include (vectors in parenthesis) the following:

Beet Curly Top Virus, BCTV (leafhoppers)
Beet Pseudo-yellow Virus, BPYV (whitefly)
Clover Yellow Vein Virus, CYVV (aphids)
Cucumber Green Mottle Mosaic Virus, CGMMV (unknown)
Cucumber Necrosis Virus, CuNV (fungus, *Olpidium* spp.)
Cucumber Mosaic Virus, CMV (aphids)
Lettuce Infectious Yellows Virus, LIYV (whitefly)
Papaya Ringspot Virus*—Watermelon strain, PRSV-W (aphids)
Squash Leaf Curl Virus, SLCV (whiteflies)
Squash Mosaic Virus, SqMV (cucumber beetles)
Tobacco Ringspot Virus, TRSV (nematodes)
Tomato Ringspot Virus, TmRSV (nematodes)
Tomato Spotted Wilt Virus, TSWV (thrips)
Watermelon Mosaic Virus-2, WMV-2 (aphids)
Zucchini Yellow Mosaic Virus, ZYMV (aphids)

This is not an exclusive listing of all viruses affecting cucurbits. Aster yellows (AY) that has viruslike symptoms is incited by a phytoplasma (previously identified as a mycoplasma).

Insect pests and mites

Beet armyworm	*Spodoptera exigua*
Cowpea aphid	*Aphis craccivora*

---

*Previously known as Watermelon Mosaic Virus 1 (WMV-1)

Cutworms	*Proxenus mindara, Feltia sub-terranea*
Fruit fly	*Dacus* spp.
Green peach aphid	*Myzus persicae*
Leafhopper	*Empoasca abrupta*
Leafminers	*Liriomyza* spp.
Melon/cotton aphid	*Aphis gossypii*
Melonworm	*Diaphania hyalinata*
Pickleworm	*Diaphania nitidalis*
Seed corn maggot	*Hylemya cilicrura, H. platura*
Silverleaf whitefly	*Bemisia argentifolia*
Spotted cucumber beetle	*Diabrotica undecimpunctata*, other spp.
Striped cucumber beetle	*Acalymma vittata, A. trivittatum*
Spider mites	*Tetranychus urticae*, other *T.* spp.
Squash bug	*Anasa tristis*
Squash vineborer	*Melittia cucurbitae*
Sweet potato whitefly	*Bemisia tabaci*
Thrips	*Thrips palmi*, other *T.* spp.
Western flower thrips	*Frankliniella occidentalis*
Whitefly, glasshouse	*Trialeurodes vaporariorum*
Wireworms	*Limonius cannus, Conoderus* spp.
Carmine mite	*Tetranychus cinnabarinus*
Two-spotted mite	*Tetranychus urticae*

A toxin of silverleaf *(Bemisia argentifolia)* and sweet potato *(Bemisia tabaci)* whitefly feeding causes silvering of squash leaves.

Nematode pests

Lesion	*Pratylenchus* spp.
Reniform	*Rotylenchulus* spp.
Root knot	*Meloidogyne* spp.
Stem and bulb	*Ditylenchus* spp.
Sting	*Belonolaimus* spp.
Stubby root	*Trichodorus* and *Paratrichodorus* spp.

A common physiological disorder is *blossom end rot.* Other abiotic disorders affecting fruit are *hollow heart* of watermelon, which is associated with rapid growth, and *sunburn,* that is often accented by inadequate leaf canopy.

IMPORTANT VEGETABLES OF THE CUCURBITACEAE

CUCUMBER, *Cucumis sativus* **L.**
WEST INDIAN GHERKIN (BUR CUCUMBER), *Cucumis anguria* **L.**

Immature cucumber fruit are eaten as salad vegetables or pickles, and also used as cooked vegetables (Fig. 24.1a). In Southeast Asia, cucumber foliage is also consumed as salads and cooked vegetables; in Japan, female blossoms are used as garnish. Immature West Indian gherkin fruit are used pickled, in some curry preparations, and as a cooked vegetable (Fig. 24.1b).

Origin

The cucumber is considered to have its origin in India where it has been grown for thousands of years. Cucumbers were also known to have been cultivated by the ancient Egyptians and Greeks. *C. sativus* var. *hardwickii,* a wild taxon native to India, is proposed as the progenitor of the domesticated forms of *C. sativus,* although it may be a feral derivative, instead.

a   b

FIG. 24.1. Various forms of cucumber fruit, *Cucumis sativus,* (a) the spherical fruit in the foreground are lemon cucumber, the long fruit at the right are slicing forms, those to the left are pickle types and the very long fruit at the top is a Armenian cucumber, which actually is a melon, *Cucumis melo;* and (b) West Indian gherkin, *Cucumis anguria.* Source (a): *Courtesy of National Garden Bureau, Downers Grove, Illinois.*

West Indian gherkin, once thought to be native to the West Indies, has since been identified as introduced from Africa as a cultigen of a nonbitter mutant of *Cucumis anguria* var. *longipes.* Cucumber and gherkin differ in chromosome number, being $2n = 14$ and $2n = 24$, respectively.

Botany

Cucumber plants are annuals, usually having a trailing or climbing growth habit, although some cultivars have a bush habit. Root systems are extensive, but usually shallow. Stems range from 1 to 3 m in length and are four angled with stiff bristle hairs; tendrils are unbranched. Leaves are triangular ovate, somewhat cordate, from 7 to 25 cm wide, with three to five angled portions or shallow lobed sinuses with a surface rough to the touch; the apex is pointed. Petioles are 5–15 cm long. Yellow flowers are bell shaped. Male flowers are in axillary clusters or singularly on slender peduncles. Female flowers, with thicker peduncles, are singular in leaf axils. The large ovary is inferior and consists of three united carpels.

Most field-grown cucumber cultivars are monoecious or gynoecious, but hermaphrodite flowers also occur. Sex expression is under genetic control but is influenced by environment and/or chemical treatment. For many monoecious forms, the usual ratio of male to female flowers is heavily skewed toward maleness. Staminate flowers are formed first, followed by pistillate and additional staminate flowers, but not necessarily in an alternating sequence. Lateral shoots tend to bear a relatively greater proportion of pistillate flowers than main stems.

Flowering is influenced by photoperiod with regard to the number and sex of flowers formed. With short days, there is a tendency for earlier and more frequent pistillate flowering; low temperatures can cause a similar response. Conversely, high temperatures and long days promote greater maleness.

Fruit are pendulous and can be spherical, blocky, oblong, or elongated in shape, and variable in size. Fruit surfaces vary in the number and size of scattered spiny tubercles (warts), which usually are more apparent on young fruit. Skin color varies from pale to very dark green; interior flesh is white to creamy white. Mature seed are flat and white, about 50 weigh 1 g.

Cucumbers with varying characteristics are used for different purposes. Fresh market types usually have large vines, fruit with a length to diameter ratio of about 4 : 1, dark green color, thick skin, somewhat pointed stem and blossom ends, and usually with white, preferable few, spines. The seed cavity of fresh market cultivars are usually larger than processing cultivars.

Cucumber types for pickling usually have small vines, fruit with a length to diameter ratio of about 2.5 : 1, light green color, thin skins, relatively blocky at the stem and blossom ends, and many surface warts, often with black spines. Black spine fruit generally have thinner skins and a lighter green color than white spine fruit, and are preferred because they result in a more attractive product after pickling. Pickling cultivars are occasionally marketed for fresh salad use; on the other hand, market types are seldom processed.

At maturity, spine (spicules) color is correlated with fruit skin coloration. The skin of mature black spine fruit typically become yellowish orange or bronze when mature. White spine fruit at maturity are light green or cream colored. Neither condition is a problem with regard to immature fruit. However, overage fruit of white spine pickle cultivars are processed as relish, but fruit of black spine cultivars usually are not because of the unacceptable color of the final product. Another association with fruit color is with regard to disease resistance. The first resistance discovered for cucumber scab and cucumber mosaic virus (CMV) was in black spine germplasm. However, similar disease resistance has since also been found in white spine germplasm. Presently, many market and pickling cultivars possess multiple disease resistances.

Parthenocarpic fruit, called seedless, English, or glasshouse cucumber, are usually grown in protective structures. These fruit are capable of being pollinated and will produce seed. However, if grown for marketing, pollination results in seeded, low-value fruit which are often misshapen. Hence, in producing parthenocarpic cucumbers, measures are taken to ensure that pollination does not occur.

The Oriental-type cucumber is thin skinned, usually long, and slender with considerable warts and spines; some cultivars are parthenocarpic. This fruit type is popular in Asian markets. In Japan, two kinds are grown: one is a day-neutral form known as the North China group; the other is a short-day or South China group that is grown for winter production. The latter produces mainly pistillate flowers, but under long days produces few pistillate and many staminate flowers.

The Sikkim cucumber, popular in India, has fruit with reddish brown skins. Smooth-skinned Beit Alpha types are most popular in the Middle East and north Africa. These are usually grown outdoors, and often in low plastic tunnels in most middle eastern countries. Beit Alpha fruit are light colored, thin skinned, and have relatively large seed cavities.

TABLE 24.1.   MORPHOLOGICAL COMPARISON OF MARKET, PICKLING, AND
PARTHENOCARPIC CUCUMBERS AT THEIR USUAL HARVEST PERIOD

	Market	Pickling	Parthenocarpic
Vines	Large	Short	Very large
Fruit length	Long	Short	Very long
Fruit length to diameter ratio	About 4 : 1	About 2.5 : 1	Greater than 5 : 1
Seed development	Immature	Very immature	Seedless
Seed cavity	Large	Small	Small
Stem and blossom end	Pointed	Blunt	Stem end pointed, blossom end often blunt
Skin color	Dark green	Light to dark green	Dark green
Skin thickness	Thick	Thin	Very thin
Skin spines and warts	Few	Many	Few, none[a]

[a]An exception are some Oriental-type cucumbers that commonly have many spines
and warts.

The Lemon cucumber is a novelty cultivar that produces a round,
creamy yellow skin fruit which is readily recognized for its large five-
carpel seed cavity. This cultivar is also unique because of its andromonoecious sex expression. Often considered a cucumber, the Armenian
or snake cucumber actually is a melon, *Cucumis melo* group Flexuous.
Very small, immature cucumber fruit are erroneously referred to as
gherkins.

West Indian gherkins are trailing monoecious annual plants. The
vines have angled stems with stiff hairs and small, simple, curling
tendrils. Leaves, borne on long petioles and resembling those of watermelon, are 4–10 cm long, deeply (3–5) lobed, with rounded sinuses.
Flowers are small, with male flowers forming in clusters. Female flowers are singular on long, slender, hairy peduncles that arise from secondary branches rather than at leaf nodes. Fruit borne on long, slender
peduncles are oval to oblong, from 5 to 10 cm in length, warty, and/or
covered with sharp soft hairs. With maturity, rind color changes from
pale green to cream. The flesh is greenish white and extremely seedy
with small, smooth, white seed; about 150 weigh 1 g. The cultivation
of West Indian gherkin is similar to cucumber; fruits are harvested
60–80 days after planting.

A somewhat similar *Cucumis* species is the African horned cucumber,
C. metuliferus, commonly called kiwano fruit. It has fewer, but longer
and harder spines than West Indian gherkin. The initially greenish
gray fruit acquires a reddish orange color when ready for consumption,
and is usually eaten as a dessert fruit for its jellylike interior.

Other *Cucumis* species of minor importance are the teasel or hedgehog gourd, *C. dipaceus,* and the gooseberry gourd, *C. myriocarpus.* An interesting species that develops fruit below ground is *C. humifructus,* known as the aardvark cucumber.

Culture

A warm climate is required for rapid and satisfactory cucumber and gherkin growth; 30°C days and 20°C nights are ideal. Temperatures less than 15°C limit growth, a period of several days of exposure to less than 10°C causes chilling injury, and plants are killed by frost.

With glasshouse cucumbers, even a short-term exposure to low soil temperatures results in leaf and fruit damage. For glasshouse-grown parthenocarpic cultivars, temperatures must be above 20°C for satisfactory growth. High humidities are conducive to foliar diseases.

Growth is rapid in well-drained fertile loam soils; a favorable pH is between 6.5 and 7.5. About 400–500 mm of water is usually adequate to produce a crop; plants are intolerant of waterlogging.

Propagation and Spacing

Direct seeding is the usual production practice; seldom are field-grown cucumbers transplanted. For rapid germination, soil temperature should be 20°C or higher; at 25–35°C, emergence can occur in as little as 2–4 days.

Common field spacings are 30–45 cm × 120 cm for row plantings, or about 90 cm equidistant for hill plantings. In the field, plants usually are allowed to trail and infrequently are trellised; in glasshouse, they are always grown on a trellis. Some producers will train trailing vines to grow in the same row direction in order to facilitate cultivation and harvesting.

Harvest and Postharvest

Fruit are usually harvested about 55–65 days after plant emergence; the time is greatly influenced by growing temperatures. Fresh market fruit (slicers) are usually harvested when about full size but before fully developed and while seed are small and soft. Pickles are harvested at an even earlier stage, and usually when very small. For purposes of processing, small-sized cucumbers have a higher value compared to larger fruit, and therefore present an incentive for early harvest of small fruit.

Most cucumber crops continue to be hand harvested. Harvest frequency is influenced by temperature, and in some situations, daily harvests are made. In the United States and several other countries, a portion of the processing crop is mechanically harvested. With most mechanical harvesters, the equipment essentially destroys the plants and, thus, only a terminal harvest results. Therefore, to maximize crop yield when the harvest is mechanized, high-density plantings are made. The use of gynoecious cultivars further enhance mechanical harvest feasibility because the trait of all-female flowering increases fruit numbers. Applications of ethephon have been used to increase femaleness and subsequent fruit yields. Mechanical harvesting results in some fruit injury which is tolerable because the damage often is not apparent in the finished product. On the other hand, machine harvest of fresh market fruit can cause unacceptable physical damage.

Generally, cucumbers and gherkins are seldom stored more than a week or two; if necessary, short-term storage is best between 10°C and 13°C and at 95% RH. Fruit should be held above 10°C to avoid chilling injury. Occasionally, fresh market cucumbers are waxed to reduce moisture losses and to enhance appearance.

Production

Cucumber production is heavily centered in Asia where almost 73% of world production occurs; China individually accounts for almost 42%. European production follows well behind at about 17%. The extremely high yield of 514 t/ha in the Netherlands is because of the intensive cultivation in glasshouses. Countries such as Japan, Spain, and Korea also produce a significant volume of cucumbers in glasshouses and other protective structures. See Table 24.2.

Glasshouse Parthenocarpic Cucumbers

The production of glasshouse parthenocarpic cucumbers is highly specialized. Glasshouses and plastic protective structures are used to manage temperature, humidity, light, CO_2, nutrition, and moisture. Such production is most prevalent in the Netherlands and several other western European counties and Japan. A very large volume of cucumbers are also produced in unheated plastic tunnels and houses during winter in many mild-climate, temperate regions.

Temperature control is especially important in these structures. At the optimum temperature of 28–29°C, seed germinate in 2–3 days. Seedlings are ready for transplanting after production of three to four true leaves. After transplanting, growing temperature is maintained at 25°C until flowering begins. Thereafter, a temperature regime of 17–18°C nights and 20–21°C days is followed. Temperatures exceeding 28°C or less than 17°C are avoided, either by ventilation or heating.

TABLE 24.2. WORLD CUCUMBER AND GHERKIN PRODUCTION, 1994

	Area (ha × 10³)	Yield (t/ha)	Production (t × 10³)
World	1,215	15.9	19,261
Africa	23	16.8	392
North and Central America	105	13.4	1,398
South America	4	15.5	54
Asia	868	16.2	14,035
Europe	214	15.7	3,362
Oceania	1	14.5	19
Leading countries			
China	468[a]	17.2	8,051[a]
Iran	96[a]	15.7	1,510[a]
Turkey	40[a]	27.5	1,100
United States	70	14.0	984
Japan	17[a]	47.1	800[a]
Ukraine	54	9.6	516
Netherlands	1[a]	514.0	514[a]
Poland	34	10.7	366
Iraq	40[a]	8.5	340[a]
Spain	7[a]	47.1	330[a]

[a]Estimated.
Source: 1994 FAO Production Yearbook, Vol. 48, FAO, Rome, 1995.

Generally, low temperatures favor vegetative development and high temperatures enhance flowering.

Humidity is mostly controlled by ventilation. High humidities encourage the development of foliar diseases, and it is important to avoid wetting the foliage. Accordingly, drip irrigation is the preferred method for providing moisture as well as supplying nutrients.

High light intensities are essential for high yields. Thus, structures should be highly transmissible to light, and plant spacing and leaf pruning manage to minimize shading. Supplemental lighting during the winter is sometimes provided when day length and light intensity are inadequate. Some producers provide 400–1500 ppm CO_2 additions to increase photosynthesis; levels greater than 1500 ppm usually are not cost-effective.

Hybrid cultivars are exclusively used for parthenocarpic cucumber production, and most are gynoecious. The complexities of hybrid seed production of parthenocarpic cultivars and the limited volume of seed used, results in very expensive seed. Nevertheless, for this specialized production, high-quality seed is an essential component. To maximize seed use, the crop is almost always started as potted transplants. Seedlings are transplanted and grow in a wide variety of growing media which may include soil ground beds, soil or peat bags, straw bales, rockwool, gravel, or sand. The nutrient film technique (NFT) procedure and other hydroponic system may also be employed. If used, soils are sterilized by steam or fumigated with methyl bromide. Use of methyl bromide is prohibited or restricted in many countries because of soil water contamination and worker safety considerations. Restrictions are likely to expand, and the use of synthetic media, often already sterile, has therefore greatly increased.

Crop nutrient requirements are high in order to support high yields and a production period usually lasting 4–6 months. Nitrogen applications range from 400 to 1000 kg/ha, and P, K, and Mg applications usually are about 80, 550, and 50 kg/ha, respectively. An adequate supply of minor nutrients should be provided, especially if soilless culture is practiced. Media pH levels should be between 5.5 and 6.0.

Many variations of vine training and support systems are employed to ensure maximum leaf exposure to light, to minimize shading, and development of straight-shaped fruit. Most spacings arrangements provide for 1.5–2 plants/m^2.

Pest and disease control are critical in protective structures. Considerable use is made of integrated pest management that can include the use of insect parasites and predators. Foliar and soil fungus diseases are frequently major problems. Good ventilation practices and selective fungicides help control diseases. Plants are sometimes grafted onto rootstocks of *Cucurbita ficifolia* which provide soil fungus resistance and cold tolerance.

The long, 20–50 cm, uniformly green fruit produced are seedless, thin skinned, and free of bitterness (Fig. 24.2). Cultivars producing smaller fruit, of about 15 cm length, are also grown. Pollination is undesirable for parthenocarpic production because seeded and misshapen fruit, both undesired, will develop. Not only are such fruit unacceptable, their occurrence interferes with the indeterminate growth of the plant and its subsequent fruiting. Although gynoecious cultivars are used, they may not always be completely female. When stressed, gynoecious plants occasionally produce staminate flowers. When male flowers occur, they should be removed immediately.

When harvested, fruit are carefully cut from the vine to avoid fruit and vine damage. Fruit are carefully handled, graded, and often covered with a shinkwrap plastic film to reduce desiccation. Storage at 13°C and 90–95% RH is optimum to achieve a postharvest life of 1–2 weeks. At temperatures less than 10°C, fruit are susceptible to chilling injury.

MELONS, *Cucumis melo* L.

Like cucumber, *Cucumis melo* is grown for the fruit; but unlike cucumber, melon fruit are larger, sweeter, and mature (ripe) when eaten.

FIG. 24.2. Parthenocarpic cucumbers, *Cucumis sativus,* in glasshouse production.

During ripening, fruit soften and fruity aromatic essences are formed. Fruit typically are eaten uncooked, although some types of *Cucumis melo* are also preserved by pickling or used when immature as cooked vegetables or fresh in salads. Some researchers are advocating that the common name of the species *Cucumis melo* should be "melon" instead of "muskmelon," an additional suggested classification places muskmelons or cantaloupes as a group within the species.

Origin

Muskmelons are reported to be indigenous to Africa. Wild forms found in India may be nondomesticated escapes. However, secondary centers of diversity exist in India, Iran, southern Russia, and China. Historical records show that muskmelon cultivation occurred in Egypt as early as 2400 B.C. The name cantaloupe is mentioned as having originated from the city of Cantaluppi in Italy or from the estate and castle of Cantalupo, also in Italy.

Botany

Muskmelons are annuals with monoecious and sometimes andromonoecious flowering characteristics. Root systems usually are extensive but relatively shallow. Stems are ridged with simple tendrils. Most cultivars are vining; the introduction of bush-type cultivars is a relatively recent development.

Leaves differ from those of cucumber in being either circular, oval, or kidney shaped, about 8–15 cm wide, and angled or with five to seven shallow lobes. Male flowers are formed in clusters of three to five on thin peduncles. Female and hermaphroditic flowers develop singularly in different axils on short thick peduncles. Flowers open only once during the early morning and are insect pollinated.

Fruit size, shape, color, and rind firmness vary greatly among melon types and cultivars (Fig. 24.3). Fruit usually are spherical or oval oblong. Surfaces are smooth, glabrous, some deeply ridged, and others are covered with a corky (reticulate) netting. Vein tracts, also called sutures, are areas of longitudinal indentations or strips on the surface without netting. These areas are associated with vascular bundles and are not readily apparent in heavily netted fruit. Fruit surfaces most often are yellow or brownish green. The flesh which is actually the

a   b

FIG. 24.3.   Muskmelon fruit, *Cucumis melo*, (a) cantaloupe type; (b) 'Honeydew,' a winter melon type.

ovary wall (pericarp) also varies considerably in thickness, color, and texture. Flesh colors can be white, green, pink, or orange. Some newly developed cultivars, when fully ripened, have flesh with two distinct colors. Melon aroma is due to the many volatile compounds, particularly alcohols, acids, and their esters formed during ripening; the amounts and ratios of these volatiles vary with the different *C. melo* groups, giving each its characteristic aroma and flavor.

A unique feature of some muskmelon cultivars is the formation of an abscission layer that coincides with fruit maturation and allows the fruit to easily detach (slip). This characteristic is a useful guide as an external indication of fruit maturation and time for harvest. Melon fruit produce relatively abundant amounts of seed which are white or buff colored and smooth. Seed range from 5 to 15 mm in length; on average, about 30 seed weigh 1 g.

Different forms of *Cucumis melo* fruit were classified as botanical varieties by Naudin in 1859. Although that classification was not recognized taxonomically, it had usefulness and was followed by tradition. With some modification, many variety designations are now recognized as groups and are indicated as follows:

Cucumis melo group:

Cantalupensis*	Netted muskmelon, cantaloupe, Persian, and some lightly or non-netted cultivars
Inodorus	Winter melon
Flexuous	Snake or serpent melon, Armenian cucumber
Conomon	Oriental pickling melon
Chito	Mango melon or vine peach melon
Dudaim	Pomegranate or Queen Anne's pocket melon
Momordica	Snap melon
Agrestis	Wild type

MUSKMELON/CANTALOUPE, *C. melo* L. group Cantalupensis

Fruit of *C. melo* group Cantalupensis are commonly called muskmelon and/or cantaloupe. Both are muskmelons, but the name canta-

---

*C. melo group Cantalupensis now includes a previous recognized group named Reticulatus.

loupe is usually associated with fruit of the Cantalupensis group produced in the arid southwestern United States. Such fruit are usually spherical or nearly so, with uniform heavy netting, indistinct ribbing, firm, thick flesh, and with a small and relatively dry seed cavity. In contrast, other muskmelons have a spherical or elongated oval shape with pronounced ribbing and vein tracts, relatively large and moist seed cavities, soft flesh texture, and strong aroma. Muskmelon cultivars are adapted to grow in many regions but have relatively poor shipping and storage characteristics. Thus, in the United States, cantaloupe and muskmelon are essentially marketing distinctions. The heavy surface netting of cantaloupes hides as well as limits abrasion and bruising, which makes the fruit better adapted to long-distance shipment and longer postharvest shelf life.

Persian-type muskmelon cultivars produce relatively large, round, hard rinded, heavily netted fruit. The fruit have a slight aroma, a moist seed cavity, and the firm flesh is a light salmon color. Many cultivars of other muskmelon types, either lightly netted or without netting but often with obvious vein tracts and thin rinds, are most popular in Europe and Asia. These fruit generally have strong aromas, moist seed cavities, and relatively short postharvest life. They are often grown in glasshouses or plastic-film-covered shelters, some of which are unheated. Some representative cultivars and types are Charentais, Ananas, Ha-ogen, and Valencia.

WINTER MUSKMELON, *C. melo* L. group Inodorus

Winter muskmelons require a growing season that is several weeks longer than that for other types of muskmelons. Best quality fruit is obtained when grown during high temperatures (30–35°C), in semiarid environments. Representative cultivars and types of winter melons are casaba, honeydew, crenshaw, Juan Canary and Santa Claus. Although morphologically dissimilar, these cultivar types readily intercross with muskmelons of group Cantalupensis.

Casaba fruit have a greenish yellow, thick, fairly hard ribby rind. Fruit are boxy or nearly round, but strongly tapered at the stem end. The white flesh is firm, sweet with a cucumberlike aroma, and the seed cavity is moist. Fruit do not abscise and have a long storage life. The name casaba is believed to be an adaptation of the Turkish name Kasaba, an area near Smyrna, Turkey.

Honeydew is the name applied by John Gauger, a Colorado farmer, to a selection from the French cultivar, White Antibes. Honeydews have hard, light green outer rinds that become creamy white and feel waxy when mature. The thick flesh is firm and a light creamy green; plant breeders have produced orange-fleshed cultivars. The seed cavity is usually relatively dry. Fruit do not abscise and have fair to good storage qualities. "Honeyloupe" is a selection from a cross of cantaloupe and honeydew made by Frank Zink in California. It is an orange-fleshed honeydew with the attributes of the cantaloupe.

Crenshaw, originated as a chance find in a muskmelon field. It has fruit which are not netted, but are relatively smooth, although slightly ribby at the stem end. Although initially dark green, the surface rind becomes patchy yellow to yellow when mature. Fruit have a waxy feel, a cucumberlike aroma, and the thick juicy flesh is moderately firm and salmon colored. Crenshaw fruit exhibit intermediate abscission, are highly susceptible to sunburning, and generally have poor storage and shipping characteristics.

Juan Canary fruit are elongated; the yellow-colored hard rind feels waxy when mature and does not have an aroma. Interior flesh is firm, crisp, and whitish green, the seed cavity is moist.

Santa Claus fruit are elongated, the hard rind has green and yellow longitudinal stripes, and the fruit is without aroma. The flesh is firm and whitish green. Fruit of Juan Canary and Santa Claus do not abscise; each have good shipping and storage features.

Sicana odorifera, commonly known as casabanana, is an unrelated cucurbit fruit that because of its strong fragrance has some resemblance to muskmelon. Casabanana fruit are edible either raw or cooked, but only when immature.

Culture

Varying with cultivars and growing conditions, most muskmelons mature within 80–120 days after planting. Optimum mean temperatures are from 18°C to 24°C. Plant growth is significantly improved with high light intensities. During cool periods, such as when the sun is low on the horizon, plant beds are shaped so as to be slanted toward the sun in order to take advantage of the additional increase of light and warmth for the growing plants; see Fig. 7.1. Various mulches, row covers, and hot caps are occasionally used to advance plant growth. In Japan and several other countries, a considerable volume of muskmelons are produced in glass and plastic houses.

Deep, well-drained soils are most suitable. Fine textured soils are potentially more productive, but usually tend to delay maturation. Melons are sensitive to acid soils; the best soil pH should be between 7 and 8.

Cultural procedures are similar to those for producing cucumbers, although nutritional requirements of melon crops are more demanding. When compared to the short production period for cucumbers, the longer growing period for melons requires more nutrients and water. Raised beds and mulches are used to minimize fruit contact with wet soil surfaces and to assist drainage; waterlogging must be avoided.

Low humidities usually reduce the incidence of most foliar diseases. Foliar diseases result in defoliation and may increase fruit susceptibility to sunburn. To reduce sunburn during periods of very high temperatures, whitewash is sometimes applied to fruit when close to maturity. Cultivars vary in susceptibility to sulfur, which is used as a fungicide for powdery mildew control; susceptible plants exhibit foliar burn.

Seed are sown 3–4 cm deep, and when soils are warm, greater than 20°C, emergence occurs within a week. Melons are not usually transplanted because bare-rooted plants do not establish well, and container-grown plants are expensive. Nevertheless, where expensive hybrid seed are used, and especially in glasshouse or other protective structure production, transplanting may become more feasible. Field plant spacings vary from 30 × 200 to 60 × 200 cm, with resulting populations of 15 thousand to 20 thousand plants per hectare. In commercial production, even at these spacings, often only one or two fruit per plant will be of market quality and suitable for harvest. Late fruit set that often abort or fail to achieve size or adequate soluble solids and sweetness are reasons for such low plant performance. Sometimes, these late fruit set are allowed to mature for seed production.

Harvest and Handling

Fruit usually mature about 6–10 weeks after anthesis. The maturity of some muskmelon cultivars can be determined by the development of the abscission zone between the fruit and peduncle. At or near maturity, fruit exhibit a groundspot and the background surface color changes from green to yellow. The noticeable aroma, detected mostly at the blossom end, is another maturity indicator. Melons that do not abscise should be cleanly cut from the vine to minimize damage to either the fruit or plant. High-quality melons have a soluble solids content of 10% or more. Most of the carbohydrates in melons are sugars, namely glucose, fructose, sucrose, and stachyose; there is no starch. Fruit sugars do not increase following harvest or during storage.

Harvest is performed by hand; containers and handling equipment are padded to minimize fruit scuffing and bruising. To reduce postharvest cooling requirements, night harvesting of muskmelons is sometimes practiced. Soon after harvest, muskmelons should be quickly cooled to 10°C, which is accomplished rapidly by hydrocooling with very cold water (0°C). Honeydew or casaba melons should also be cooled to the same 10°C level, but the rate of cooling need not be as rapid.

Accurate field determination of honeydew melon maturity is difficult, and harvested fruit are often treated with ethylene to accelerate and achieve ripening. Ethylene exposure for 12–24 h at 1000 ppm is made at 20–30°C. The exposure time is shortened at higher temperatures. Other types of winter muskmelons can be similarly treated, although less frequently.

Storage

For short-term storage of 5–10 days, melons fruit can be held at temperatures of 5°C and high relative humidity, (95%). A long storage period at this temperature can result in chilling damage. Winter-type melons are better suited to a longer period of storage; they can be held at about 10°C for 1–2 months.

Production

Asia, with more than 59% of the tonnage produced, dominates world production of cantaloupe and other melons. Europe, and North and Central America follow with more than 17% and almost 13%, respectively. See Table 24.3. Among the leading countries, China has the largest production, with more than 25%. Developed countries produce less than one-third of total production, suggesting that the potential for export to these countries is high and that international trade could increase. Considering the large volume of home-grown melons produced, the preceding statistics shown are much lower than the actual annual production.

Other melon groups of *C. melo* may or may not be as important as those of Cantalupensis and Inodorus. The importance depends on the region of production and usage. Described below are some of these melons.

SNAKE MELON/ARMENIAN CUCUMBER, *C. melo* L. group Flexuosus

The pale green fruit are long and slender with a smooth longitudinally slightly ridged surface (Fig. 24.1). Fruit sizes range from 20 cm up to 1 m in length with diameters from 4 to more than 10 cm. The fruit

TABLE 24.3.   WORLD PRODUCTION OF CANTALOUPE AND OTHER MELONS, 1994

	Area (ha × 10³)	Yield (t/ha)	Production (t × 10³)
World	803	17.3	13,894
Africa	55	18.4	1,014
North and Central America	116	15.4	1,793
South America	30	8.7	262
Asia	448	18.5	8,304
Europe	150	16.2	2,440
Oceania	4	21.6	80
Leading countries			
China	133[a]	26.6	3,540[a]
Turkey	100[a]	17.0	1,700
Iran	86[a]	13.8	1,185[a]
Spain	50	18.2	916
United States	42	20.5	859
Romania	50[a]	13.2	660[a]
Mexico	50[a]	13.0	650[a]
Egypt	24[a]	18.7	450[a]
Morocco	20[a]	20.8	415
Japan	18[a]	22.4	410[a]

[a]Estimated.
Source: 1994 FAO Production Yearbook, Vol. 48, FAO, Rome, 1995.

often is curved, but less so when grown on a trellis. The crop is popular in the Middle Eastern countries where the fruit is known as Armenian cucumber. They are harvested immature for fresh use in salads and also are cooked like summer squash; mature fruit are also consumed.

ORIENTAL PICKLING/SWEET MELON, *C. melo* L. group Conomon

The origin of the oriental pickling melon is obscure. Its presence is referenced in Chinese literature from A.D. 560. In Japan, the crop is known as "uri" and is grown to make pickles called "tsukemono" or "koko." The group name Conomon is believed to be a corruption of "koko no mono," meaning material for making koko. Fruit are usually smooth and cylindrical, 20–30 cm long, and 6–9 cm in diameter. The flesh is medium thick and white. Ao uri, shiro uri, and shima uri are representative cultivar types, and are green, white, and striped green and white, respectively (Fig. 24.4). Some fruit within this group have a high sugar content and, when mature, are eaten as dessert fruit.

Growth is optimum at 25–30°C, and greatly limited below 13°C; cultivation is similar to that for cucumber and melons. Plants flower about 40 days after planting. Fruit for pickling are harvested when full size but still immature. After seed removal, the rind is preserved,

FIG. 24.4. Immature fruit of pickling melon, *Cucumis melo* gp. Conomon.

either pickled or candied. Small fruit are used much like summer squash.

MANGO MELON, *Cucumis melo* L. group Chito

Grown for its fruit, eaten cooked or pickled, the mango melon has many common names such as vine peach, lemon melon, and orange melon. Plants produce spreading vines with foliage that resembles muskmelon. The small round fruit look like small honeydews or lemon cucumbers that have a yellow or pale green color and often are lightly striped lengthwise. In about 80–90 days after seeding, the mature fruit will slip. The firm flesh is yellowish white with a cucumberlike texture; seed are embedded in a gelatinous medium.

POMEGRANATE/QUEEN'S POCKET MELON, *C. melo* L. group Dudaim

Fruit are small, globular, and pubescent with alternating dark and lighter green surface striping that turns brownish orange when mature.

FIG. 24.5.   Queen Anne's melon, *Cucumis melo* gp. Dudaim.

The rind is thin and the seed cavity large and moist (Fig. 24.5). It is used as an odoriferous fruit and as an ornamental, although some accessions lack fragrance. The popularity is highest in areas of Asia Minor and norther Africa. Some taxonomists believe this group and the Chito group should be combined.

SNAP/PHUT MELON, *C. melo* L. group Momordica

The smooth fruit has a light orange or white mealy flesh. The snap name apparently has to do with the cracking of the fruit surface and its disintegration with advancing maturity. The production and popularity is largely limited to India, where it is called "phoot," but the crop is also grown in other Asian countries.

WATERMELON, *Citrullus lanatus* (Thunb.) Matsum & Nakai (*C. vulgaris*)

Watermelons are grown for their fleshy fruits, which are juicy and sweet. Immature fruit are occasionally used like summer squash; the pickled rind and the seed are also edible. Some watermelons are used for livestock feed, and the extracted juice from others is fermented to produce an alcoholic drink. Fruit have also served as a reserve water source (botanical canteen) in drought areas or where drinking water is contaminated.

Origin

The origin of watermelon is still unknown. The crop has an ancient cultural history and was known to the Egyptians before 2000 B.C. One theory is that it is derived from a perennial relative, *C. colocynthis,* which is endemic to Africa. That species can cross with watermelon, and colocynth seed have been found in early archaeological sites preceding findings of watermelon remnants. Another theory is that watermelon was domesticated in Africa from putative wild forms of *C. lanatus.* South central Africa near the Kalahari desert is a region where wild forms are still found, some of which do not have bitter fruit. The practice of "plugging" watermelons, where a small triangular piece (plug) is removed for tasting, was possibly developed by natives to avoid carrying bitter fruit back to their villages. The related species, *C. lanatus* var. *citroides,* known as citron or preserving melon, appears to have a similar origin. Early explorers in central Africa (David Livingston, in particular) found huge areas covered with wild watermelon vines. The Equsi melon, a primitive form of watermelon, is grown in Africa for seed production; both bitter and nonbitter fleshed fruit types exist. In addition, this plant possess some important diseases-resistant traits that are being exploited in current breeding.

A similar fruit, known as squash melon, round melon, or tinda, *Praecitrullus fistulosus* (previously identified as *C. lanatus* var. *fistulosus*), is grown in India where the small round fruits are used as fresh vegetables or pickled; seed are also eaten. Fruit of this species are reported to have a higher nutritive value than most cucurbits. However, plants differ in chromosome number and are not closely related to watermelon.

Botany

Watermelons are annual monoecious (some andromonoecious) plants with long trailing stems, frequently easily exceeding lengths of 5–6 m. Stems are thin, angular, grooved, and hairy, with branched tendrils. Root systems are extensive, deeper than some other cucurbits, although still relatively shallow. The majority of the roots are within 60 cm of the surface.

Leaves are large, 5–20 cm long, but differ from *Cucurbita* and *Cucumis* species in being deeply and many lobed. Flowers open for only

1 day, are small, solitary, pale yellow and insect pollinated. When inadequately pollinated or if fruit load is excessive, flowers may abort. Poorly pollinated fruit that do not abort usually are misshapen.

Fruit of different cultivars vary considerably in size, shape, and surface color (Fig. 24.6). Sizes range from 1 to 3 kg, "icebox" types, to those weighing more than 25 kg. Fruit shapes are round, oblong, or elongated with blocky or pointed ends. The usually glabrous rind is pericarp tissue, which is firm but not hard or strong. Rind thickness can vary from less than 1 cm to 4 cm. Exterior rind colors range from blackish green to light grayish green with solid, striped, or mottled coloring. Rind surface wax accumulation increases with maturity. The fruit does not have a cavity; instead, the seed are imbedded within the placental tissue, which is the primary edible portion. Small, flat, and smooth seed are of many colors, such as white, tan, green, red, or black; about 10–15 seed weigh 1 g. Some watermelon cultivars and *Citrullus colocynthis,* a related species, are grown exclusively for their abundant and large seeds, which are eaten after roasting, mostly as a snack food.

Cultivars differ with regard to flesh texture, color, and sugar content. With some fruit, the flesh texture can be fairly stringy or fibrous. Flesh colors include shades of red, orange, pink, and yellow and also white. Red flesh color is due to the pigment lycopene; yellow mostly from beta-

FIG. 24.6.   Watermelon fruit, *Citrullus lanatus,* exhibiting size, shape and color variations. Source: Courtesy of National Garden Bureau, Downers Glove, Illinois.

carotene and xanthophylls. Occasionally, flesh bitterness occurs due to the presence of cucurbitacins, but is seldom found in present-day cultivars. The sugar content increases as fruit mature and some cultivars achieve levels greater than 12%.

Seedless Watermelon

Seedless watermelons, first developed by Kihara in Japan, are triploids. Being triploid, they normally are sterile and fertilization does not occur; therefore, no seed develop. If fertilization, which is extremely rare, does occur, the embryo usually aborts. However, the process of pollination stimulates the ovary to enlarge and develop parthenocarpically. Small, empty ovules (vestigial seeds) can sometimes be found in the placental flesh, but they are soft, without an embryo, and usually not readily detectable. To further reduce detection, plant breeders use germplasm that provides white or light colored seed coats.

Chromosome imbalance because of triploidy does not significantly affect plant or fruit development, although fruit defects such as triangular shape, large blossom scars, hollow heart, light flesh color, and delayed maturity of varying levels are occasionally observed.

Triploid $(3n)$ seed is produced from crossing a tetraploid $(4n)$ plant with a diploid $(2n)$ plant. The tetraploid plant is obtained by treating a diploid plant with colchicine,* which is used to double chromosome number. In performing the $4n \times 2n$ cross, pistillate flowers of the $4n$ plant are pollinated with $2n$ pollen. Fertilization results in fruit with $3n$ seeds; these are the seed that are used for planting the seedless crop.

When plants grown from triploid seed flower in the field, they are pollinated with diploid pollen. Usually, one row of a diploid cultivar is planted to pollinate two to four rows of triploid plants. The pollination of triploid flowers produces seedless fruit. The pollinating cultivar should be of a contrasting color or shape to distinguish selfed diploid fruit from triploids at the time of harvest. Fruit of the pollen supplying diploid cultivar are harvested as conventional seeded fruit.

Culture

Productive growth of many watermelon cultivars requires a relatively long growing season, 100–150 days during warm to hot weather; optimum day and night temperatures are 30°C and 20°C, respectively. Hot caps, row covers, or low tunnels are occasionally used to obtain early

---

*Colchicine is an alkaloid compound extracted from the autumn crocus, *Colchicum autumnale,* that has the property of inducing polyploidy because of interference with mitosis.

production. Watermelons are tolerant of low humidity and have fairly good drought tolerance, but are sensitive to waterlogging. Early-maturing cultivars, many of which are short internode, bush-type plants are being grown with greater frequency. Long vines require wide row spacing and are susceptible to wind damage.

Soil compaction severely limits root growth, thus friable, deep, and well-drained sandy loam or loam soils are preferred. The ideal soil pH is between 6.0 and 6.5, but a range from 5 to 7 is usually acceptable. Watermelon root systems are efficient in water uptake and use. From 400 to 700 mm of rainfall or irrigation will adequately support most crops. Crop rotation is recommended if soils are infected with *Fusarium* or nematodes. In addition to other plant nutrients, potassium is also important because of its role of increasing rind thickness and cracking resistance.

Seed are planted 2–4 cm deep. Most cultivars have a seed germination optimum between 25°C and 35°C, which will result in emergence in less than 1 week, but between 15°C and 20°C, germination may require 2 weeks. Seed fungicide treatments are often recommended, especially when planting into cool soils. Because triploid seed germinate poorly and seedlings are not vigorous, they should not be planted unless conditions are ideal. Also because seed is expensive and germination requirements exacting, transplants are used to reduce establishment costs. For producing transplants, triploid seeds are germinated at a high temperature (30°C) and the temperature is then lowered to 22–23°C for seedling growth. Another difficulty common to triploid seed is the chronic adherence of seed coats to emerging cotyledons, which can result in distorted and poor development of seedlings. This occurrence can be reduced by orienting seed with the radicle end up when planted.

Watermelon vines spread rapidly, and because of their length, they utilize considerable surface area. Therefore, wide row or hill spacings are typical for watermelon production. Except for bush-type cultivars, spacings ranging from 1 to 2 m in the row or between hills, and from 2 to 3 m between rows are usual. Plant populations can range from 3200 to 8000 per hectare. Vines of some cultivars are sometimes trained toward row centers to facilitate cultivation and harvesting. Bush-type cultivars can be spaced closer; distances vary but are usually half that of vining types.

Early-developing fruit have an inhibitory influence on the development of later-formed fruit. Most plants do not adequately support more than two or three fruit, and late-set fruit seldom achieve acceptable size or maturity unless the growth period is extended. Fruit thinning is sometimes practiced to improve the size and sugar accumulation of remaining fruit.

Harvest and Postharvest

Cultivars differ greatly as to time from planting to fruit maturity; some are harvested about 3 months after planting, others may require 5 months. Ideally, fruit are harvested when the sugar concentration is highest. Soluble solids readings taken from the center of the fruit should be about 10%. Soluble solids of 12% to more than 13% can be achieved by some cultivars. Harvest delay may result in flesh texture that becomes mealy and stringy. Maturity determinations, such as groundspot color change, tendril drying, and tapping or thumping are subjective and not as reliable compared to soluble solids measurement or direct tasting. When sampled fruit are found to be suitable, it is assumed fruit of similar size and age are of equivalent maturity and, therefore, other fruit in the same field can be harvested.

Fruit should be cleanly cut at the stem end rather than pulled from the vine. Fruit rinds may give the appearance of strength but actually are susceptible to cracking from compression or impact shock and should never be stacked on stem or blossom ends. Fruit are most turgid and susceptible to cracking during early morning hours. Flesh firmness and rind toughness are important cultivar characteristics for resisting damage. To maintain harvest quality, fruit should be quickly cooled to about 15°C.

Storage

Watermelons are not suited to long-term storage, but are easily stored at 13–16°C and 80% RH for 2–3-week periods with little quality loss. The somewhat waxy rind limits desiccation. Extended exposure to 10°C or less results in quality loss and possible chilling injury, although high-temperature preconditioning improves subsequent low-temperature storage. Normally, fruit ethylene evolution is slight, but exposure to other ethylene sources accelerates quality deterioration.

Production

Asia clearly leads the world in watermelon production with two-thirds of the volume, followed well behind by European (13%) and African (6%) production. China with 23% of the world's watermelon production provides most of the Asian volume. See Table 24.4.

TABLE 24.4.   WORLD WATERMELON PRODUCTION, 1994

	Area (ha × 10³)	Yield (t/ha)	Production (t × 10³)
World	1,824	16.1	29,360
Africa	125	15.8	1,978
North and Central America	133	17.8	2,364
South America	129	9.0	1,162
Asia	1,104	17.9	19,783
Europe	339	11.8	3,993
Oceania	5	17.3	80
Leading countries			
China	363[a]	18.6	6,760[a]
Turkey	135[a]	25.9	3,500[a]
Iran	140[a]	18.4	2,580[a]
United States	84	21.6	1,814
Korea, Rep.	39[a]	23.2	898[a]
Georgia	60[a]	13.3	800[a]
Japan	22[a]	32.7	710[a]
Egypt	33[a]	21.2	700[a]
Uzbekistan	62[a]	11.3	700[a]
Greece	18[a]	36.4	656[a]

[a]Estimated.
Source: 1994 FAO Production Yearbook, Vol. 48, FAO, ROME, 1995.

CITRON, *Citrullus lanatus* var. *citroides* (L.H. Bailey) Mansf.

Other name: Preserving melon

The leaf blades of citron plants are more broad and less lobed than those of watermelon, but the fruit resemble and are easily mistaken for watermelon. However, the rind is hard and very tough; immature fruit are very bitter. Flesh color is white or pink tinged; seed are usually a greenish tan color. The rind of mature fruit is fed to animals; and also is sweet pickled and used for candies and in baked products.

PUMPKINS, SQUASHES, and GOURDS, *Cucurbita pepo* L., *C. moschata* Duch. ex Poir., *C. maxima* Duch., *C. argyrosperma* Huber *(C. mixta)*, and *C. ficifolia* Bouche

Cucurbita is a New World genus of about 20 species, most are mesophytic, and a few xerophytic. They have been and remain important in diets of Western Hemisphere populations ranging from the tropics to warm, temperate regions. Additionally, cultivation of the domesticated *Cucurbita* species has spread beyond their New World origin with the result that Asia, and China in particular are major producers. *Cucurbita,* especially summer squash, production in the Mediterranean region, the Middle East and many European countries is also significant.

Five *Cucurbita* species, *C. pepo, C. moschata, C. maxima, C. argyro-sperma,* and *C. ficifolia,* comprise the principal cultivated squash and/or pumpkin crops (Fig. 24.7). Both mature and immature fruit are the most important edible plant parts, although for some species, seed, flowers, roots, and even leaves are consumed. Wild *Cucurbita* fruit are bitter; however, the seeds usually are not. Undoubtedly, early domesticators found nonbitter mutant fruits and practiced selection for plants producing nonbitter fruits.

In regard to production volume, summer squash *(C. pepo)* is the most important *Cucurbita* species, with world production estimated to exceed 7 million tons. Summer squash have a mild flavor, eaten raw or cooked, and have a short storage life compared to the strongly flavored winter squash or pumpkins, which include all five of the previously mentioned species. In contrast to summer squash, the postharvest life of winter squashes and pumpkins is much longer and the fruit are not eaten raw.

Both squash types, summer and winter are processed by canning, freezing, and by dehydration. Roasted seed of some species are a favorite and highly nutritious food. "Flor de calabaza," the large, usually male flowers of some cultivars, are another popular specialty food. Some pumpkin and winter squash cultivars are grown for livestock feed.

The word squash may have been derived from an American Indian word for the *Cucurbita* fruit. The word pumpkin is believed to have originated from the Greek "pepon," meaning large melonlike fruit. The French name "potiron" was modified to pampion and to pomkin by the English, and finally as pumpkin by North American colonists. Squash and pumpkin have no precise botanical identity and the terms are used interchangeably and with some confusion. One generalization is that pumpkin flesh is more coarse and stronger flavored than that of winter squash. Another is that squash have a traditional vegetable usage, consumed after baking or boiling, whereas pumpkins are used for pie making, for ornaments, and for livestock feed. In Europe and other parts of the world, the terms "courgette" and "vegetable marrow" identify specific forms of summer squash.

Origin

Species of *Cucurbita* are natives of tropical and subtropical America, with some having relatively wide environmental adaptation. Archaeological evidence indicates they were cultivated in pre-Columbian times.

Cucurbita pepo is believed to be the oldest of the domesticated species; its cultivation ranged from southern Mexico into southwestern United States about 8000 B.C. *Cucurbita pepo,* the most diverse *Cucurbita* spe-

FIG. 24.7.   Composite of different *Cucurbita* fruit: (a) pumpkins, (b) winter squashes, and (c) summer squashes, and (d–f) various gourds of *Cucurbita pepo*. Source (a–c): Courtesy of National Garden Bureau, Downers Grove, Illinois.

cies, has slightly more cold temperature tolerance than other related species, with the exception of *C. ficifolia.*

Evidence has been found for the presence of *C. moschata* in southern Mexico about 5000 B.C., and Peru about 3000 B.C. *C. moschata* tolerates high temperatures better than the other cultivated species and is probably the least cold tolerant. *C. argyrosperma* domestication in southern Mexico probably occurred about 5000 B.C. Until recently, plants of *C. argyrosperma* were classified as *C. mixta,* and previous to that were not identified as a different species but rather as variants of *C. moschata.* Wild species, *C. lundelliana* and *C. okeechobeensis* subsp. *martinezii,* are also found in areas of Mexico and Guatemala. Domestication of *C. maxima* appears to have occurred in southern South America as evidenced from excavated seed that were dated to 1200 A.D.. A related wild species, *C. maxima* var. *andreana,* is also found in South America.

Cucurbita ficifolia, indigenous to the highlands of Mexico and Central and South America, has good cold tolerance. Flowering is short-day responsive, but some forms are day neutral. Buffalo gourd, *C. foetidissima,* that is native to southwestern United States and northwest Mexico is discussed in Chapter 14.

Plants, classified as *C. pepo* var. *ovifera,* produce hard-shelled, bitter, inedible, often multicolored fruits of various shapes and surface textures. These are used as ornamentals, although their bright-colored patterns fade in less than a year. Also producing similar hard-shelled inedible fruit is *C. pepo* var. *texana.* This wild form occurs in the south central United States.

Some wild species have desirable traits, such as disease resistance, which can be transferred to cultivated species. *C. lundelliana* has powdery mildew resistance and *C. ecuadorensis* and *C. foetidissima* have resistance to CMV, PRSV-W, and WMV. Some wild species can also be useful as bridges or intermediates in crosses that otherwise are difficult to achieve. For example, *C. okeechobeensis* subsp. *martinezii* can be crossed with *C. moschata* in order that the resulting progeny are then able to cross with *C. pepo.*

Taxonomy

The domesticated species of *Cucurbita* are generally isolated from one another. Although there are barriers to hybridization of *Cucurbita* species, none is completely isolated from all others. Although difficult, crosses among them can be made, but the resultant interspecific hybrids are usually self-sterile or sparingly fertile. New cultivar forms have been developed from certain interspecific crosses; an example is the cultivar, Iron Cap, a hybrid of *C. maxima* × *C. moschata.*

The following listings shows the common names and cultivar groups of cultivated *Cucurbita species,* and a brief description of the different summer squash groups.

Cucurbita species	*Common names and cultivar groups*
C. pepo subsp. *pepo*	Pumpkins, winter squashes (acorn), summer squashes (marrow, zucchini, cocozelle), and ornamental gourds
C. pepo subsp. *ovifera*	Summer squashes (scallop, crookneck) and ornamental gourds
C. maxima	Winter squashes and pumpkins (Boston marrow, hubbard, banana, Turk's turban, delicious)
C. argyrosperma (C. mixta)	Winter squashes and pumpkins (green striped cushaw, Japanese pie, Tennessee sweet potato)
C. moschata	Winter squashes and pumpkins (cheese, golden cushaw, Canada crookneck, butternut)
C. ficifolia	Figleaf/malabar gourd
C. foetidissima	Buffalo gourd

Summer squash group	*Fruit characteristics*
Cocozelle	Long, slightly tapered, and cylindrical, broadened or bulbous near-distal portion; wide range of green color, smooth surface
Crookneck	Elongated, narrow, and slight or strongly curved neck, broadened distal half; yellow skin intensifies and surface become warty with maturity.
Marrow (vegetable marrow)	Relatively short, strongly tapered toward blossom end, usually pale green, smooth surface
Scallop	Flat with scalloped margins, yellow and green cultivars, surface generally smooth
Straightneck	Cylindrical, with constriction at stem end, with a tendency to

	become broadened at distal half, smooth surface, yellow
Zucchini	Long, uniformly cylindrical, relatively blunt at stem and blossom ends, smooth surface, colors range from light to very dark green, and also yellow

Botany

Cultivated *Cucurbita* are monoecious plants, most of which have long trailing vines and a prostrate growth habit unless trellised. Certain cultivars of *C. pepo* (most being summer squashes) and some *C. maxima* plants have short internodes and a bushy growth habit. Taproots are moderate to deeply rooted, with most of the rapidly established root system horizontally extensive, although relatively shallow. Flowers are bright yellow, borne singly in leaf axils, and seldom open more than 1 day. Most *Cucurbita* species are day neutral, although a few have photoperiod sensitivity.

Cucurbita fruit are among the largest in the plant kingdom. Several large-fruited cultivars of *C. maxima* and *C. argyrosperma* bred expressly for exhibition purposes have achieved weights greater than 440 kg; many people believe that a 500-kg fruit can be achieved.

Ornamental use is made of both pumpkin and gourd cultivars of *C. pepo*. The gourds have distinctive shapes and colors, but are not edible because of very hard rinds. The pumpkins are edible and do not have lignified rinds. The texture of the edible portion, which is pericarp tissue, can vary considerably, and the color can range from white or varying shades of yellow to dark orange. Seed colors can be white, tan, brown, or black.

Some *C. pepo* cultivars produce what is sometimes called "naked seed" (Fig. 24.8). The result of this inherited variant characteristic is very thin nearly transparent seed coat in place of the normally thick coat. Because the seed coat is formed from maternal tissue, cross-pollination of these cultivars does not affect seed coat production in the current crop. Although the feature improves seed edibility, the characteristic makes seed more susceptible to decay after planting. Another unusual trait is seen in the spaghetti squash, also a cultivar type of *C. pepo*. After cooking, the edible pericarp tissue of this squash can easily be separated into loose strands resembling spaghetti. In Japan, spaghetti squash is used fresh or pickled in salad because of its self-shredding quality.

FIG. 24.8.   Naked seed of *Cucurbita pepo.* Note: the dark appearance of seed in the center are those with a thin parchment-like seed coat (naked), the other seed have a normal thick seed coat.

Although leaf shape and leaf surface markings can vary within a species, a combination of stem, androecium, peduncle, flesh texture, and seed features are sufficiently distinct to be useful for species identification. These features are indicated in Table 24.5.

Culture

Most cultivated *Cucurbita* are well adapted for growth at temperatures from 18°C to 30°C; most grow poorly at lower temperatures and are damaged or killed by frost. Plants generally do not grow well in the wet tropics, although certain forms of *C. moschata* are adapted to that region. Fertile, well-drained soils with a pH range from 6.5 to 7.5 are best suited for good growth and yields.

The large leaf area of many *Cucurbita* species can result in high evapotranspiration. However, many cultivars because of moderately deep to deep root systems possess relatively good drought tolerance. Nevertheless, because of the high moisture requirements of these crops, soils with high water holding capacity and well supplied with moisture are suited for high crop yields. Summer squashes having less extensive rooting are more readily stressed by low soil moisture. In general, vining *Cucurbita* crops utilize about 500–900 mm of water during growth.

TABLE 24.5.   IDENTIFYING CHARACTERISTICS OF
DOMESTICATED *Cucurbita* SPECIES

Species	Stems
C. pepo	Hard, 5 angled, no tendrils in bush cv.
C. maxima	Soft, round
C. argyrosperma	Hard, 5 angled
C. moschata	Moderately hard, smoothly 5 angled
C. ficifolia	Hard, smoothly angled
	Leaves
C. pepo	Deeply lobed, spiculate
C. maxima	Slight lobing, moderately spiculate
C. argyrosperma	Moderate lobing, nonspiculate
C. moschata	Shallow lobing, nonspiculate
C. ficifolia	Moderate lobing, moderately spiculate
	Androecium
C. pepo	Short, thick, conical
C. maxima	Short, thick, columnar
C. argyrosperma	Long, slender, columnar
C. moschata	Long, slender, columnar
C. ficifolia	Short, thick, columnar
	Peduncle
C. pepo	Hard, angular, ridged
C. maxima	Soft, rounded, enlarges by soft cork
C. argyrosperma	Hard, angular, enlarges by hard cork
C. moschata	Hard, angular, flared at fruit attachment
C. ficifolia	Hard, smoothly angled, slight flaring
	Fruit flesh
C. pepo	Coarse grained
C. maxima	Fine grained
C. argyrosperma	Coarse grained
C. moschata	Fine grained, gelatinous fibers
C. ficifolia	Coarse grained
	Seed attachment
C. pepo	Obtuse, symmetrical
C. maxima	Acute, asymmetrical
C. argyrosperma	Obtuse, slightly asymmetrical
C. moschata	Obtuse, slightly asymmetrical
C. ficifolia	Obtuse, slightly asymmetrical
	Seed margin
C. pepo	Smooth, obtuse
C. maxima	Smooth, obtuse
C. argyrosperma	Slightly scalloped, acute
C. moschata	Scalloped, obtuse
C. ficifolia	Smooth, obtuse

Seed are planted about 2 cm deep in heavy soils and about 5 cm deep in sandy soils. Soil temperatures should be above the minimum of 15°C for seed germination; at 30–35°C, emergence can occur within a week. In some tropical areas, plants occasionally are propagated from cuttings.

Wide plant spacings are made to accommodate spreading vine growth. Spacing within rows can vary greatly depending on plant type and cropping objective with regard to fruit size, number, and yield.

Thus, spacings from 50 to 150 cm in rows and from 2 to 3 m between rows are not unusual. Wide spacings are also used to facilitate inter-cropping cultivation, frequently practiced in the tropics. Obviously, bush-type squashes are spaced closer, so that populations may be 100–200% greater than a vining crop. *Cucurbita* crops are not self-pollinating and when native insect pollinators are inadequate, bee colonies can be placed in or near the field.

Harvest, Postharvest, and Storage

Most pumpkins and squashes require 80–150 days of growth before being harvested; the period is strongly cultivar influenced. Pumpkins and winter squashes are allowed to fully mature before harvesting. However, summer squash are harvested as immature fruit, often as soon as 40–50 days after planting. Some summer squash fruit are harvested early in their development, often just a few days after anthesis and with the corolla still attached. In certain markets such fruit are in high demand and accordingly have a high value. Developing fruit tend to suppress subsequent pistillate flowering. Often in order to favor development of early set fruit in pumpkins and squashes, the late fertilized fruit are removed.

Essentially, all *Cucurbita* fruit are hand harvested, and except for summer squash, rind hardness is a usual indication of maturity, sometimes accompanied by vine senescence. When ready for harvest, fruit are carefully cut free from the vine with a sharp knife to minimize peduncle injury, a possible site for disease entry. Harvesting of summer squashes is more difficult than winter squash because the short internodes result in closely spaced fruit that can interfere with removal. Summer squash fruit are cut or twisted from the soft succulent peduncle. Having soft rinds, the fruit are easily scratched by the stiff foliar trichomes and also are very susceptible to other physical handling injuries as well as rapid moisture loss.

Pumpkins and winter squashes, despite having a relatively hard rind, can be damaged by rough handling. Another consideration is to avoid exposure of fruit to bright sunlight or frost. Some cuts and bruises can be healed (suberized) to restrict pathogen entry into wounds. This is accomplished by subjecting fruit to temperatures between 27°C and 30°C at 80% RH for about 10 days. Following the curing period, further storage is maintained at temperatures between 10°C and 15°C and 55–60% RH. Under these conditions, healthy fruit can be stored for 1–6 months; the period varies with the specific commodity. On the other hand, satisfactory quality of summer squashes can be maintained

TABLE 24.6.   WORLD PUMPKIN, SQUASH, AND GOURD PRODUCTION, 1994

	Area (ha × 10³)	Yield (t/ha)	Production (t × 10³)
World	668	12.6	8404
Africa	77	14.7	1136
North and Central America	74	6.0	443
South America	71	11.0	786
Asia	279	14.2	3951
Europe	154	12.5	1925
Oceania	12	13.4	163
Leading countries			
China	94[a]	17.2	1616[a]
Ukraine	44	18.0	790
Argentina	38[a]	9.8	368[a]
Romania	80[a]	4.5	360[a]
Turkey	19[a]	17.9	341
Mexico	41[a]	7.8	320[a]
Egypt	20[a]	15.0	300[a]
South Africa	16[a]	18.6	298
Japan	19[a]	15.3	285[a]
Spain	7[a]	35.7	250[a]

[a]Estimated.
Note: The production figures are not separated by species or commodity groups and do not include a considerable volume of home garden production.
Source: 1994 FAO Production Yearbook, Vol. 48, FAO, Rome, 1995.

for about 7–10 days at 10°C and high humidity; temperatures below 10°C, even for a few days, can cause chilling injury.

Production

With 47% of world production, Asia is the leading continent and China the leading country with 19%. Production in developing countries is nearly four times that of the developed countries, a clear indication of the staple food use of the *Cucurbita* species. See Table 24.6.

FIGLEAF GOURD, *Cucurbita ficifolia* Bouche

Other names: Malabar gourd, chilacayote

Figleaf gourd is a cultivated *Cucurbita* species and is of some importance in Mexico and highlands of South America. The plant is an annual or short-lived perennial with figlike leaves and fruit having edible black-colored seeds; some accessions have white seed. The dry, fibrous, white flesh of mature fruit is candied or fermented to produce an alcoholic drink. Mature fruit are often boiled with sugar and eaten as a dessert food. Tender immature fruit are used like summer squash, the young leaves and stem tips for greens, and the seed are roasted.

Flowering is monoecious and requires a short photoperiod. The taproot is long and deep, and lateral roots are several meters long. Plants are sometimes used as root stock for glasshouse cucumber because of their disease resistance and cold tolerance. The cold adaptation enables the crop to grow at high elevations. Leaf blades are nearly round, but lobed and about 20–25 cm in diameter. Light orange flowers are solitary and insect pollinated. Fruit are globular or cylindrical, from 15 to 50 cm long with white stripes or blotches on a green background; fruit color can also be entirely white or green. The rind of mature fruit is hard and relatively smooth. Mature fruit have an excellent storage life, as long as a year without refrigeration.

CHAYOTE, *Sechium edule* (Jacq.) Swartz.

Other names: Vegetable pear, mirliton, chuchu, chayotl, mango squash

Chayote is primarily grown for its edible fruit. The plant is an herbaceous, short-day, perennial, climbing vine; some vines are as long as 15 m. Indigenous to southern Mexico and Central America, chayote was well known to the Aztec civilization. The name chayote was derived from the Aztec word chayotli. Plants are best adapted to a warm tropical climates and least adapted to arid or very high elevations.

Stems are longitudinally furrowed with large branched tendrils and are nearly hairless. Leaves are large and ovate, with shallow lobes. Separate male and female flowers are produced in the same leaf axils; staminate flowers occur in small clusters, and pistillate flowers are found singularly on a short peduncle. Flowers are a greenish cream color.

Fruit often are pear shaped, fleshy, and soft skinned, with longitudinal grooves and a whitish green color (Fig. 24.9). The blossom end characteristically appears incompletely closed. Another unique fruit characteristic is its single, large (3–5 cm), flat, white seed; occasionally, parthenocarpic fruit occur. Parthenocarpy can also be induced by gibberellic acid (GA_3 and $GA_{4/7}$ at 1000 ppm), applied to pistillate flowers.

Cultivars are classified by fruit shape and color features. Some cultivars have spines or thorns on the fruit; the fruit peel can cause skin irritations to some people. Among cultivars, fruit show considerable variation and can range between 10 and 20 cm in length, with diameters from 5 to 15 cm; fruit usually are not cylindrical but slightly flattened. Fruit weight usually ranges from 200 to 400 g, but some fruit weigh more than 1 kg.

FIG. 24.9. Chayote fruit, *Sechium edule,* growing on a trellis in Brazil.

Under favorable day lengths, slightly over 12 h or shorter, flowering can begin 3–5 months after planting. Flowering may not occur unless these conditions are met. Fruit are harvested 30–35 days after anthesis. In the tropics, flowering is continuous, and the prolific plants often are productive for 3–4 years, and very high yields are achievable. As many as 500 fruit have been obtained from a individual plant. An astonishing yield obtained over a 3-year period in Georgia (formerly part of the USSR) reported 300 t of fruit, 22 t of tuberous roots, and 90 t of green shoots per hectare. However, more common are crop yields of 8–10 t/ha per year.

Propagation is by transplanting well-rooted cuttings obtained from young shoots, or by planting intact mature fruit. Being viviparous, seed sprouting may occur while fruit are still on the vine. Ungerminated seed cannot survive outside the fruit; the seed are intolerant of desiccation and low temperatures.

When used for propagation, the fruit are placed about 5–8 cm deep and angled so that the stem end is higher than the blossom end. The sprouted shoot exits from the blossom end and turns upward before emerging from the soil. The fleshy tissue of the fruit functions like an endosperm to nourish the seedling.

The crop grows best in fertile, well-drained soils. The vining growth habit requires wide plant spacings that commonly provide 2 m between plants in the row and 3–4 meters between rows. Plant densities range from 1200 to 1500 per hectare. Vines should have a strong trellis to support growth and to avoid fruit contact with the soil. Shelter from wind is sometimes necessary. Foliage is damaged by temperatures less than 5°C and is killed by frost. In some temperate areas, mulches are used to protect roots against low-temperature injury and are also useful for shoot regrowth after soils warm in the spring.

Careful harvest and postharvest handling are necessary to avoid physical injury of the thin-skinned, tender fruit. Storage at 10–15°C and 90% RH can maintain fruit in good condition for 2–4 weeks. Fruit are chilling sensitive, and if intended for storage or for propagation, they should not be exposed to temperatures less than 10°C.

Chayote are eaten cooked, and sometimes are pickled. The plant's enlarged yamlike starchy roots are also eaten, prepared as a boiled or fried vegetable, and tender shoots and leaves are consumed in soups or as a potherb. Foliage and fruit are also used for animal feed. The flavor of the fruit resembles that of artichoke hearts or new (young) white potatoes, but also may have a sweet watery taste. The latter was reported to be used as an apple substitute for pies and as a food filler. The tuber contains a very good quality starch, reported to be superior to that of arrowroot.

SMOOTH LOOFAH, *Luffa aegyptiaca* Miller *(L. cylindrica)*

Other names: Vegetable sponge, dishcloth or sponge gourd, ghia tori

ANGULAR LOOFAH, *Luffa acutangula* (L.) Roxb.

Other names: Ridged luffa, kali tori, mo kua (mo kwa), Chinese okra

Both *Luffa* species are indigenous to tropical Asia; exactly where is uncertain, as is knowledge of their transfer to the New World, although several theories have been proposed. Loofah is an Arabic name, and the plant and fruit were mentioned in Sanskrit and Egyptian writings. For centuries in India, and other parts of Asia and Africa, the immature fruit of these plants has been used for food and mature fruit for the strong fibrous spongelike tissues. Wild forms of angled loofah, still found in India, are extremely bitter; domesticated cultivars are much less bitter or free of bitterness.

Each species is an annual with vigorous climbing vines. Stems are four or five angled with branched tendrils. Both *L. aegyptiaca* and *L.*

acutangula plants are monoecious, but some *L. aegyptiaca* plants exhibit andromonoecious flowering.

Leaves have rough textured surfaces much like those of cucumber but are larger and many angled with varied lobing. Flowers are yellow and about 5 cm in diameter. Staminate flowers, between 5 and 20, are grouped in racemes in leaf axils. The pistillate flower is solitary and also borne in the same axil.

Luffa aegyptiaca flowers open in the early morning, those of *L. acutangula* in the afternoon. Stigmas of smooth loofah are receptive about a day longer than those of angled loofah, or a total of about 60 and 36 h, respectively. Even though staminate flowers greatly outnumber pistillate, pollination can sometimes be poor, which can result in misshapen fruit. Treatment with the growth regulator, indole acetic acid (IAA), reduces the ratio of male to female flowers, and short days tend to increase pistillate flowering. When pollination is adequate, many black, flat seed, about 12 mm long and 8 mm wide, are produced. Smooth loofah seed are slightly narrower and slightly pitted compared to those of angled loofah.

Seed are planted 1–2 m apart in hills or rows and up to 3 m apart when not trellised. When grown with trellis support, spacings are generally about 1 m × 1 m. Vines are supported in order to prevent the fruit contacting the soil and to obtain well-shaped and straight fruit.

Mature *L. aegyptiaca* fruit are 30–60 cm long, 8–15 cm wide, and smooth and cylindrical with vein tracts (fibrous bundles) which are visible but are not prominent. Mature *L. acutangula* fruit are almost the same size and shape but have 10 distinctive longitudinal acute ribs extending the length of the fruit (Fig. 24.10). Both loofah types have a thin skin. The edible fleshy portion is endocarp tissue.

Loofah plants are capable of producing many fruit. The number is a function of how many fruit are harvested and the length of the growing period. Frequent removal of fruit increase pistillate flowering. For vegetable use, fruit are harvested about 2 months after seeding and while still immature and tender and well before fully elongated; the fruit of angled loofah are preferred to those of smooth loofah. Whichever type fruit are used, they are generally prepared as a boiled or fried vegetable. About 4 or 5 months is required after planting for seed to mature.

For fiber sponge qualities, mature, smooth loofah fruit are preferable to those of angled loofah because their fiber and vascular bundle development is superior. To extract the fibrous tissue, fruit are soaked in water for 1–2 weeks, during which time the outer fruit wall and inner pulp rot and disintegrate; this procedure is called "retting." Subsequent

FIG. 24.10.   Loofah fruit: (a) smooth loofah, *Luffa aegyptiaca,* and (b) angled loofah, *Luffa acutangula.*

washing removes seed, skin, and pulp from the fibers. The fibers are dried, usually by the sun; normally, they are a light tan color and frequently are bleached. In addition to sponge uses, the fibers are used as coarse filters, insulation, padding, and packing materials.

BITTER MELON, *Momordica charantia* L.

Other names: Balsam pear, bitter cucumber, bitter gourd, alligator pear, karela, fu kua (fu kwa), niga uri

Bitter melons are believed to be indigenous to the tropics of India or Southeast Asia, where their cultivation as a food crop and for folk medicines has a long history. It has been reported that a seed protein of bitter melon inhibited growth of the immunodeficiency virus (HIV-1) in human cell cultures.

Bitter melons are perennials, usually monoecious plants, commonly grown as annuals. Plants develop tuberous storage roots and vigorous herbaceous vines, sometimes as long as 10 m. Stems are thin, five angled, and grooved with simple and forked tendrils. Palmate leaves, showing multiple deep lobes, are borne on long petioles.

Small, usually yellow, flowers are borne singly on slender peduncles in leaf axils. Male flowers appear first and greatly exceed the number of pistillate flowers often by 20 or 25 to 1. Flowering usually occurs about 30–35 days after planting, and individual flowers open at sunrise for only 1 day.

The pendulous fruits develop about 10 irregular, longitudinally rounded ridges on a warty surface (Fig. 24.11). The skin of immature fruit is soft and, depending on the cultivar, varies from a dark to light green color; some are white. Fruit are slender, ranging from 5 to 25 cm in length. There are two important cultivar types: one produces fruit that are tapered at each end, the other produces fruit that are not tapered.

With advancing maturity, fruit surface color changes from green to a greenish yellow color, and at full maturity, it is bright orange. Interior flesh color also changes from white to yellow or orange. Mature fruit typically rupture, revealing a somewhat hollow center, and the seed are oval, flat, and gray or brown and are coated with a scarlet-colored pulpy tissue. The seed coating tissue (aril) is edible, sweet, and an attraction to birds, as well as children. With the coating removed, about six seed weigh 1 g.

Seed are planted 1–3 cm deep; when planted into warm soil, emergence can occur in about 1 week. For successful transplant establishment, seedlings should have an undisturbed root system; bare-rooted transplants do not survive very well. Typical plant spacings are 45–60 cm within rows and 120–150 cm between rows, generally resulting in a population between 13 thousand to 17 thousand plants per hectare.

Bitter melons are usually grown on supports similar to other climbing cucurbits to avoid fruit contact with soil. Fruit are often protected from fruit flies with paper bags that are open at the bottom. Interestingly, flies do not enter the bags when fruit are covered in this manner.

a   b

FIG. 24.11. Bitter melon, *Momordica charantia:* (a) plants growing on trellis to make fruit harvest easier, and (b) harvested fruit marketed, Taiwan.

Production of 10–12 fruit per plant or 10–15 t/ha is considered a good yield.

Immature fruit are the primary edible product and these are usually harvested 2–3 weeks after anthesis, which generally is about 55 days after planting. Fruit are cut free of the vines and carefully handled because they are tender and easily damaged. At this harvest stage, individual fruit weigh 80–100 g. Frequent harvests are made to encourage new fruiting, as persisting fruit tend to inhibit flowering. Immature fruit are less bitter than older fruit; fully mature fruit are extremely bitter, and consequently undesirable. The bitterness is due to the alkaloid momordicine.

Exposure to even minute quantities of ethylene accelerates maturation and quality loss. Fruit should be stored at moderate temperatures, and preferably above 12°C to avoid chilling injury. At 12–15°C, fruit can be held for 1–2 weeks.

Fruit are peeled and parboiled or steeped in salty water to reduce the level of bitterness before preparation for consumption. Fruit are also prepared as pickles and as ingredients in curries. Tender shoots and leaves are eaten as potherbs after parboiling to leach out the bitter alkaloid. Bitter melon fruit are a good vitamin C source and provide a fair amount of pro-vitamin A, phosphorus, and iron. Vine tips are an excellent source of pro-vitamin A, and a fair source of protein, thiamine, and vitamin C.

Balsam apple, *M. balsamina,* is similar to bitter melon, but plants are slightly smaller, and fruits are also smaller and have different shapes. Fruit of *M. dioica,* also occasionally known as Balsam apple and as kartoli in India, are small, 3–5 cm long, and not bitter. These and several other *Momordica* species are also used for medicinal purposes.

SWEET GOURD, *Momordica cochinchinensis* (Lour.) Spreng.

Other names: Chinese cucumber, spiny bitter cucumber, kakur, kheksa

Like bitter melon, sweet gourd originated in tropical India and Southeast Asia. The perennial climbing plants have three to five lobed, dark green leaves. Sweet gourd plants are dioecious; flowers are straw colored with a dark purple area at the base.

Fruit are ovate, 6–10 cm in diameter, and up to 20 cm in length. When mature, fruit are bright orange-red, warty, and covered with short sharp spines; the spiny skin is removed before cooking. The light yellow flesh is about 1 cm thick, and the seed are embedded in the orange-red aril. Fruit are less bitter and have a greater proportion of edible flesh compared to bitter melon, *M. charantia.*

Of the two major cultivar types, those with yellowish green oblong fruit are preferred to those that are small, round, and dark green. Plants are propagated by seed or vegetatively with tuberous stem cuttings. Vegetative propagation is preferred because propagating materials are taken from female plants, which assures that the cloned plant produces female blossoms. Nevertheless, a few male plants (10%) should be planted to provide pollen for fertilization.

The plantlets are planted into holes, about 10 cm deep, covered with soil, usually fertilized with about 40 kg P 60 kg K per hectare, and watered. Nitrogen, about 25 kg/ha is applied after establishment and again at fruit set. Plant densities range from 25 thousand to 50 thousand per hectare. Favorable growing temperatures are between 20°C and 35°C and preferably with high humidity. The crop is usually grown during the rainy season and becomes dormant during the dry season unless irrigated.

Flowering begins about 55–65 days after planting. Marketable fruit are harvested 10–12 days following anthesis and when the skin is light green and seed are immature. Sweet gourd is relatively free of diseases, although fruit fly maggots damage fruit and *Epilachna* beetles feed on the foliage.

CHINESE WINTER MELON/WAX GOURD, *Benincasa hispida* (Thunb.) Cogn.

Other names for mature fruit: Winter melon, Chinese preserving
    melon, white and ash gourd, petha, and don kwa
Other names for immature fruit: Chinese squash, mo kwa

Although wild types have not been found, the Indo-Malayan area of Southeast Asia is the likely center of origin. Indochina and India are centers of greatest diversity. The names wax, white, or ash gourd are suggested by the layer of wax that develops on the mature fruit surface of some cultivars. Chinese names for these and other vegetables often vary due to differences in phonetic spellings or different regional pronunciations. Thus, don kwa may be also known as dong kua, dong gua, tong kwa, and other variations.

Generally, plants are vigorous, spreading, herbaceous, stout vines with hairy, whitish green stems reaching several meters in length, and with branched tendrils that are spirally coiled at the tip. Large, broad, heart-shaped, irregularly lobed, obvate leaves are borne on tall petioles.

Plants are annuals and usually monoecious. Staminate flowers have long peduncles, pistillate flowers have short peduncles, and both are solitary in leaf axils.

Flowering begins about 60–80 days after planting; flowers are insect pollinated. The ratio of female to male flowers is increased by cool weather and short days. Young, immature fruit can be harvested within 7–8 days after anthesis (Fig. 24.12a) Harvest of young fruit prolongs the flowering period. Stem tip and flower pruning are sometimes performed to improve the growth of fruit intended to achieve full maturity, which requires 60–70 days or more after anthesis.

Four major cultivar groups are recognized: (1) a late-maturing winter melon group that have unridged seeds and cylindrical (50–100 cm long), dark green fruit with sparse wax; (2) a ridged winter melon group that is similar except the seed are ridged; (3) a fuzzy gourd group with ridged seeds and narrow cylindrical (20–25 cm long), green, hairy, sparsely waxed fruit; and (4) the wax gourd group that also has ridged seeds, and an oblong (10–60 cm in diameter), light green, waxy, and sometimes hairy fruit.

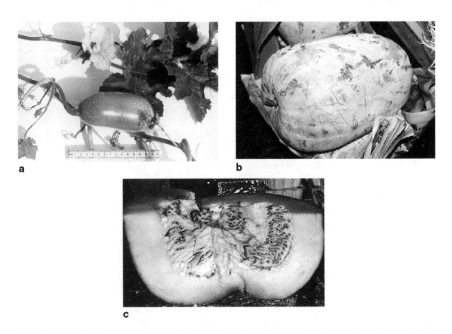

a b c

FIG. 24.12. Wax gourd, *Benincasa hispida:* (a) immature (mo kwa) fruit, (b) mature (don kwa) fruit, and (c) cut-open mature fruit.

The initially hairy fruit become glabrous, and depending on cultivar, become darker green and covered with a dusty white, waxy bloom as they mature; hairy surfaces consist of bristlelike trichomes. The wax layer continues to thicken even after harvest. It is reported that the wax was scraped and used to make candles. The fruit rind is firm but not durable. The juicy, mild-tasting, white flesh is endocarp tissue. Seed are oval, flat, and smooth and are contained in the spongy placenta tissue. They are about 15×5 mm in length and width; about 13–15 seed weigh 1 g.

Specific cultivars are used to produce the immature fruit known as mo kwa. Their immature fruit are harvested when 12–15 cm long (sometimes smaller) and are used like summer squash. Of the several cultivar types grown, the large mature fruit, called don kwa, are usually about 30–50 cm long with a diameter of 15–25 cm, and normally weigh about 10–15 kg, although some weigh as much as 45 kg (Fig. 24.12b). Crop yields as much as 20 t/ha are achieved.

Wax gourd growth is best suited to moderately dry areas of the lowland tropics. Growth is optimum at 24–27°C and poor at lower temperatures. Plants grow well in fertile, well-drained, light soils, preferably when soil pH is slightly acid, 5.5–6.5. Plants have relatively good drought tolerance. Most cultural procedures are similar to those for *Cucurbita* winter squashes. Seed-propagated field populations range from 8 thousand to 10 thousand plants per hectare when trellised, and about half that when vines are allowed to trail over the ground.

The long storage life of mature melons is a valuable feature. When stored at 13–15°C and 70–75% RH, fruit remain in good condition for 6 to as long as 12 months. Suberization of bruises and cuts helps reduce deterioration and decay. Stored fruit must be carefully handled and often are placed in storage so that individual fruit are not in direct contact with each other. Immature fruit do not store well and, therefore, like summer squash are consumed soon after harvest.

During food preparation, the skin of mature fruit is scraped or peeled and the seed cavity is removed and discarded. The white interior flesh is cut into pieces and cooked, used for soup flavoring, and sometimes candied in sugar syrup (tangkwe), or dried for later use. The Chinese use the fruit as a seasonal and festival specialty eaten in the winter and during special holidays. At such occasions, the fruit serves as a consumable soup tureen with the seed cavity containing the cooked ingredients. Additionally, seed are fried, and young leaves and flowers can also be eaten.

BOTTLE GOURD/WHITE FLOWERED GOURD, *Lagenaria siceraria* (Mol.) Standl.

Other names: Calabash gourd, cucuzzi, zucca melon, lauki, Chinese squash, fu kwa

Lagenaria siceraria is an ancient, widely distributed plant especially well suited to semidry tropical and subtropical conditions. Its assumed origin is Africa, and early domestication occurred in the tropical lowlands of south central Africa. Plant remains found in Egyptian tombs were dated to 3500 B.C. Interestingly, rind fragments of similar or possibly older age are evidence of the presence of *Lagenaria* in Peru; specimens found in Mexico were predated to 5000–7000 B.C. This species has the distinction of being the only crop known to have been cultivated in pre-Columbian times in both the Old and New Worlds. How this bihemispheric distribution occurred is speculative and remains unresolved. A leading theory is that mature fruit with viable seeds were dispersed either naturally or with human intervention after drifting across the ocean from Africa to South America. In support of that theory, *Lagenaria* gourds were demonstrated to be able to tolerate seawater immersion for long periods without loss of seed viability. Some taxonomists consider the African and American variants as subspecies, *siceraria* and *asiatica,* respectively.

Bottle gourd is a monoecious annual with long, ribbed, well-branched, climbing vines of lengths from 3 to 15 m. Stems are longitudinally grooved with soft hairs and branched tendrils; one being short, the other long. Leaves are simple, very large, cordate or oval measuring 15–30 cm across with a pubescent velvety surface and musky odor.

Large white flowers are borne singularly in leaf axils, pistillate flowers on short peduncles, staminate on tall slender peduncles; most of these often standing well above the foliage. Flowers open in the evening and usually remain open until the next afternoon. Bees are the usual pollinators, although flowers are adapted to hummingbird and bat pollination. Flowering occurs about 2 months after seedling emergence. Seed mature about 90 days after fertilization. Their shape is rectangular, with ridged grooves at the attachment end, although unridged forms occur. White to tan colored, the seed are flat and between 1 and 2 cm in length. Seed are edible and are used as a snack food.

The developed fruit is a pepo that can vary greatly in shape, size, and color. Cultivar identification is imprecise, relying mostly on fruit shapes and sizes (Fig. 24.13). Globe to elongated clublike and numerous intermediate forms occur; some are long, cylindrical, and coiled.

a

b

c

FIG. 24.13. Bottle gourd, *Lagenaria siceraria:* (a) immature (fu kwa) fruit, (b) mature fruit, and (c) kanpyou (kanpyo) shavings.

Lengths range from 10 cm to over 1 m with some slender and others with diameters greater than 35 cm. Fruit rind color is greenish white to dark green. The interior flesh is white and becomes pulpy as the fruit matures.

Immature fruit are tender and covered with soft hairs. They are harvested about a week or 10 days after anthesis and are used like summer squash; fruit grown for other purposes are allowed to develop further. Frequent harvesting tends to prolong productivity of the plant. Although not affecting sex expression, short days tend to promote flowering.

Propagation and cultivation is similar to that of Chinese winter melon; depending on intended purpose, the cropping period is often extended to 6 or 7 months. Excessive soil moisture should be avoided because of the waterlogging susceptibility of the relatively shallow root system. Light textured, well-drained soils with a near-neutral pH are most favorable. Seed are directly sown; plant densities usually are about 5 thousand to 7 thousand per hectare for unsupported plants and about 10 thousand if trellised. Bottle gourd vines should be supported in order to obtained uniformly shaped fruit.

Wild forms of *Lagenaria* are bitter; cultivated vegetable forms are not. Some types cultivated for utensils and other nonfood uses may be bitter. Immature fruit are prepared and consumed like summer squashes, and are popular in Italy, China and Southeast Asia, and to a lesser degree in Latin America. Immature starchy fruits called kashi in Brazil are used as a vegetable, although their insipid taste is usually overcome with spicy ingredients. The flesh of young fruit, although firm and still moist, is used in the preparation of various vegetable and soup dishes. In Japan, after peeling of the rind, the flesh is thinly sliced and air dried to produce kanpyo which after rehydrating is used in various cooked dishes, and as an ingredient in sushi. *Lagenaria* flesh is also made into candied and sweet spiced glace, used as a moist filler in bakery products, particularly fruit cakes. Young tender leaves are used as a potherb in India and some Mediterranean countries.

Fully mature fruit require 4–5 months of growth and have a hard durable rind; the interior flesh is well desiccated. The seed will rattle when the dry fruit is shaken. Such fruit have a wide variety of uses as containers, fishnet floats, musical instruments, and carved and decorated ornaments.

SNAKE GOURD, *Trichosanthes cucumerina* L. var. *anguina* L. *(T. anguina)*

Other names: Viper gourd, serpent cucumber, chichinda

The center of origin is probably Southeast Asia, where wild forms are found. Wild forms are also found in India and tropical Australia; domestication may have begun in India.

Plants are annual and monoecious, having rapidly growing climbing vines that thrive in warm and humid climates. Optimum mean temperatures are between 30°C and 35°C; temperatures below 20°C restrict growth. Although sensitive to waterlogging, plants require an abundant supply of soil moisture to be productive.

Stems are five angled and grooved with tendrils that are branched. Simple leaves are angular with moderately deep five to seven lobes and are dark green on upper surfaces and lighter green on lower sur-

faces. The pubescent raceme inflorescence consists of five or more staminate flowers on long peduncles. A single sessile pistillate flower is found
in the same leaf node. The single carpel is long and hairy, with many
ovules. Both flower types are fragrant, with showy white delicately
fringed hairlike petals. Flowers open in the early evening and are insect
pollinated. Male flowers form relatively soon after planting and are
the first to appear.

Fruit are long, cylindrical, slender, and tapered (Fig. 24.14). At the
widest portion, fruit diameter ranges from 4 to 10 cm. Fruit lengths
commonly range from 30 to 70 cm, although fruit lengths greater than
150 cm are obtained. Surface colors are greenish white or white and
may be striped; interior flesh is white. Mature fruit become orange or
red and are inedible because of extreme bitterness and fiber development. Mature seed are surrounded with a red pulp. They are flat,

FIG. 24.14. Snake gourd fruit, *Trichosanthes cucumerina.*

thick, about 10–15 mm long, brown, and sculptured, and are used for propagation. Vines are supported on a high trellis to avoid fruit twisting or coiling. Sometimes a small weight is attached to the end of the pendent fruit to assure straight growth. In most other respects, culture is similar to that of other climbing cucurbits. Plant populations are between 7 thousand and 8 thousand per hectare.

Fruit can be harvested as early as 15–20 days after anthesis, and flowering and harvest usually continues for 1–2 months. Most often fruit are harvested about 3 months after planting, and typically weigh 400–500 g. With individual plants providing about 5–10 fruit, and sometimes many more, 15-t/ha yields can be realized. Being succulent, fruit have a short postharvest life, but if stored at 16°C and 90% RH, they will keep well for 2 weeks.

The immature fruit are used as cooked vegetables in salads and in various curry dishes. Young shoots and leaves are also edible. Although all plant parts taste bitter and have an unpleasant odor, these characteristics disappear after cooking. Other *Trichosanthes* species, such as *T. celebica, T. ovigera,* and *T. villosa,* have minor vegetable usage. The Japanese snake gourd, *T. cucumeroides,* is grown in Japan and China as a source of starch. Fruit of wild *Trichosanthes* species are very bitter and inedible but have some medicinal uses. A protein from the roots of snake gourd, *Trichosanthes cucumerina,* was reported to inhibit growth of immunodeficiency virus (HIV-1) in human cell cultures. An additional compound with similar apparent anti-HIV activity is extracted from *Trichosanthes kirilowii* var. *japonica.*

POINTED GOURD, *Trichosanthes dioica* Roxb.

The pointed gourd is similar to snake gourd but is a dioecious perennial, and although popular in India where it is known as "parwal," the crop is infrequently cultivated elsewhere. Propagation is with stem cuttings taken from female plants. Cuttings about 60 cm long with several nodes are formed into a coil and planted to produce several rooted seedlings, which are then used for transplanting. Another method uses rooted suckers removed from the mother plant. When transplanting, female plants are interspersed with male plants. Fruit are smaller than those of snake gourd and also become orange when mature. Peeled fruit are boiled with spices, made into curries, or pickled; tender shoots are sometimes cooked as a potherb.

CAIHUA, *Cyclanthera pedata* (L.) Schrader var. *pedata*

Other names: Wild cucumber, achoccha, korilla, meetha karela

A native of Andean South America, with likely domestication in Central America, this plant is grown in limited volume in the Caribbean region, Mexico, Central America, and western South America, and now occasionally in the Old World tropics, namely India. The plant has a broader adaptation to climates than many other cucurbits and will tolerate some cold and, therefore, is suitable for higher elevations in the tropics. A related species, *C. brachystachya (C. explodens),* having similar usage is slightly more cold tolerant. It is mostly grown in the Andes but is also cultivated in Japan.

Caihua is an annual, monoecious vigorous spreading plant with climbing vines greater than 3 m in length. Plants are more productive when vines are supported. The crop is grown in hill plantings generally spaced about 1 m or more apart. Stems are glabrous with both simple and branched tendrils and emit a noticeably strong odor. Leaves are 8–18 cm wide, about 20 cm long, and show considerable lobing. Flowers are inconspicuous, small, and greenish white or white. Staminate flowers are formed in racemes; the pistillate flower is solitary. Fruit shapes are ovoid and oblique, and they are often spiny or warted, and beaked (Fig. 24.15). The pale green fruit are about 5–15 cm long with a diameter between 5 and 10 cm. The fleshy rind is 3–4 mm thick and the seed cavity is spongy and partly hollow. The flat, black seeds (6–8 mm long) tend to remain attached to the white, spongy, placental tissue.

Fruit are harvested about 3 months after planting and remain productive with abundant fruiting usually continuing for several months. Four to six harvests are possible during a crop year. Full size, but still immature, fruit are harvested for use either raw in salads or cooked; shoots are also edible. One preparation method involves stuffing the fruit cavity after seed removal.

FLUTED GOURD/FLUTED PUMPKIN, *Telfairia occidentalis* Hook. f.

The fluted pumpkin is indigenous to western Africa; wild types are not found. The plant is cultivated extensively in Nigeria and Ghana for its edible leaves and large seeds.

The crop is well adapted to hot and moist environments, yet has

FIG. 24.15. Caihua fruit, *Cyclanthera pedata*.

good drought tolerance. Plants are perennial, dioecious, climbing vines. During culture, the long vines are often supported by nearby trees. The five-angled, ridged stems, petioles, and leaf blades are covered with multicellular hairs. Compound leaves have three to five leaflets. Roots contain toxic alkaloids and saponins.

Flowers are white with a dark purple mark at the inside base of the five petals. Male plants produce flowers on racemes about 30 cm in height; flowers of female plants are solitary. Fruit are 10-ribbed, usually about 50–60 cm long, and covered with a white waxy bloom. Seed are large, somewhat circular, flat, red, 35–45 mm in diameter, 15 mm thick, and rich in oil and protein. They are peeled and cooked whole and eaten like beans, roasted, and as a powder after processing. Oil extracted from seed has nondrying properties and characteristics similar to olive oil.

Growing on a trellis results in superior yields compared to cultivation that permits vine trailing on the ground. Tender leafy shoots, 50 cm long, are cut for marketing or home use at 3–4 week intervals. Shoots and leaves contain high levels of protein. Frequent harvests tend to encourage new shoot formation and thus extends the harvest period.

The crop is seed propagated, but seed are short lived and sometimes viable for only a few days. Seed occasionally are found already germinated *(vivipary)* in the fruit. Shoot emergence is slow, often requiring

2 weeks. Ratoon propagation is sometimes used in order to ensure a high population of female plants.

OYSTER NUT, *Telfairia pedata* (Smith ex Sims) Hook. f.

Oyster nut, possibly an east African native, is a perennial and prolific climber with extensive branching and vine length that enables the plant to climb and spread over a large area. The stem base sometimes exceeds 15 cm in diameter. The plant is dioecious; male flowers form on a raceme; flowers of female plants are solitary. Propagation is usually with cuttings.

The plant is grown for its large, 35–40-mm diameter, 10-mm thick seed which are contained in a large, oval-shaped, ridged gourdlike fruit. Fruit length ranges from 45 to 90 cm and are about 20 cm in diameter. The rind is about 1 cm thick and large fruit weigh about 10 kg. When ripe, the fruit splits, and as many as 200 seed can be released. The harvested seed are dried, and if kept dry, have a long storage life.

The seed surface is covered with corklike fibers, and after the bitter-tasting seed covering is removed, the kernel are eaten raw or roasted. They are also eaten in soups and sometimes are pickled. An edible oil can be extracted from the seed. Leaves and shoots are also edible.

IVY GOURD, *Coccinia grandis* (L.) Voigt *(C. cordifolia, C. indica)*

Coccinia grandis, also known as tindora or tondli and as kundsru in India, occurs wild from Africa to Malaysia. Its cultivation occurs mostly in India and the southeastern Asian tropics. The glabrous climbing perennial vining plant produces a tuberous root. The stem is longitudinally ribbed when young and becomes round and woody when older. A tendril is formed at each node. Leaves are simple, alternate, and broadly ovate to almost pentagonal in shape; the largest leaves are about 15 cm long and 12 cm wide.

Plants are dioecious, with large white flowers. Male flowers develop singularly or paired in the axils of a short raceme; female flowers also develop axillary and are solitary. Pollination is accomplished by insects.

Growth is best suited to well-drained soils in areas of high rainfall and humidity. The crop is generally trellised and its cultivation is similar to that for bitter melon, *Momordica charantia.* Seed is seldom used for propagation because the dioecious character of the plant results

FIG. 24.16.   Ivy gourd fruit, *Coccinia grandis.*

in a high proportion of nonproductive male plants. The preferred ratio of female to male plants is 10 : 1. For propagation, stem cuttings of 10–15-cm length, are commonly used. These are planted at spacings to provide about 8 thousand to 10 thousand plants per hectare.

Fleshy fruit are elliptical in shape, about 3–7 cm long and 1.5–4 cm wide, and green with white striping when young, and become red when mature (Fig. 24.16). The white seed are small and flat with a feltlike surface.

Young fruit are used in soups and curries. Mature fruit of sweet cultivars can be eaten without cooking and are often candied. Fruit of wild or semidomesticated forms are very bitter but can be used as dried chips that are rehydrated before consumption. The young shoots and leaves are also popular and often consumed as a potherb or fried greens.

CHINESE LARD FRUIT, *Hodgsonia macrocarpa*

A dioecious perennial, Chinese lard fruit is native to northern India and southern China tropical forests. Vines 30 m long or more produce nearly spherical large fruit, 20–25 cm in diameter. Large (6–8 cm long) seed are eaten raw or roasted and have a lardlike taste. Seed have a high content of extractable edible oil; foliage is also edible.

WHITE SEEDED MELON, *Cucumeropsis mannii* Naudin

Native to west tropical Africa, this monoecious plant is cultivated mainly for the seed which have a high protein and oil content. The seed are white and about 2 cm long, and are used in soups, ground into a vegetable paste, and for cooking oil. Fruit and leaves are also edible.

Future Improvements

Cucurbitaceae remains an important family for several vegetables having a status close to that of a food staple, particularly the various squash species. Additionally, some members, such as melons and watermelons, are likely to remain favorites for their refreshing taste.

Although some melon and watermelon products are bulky, there is considerable international transfer, primarily from less developed to developed countries. This is an export opportunity developing countries are likely to pursue further.

Nearly a half century since its introduction, the seedless watermelon only recently has moved from obscurity to some level of prominence that promises to expand. Watermelon male sterility could replace the difficult production of triploid seed, presently used for seedless watermelon production. Continued use of dwarf cultivars can make land use more efficient in the production of cucumbers, melons, pumpkins, and squash.

The ability to manipulate floral sex offers promise for the production of earlier and higher yields. Plant breeders utilizing molecular technologies have successfully transferred virus resistance into summer squash cultivars. Other similar advances for disease and pest resistance will reduce some of the effort and pesticides required for pest management. Product-edible quality improvement are additional goals toward which biotechnology will be applied.

SELECTED REFERENCES

Akoroda, M.O., and Adejoro, M.A. 1990. Pattern of vegetative and sexual development of *Telfairia occidentalis* Hook. f. Trop. Agric. (Trinidad) *67*, 243–247.

Aung, L.H., and Flick, G.J. 1976. Gibberellin induced seedless fruit of chayote (*Sechium edule*, Schw.) HortScience *11*, 460–465.

Aurin, M.T.L., and Rasco, E.T., Jr. 1988. Increasing yield in *Luffa cylindrica* L. Roem. by pruning and high density planting. Philippine J. Crop Sci. *13*, 87–90.

Bates, D.M., Robinson, R.W., and Jeffrey, C., eds. 1990. Biology and Utilization of the Cucurbitaceae. Cornell University Press, Ithaca, NY.

Bianco, V.V. 1986. Variations in winter melon (*Cucumis melo* L. var. inodorus Naud.) germplasm collected in Southern Italy. HortScience *21*, 891.

Bush, A. 1978. Citron melon for cash and condiment. Econ. Bot. *32*, 182–184.

Cantliffe, D.J. 1981. Alteration of sex-expression in cucumber due to changes in temperature, light intensity and photoperiod. J. ASHS *106*, 133–136.

Chisholm, D.N., and Picha, D.H. 1986. Distribution of sugars and organic acids with ripe watermelon fruit. HortScience *21*, 501–503.

Cutler, H.C., and Whitaker, T.W. 1961. History and distribution of the cultivated cucurbits in the Americas. Am. Antiquity *26*(4), 469–485.

Da Costa, C.P., and Jones, C.M. 1971. Cucumber beetle resistance and mite susceptibility controlled by the bitter gene in *Cucumis sativus* L. Science *172*, 1145–1146.

Dane, F., Denna, D.W., and Tsuchiya, T., 1980. Evolutionary studies of wild species in the genus *Cucumis*. Z. Pflanzenzucht *85*, 89–109.

Decker, D.S. 1988. Origin(s), evolution, and systematics of *Cucurbita pepo*. Econ. Bot. *42*, 4–15.

Engles, J.M.M. 1983. Variation in *Sechium edule* in Central America. J. ASHS *108*, 706–710.

Heiser, C.B., Jr., and Schilling, E.E. 1988. Phylogeny and distribution of *Luffa* (Cucurbitaceae). Biotropica *20*, 185–191.

Heiser, C.B., Jr., Schilling, E.E., and Dutt, B. 1988. The American species of *Luffa* (Cucurbitaceae). Syst. Bot. *13*, 138–145.

Heiser, C.B., Jr. 1979. The Gourd Book. University of Oklahoma Press, Norman, OK.

Herklots, G.A.C. 1972. Vegetables in South-east Asia. George Allen & Unwin Ltd., London.

Jeffrey, C. 1980. A review of the Cucurbitaceae. Bot. J. Linnean Soc. *81*, 233–247.

Kihara, H. 1951. Triploid watermelons. Proc. ASHS *58*, 217–230.

Lester, G.E., and Dunlap, J.R. 1985. Physiological changes during development and ripening of "Perlita" muskmelon fruits. Sci. Hortic. *111*, 323–331.

Lower, R.L., Pharr, D.M., and Horst, E.K. 1978. Effects of silver nitrate and gibberellic acid on gynoecious cucumber. Cucurbit Genet. Coop. Rep. *3*, 15–16.

Maynard, D.N. 1989. Triploid watermelon seed orientation affects seedcoat adherence on emerged cotyledons. HortScience *24*, 603–604.

Okoli, B.E. 1984. Wild and cultivated cucurbits in Nigeria. Econ. Bot. *38*, 350–357.

Paris, H.S. 1988. Historical records, origins, and development of the edible cultivar groups of *Cucurbita pepo* (Cucurbitaceae). Econ. Bot. *43*, 423–443.

Paris, H.S. 1996. Summer squash: History, Diversity, and Distribution. HortTechnology *6*, 6–13.

Pike, L.M., and Peterson, C.E. 1969. Inheritance of parthenocarpy in cucumber (*Cucumis sativus* L.). Euphytica *18*, 101–105.

Provvidenti, R. 1990. Viral diseases and genetic sources of resistance in *Cucurbita* species. In Biology and Utilization of the Cucurbitaceae, D.M. Bates, R.W. Robinson, and C. Jeffrey, eds. Cornell University Press, Ithaca, NY, pp. 427–435.

Richardson, J.B. 1972. The pre-Columbian distribution of the bottle gourd (*Lagenaria siceraria*): a re-evaluation. Econ. Bot. *26*, 265–271.

Robinson, R.W., Munger, H.M., Whitaker, T.W., and Bohn, G.W. 1976. Genes of the Cucurbitaceae. HortScience *11*, 554–568.

Shannon, S., and Robinson, R.W. 1979. The use of ethephon to regulate sex expression of summer squash for hybrid seed production. J. ASHS 104 674–677.

Shifriss, O. 1985. Origin of gynoecium in squash. HortScience *20*, 889–891.

Walters, T.W., and Decker-Walters, D.S. 1988. Balsam-pear (*Momordica charantia,* Cucurbitaceae). Econ. Bot. *42,* 286–288.

Walters, T.W. 1989. Historical overview on domesticated plants in China with special emphasis on the Cucurbitaceae. Econ. Bot. *43,* 297–313.

Wehner, T.C., and Miller, C.H. 1985. Effect of gynoecious expression on yield and earliness of a fresh market cucumber hybrid. J. ASHS *110,* 464–466.

Whitaker, T.W., and Carter, G.F. 1961. A note on the longevity of seed of *Lagenaria siceraria* (Mol.) Standl. after floating in sea water. Bull. Torrey Bot. Club *88,* 104–106.

Whitaker, T.W., and Davis, G.N. 1962. Cucurbits: Botany, Cultivation and Utilization. Interscience, New York.

Zitter, T.A., Hopkins, D.L., and Thomas, C.E., eds. 1996. Compendium of Cucurbit Diseases. APS Press, St. Paul, MN.

25

Other Succulent Vegetables

FERNS

FIDDLEHEAD FERN, *Osmunda cinnamomea*, L. *(O. japonica)*

Family: Osmundaceae
Other names: Zen mai (Japan)

Fiddlehead fern is a perennial, native to temperate and tropical areas of Asia and the Americas, and grown for the edible young *croziers* that are used like asparagus or greens (Fig. 25.1). The astringent tannins contained are removed by parboiling in alkaline water.

BRAKE FERN, *Pteridium aquilinum*, (L.) Kuhn *(Pteris aquilina)*

Family: Polypodiaceae

Young croziers and rhizomes are eaten, although their food safety is somewhat questionable. The amount consumed must be considered with regard to possible health effects. They are reported to contain thiaminase, an enzyme which hydrolyzes thiamine, making it inactive as a vitamin.

LOPLOP, *Blechnum capense* L.

Family: Polypodiaceae

This fern, found mainly in the Southern Hemisphere, is used as a leafy vegetable in New Zealand. Plants are best adapted to temperate conditions, moist soils, and a dry atmosphere.

FIG. 25.1.   Fiddlehead ferns, *Osmunda cinnamomea.*

GYMNOSPERMS

GINKGO, *Ginkgo biloba* L.

Family: Ginkgoaceae

Ginkgo is a large, tall, dioecious tree indigenous to southeast China. Its foliage has a characteristic fan shape with a notch in the middle of the leaf blade. Another identifying feature of the ginkgo is the foul butyric acid odor of ripe, fallen fruit from female trees. In the plumlike fruit is a ivory-colored seed kernel used as a cooked vegetable in the Orient; kernels have a cheeselike texture (Fig. 25.2). The outer fleshly layer of the seed and seed oil can cause dermatitis, and excessive consumption of kernels may cause food poisoning. Trees can be seed propagated, but to guarantee the sex of the plant, cuttings are used. The male plant is often grown as an ornamental tree.

FIG. 25.2.   Ginkgo kernels, *Ginkgo biloba.*

ANGIOSPERMS-MONOCOTYLEDONS

SPANISH BAYONET, *Yucca elephantipes,* Regel Y. *aloifolia*
Family: Agavaceae
Other names: Itabo, ozote

In Mexico and Guatemala, flowers of this treelike plant are used in soups and other dishes. Flowers have a high vitamin C content. The tender potion of the shoot apex is also edible and utilized much like heart of palm and is a rich source of calcium.

CABBAGE TREE, *Cordyline australis* (G. Forst.) Hook. f. *(Dracaena australis)*
Family: Agavaceae
Other name: Palm lily

BLUE DRACAENA, *C. indivisa,* (G. Forst.) Steud. *(Dracaena indivisa)*
Family: Agavaceae

These minor vegetables are cultivated in New Zealand. The cabbage tree is 10–12 m tall with thickened, meter-long, narrow leaves. Young tender leaves are cooked, usually boiled, and used in soups and stews. The sword-shaped leaves of *C. indivisa* are twice as long as *C. australis,* but the tree is shorter, about 7–8 m tall, when mature. Both are propagated vegetatively or by seed.

HEART OF PALM, PALM CABBAGE, several genus of the family Arecaceae (Palmae), usually *Bactris gasipaes* Jacq. ex Scop., *Euterpe edulis* Mart. and *E. oleracea* Mart.

Other names: Pejibaye palm, peach palm, palmito, Assai palm

Palms are treelike plants native to the tropics and warmer temperate areas of both hemispheres. Most palms are monoecious, with heights varying from several to more than 30 m. Those used for heart of palm production are usually in the range 2–5 m. The best known edible vegetable portion, called heart of palm, is the central shoot (crownshaft) consisting of the growing point and associated compressed tightly enfolded leaf blades. This portion is commonly harvested as a cylindrical bundle from 60 to 100 cm long and 3 to 10 cm thick (Fig. 25.3). Harvest of the "heart" removes the apical bud, and as no further growth can occur, the tree or shoot is destroyed.

Bactris gasipaes (Guilielma gasipaes), Euterpe edulis, and *E. oleracea* are major species, although several other species have food uses. Additional genera from which heart of palm are obtained include *Prestoea, Roystonea, Sabal, Acrocomia, Caryota,* and *Cocos.* However, not all species have edible hearts and some are not used because they are difficult to harvest. *Euterpe edulis* and *E. oleracea* known as Assai palms, are tall, spiny, and difficult to harvest. *Bactris gasipaes,* known as peach palm or pejibaye palm, are shorter trees, some being spineless, and therefore easier to harvest. *B. gasipaes* palms typically produce multiple stems (suckers). Suckers can be harvested for vegetable use or used as rooted propagating material. Generally, 6-month-old seedlings are used for propagation. Although seed produced are viable, they are seldom used for propagation.

Hearts of palm are regarded as a gourmet vegetable and were extensively harvested from the wild, but that source has been greatly depleted. Replacement plantations have been and continue to be established. Major production is centered in Brazil, with a sizable volume from Ecuador, Colombia, and several Central American countries. Some plantings have recently been established in Hawaii. Hearts of

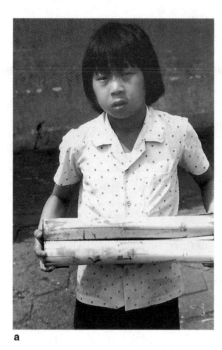

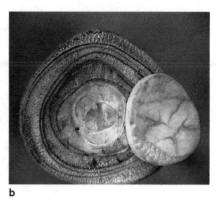

FIG. 25.3.   Heart of palm:
(a) Brazilian boy holding several
crownshaft sections; (b) cross sections
of the harvested heart of palm. The
front portion has been trimmed and is
edible, the underlying rearmost
section has not been trimmed.

palm are processed by canning and are used mostly for salads. Also used as fresh vegetables are the tender stems just below the apical meristem and tender young leaves just above the heart tissue. *Bactris* palms also produce edible, small (2–5 cm), single-seeded orange or yellow fruit.

Slightly more than 2 years of growth after seedling establishment are permitted before harvest occurs. Plants at this time are usually about 2–3 m in height as measured from the base to the fork of the first fully expanded leaf. During harvest, the crown is cut and the fibrous petiole sheaths are stripped away. This reveals the following: the edible portion which is composed of the active differentiating and expanding stem just below the apical meristem; the heart of palm composed of the expanding leaves above the meristem which are enclosed within the petiole sheath of the flag leaf; and the expanded leaf not enclosed by the petiole sheath. Each portion differs in texture and usage, but only the heart of palm is processed by canning, principally for an export product.

SWAMP CABBAGE, *Sabal palmetto* (Walt.) Lodd. ex Schult. & Schult. f.

Family: Arecaceae
Other name: Palmetto palm

Sabal palmetto and other *Sabal* species are native to the Caribbean region. Although not usually cultivated, young 2–3-m-tall plants are sacrificed for the central shoot (heart) which is removed from the upper trunk. The cylindrical-shaped, creamy white, heart portion of about 60 cm length is comprised of many layers of tightly wrapped leaves that have a texture somewhat like cabbage. The product is thinly sliced and used raw in salads or is cooked, usually with meat dishes for the smoky flavor it provides. Mature trees achieve a height of 25 m and are not used because they are difficult to harvest.

SAGO, *Metroxylon rumphii* **Mart.,** *M. bougainvillii*

Family: Arecaceae

Sago is a starchy food staple cultivated adjacent to swampy areas in New Guinea and some other South Pacific Islands. Each tree is capable of providing about 100–150 kg of crude starch obtained from the central stem. Sago should not be confused with sago palm, a cycad.

ASPARAGUS, *Asparagus officinalis* **L. var.** *altilis* **L.**

Family: Liliaceae

Origin

The region between the eastern Mediterranean and eastward to the Caucasus mountains is generally believed to be the center of origin of asparagus. The plant is known to have been cultivated for medicinal and food use for more than 2000 years, and its cultivation is now extensive in many worldwide locations. Related species, *A. cochinchinensis,* produces edible tubers, and *A. racemosus* produces edible roots.

Botany

Asparagus is a dioecious perennial monocot grown for its edible, tender, unexpanded shoots, commonly called spears (Fig. 25.4a). Underground portions of the plant consist of the rhizome with its cluster

FIG. 25.4.   Asparagus, *Asparagus officinalis:* (a) green asparagus bundled for market; (b) cluster of excavated crowns revealing thickened storage roots, rhizomes, buds and emerging spears; (c) seedlings growing in a shallow trench which will be filled in as the plants enlarge; and (d) hand harvesting of asparagus spears.

of buds and associated storage and fibrous roots, collectively called the "crown" (Fig. 25.4b). Upper portions of the horizontal rhizome contain the buds from which shoots emerge and elongate to form spears. Fleshy and fibrous roots develop from the lower portion of the rhizome. Fleshy roots are unbranched, act as storage organs, and have little absorptive activity, whereas the role of fibrous roots is absorption. Storage roots are unbranched, usually radiate equidistantly from the crown, and function for several years. On the other hand, fibrous roots are short lived; they die and are replaced each year. Fibrous roots develop from the stele tissue of storage roots and most often are found toward the apical end. Each generation of storage roots is also initiated at the crown.

The rhizome is comprised largely of vascular tissues and stores relatively little carbohydrate. It functions as a conduit for the translocation of carbohydrate between storage roots and aboveground tissues. During each growth cycle, lateral growth of the rhizome and storage roots occurs. Storage roots can grow laterally for 2–3 m before turning downward and can achieve lengths up to 5 m. Fibrous roots that are produced at the apical ends of the rhizome absorb nutrients. They also are deep and can reach depths of more than 80 cm.

During the fern growth period, buds are differentiated in clusters near the growing point of the rhizome. Root radicles are also initiated near this meristematic area. Spear growth begins when buds on the crown sprout and elongate. The sprouting and elongation of each bud is accompanied by the development of several storage roots. The primary bud is the largest and innermost bud of a bud cluster and is followed by development of progressively smaller secondary buds. The primary bud exhibits apical dominance, which restricts the development of younger buds. The dominance is progressive, whereby older buds continue a level of dominance over younger buds. Ethylene is known to negate apical dominance.

Spear size is positively correlated with bud size: Large buds produce large spears. Bud size tends to vary more than bud numbers. With each growth cycle, the crown exhibits a lobelike, outward, and slightly upward enlargement. After several years, lateral expansion of the rhizome to 50 cm or more is possible, as well as its upward movement toward the soil surface. In old plantings, the initial crown often become segmented and portions function as separate plants.

Disease and cultivation damage are primary reasons for plant decline. New plantings of asparagus should not immediately follow in the same field, especially if *Fusarium* is present and also because of allopathic characteristics of asparagus plants. However, the allopathic agent that is inhibitory to seed germination is short lived and can be leached by water. It has been suggested that the chemical is autotoxic and might also predispose plants to *Fusarium* infection.

The spear is composed of the apical meristem with many tightly closed lateral buds covered with tightly formed bud scales. With continued spear growth, lateral branches (feathering) elongate from nodes under the bud scales along the spear. Lateral branches produce fernlike cladophylls, which appear and act like leaves, but actually are modified stems. The compressed triangular-shaped bud scales that cover the lateral buds are the true leaves. Primary and secondary branches and flower primordia are differentiated early in the development of the spear.

Female plants ordinarily do not bloom the first year when grown in temperate regions. Male plants will flower about 4–5 months after seeding. Flowers are borne in axils of secondary branches. The bell-shaped greenish yellow flowers, usually two per node, develop about the same time as the cladophylls, and generally flowers mature before the fern is fully developed.

Flowers on staminate plants have six stamens and functional anthers, and a rudimentary stigma and ovary. Pistillate flowers have a functional three-lobed pistil, stigma, a three-carpel ovary, and rudimentary anthers. Flowers have nectaries and are insect pollinated. Seed mature about 90 days after anthesis. Fruit, initially a green berry, become red when mature. Black seed, potentially two per carpel, are round with a flattened side; about 40 seed weigh 1 g.

The sexual state of the plant is controlled by a single gene, maleness being dominant. Because all flowers are potentially hermaphroditic, there can be a gradation between male and femaleness in flower development with some andromonoecious plants having small but functional pistils. Although rare, hermaphroditic plants occur. Plants homogametic for maleness are called super males. Growers often prefer all-male field populations, because growth performance differs with plant sex. With many, but not all cultivars, male plants are more productive because they produce more spears, although average spear diameters are slightly smaller. Male plants typically have a longer productive life and exhibit better disease tolerance. A probable reason is that male plants, not producing seed, can direct more stored carbohydrate to spear growth. Cloning via tissue culture is used for the production of parental plants for seed production fields. Super males are made by selfing of male plants or tissue culture of haploids from anther culture, and these are used for production of stock seed.

Climatic Requirements

Asparagus, being a broadly adapted cool season plant, is grown in most temperate and some subtropical regions. Generally, plant longevity is greater in cool, temperate areas. Mean day temperatures of 25–30°C and 15–20°C at night are ideal for spear and fern growth, also for the storage of accumulated photosynthate reserves. Soil temperatures above 10°C in the vicinity of the crown is sufficient to initiate bud sprouting. Aboveground plant parts are injured or killed by frost; the below-ground crowns can escape freezing injury if temperatures are not severe. For areas with below-freezing temperatures, crowns are protected from freezing by mulching. Cultivars vary in their cold hardiness.

Production and longevity are enhanced when plants have a dormant period. However, dormancy is not necessary for commercial production of asparagus. Where fern is killed by frost and soil temperatures are low, plants enter a dormant state because of diminished respiration. This situation tends to preserve much of the stored carbohydrate reserves, which are then available for subsequent spear production. An imposed moisture deficit can partly substitute for the effect of dormancy.

In regions of mild winters or the tropics, foliage growth tends to be continuous. Under these conditions, it is difficult to arrest growth and reduce respiration. Therefore, plants do not become dormant and relatively little accumulation of food reserves occurs. The "mother-stalk" cultural procedure is used to overcome the lack of a dormancy period and to manage the plant's carbohydrate reserves. This procedure permits four or five shoots to grow into fern in order to produce photosynthates while other shoots are harvested. When these ferns become senescent, others are allowed to grow as replacements. With this procedure, a long period of spear production and higher yields can be realized, but at the cost of high labor inputs and reduced plant longevity. In the low-elevation tropics, production is also limited by heat stress.

Soil, Moisture, and Nutrition

Being a perennial, asparagus land preparation that assures stand longevity as well as productivity is important. Productive plantings can be maintained for many years. Friable, deep, well-drained, and fertile soils are conducive to longevity. Light textured soils are preferred for spear development and ease of harvest. Organic (peat) soils, because they are easily tilled, are an advantage, especially for white (blanched) spear production. However, such soils are prone to wind erosion that can uncover crowns. Raised-bed culture is often preferred because of drainage and soil-warming features while providing adequate soil cover and protection to the crowns.

Although a neutral pH is ideal, asparagus plants are tolerant to slightly alkaline or acidic soils. Mature plants have a high tolerance to salinity, but seedlings are very sensitive to salt.

Asparagus is a moderate consumer of moisture and has fair drought tolerance during minimum fern growth or when dormant. However, adequate soil moisture is critical during rapid spear development.

When soil moisture is inadequate, water should be supplied; an annual supply of 1.2 times evapotranspiration is recommended. Surface-applied irrigation is often preferred to overhead sprinkling in order to minimize the incidence and severity of asparagus rust, *Puccinia asparagi,* and most other foliar diseases. Poor drainage and waterlogging cause damage to rhizomes and storage roots and is conducive to a greater incidence of *Phytophthora* disease.

An adequate initial source of phosphorus and potassium is needed for good plant establishment because later additions of phosphorus and potassium are difficult to apply effectively without inflicting some root injury. A small quantity of nitrogen is usually applied in advance of crown plantings. Later nutrient applications should be based on fern tissue analysis and kept at or above critical nutrient levels. The nitrogen should be applied during the fern development when nutrients can be utilized for growth rather than spear harvest when little absorption takes place.

Propagation and Cultural Practices

Asparagus is propagated by direct sowing, transplanting rooted seedlings, or transplanting crowns. The procedure of dividing large crowns for propagation is seldom used because of subsequent poor yield and plant longevity. Precision seeding has greatly improved direct field sowing (Fig. 25.4c). Seed are uniformly placed in the bottom of a 15–20-cm-deep furrow and initially covered with 3–5 cm of soil. As the seedlings develop, soil is gradually filled, often accomplished during mechanical weed cultivation.

The initial primary shoot and root of the newly emerged seedlings does not persist. An axillary shoot soon develops at the transition of the primary shoot and root, and this site is the origin of the future bud cluster. Secondary bud clusters develop in the axils of some of the primary buds. Additional shoots and roots are quickly formed and continue to grow. By the end of the growing season, the furrows will have been filled in so that the developing crowns are about 15–20 cm below the surface. Placement deeper than 20 cm adversely affects spear quality. Shallow placement results in early production but tends to decrease plant longevity and spear diameter.

To ensure full stands, extra seed are sown and the excess seedlings are thinned. Thinning is difficult because of root entanglement and because stems easily break when pulled, leaving some crown portion of the seedling in the soil that is capable of regrowth. Uniformly spaced seed helps to reduce this problem. Germination is optimum between

25°C and 30°C. Seed soaking prior to sowing can accelerate germination, which is slow in cold soils. Although establishment is less costly, directly seeded fields require more time for plant growth before production occurs.

Field establishment with seedling transplants is also practiced, in part because of the high seed cost of new hybrid cultivars. The advantage of using seedling transplants is that full stands of uniform plants at accurate spacings with minimum seed use is achieved. Transplants are usually 10–12 weeks old when put into the field. Transplants are also placed into shallow trenches and handled like seed-sown plantings, with gradual filling of the trench during growth.

Crowns are the traditional propagating material and are widely used. Although a more expensive and labor-demanding practice, crown plantings do provide full plant stands, and allow for more efficient weed management. However, the primary advantage is rapid crop establishment and earlier production.

In temperate regions, crowns are usually removed from the nursery after 1 year's growth. Two-year-old crowns are not used because subsequent growth and production tends to be less than that of 1-year-old crowns. In subtropical areas, crowns are grown for 3–4 months before transplanting. Crowns are dug at the conclusion of fern growth and, after removal, are individually separated and if necessary can be held in cold storage at 5°C and 90–95% RH until planted. When dug, crowns are often dormant; therefore, they need not be planted immediately. When out of dormancy and planting conditions are favorable, crowns are placed into a 15–20-cm-deep trench. Preferably, the crowns should be planted with buds up and with storage roots spread outward, and then covered with soil. Crowns randomly placed or placed upside down will exhibit delayed and variable spear emergence.

Populations established directly from seed require a year for adequate seedling growth. Using transplants reduces the nonproductive period, but not substantially. In the second year, harvest is either omitted or restricted in order to build up storage root and crown carbohydrate reserves. Even during the third year, it is sometimes prudent to restrict the harvest period so as not to exhaust storage carbohydrates, and thereby limit fern growth. The storage carbohydrate is a high-molecular-weight fructose oligosaccharide of about 90% fructose and 10% glucose.

Crown-established plantings have a shorter period before the harvest cycle begins, and limited harvests usually can be made during the first year following planting, without affecting subsequent spear yield and quality. In following years, a normal harvest cycle is followed. Further cultural and harvesting practices resemble those used for direct seeding. Regardless of propagation method, the period from initial seed to when a significant yield of harvested spears occurs is about 3 years.

Plant spacings range between 20 and 50 cm in row and from 100 to 200 cm between rows. Double-row plantings are sometimes made at slightly wider row spacings. For white asparagus production, wide row spacings are commonly used to allow for the construction of high beds that cover the developing spears. Plant populations range from 15 thousand to 25 thousand per hectare for white asparagus and from 25 thousand to 50 thousand per hectare for green spear production.

At the end of the harvesting period, newly formed spears are allowed to continue growth and develop fern. The fern is grown as long as possible or until senescence, in order to replenish and store carbohydrate reserves for the next season's production, or is grown concurrently with harvesting as in the "mother-stalk" procedure. In some areas, the fern is killed by frost. Prior to the start of the next harvest period, any remaining fern growth is destroyed or removed so as not to interfere with harvesting. Destruction is by flailing, disking, or burning.

Effective weed management is important in order to limit weed competition for nutrients, moisture, and light. Mechanical cultivation is effective during the dormant or postfern period when bed tops are worked at a shallow depth to avoid damage to crowns. Use of male-only plant populations eliminates volunteer seed from becoming weedlike competitors. A cultural practice no longer followed actually applied salt (NaCl or KCl) to manage weeds in established plantings. Several effective chemical herbicides are available and are frequently used for weed control.

Asparagus Diseases and Pests

Diseases

Bacterial soft rot	*Erwinia cartovora*
Botrytis	*Botrytis cinerea*
Cercospora leaf spot	*Cercospora asparagi*
Fusarium wilt	*Fusarium oxysporum* f. sp. *asparagi*
Phytophthora	*Phytophthora megasperma*
Rust	*Puccinia asparagi*
Seedling crown rot	*Penicillium martensii*
Stem and crown dry rot	*Fusarium moniliforme*
Stem blight	*Phoma asparagi*
Stemphyllium leaf spot (purple spot)	*Stemphyllium vesicarium*

Asparagus virus I and II diminish plant vigor and longevity.

Insect pests

Asparagus beetle	*Crioceris* spp.
Asparagus fly	*Platyparea, Delia, Phorbia* spp.
Asparagus leafminer	*Melanagromyza simplex*
Carpenter moth	*Hypopta caestrum*
Centipedes	*Scutigerella* spp.
Cutworms	*Noctuidae, Euxoa, Agrotis, Crymodes, Feltia* spp.
European asparagus aphid*	*Brachycolus asparagi*
Tarnish plant bug	*Lygus lineolaris*
Thrips	*Thrips tabaci,* other *T.* spp.
Wireworms	*Alelanotus* spp.

Nematode species of *Meloidogyne, Pratylenchus, Trichodorus,* and *Heterodera* also are important pests.

Harvest

Bud break and spear elongation occur when soil temperature at crown depth is greater than 10°C. Crowns can be forced into early spear development by increasing soil temperatures. Spears are usually harvested before excessive elongation occurs. Reduced quality is caused by internode lengthening and loss of tip compactness (feathering). Large spear diameter and length with compact tip, and spear tenderness are quality factors.

Spear diameter is strongly influenced by plant vigor. Differences in spear cell size rather than cell number are responsible for variation in spear diameters. The amount of soil cover over the crown is also correlated with spear diameter. At deeper planting depths, developing spears are thicker. Low concentrations of ethylene cause cell expansion, and at greater depths, ethylene levels are higher. Thus, white asparagus spears, subjected to a greater depth of soil cover, produce larger-diameter spears.

Plant vigor and seasonal temperatures affect spear length and growth rates, as well as harvest frequencies and the length of the harvest period. At high temperatures, the internodes of the spear elongate rapidly (feather) and axillary buds may also elongate, both leading to reduced quality. During temperatures greater than 30°C, spear elon-

*European aphid feeding toxin causes severe bushy and stunted plant growth.

gation of 10–15 cm per day can occur. In such conditions, daily or twice-daily harvest may be necessary.

Harvests are usually terminated either by calendar schedule or when it is observed that spear diameters and spear numbers rapidly decrease. This observation suggests carbohydrate reserves have decreased and further harvesting may reduce reserves to a level that will affect fern growth vigor and thereby decrease the number and spear size of subsequent crops, and also may reduce plant longevity. Where fern growth is greatly restricted, harvest periods may be limited to as little as 2–3 weeks. Similarly, in tropical areas where a dormancy cannot be achieved, plant life span is often limited. Carbohydrate reserves are usually at their lowest level at the time cladophylls are beginning expansion. The harvest period can be lengthened to as much as 15–20 weeks where adequate time and temperatures for subsequent fern growth will produce sufficient photosynthates to replenish storage root reserves.

Most asparagus is hand harvested (Fig. 25.4d) although a limited amount of mechanical harvesting occurs; machine harvested spears are usually processed. For green spear production, spears are hand snapped or cut below soil level with a long-handle, narrow-blade knife. Damage to young spears and latent buds below the surface is possible, so careful insertion of the knife is important. Spears cut below the surface have a white base. Occasionally, purple pigmentation occurs at the tip of the spear, usually considered a cosmetic market defect. However, some newly introduced cultivars that produce purple spears are grown as a novelty.

The longevity of an asparagus planting varies from as little as 3 or 4 years to more than 15 years. When the percentage of large spears shows a continuous and significant decline, the production is discontinued. The reason for stopping production is that profitability declines along with the decline in large-spear size (Fig. 25.5). The rate of the decline of large-spear production are usually a result of an overly extended cutting period, disease, pest infestations, and other damage to the crown.

White or blanched spear harvesting is more difficult. Spears are harvested when the tip is about to emerge, or just protrudes through the soil surface. Even a slight amount of tip greening of white asparagus diminishes its perceived market quality. The harvester must judge where the location of the spear base is to properly insert the cutting knife well below the surface to cut just above and not into the rhizome. The difficult and costly production of white asparagus is steadily being

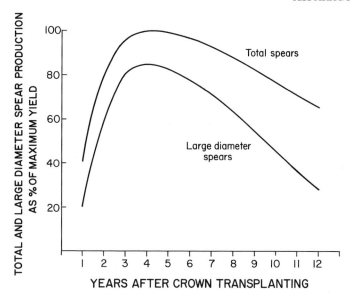

FIG. 25.5. Relationship of large-diameter spear production compared to total spear production over time.

displaced by green production. Nevertheless, the demand for white asparagus persists in European and Japanese markets, probably because of flavor differences. Cultivars are specifically bred for white, others for green, and some for both white and green production. Home gardeners have used clay bell jars placed over developing spears to blanch them as well as to advance their rate of development.

Individual spear growth is influenced by the apical dominance of preceding spears as well as its own. The apical meristem of the most advanced spear retards development of adjacent buds of the same crown. Spear removal reduces the suppressive influence of apical dominance, so that subsequent buds can elongate. The net effect is a pronounced indeterminate or cyclic emergence of spears, which limits the feasibility of efficient mechanical harvesting. Nonselective machine harvesting results in spears of differing lengths, many not suitable for fresh market but suitable for processing. Several harvests, a week or 10 days apart, are commonly made. Although this method results in lower harvesting costs, total yields of marketable product are also reduced. Selective mechanized harvesting has not been economically feasible because spear emergence is not adequately synchronized.

Postharvest and Storage

Asparagus spears have a high respiratory rate, and storage life is markedly shortened at high temperatures. Hydrocooling is an excellent procedure to rapidly reduce spear temperature. The recommended handling temperature is 2°C and with relative humidity greater than 95%. Under these conditions, high-quality asparagus can be maintained in good condition for 10–14 days. At temperatures less than 2°C, chilling damage occurs with prolonged holding. At high temperatures, rapid spear elongation, sugar loss, and senescence result; ethylene exposure also causes senescence. An increased rate of fiber (lignin) development is directly correlated with spear length and age and also with high postharvest temperatures. The area where spears easily snap coincides with the beginning of the fibrous region.

Fresh market white or green asparagus is often marketed as a bundle of uniformly sized and trimmed spears. Spears are graded for length and diameter and packed upright with the basal end down. If placed horizontally, spear tips exhibit negative geotropism and turn upward. Specially shaped containers, wide at the base and narrow at the top, are often used to accommodate the long tapered spears. The basal portions of packed spears are frequently placed on a water-moistened pad to prevent wilting. Spears continue to elongate after harvest and headspace is provided in packaging to avoid tip damage.

Uses and Composition

Fresh green asparagus consumption continues to increase relative to that of white asparagus and has surpassed total white asparagus production. Both kinds are processed by canning, freezing, or dehydration. White asparagus is a more attractive canned product than canned green asparagus, and presently exceeds green canned consumption. Intact spears are a premium canned product in contrast to cut portions. Whole peeled spears receive an extra market premium. Although canned in greater volume, the quality of the frozen product is considered superior and the frozen volume is increasing. Asparagus seeds have been used as a coffee substitute.

On a weight basis, the nutritional value of asparagus is good (see Appendix Table C). However, asparagus nutrient production on a per hectare basis is low because crop yields are low; the world average is about 3 t/ha. Spears contain asparagine aminosuccinic acid monoamide, which when ingested causes the noticeable methyl mercaptan odor in urine. Asparagus contains rutin, a compound useful in hemorrhage prevention, and also has diuretic properties.

Production

Accurate production statistics for asparagus are difficult to obtain. An estimate of the world's 1996 production area of asparagus is about 208,500 hectare. From that area, about 600,000 tons of asparagus are produced (Table 25.1).

It should be recognized that area of production is not necessarily correlated with volume as yields vary considerably among countries. Most of the white asparagus production is centered in Europe, South Africa, and China. The European market prefers white asparagus and consumes most of the domestic production and imports a large proportion of white asparagus from other producers. However, utilization of green asparagus is increasing in the European countries. The United States is the major green asparagus producer and also the largest consumer; many western European countries and Japan are also large consumers of asparagus. Countries such as Peru, Mexico, Chile, Taiwan, and Ecuador are major exporters of fresh and processed asparagus.

TABLE 25.1. ESTIMATED AREA OF ASPARAGUS PRODUCTION FOR VARIOUS REGIONS AND COUNTRIES, 1996

	Hectares
World	208,500
Africa	3,700
South Africa	3,000
Europe	53,800
France	17,000
Spain	13,000
Italy	6,000
Germany	6,000
Greece	6,000
Asia	67,900
China	50,000
Japan	11,000
North America	47,700
United States	38,500
Mexico	8,000
Central and South America	27,800
Peru	17,800
Chile	5,500
Oceania	7,300
Australia	4,500
New Zealand	2,800

Source: Benson (1996).

BAMBOO, Several Poaceae genera

Family: Poaceae (Gramineae)

Bamboo is a general term identifying certain genera and species of the Poaceae family that are widely distributed throughout the world's tropical and many temperate regions.

Bamboo plants are highly regarded and utilized in many areas, especially in the Orient for its food, timber, and many other uses. Writings about bamboo in ancient Chinese literature date back many centuries. The Yunnan Province in southwest China has the most diversity with at least 28 primitive and 5 semiprimitive types, and therefore can be considered a center of origin. China has about 20% of the world's area of bamboo forests. Malaysia, Indonesia, and South America are also considered centers of diversity or origin. In contrast, most bamboo species in South America are highly advanced and only one species is considered primitive. The genera most important for vegetable use are *Phyllostachys, Bambusa,* and *Dendrocalamus.* Important species are *P. dulcis, P. edulis, P. bambusoides, P. pubescens, P. nuda, P. viridis,* and other more temperature-hardy *Phyllostachys* species. *Bambusa beecheyana, B. vulgaris, Dendrocalamus strictus,* and *D. latiflorus (Bambusa oldhami)* also have vegetable usage.

The bamboos are woody-stemmed, evergreen perennials, typically consisting of aboveground jointed stems *(culms)* and below-ground rhizome. Those exhibiting a clump growth habit are usually tropical; those with a runner-type growth are usually temperate.

The edible vegetable portions of bamboo are the young emerging shoots that are harvested before significant fiber development occurs (Fig. 25.6). Shoots are progressively formed from latent buds on rhizomes. To maximize development of tender young shoots, mulch or soil is placed around the base of the plant into which the shoots grow. Exposure to light during shoot elongation is prevented in order to limit formation of bitter-tasting cyanogenic glucosides. Harvesting occurs when the shoot tip first emerges above the soil. The shoot is cut free below the soil and near the base of the rhizome.

Growers in temperate regions sometimes accelerate shoot emergence by warming the soil surrounding the rhizomes. Heat forcing can advance production by a month. New shoots continue to emerge from lateral buds on the rhizome. However, the period of shoot-sprouting activity usually lasts only a few weeks. Toward the end of that period, the late-formed shoots are not harvested but allowed to fully developed. Their subsequent growth produces the storage carbohydrate reserves

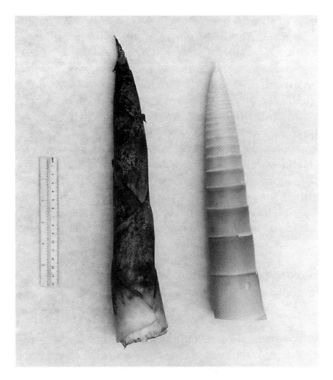

FIG. 25.6.   Bamboo shoots tips, species unidentified. The left shoot tip has not been trimmed of sheath leaves; the right shoot tip is edible after trimming, cooking and leaching of toxic substances.

for the rhizome and the next season of shoot growth. Thus, it is important that growers avoid excessive harvesting that can reduce the vigor and future productivity of the parent plant.

Propagation is made with rhizome divisions, culm segments, or seed; rhizomes are preferred. An interesting feature of some species is the synchronous timing of flowering even when portions of the rhizome of one plant are planted at different locations. The life cycle of some bamboo species is also interesting in that once flowering occurs, stems and foliage die. Whole forests have been observed to die off after flowering and seed maturation. Nevertheless, with some species, regrowth can occur from basal buds or from sprouted seed.

Some species flower within 3 or 4 years, others may continue to be vegetative and not flower, even after 40 years. When flowering does occurs, florets are formed on spikelets typical of Poaceae plants, and viable seed are produced. In times of famine, the grain (seed) has been used for food.

Shoots contain homogentisic acid which is responsible for a pungent offensive flavor; called "egumi" in Japan. The homogentisic acid content ranges from 100 to 250 μg/100 g fresh weight in early harvested shoots and decreases in later harvests; the content can also vary with production locations.

Before food use, leaf sheaths of harvested shoots are removed and the tough basal portion is cut away and discarded. Raw shoots are acrid and the acridity is removed by boiling in water for about half an hour, sometimes longer to remove the bitterness. Bamboo shoots, while largely devoid of flavor after the bitterness is removed, do provide a crisp texture to foods that remains even after prolonged cooking.

MYOGA, *Zingiber mioga* (Thunb.) Roscoe
Family: Zingiberaceae

Myoga is a perennial that is native to temperate regions of eastern Asia and is grown in Japan, China, and Korea. It is known to have been cultivated in Japan since the 10th century. Plants are killed by frost, but unless the low temperatures are severe. the hardy underground stems (rhizomes) survive to permit regrowth. Favorable growing temperatures range from 20°C to 30°C. Mature plants grow to heights from 50 to 100 cm, with lanceolate foliage about 6–7 cm long and 3–4 cm wide. Flowers rarely set seed and thus propagation is with rhizomes.

Myoga is grown for the condiment use made of the immature inflorescence (hanamyoga) and blanched stems (myogatake) (Fig. 25.7). The inflorescence are used to a greater extent than blanched stems, possibly because production of the former is easier. Many clones, identified by maturity characteristics, are grown. Early-maturing clones are used for inflorescence production, and the intermediate- and late-maturing clones are preferred for blanched stem production.

Plants are managed differently, depending on whether the inflorescence or stems are to be produced. For inflorescence production, rhizomes are planted in the spring for autumn harvest. Inflorescence are produced near the base of the stems, thus some are formed below and at the soil surface, as well as slightly above. Exposure to sunlight results in the tips of the inflorescence developing a reddish purple color.

For production of blanched stems, plants are grown in pits, or soil

FIG. 25.7.   Myoga, *Zingiber mioga:*
(a) plant (note floral bud at base of stem);
and (b) floral buds as marketed in Japan.

is repeatedly mounded against the stems as they grow. Another method is forcing, in which plants are removed from the soil and put into heated soil beds and mulched with straw to exclude light. Stems are harvested about 2 months after replanting. At that time, the marketable portions of the stems usually have a length of 50–60 cm. Shortly before harvest, the mulch is removed, and light exposure results in development of reddish color.

GINGER, *Zingiber officinale* Roscoe
Family: Zingiberaceae

Ginger is a tropical perennial of uncertain origin and unknown in the wild state. India, south China, and possibly some of the Pacific Islands may be regions of origin. Ginger has a long history of cultivation in southeast China and India as noted by the explorer Marco Polo. The crop was introduced into the West Indies by Spanish traders. Its early use was medicinal as well as for flavoring.

Ginger is an important product in food preparation, especially throughout Southeast Asia, grown in significant volume in China, as well as Taiwan, India, Nigeria, Australia, and Jamaica. The crop is also grown in warm, temperate areas. Ginger has numerous flavoring uses in fresh, dried, or dehydrated powdered form in addition to being candied in crystallized sugar, prepared as a syrup, and also has some perfume use.

From nodal buds, the modified stem develops highly branched and laterally compressed rhizomes, 2–4 cm thick, that grow in a palmate fashion close to the surface (Fig. 25.8). Each portion of the rhizome develops from one of these buds and, in turn, produces additional buds. However, not all develop additional rhizome portions. The thin outer skin is scaly, light brown or tan; the interior pale yellow. Each portion of the rhizome is capable of producing aerial (pseudostem) shoots. The closely grouped, unbranched, canelike shoots grow to a height of 50–100 cm. Leaves are opposed and alternate, simple, smooth, and thin with long (10–25 cm) and narrow (2–3 cm) blades with ligules. The first roots emerge from the first developed portion near the neck of the rhizome, but, later, adventitious roots develop from the underside of terminal portions. Roots are fibrous and remain relatively close to the surface.

Ginger grows best in a warm and moist climate; an average temperature between 25°C and 28°C is ideal. Above 30°C, the rate of photosynthesis decreases, and the potential for sunburning of the rhizome is increased. Ginger grows well in full sun but is often interplanted because it is also adapted to partial shade. Although rhizomes tolerate light frost, temperatures below 15°C suppress foliage growth. Low temperature or freezes tend to induce dormancy. Vegetative growth is

a   b

FIG. 25.8.   Ginger, *Zingiber officinale:* (a) bunched plants marketed with immature rhizomes, and (b) mature fresh rhizomes.

enhanced under long (16 h) day conditions, whereas rhizome enlargement is promoted under short (10 h) days. Flowering is promoted by increasing day length (14–16 h). The day length response varies with clones, and some appear to be insensitive.

Well-drained, friable, silty loams are preferred to heavier soils which interfere with rhizome shape. The preferred pH is between 5.5 and 6.5. Ginger is a heavy user of nutrients and moisture. For commercial production, fertilization is necessary to obtain satisfactory yields. The required moisture is about 1500 mm per year. Waterlogging often results in root rot and must be avoided.

Although a perennial, most production is as an annual crop. Ratooning generally is unsatisfactory. Because some clones exhibit a short dormancy period, unharvested rhizomes may resume growth and sprout. Propagation, usually in the spring, is with rhizome divisions. Hot water and fungicide treatment is commonly made to provide nematode- and disease-free rhizome segments (setts). Nematode eradication is accomplished by hot-water (51°C) immersion of the rhizomes for 10 min. Propagules are placed 8–10 cm deep at spacings usually 30 × 50 cm. These segments having several buds usually weight about 75 g. Roughly 20–25% of the production is utilized for propagation. The best temperature for sprouting is 25–30°C. Shoot and root growth can be accelerated by soaking rhizomes in a 750-ppm ethephon solution for 15 min at 21°C.

Clones are relatively distinct with different production regions usually growing a preferred clonal type. Clonal types are generally identified by the region where first principally grown and are known as Chinese, Canton, Malay, Jamaican, and Fijian. In Japan, clones are also classified according to rhizome size and tiller numbers. Clones vary regarding pungency, flavor, aroma, color, amount of fiber and firmness, and whether intended for fresh, dried, or confectionery use.

Harvests are made when aerial parts become senescent and dry. Senescence usually occurs 9–11 months after planting. At this stage, rhizome skin becomes set. Even so, careful handling during harvest and postharvest is needed to avoid skin damage. Early harvesting is made in order to obtain a low fiber; however, rhizome pungency is low. Delayed harvest tends to increase rhizome weight and pungency. For processing, such as pickling and preserves, tender rhizomes are preferable to the high fiber and strong pungency of late-harvested crops; early-harvested rhizomes are more easily peeled. Old and fibrous rhizomes are usually dried and ground into a powder for flavoring uses.

Flavor is contributed by the complex essential oil of which zingiberene is an important aromatic component. The pungent agent in the rhizome is due to the oleoresin, gingerol. Yields of 40–50 t/ha of fresh ginger are obtained by large-scale commercial producers. In large-scale production, mechanization is used for topping and digging rhizomes.

Postharvest life is generally good if rhizome skin is well developed and if moisture loss and/or chilling injury is avoided. For long storage, ginger is maintained at 12–13°C and 65% RH.

Some diseases affecting ginger production are bacterial wilt, *Pseudomonas solanecearum,* and soft rots caused by *Sclerotium rolfsii* and species of *Pythium, Pseudomonas,* and *Erwinia.* Leaf spot *(Colletotrichum zingiberis)* and *Fusarium* rot *(Fusarium oxysporum* f. *zingiberi* are other important diseases. Additionally, root knot nematode (*Meloidogyne* spp.), the burrowing nematode, *Radohpolus similis,* and shoot borer *Dichocrosis punctiferalis* are other crop damaging pests.

DICOTYLEDONS

ICE PLANT, *Mesembryanthemum crystallinum* L. *(Cryophytum crystallinum)*

Family: Aizoaceae (Mesembryanthemaceae)

Ice plant is a perennial that thrives in a hot and dry climate and produces many semiprostrate, 30–35-cm-long stems. Stem tips and the flat, spatulate succulent and the slightly acid-tasting leaves are used as a potherb. The plant is mainly used as an ornamental and is killed by freezing temperatures.

AMARANTHUS, *Amaranthus* species

Family: Amaranthaceae
Other names: Chinese spinach, tampala*

Amaranths plants are short-lived, usually monoecious annuals. Within this large genus, considerable variability exists relative to growth habit, leaf shape, colors, inflorescence characteristics, and uses.

The *Amaranthus* genus is better known for the grain (quinoa) producing species, but there are cultivars specifically grown for the spinachlike vegetables (Fig. 25.9). The important grain amaranth species originated in western Central and South America. Centers of diversity for many of the leafy vegetable *Amaranthus* species are Central and South America,

---

*Tampala actually represents a green leafed cultivar of *A. tricolor.*

FIG. 25.9.   Amaranthus, a vegetable type produced for its foliage rather than the seed.

India, and Southeast Asia, with secondary domestication in west and east Africa. The most diversity of leafy amaranths is found in India. The leafy vegetable species, usually consumed as a potherb, present a low-cost, good protein source to many tropical, subtropical, and temperate region populations. These are especially popular vegetables in the Orient. Some general disadvantages are the early short-day flowering response and low-temperature sensitivity of some species and the high calcium oxalate content in leaf tissues. Nevertheless, these plants supply large amounts of pro-vitamin A, vitamin C, protein, and fiber.

Major leafy vegetable amaranth species are *Amaranthus tricolor, A. lividus, A. dubius, A. gangeticus, A. blitum,* and *A. hybridus. A. caudatus* is the most important grain (quinoa) species, but can also are eaten as a leafy vegetable when harvested as young seedlings. Several weedy species are also used as vegetables; one, *A. spinosus* known as *uray,* is a vegetable of some importance in the Philippines.

Amaranthus lividus, known as *bondue,* is grown for vegetable uses in tropical Africa; young plants of *A. leucocarpus* are potherbs in Algeria; in addition, the seed is made into candy. Another crop of the amaranth family grown in tropical Asia for its edible leaves is *Celosia argentea.* The many cultivars in Southeast Asia are usually classified by leaf color and shape.

The amaranths grow on a wide range of soils; slightly acid sandy loams and good drainage are preferred. Root systems are generally sparse, but because of the plant's C_4 physiology, photosynthesis under high temperature and moisture stress is relatively efficient, and drought tolerance is good.

Most leafy-type amaranth plants are erect, about 30–90 cm tall, and produce numerous small flowers on terminal and axillary spikes. Although individual seed are very small, plants produce an abundance of the edible seed which have high protein and oil content. The grain amaranths, selected and bred expressly for seed production, are more productive seed producers than those grown for vegetable use.

The amaranths are usually seed propagated, but occasionally ratooned. However, a common practice is to thinly sow or broadcast seed, and about 20–30 days later, thinned and surplus seedlings are used for transplants or eaten as potherbs. Final populations typically range between 15 and 50 plants/m². Periodic applications of fertilizer are recommended to encourage vegetative growth and high yields.

When harvested, plants are pulled with roots left on to facilitate bunching. In another method, partial leaf removal is made with regrowth permitted for successive harvesting. Frequent harvesting, every 7–10 days, tends to delay flowering and encourages new shoot and leaf growth. Total yields of 25 t/ha or more can be obtained, especially with intensive culture of indeterminate, late-flowering species. Postharvest life is relatively short because of rapid wilting of the tender foliage.

Disease concerns center around damping off, leaf spot, and white rust caused by species of *Pythium, Cercospora,* and *Albugo,* respectively. Wet rot due to *Choanephora cucurbitarun* is another problem. Many chewing insects and some nematodes also attack amaranth plants.

AMERICAN GINSENG, *Panax quinquefolius* L.

Family: Araliaceae

CHINESE GINSENG, *Panax ginseng* L.

Family: Araliaceae

Since ancient times, the curative powers of ginseng have been reported, especially in the Orient. Even today, ginseng is still highly appreciated in China, Korea, and Japan, although there are patrons in other countries. The putative, cure-all, pharmacological properties of reducing stress and increased physical stamina are sometimes exceeded by claims of ginseng's aphrodisiac benefits and reduction of short-term memory loss, as well as other claimed benefits.

American ginseng, *P. quinquefolius,* is a perennial plant indigenous to North America, mostly the eastern and south central regions in Canada, and eastern and north central regions in the United States. *P. ginseng,* known as Chinese or Oriental ginseng, is native to wooded areas of northeastern Asia, where it has been cultivated since 2000 B.C. China and South Korea are the major producers of Chinese ginseng. Wisconsin is the primary U.S. producer of cultivated American ginseng. Cultivated production in Canada has increased substantially, especially in British Columbia, and possibly has surpassed United States in volume. More than 90% of the United States and Canadian production is exported, primarily to Asian counties. Some other *Panax* species are *Panax pseudo-ginseng, P. trifolium,* and *P. japonicum.*

The saponin glucosides (ginsenosides) are recognized as the primary biologically active ingredients and the major components determining the value. They vary with species and growing environment, and they increase with root age. Other active principles are identified as ginsenin, panaxic acid, panacen, and panaxin. Chinese ginseng contains the glucoside panaquilon.

American ginseng roots from native noncultivated wood lot or forest populations were collected by gatherers, much like wild mushrooms. Traditional gathering from indigenous native growth became limited and essentially has stopped because noncultivated American ginseng is an endangered species. Short supply and the high demand for ginseng has encouraged cultivation, which was started during the late 1800s in North America. However, ginseng root as a cultivated crop is not as highly regarded as wild ginseng by its consumers.

The preferred natural habitat of ginseng is a moderate to rich, well-drained soil of high organic matter content shaded by hardwood trees; shade of about one-third normal sunlight is very important. An optimum soil pH is between 5.5 and 6.5. Moisture in the amount of 1000 mm annually is desirable, preferably supplied by rain. Soil moisture at about 50% of field capacity is most favorable for plant growth. If soils are fertile, additional fertilizer may not be necessary. It is important to avoid excessive soil nitrogen levels because vegetative growth will be promoted to the disadvantage of root development. Mulching is an important production factor by acting as a buffer to moisture and temperature fluctuations. Cultivated crops are commonly grown on raised beds and are provided shade by wood lath or fabric screens usually made from synthetic materials.

The best overall growth occurs at 18–20°C; root growth is optimum at 15–18°C. No leaf growth occurs at less than 5°C and is suppressed above 30°C. Plants are relatively temperature hardy, and well-mulched plants tolerate temperatures as low as −10°C.

Two recognized types of cultivated *P. ginseng* are grown in Asian countries. One type produces large, round rhizomes with short, thick, main roots having many side branches. This type is fast growing and high yielding, and form seed in the third year. The other type produces long, thin rhizomes with few side branches on the short main root. It is slower growing and lower yielding than the large round types, and seed are usually formed in the fourth year.

Ginseng grows very slowly, and even after 5 years, the spindle-shaped, fleshy taproot (25–30% dry matter) may only be 10 cm long and 2–3 cm thick (Fig. 25.10). Root weight increases with continued

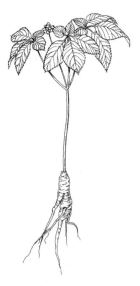

FIG. 25.10.   American ginseng, *Panax quinquefolius.*

growth and is affected by plant spacings. Beyond 5 years, considerable root branching occurs. Although a cosmetic factor, root branching has a strong effect on perceived value, in that the more the root and lateral branches resemble the human figure, the greater its value. Some growers intentionally promote root branching by replanting.

The rhizome grows upright or horizontally; upright growth is more common with cultivated crops. A new terminal bud is initiated on the rhizome during midsummer and enlarges during the rest of the active growth period. In the spring, the bud elongates with the occurrence of warm temperatures. Annual abscission scars appear on the rhizome.

Plants are deciduous and foliage growth is slow and sparse. Usually, only a single compound leaf is formed which has three to five palmate-like leaflets. Only older plants produce a whorl of foliage with four or five leaves, and those with five leaflets do not appear until the fourth or fifth year, and sometimes longer. The limited leaf growth partly explains the tediously slow plant and root development. Native plant growth is much slower than those of cultivated plants, and after the fourth or fifth year, may only achieve a height of 50 cm.

Flowering varies with plant type but can begin as early as the end of the second or third season's growth, and generally continues for about 3 weeks. Flowers are polygamous and self-pollinated. From 15 to 50 small greenish white flowers form in a cluster on a relatively tall seed stalk. The fruit is a pea-size berry, bright red when mature, and usually contains two small (5 mm), flat, wrinkled seed; about 20 seed weigh 1 g. Seed from plants older than 4 years are preferred to seed from younger plants. A 4-year-old plant can be expected to provide 30–40 berries.

When the production emphasis is root growth, growers may remove flowers. Removal has been shown to enhance root yields about 30%. A portion of production is allowed to produce seed which are used for propagation. The berries develop unevenly, and thinning is used to increase the size of remaining berries and to select for those of similar maturity.

Harvested seed is washed after removal from the fruit, but for seed to remain viable, they cannot be allowed to dry. Because of embryo immaturity, seed are stored at 2–4°C in slightly damp sand. This procedure, called stratification, completes embryonic development along with physiological after-ripening. Freshly harvested seed normally require about 18–20 months of stratification before being capable of germination. Before sowing, seed should be pregerminated. Seed stratified for propagation commonly sells for twice that of fresh (green) seed.

Most plantings are made in the fall, either by direct seeding or with year-old nursery transplants. Plant populations range from 15 to 20 per m^2, whereas native or noncultivated populations are variable and usually at low densities.

Ginseng is susceptible to diseases that include *Botrytis* and *Alternaria* leaf blight, *Phytophthora* root rot, *Verticillium* wilt, and damping off. Insect pests are leafhoppers, aphids, and cutworms. Additional pests are root knot nematode, snails, slugs, mice, and weeds.

Harvest of cultivated ginseng usually is performed after 4 years of growth. Individual harvested roots are 7–13 cm in length, 2–5 cm in diameter, and weigh between 40 and 80 g, with yields of dried roots ranging from 1800 to 2500 kg/ha depending on number of seasons grown. Some growers harvest after 3 years' growth. Although yields are lower, their motivation is to recover the high production costs associated with this crop as soon as possible. Ginseng growers, much like truffle producers, are secretive about their activities, especially regarding access to rapidly disappearing noncultivated populations. Relatively few producers and limited supplies tend to keep prices high.

For small holdings, harvest is usually accomplished with hand labor and involves digging out roots following leaf fall in autumn. With large plantings, digging is mechanized, using potato harvester or similar equipment. After being washed and cleaned free of soil, roots receive postharvest curing by drying. Roots are not washed if they are to be placed in cold storage for drying at a later time. Roots can be held for 6 weeks at 5–8°C and 80–90% RH and then washed, size graded, and dried.

Ginseng is not used until it has been dried. Drying is a critical procedure for high-quality production. Natural drying is preferred but cannot always be practiced. Generally, 25°C and high air movement is used for the initial stage of drying and then ventilation is slowed and the temperature is increased to 32°C. An alternative procedure applies 38–43°C initially, which is then decreased to 32°C. The drying procedure is usually carried out over a 2-week period. Properly dried roots have a moisture content between 10% and 13%. Good quality after drying is judged by a creamy yellow exterior and white interior root color. Excessively high drying temperatures and long duration are avoided because they result in a dark brown discoloration. Drying at low temperatures results in a undesirable green color.

Dried roots are aromatic, have a licoricelike taste, and most often are used in a thinly sliced, shaved or powdered form in teas, various drinks, candies, and many other products, as well as numerous medicinal preparations.

UDO, *Aralia cordata* Thunb.

Family: Araliaceae

Udo is native to Asia and the Malaysian peninsula. It is a tall (2.5-m), perennial plant with small greenish white flowers. Leaves are two to three pinnate, each having five to nine leaflets. Propagation is either with seed or root cuttings. In the spring, the storage roots are placed in a warm dark cellar and the blanched shoots are forced much like witloof chicory. The shoots which are popular in Japan have an aromatic turpentinelike taste; they are sliced and used raw as garnish or put into soups.

The aromatic shoot tips and foliage of *Nothopanax fruticosum (Polyscias fruticosum)*, a related Araliaceae species, are also eaten as vegetables. This tall (2-m) shrub is native to tropical Asia and Polynesia.

MALABAR SPINACH, *Basella alba* L.

Family: Basellaceae
Other name: Malabar nightshade

CEYLON SPINACH, *Basella rubra* L.

Family: Basellaceae
Other name: Indian spinach

These *Basella* species are succulent annual or biennial (perennial in the humid tropics) plants that exhibit a rampant, highly branched, long (3–5 m), twining vine growth, often grown on supports. The plants, having glabrous stems and leaves, are adapted to many soil types and to tropical and warm, temperate climates. Leaves are more tender when grown in partial shade. Temperatures most favorable for growth are between 25°C and 30°C during the day and 17–20°C during nights; the crop is very frost sensitive. Flowering occurs during days of 12 h or less; day lengths greater than 13 h inhibit flowering. Water-stressed plants will often flower.

Basella alba had an African or southeast Asian origin. Leaves are narrow and lance shaped; stems and petioles are not pigmented. White flowers are borne in loose clusters on elongated spikes.

Basella rubra possibly originated in India or Indonesia. Leaves are slightly pointed, nearly as wide as long. Red flowers are sessile in small clusters on short axillary spikes. Stems and petioles have red

FIG. 25.11.   Malabar spinach, *Basella alba.*

pigmentation. Hortus Third places *B. rubra* as a cultivar of *B. alba;* other authorities give each a species status. A cultivar with heart-shaped leaves is sometimes identified as *B. cordifolia.*

Propagation is by direct seed, transplants, or stem cuttings. Common plant spacings are as close as 45 × 45 cm and are up to 1 m apart if hill planted and trellised; trellised plants are more productive.

Stems tips, about 20–30 cm long, are harvested about 55–70 days from seeding. Repeated harvests are made of new stem growth so that yields of 15–20 t/ha can be obtained. The succulent leaves and shoots are widely used as potherbs (Fig. 25.11). The mucilaginous texture is especially useful as a thickener in soups and stews. The red pigment in fruit was used for a dye and ink in ancient China. *Pythium* caused damping off and *Fusarium* root decay are diseases most often responsible for crop losses.

BORAGE, *Borago officinalis* L.

Family: Boraginaceae

Native to the Mediterranean region, borage is a hardy annual which often exhibits biennial properties. The leaves are gray-green and ob-long–lanceolate in shape with bristly surfaces that can be irritating and cause dermatitis for sensitive people. The bristly stems are hollow, and the bluish purple flowers are star shaped. Plants grow to a height of 45–60 cm. Propagation usually is with seed, although crown divisions are also used.

Foliage, tender stems, and the flowers are edible. The leaves have a cucumberlike flavor and odor and usually are prepared as a potherb, and fresh in salads. The foliage has a high tannin, calcium, and malic acid content, and very good levels of beta-carotene and vitamin C. Borage has been used for multiple medicinal uses for centuries.

COMFREY, *Symphytum officinale* L. *(S. peregrinum)*

Family: Boraginaceae

Comfrey is a hardy herbaceous perennial of European and west Asian origin, growing to a height of 1 m when fully mature. Leaves are used for their flavor and herbal properties, and occasionally as a potherb Crowded onto the compressed stem, leaves are about 30 cm long and 10 cm wide, with many short bristly hairs on the surface. With continued growth, the stem forms new branches. White, light yellow, or purple, bell-like flowers are produced on long pedicels. Seed, root, or crown divisions are used for propagation.

CACTUS, *Opuntia ficus-indica* (L.) Mill.

Family: Cactaceae
Other names: Prickly pear, Indian fig, nopales

Within the Cactaceae, the *Opuntia* genus is important as a food crop; some species of *Nopalea* also have food use. These plants are perennials grown in tropical and subtropical areas for their edible fruit and clad-odes (stems). Of about 300 species, only a few are cultivated expressly for food.

Cacti are of New World origin. Columbus introduced cacti to Spain, and their spread into parts of the Mediterranean and Africa was extensive. Continued introductions find cacti in many other regions. Cacti presently occupy significant areas in Italy and North Africa. It appears that the center of diversity for *Opuntia* is Mexico. Its use, both wild and cultivated, has a long history in the Americas.

Species of *Opuntia* commonly used for vegetable production of young tender cladodes are *O. ficus-indica,* which produces long slender cladodes, and *O. inermis,* which produces rounder, heavier cladodes. Plants are productive throughout the year, although growth rates diminish with cool temperatures. Fruit of several species are eaten, but *O. ficus-indica* is the species most often grown for commercial fruit (tuna) production. Plants of these and other species are also used for livestock forage, and as hedges and for ornamental purposes.

Cacti vary greatly in size and growth habit; some are tree like, others exhibit short prostrate growth. Roots are fibrous, thick, and fleshy. Cladodes are usually very fleshy and become woody with aging. Vegetative buds on cladodes produce additional cladodes and the process continues to be repeated, resulting in the extension of steplike branching. Spines (glochids), sometimes many, are solitary or clustered and found on all aboveground plant parts. Spines vary in size; some are large and prominent, others are minute. Cladodes, also known as "pads," "nopales," or "cactus leaves" are not leaves but flattened, elliptical- or oblong-shaped stems. Plants have true leaves; however, these are vestigial and abscise during early cladode growth.

Seed propagation of cacti is possible; however, early growth is much slower than vegetative propagation, where mature "mother" cladodes are cut from donor plants. These cut cladodes are held under conditions favoring wound callus development. After about 2 weeks, the healed surface of the cladode is placed about 3 cm deep into moist but not wet soil. Cladodes have sufficient moisture to support initial root development and to become established without added water. Once established, water, if needed, can be applied. For commercial vegetable production, plant populations approach 40,000 plants/ha, spaced at 30×80 cm for hand labor or slightly wider when equipment is used. Plantings for cladode production are renewed more frequently than those for fruit production. New cladode growth usually is fairly rapid, and within 2–3 months after planting, several tiers of cladodes can form.

Cactus plants have a process for photosynthetic carbon fixation differing from C_3 and C_4 plants. Cactus are Cassulacean acid metabolism (CAM) plants. CAM plants have stomata that open at night, allowing carboxylase enzymes to fix CO_2 into four carbon organic acids, such as malic. Malic acid is later converted into sugars during daylight. As a result, acid content of the cladodes varies, exhibiting a diurnal variation, where acidity decreases during the day and increases during the night. Being CAM plants, cactus species are several times more efficient in utilizing water for dry matter production than the best of efficient C_4 species, such as *Zea mays.* This partially explains why cacti are well adapted to arid conditions.

Although plants survive without much rainfall or irrigation, for commercial production, between 300 and 600 mm of moisture is required annually. Good soil drainage is essential, as even brief periods of flooding can damage roots. Fertilizer is beneficial, and more is applied for producing the vegetative crop than the fruit crop. Little, if any, nitrogen is supplied to plantings producing fruit because nitrogen increases cladode growth preferentially to fruit development.

Cladode Production

Harvests are typically delayed until at least three or four cladodes have fully developed. Cladodes are harvested by cutting at the articulation with the subtending cladode. Careful cutting reduces damage because cut areas are susceptible to decay and tender cladodes are also susceptible to bruising. Plants are capable of producing more than 20 harvestable cladodes per year and yields of 90 t/ha are attained.

The preferred size for vegetable use are about 20 cm long, 10 cm wide, and 2 cm thick. Some cultivars are spineless (Fig. 25.12a). Fresh cladodes are turgid and bright green. Cladodes are composed largely of water, about 92%, the remainder being fiber (4–6%) and protein (1–2%). They contain good levels of pro-vitamin A and vitamin C. After trimming, cladodes are consumed as a fresh or cooked vegetable, tasting similar to fresh green bean pods. They have a mucilaginous texture that also makes them well suited for use in soups and stews.

Generally utilized soon after harvest, cladodes are infrequently stored. However, storage at 5–10°C and high humidity can provide about 2–3 weeks of postharvest life, after which chilling injury will usually become evident. Storage at 20°C avoids chilling but reduces shelf life. Storage at the higher temperature will decrease cladode acid content, whereas low temperatures may increase acidity.

Fruit (Tuna) Production

Fruit production differs considerably from that for vegetable cladodes. For example, flowers are borne on upper surfaces of cladodes from buds developed in the previous season and form on mature cladodes which are no longer palatable for vegetable use. Large plants are grown to provide more fruit-growing surfaces; 10–15 fruit can form on a large cladode, but excess fruit are usually thinned. To produce large plants, wide spacings are used and average about 2000 plants per hectare.

Some chilling is needed for differentiation of floral primordia. Flower buds are cylindrical in shape; vegetative buds are flat. Flowers develop about 4 weeks after bud emergence and are showy with yellow, red, or

FIG. 25.12.   Cactus, *Opuntia ficus-indica:* (a) cladodes (pads), and (b) fruit (tuna).

green corollas; they are self-pollinated. The edible fruit is a globe- or ovoid-shaped berry containing many flat, white seed.

Fruit exterior and interior colors vary from pale green or yellow to dark purple. The pear- or fig-shape fruit are about 3–7 cm wide and 8–10 cm long, and typically have a flattened or sunken floral cavity (Fig. 25.12b). The thick and fleshy skin may be a third of the fruit weight and is not consumed, nor are the plentiful seed that contribute 5–10% of the fruit weight. The remainder is the usable sweet juicy pulp.

Sugar content increases with maturity, whereas acidity decreases; the acid content is highest in the peel.

Fruit are harvested by hand, by twisting from the cladode. Careless removal tears tissues and can lead to rapid decay and loss. Preferably, fruit should be carefully cut free. Harvesting is made difficult by the numerous minute spines (glochids) on the fruit that can become airborne, especially when humidities are low. Harvest during periods when dew is present is helpful. Heavy gloves are worn for fruit removal and handling. Various methods of brushing the harvested fruit are used to remove the remaining glochids. Yields of 8 t/ha can be obtained by the third year of production and mature plantations (10 years) can produce 20 t/ha.

Tuna are usually consumed soon after harvest and are eaten fresh or cooked. They are also dried or used for syrups and jams. These are nonclimacteric fruit and have a low respiration rate and very low ethylene production. However, to limit wilting and quality loss, fruit may be waxed. Somewhat similar to cladodes, fruit can also be held at 5–7°C for up to 3 weeks, at which time chilling injury becomes apparent.

Although spines can protect cactus from predatory animals, plants are attacked by disease, insects, and other pests. Several important diseases are soft rot *(Erwinia aroideae)*, rust *(Phillostica opunitae)*, black rot *(Phytophthora cactorum)*, and black spot *(Perisporium wrightii)*. Problem insects are thrips species, and the yellow-stripped armyworm *(Spodoptera latifascia)*. Other pests are spider mites *(Tetrancychus opuntiae)* and snails *(Helix tortensis)*.

RAMPION, *Campanula rapunculus* L.

Family: Campanulaceae

Rampion is a erect, 60–70-cm-tall, biennial plant with a rosette growth habit. It is a vegetable well known in Britain and also grown in some parts of northern Europe. Leaves are entire, long, and oval and about 15 cm long; the bell-shaped corolla is sharply lobed and light purple. Fleshy, slender, first-year, white roots and basal leaves are eaten raw or cooked. Short and unbranched roots are preferred to old roots that can reach more than 25 cm in length. Although usually seed propagated, cuttings and root divisions are also used. Cultivation is similar to garden radish.

CAPER, *Capparis spinosa* L.

Family: Capparaceae

The caper plant, a shrubby deciduous native of the Mediterranean area, is grown for it flower buds *Caper sodala,* a related caper known as timbuctoo, is grown in northern Africa and another, *C. corymbifera,* is grown in South Africa.

Unsupported plant height is about 60–75 cm with branched horizontally spreading vines reaching 2–3 m in length. Having a deep root system, the plant is resistant to drought. Although relatively little cultural care is required, good soil drainage is important. Plants are relatively free of disease and insect pests and when dormant, are hardy enough to withstand freezing temperatures.

Leaves are alternate, oval to round in shape, about 5 cm in diameter, leathery, and shiny green. A pair of spines located at the base of each petiole can make harvesting difficult; some cultivars are spineless. Numerous flower buds are formed, one in every leaf axil. Flowers are bisexual with four white petals, each 2–3 cm long, with many prominent violet stamens, all much longer than the style. Flowers are open for 1–2 days.

Unopened flower buds are the primary product. These are hand harvested during the summer flowering period, often daily because the smallest buds have the highest value. A mature plant is capable of producing 8–9 kg of fresh buds annually. Fresh caper buds are not palatable and until processed, do not exhibit their unique pungent flavor. Buds are processed by pickling or brining. The principal use of capers is as a condiment.

The pollinated flower produces a fruit that is an elongated (5 × 1.5 cm), green berry, containing many seed. Seed are used for propagation, but germination and seedling growth is slow. Cuttings are the preferred propagating material, exhibiting less plant-type variation, and are productive earlier. One-year-old rooted cuttings are commonly used. Fields are established in late winter. Plant densities vary; some spacings are 5 × 5 m or closer in intensive cultivation. During the first season, few buds are produced, but by the fourth year, full production potential is reached.

Favorable growing temperatures range from 13°C to 27°C. Capers are often found growing on poor soils and with minimum moisture. However, production from such plantings is low. Growth is enhanced from annual applications of about 100 kg/ha of a mixed fertilizer, and the plant should not be subjected to moisture stress. Plants are dormant after leaf fall and any necessary pruning is done during this period.

Tender new shoots that emerge in the spring are sometimes harvested and used as a vegetable, much like asparagus. Excessive shoot harvesting results in reduced bud production.

CASTOR BEAN, *Ricinus communis* L.

Family: Euphorbiaceae

Castor bean, endemic to Africa, is grown throughout the tropics and many other warm-temperature regions mainly for its inedible seed oil that has some medicinal and many industrial uses. The plant grows as a large, almost treelike shrub with large-palmate, multilobed leaves. Plants are annuals and monoecious, with flowers forming on different parts of the panicle. Seed borne in three-celled spiny capsules have an oil content of about 50% and also contain ricin, a *extremely toxic* alkaloid. Vegetable use is made of tender shoots and leaves after detoxification, which involves boiling with several changes of water.

CHAYA, *Cnidoscolus chayamansa* McVaughan

Family: Euphorbiaceae

Chaya, also called tree spinach, is a large, fast-growing leafy, perennial shrub, 2–3 m tall when mature. The foliage is used as a green vegetable in countries from Mexico to Brazil. The plant resembles cassava, with foliage similar to that of okra. Leaves are alternate, simple, and palmately three or five lobed, about 15–20 cm wide, and supported on a long, thin petiole (Fig. 25.13). The semiwoody stems contain latex. Small (1 cm) male and female flowers are borne together on a long pedicel. Round seed pods are about 2.5 cm in diameter. Plants are tolerant of heavy rains and have some drought tolerance, but are cold sensitive. Propagation is with stem cuttings.

Tender shoots and leaves are used much like spinach or other potherbs. Domesticated cultivars have little or none of the irritant features found in leaves or stem spines and hairs common to wild-type chaya, *C. aconitifloius*. Plants can be harvested continuously, but not more than half of the foliage should be removed at any one time to avoid impeding subsequent growth. Raw tissues are poisonous because of their hydrocyanic glucoside content, but cooking and leaching renders the product safe to eat.

FIG. 25.13.   Chaya, *Cnidoscolus chayamansa,* shoot tip and leaves.

KATUK, *Sauropus androgynus* (L.) Merr.
Family: Euphorbiaceae

Also known as chekurmanis, katuk plants are widely grown for their tender shoot tips. This vegetable is consumed extensively in Indonesia, especially in Borneo and throughout areas from India to Southeast Asia. The perennial shrub has tropical and subtropical adaptation and provides year-round productivity, although plants tend to become somewhat dormant in cool weather.

Plants exhibit prolific growth of long, upright stems, which often tip over. Consequently, plants are usually pruned and grown as a hedge. Pruning encourages lateral growth of shoots. Edible quality is improved when plants are grown in partial shade.

In addition to the tender shoot tips, the leaves and flowers are also eaten, either raw or cooked. The older, less tender leaves are usually cooked. Shoot tips resemble asparagus and have a pealike taste. Foliage protein content is about 7% and katuk shoots are a very good source of pro-vitamin A, vitamin C, and also calcium, iron, and magnesium.

PERILLA, *Perilla frutescens* (L.) Britt. *(Ocimum frutescens)*
Family: Lamiaceae (Labiatae)
Other name: shiso, beefsteak plant

Perilla is an annual aromatic herb widely cultivated in throughout eastern Asia and in Japan for its foliage and edible seed oil. Perilla

FIG. 25.14.   Perilla, *Perilla frutescens.*

originated in Southeast Asia but is now widely domesticated and is also known as red shiso and beefsteak plant. The plants has squarelike stems that grow erect to a height of 90 cm. The broadly ovate leaves are about 12 cm long on long petioles. Leaf margins are toothed and usually a brownish purple or bronze color or may be variegated; some cultivars are green (Fig. 25.14). The leaves are appreciated for their curry-like flavor.

OKRA, *Abelmoschus esculentus* (L.) Moench. *(Hibiscus esculentus)*

Family: Malvaceae
Other names: Bhindi, gumbo, and lady's finger.

Origin and Botany

Okra is an erect, moderately branched, warm-weather annual, well adapted to the tropics and to warm, temperate areas. It is grown primarily for the immature fruit (Fig. 25.15). The plant explorer Vavilov indicated that the domestication of okra was in the area of Ethiopia; others identify its origin as India. However, genetic studies revealed discrep-

a

b

FIG. 25.15.   Okra, *Abelmoschus esculentus:* (a) foliage, flower, cross-sectioned pods, and seed; and (b) intact immature pods. Source (a): Courtesy of National Garden Bureau, Downers Grove, Illinois.

ancies which indicate that the crop might be composed of multiple species, origins of which might have been Southeast Asia, India, west Africa, and/or Ethiopia. Ploidy number varies greatly, with the diploid ranging from $2n = 66$ to $2n = 144$. Some investigators believe okra to be of amphidiploid origin exhibiting genomic contributions from hybridization of species of different ploidy. Cultivated species of *Abelmoschus* in addition to *A. esculentus* are *A. manihot* and *A. moschatus*. The wild species, *A. crinitus*, *A. angulosus,* and *A. tuberculatus* are confined to Asia; *A. ficulneus* is present in west Africa. A perennial treelike okra grown in west African villages is believed to be an intermediate between *A. esculentus* and *A. manihot*.

The taproot of okra is deep with many shallow laterals. Cultivars range in height from less than 1 m to more than 2 m; stems of some have a reddish tinge. With increasing age, stems rapidly become woody. Leaves are spirally arranged, large, and ovate with three or five shallow lobes. Single, large, 5–8-cm-long, yellow flowers with dark red or purple centers are formed in leaf axils. Self-fertile flowers have five petals and many stamens and stigmas; the ovary is superior. Flowering begins near the base of the plant and progress up the stem. Floral development is simultaneous with stem elongation. Seldom does more than one flower bloom at a time on an individual stem. Flowers open only once in the morning and are receptive for a short time. Flower buds are evident in advance of the first open flower, but these do not bloom until after subtending flowers have bloomed. Flowering occurs 35–60 days after planting; an additional 40 or more days after anthesis are required for seed to fully mature. Normally self-pollinated, flowers also can be pollinated by insects.

The fruit is an erect pod (capsule) formed on a short pedicel. Pods are smooth or ridged and beaked; surfaces are almost glabrous or with light hairs, colors vary from pale green to dark red. Fruit length can range from 5 to 50 cm and contain 30–80 seed. Pods rapidly lose tenderness and become woody with advancing maturity. Mature pods dehisce, releasing round, very hard, dark green or brown seed. Seed sizes vary, such that a 1000 seed sample may weigh between 30 and 80 g.

Domestication resulted in smooth fruit being selected from spiny wild types. Modern cultivars are essentially spineless, and although some have soft bracts, these usually abscise from the base of the pod.

Okra cultivars are short-day responsive, although some are day neutral. Temperature and photoperiod interact to influence flowering. Flower initiation and flowering are delayed at high temperatures. Temperatures above 20°C are needed for normal development, 30–35°C being the most favorable for optimum growth. At temperatures less than 15°C, growth is poor, and at 10°C, chilling injury can occur. At temperatures greater than 42°C, flower abortion can occur.

Culture

Fertile, well-drained, sandy to clay, loam soils are preferred for cultivation. Okra is slightly sensitive to soil acidity; a pH range from 6.0 to 7.0 is satisfactory. Plants have good drought tolerance; however, yield is adversely affected when plants are stressed for moisture during flowering. The crop's moisture requirement is about 50 mm per week. Nutrient uptake is high; therefore, okra plants are responsive to fertilization. For effectiveness, the fertilizer is usually divided among two or three application periods.

The crop is seed propagated. Germination is negligible at temperatures less than 17°C; about 29–30°C is optimum. Because of a hard seed coat, emergence is improved by soaking in water for about 24 h before planting. Acid scarification can also be used. Seed are planted 2–3 cm deep, often into a raised seedbed, and normally emergence occurs within a week. An excess of seed is often sown with subsequent thinning to spacings of 20–30 cm within rows and 100 cm between rows. However, populations vary considerable and can range from 50 thousand to 150 thousand plants per hectare. High plant densities tend to limit lateral branching and also impede harvesting.

Diseases and Pests

Being in the same family as cotton, okra is subject to many of the diseases and pests of that major world crop. Some important diseases include the following:

Alternaria leaf spot	*Alternaria hibiscinum*
Cercospora blight and leaf spot	*Cercospora abelmoschi, C. hibiscina, C. malayensis*
Damping off	*Pythium, Rhizoctonia,* and *Fusarium* spp.
Dry rot	*Macrophomina phaseolina*
Pod rot	*Phytophthora parasitica*
Fusarium wilt	*Fusarium oxysporum* f. *vasinfectum*
Pod spot	*Ascochyta abelmoschci*
Powdery mildew	*Erysiphe cichoracearum*
Verticilium wilt	*Verticilium dahliae*

Some of the important insect pests include the following:

Corn earworm	*Heliothis armigera*
Flea beetle	*Nisotra gemella*
Leafhopper	*Amrasca biguttula*
Melon/cotton aphid	*Aphis gossypii*
Pink bollworm	*Pectinophora gossypiella*
Spider mite	*Tetranychus* spp.
Stemborer	*Earias biplaga, E. insulana*
White fly	*Bemisia tabaci, B. argentifolia*

Yellow vein mosaic virus causes some crop losses, and *Meloidogyne* species of root knot nematode often parasitize okra plants and also reduce yields.

Harvest and Postharvest

Harvest often begins about 7 weeks after planting. Okra harvesting is a disagreeable task because irritant hairs on all plant surfaces can cause skin irritation and dermatitis. Pods are hand harvested by slightly twisting and snapping pods free from the stems, although cutting through the tough pedicel results in less pod and stem damage.

As pod size increases, tenderness decreases. The largest pod size with acceptable tenderness usually occurs about 4–7 days after anthesis. Rapid pod fiber development starts about the same time as maximum rate of expansion. Harvesting is usually performed at 2–3 day intervals, but high temperatures may mandate daily harvesting. Delay or failure to harvest results in increased pod fiber and also slows subsequent flower and plant development. Frequent harvesting of young pods tends to sustain vegetative growth and prolong the harvest period. Most of the yield is achieved in the first half of the typical 50–55-day harvest period. A well-grown, vigorous plant is capable of producing as many as 100 pods. Nevertheless, yields vary greatly. The current world average is 6 t/ha, a yield of 10 t/ha is considered very good, but the potential is three or four times greater.

Because of its high respiration rate, sensitivity to bruises and rapid desiccation, okra requires careful handling. Marketing of fresh product is usually near production areas. Pods should be cooled to 10°C as soon as possible to maintain quality. Refrigerated cold rooms are generally used; hydrocooling is seldom utilized. Holding at less than 7°C is not recommended because of chilling injury; pod wilting can occur at temperatures above 10°C if humidity is low.

Uses

Okra has many uses, being consumed fresh or processed by canning, freezing, and pickling. Because of its mucilaginous properties, it is a common ingredient in soups and sauces. Processors prefer short, less than 10-cm pods, because they are easier to process and produce a more attractive product. Okra is also dried for later rehydration, and some of the dried okra is ground for use in a powdered form. In some countries, okra seed are roasted and used as a coffee substitute. In Southeast Asia and central Africa, leaves and tender shoots, having a slightly acid taste, are eaten much like spinach. Sometimes, leaves and shoots are sun dried for later use.

If available, a greater use can be made of hybrid cultivars along with the introduction of broader and better-adapted germplasm that also incorporates pest resistance. Product improvement goals are to maintain pod tenderness longer while pods continue to enlarge to obtain high yields. Additional objectives are to slow the rate of seed hardening and to introduce cultivars with fleshy, thicker, and shorter pods. Because of its 20% protein and 20% high-quality oil content, the seed has a high potential for expanded food usage. However, the presence of cyclopropenoid fatty acids causes some toxicity concerns. A related, somewhat more robust species, *A. caillei,* also of African origin, is known as west African okra. It is cultivated much like okra in west and central Africa.

SUNSET HIBISCUS, *Abelmoschus manihot* (L.) Medic. *(Hibiscus manihot)*

Family: Malvaceae

Resembling okra somewhat, sunset hibiscus, also known as aibika in Southeast Asia, is a perennial shrub and possibly a native of tropical Asia. It is widely cultivated in many South Pacific islands, Indonesia, and other parts of tropical Asia. The highly palatable young leaves and stem tips are used raw or cooked and have a sweet taste; they have a fairly high protein content and are nearly fiber free. Plant height varies from 2 to 3 m, but some grow as tall as 5 or 6 m. The alternate leaves on long petioles are mostly narrow, although sizes and shapes vary. The foliage of some plants have a red tinge; flowers resemble those of okra. The crop is most productive when grown in full sun at temperatures greater than 25°C; shading reduces productivity. Plants have a relatively high moisture requirement.

Plants are propagated with stem cuttings and are spaced about 25 cm apart in rows about 100 cm apart, resulting in 40,000 plants/ha. Harvests start about 3 months after planting. With regrowth, harvest is continued for as long as 9 months, but annual replanting is commonly practiced. Harvesting encourages branching, which results in a compact, bushy growth that tends to delay flowering. About 15 cm of the tender shoot tips and attached leaves are harvested and are generally consumed the same day because postharvest life is very short. Yields from multiple harvests range from 15 t/ha to as much as 60 t/ha.

ROSELLE, *Hibiscus sabdariffa* L.

Family: Malvaceae
Other names: Jamaican or Indian sorrel

A central or west Africa native, roselle is grown in many parts of Africa, India, Southeast Asia, and the South Pacific for its edible fleshy enlarged *calyces* used fresh or dried (Fig. 25.16). The young tender shoots and leaves are used as a potherb, and the seed are also eaten, usually after roasting.

Roselle is well suited for tropical conditions, showing rapid growth in warm temperatures. Because of sensitivity to low temperatures, roselle is seldom grown in temperate regions. Additionally, flowering is initiated by short-day length, which is associated with decreasing temperatures in temperate zones. Plants are responsive to rich soils, although they are often grown on less fertile soils. The crop is seldom fertilized except when grown for intensive, leafy, vegetable production. Plants have fairly good drought tolerance.

Roselle is a tall (2–4 m), bushy, branched annual with a deep root system. Stems are green; some have a red tinge. Foliage appears similar to okra but is lancelike with deeper lobes. Flowers are similar to those of okra.

Plants are usually seed propagated, but cuttings are also used. For calyx production, plants are spaced equidistant about 1 m apart, and for leafy, vegetable production at spacings about 60 cm × 90 cm. Mound or hill plantings are also made instead of rows.

In subtropical and tropical areas flowering starts 80–140 days after plant establishment. Calyces are harvested about 3 weeks after bloom and before they become woody. When mature, calyces turn red; light yellow types are also known. The calyces have a high citric acid content and are usually boiled with sugar for use in preserves, jelly, juices, and

FIG. 25.16.   Roselle calyces, *Hibiscus sabdariffa.*

teas. The short, egg-shaped, hairy fruit capsules are not harvested, unless ripened for seed. Seed capsules contain about 25 small, dark brown seed; about 40 weigh 1 g. Leaves and stalks have use in salads, cooked potherbs, and as seasoning for curry preparations. Leaves have a sour taste and a mucilaginous texture. Leaf and shoot harvesting often begins about 3 months after planting, well before calyces are harvested. Representative yields for calyces production are about 8 t/ ha, and about 10 t/ha for leafy shoot production. Some roselle cultivars are grown specifically for the jutelike fiber and for paper manufacturing. A closely related species, *H. acetosella,* known as false roselle, is also of African origin; the leaves and shoots have a similar taste and texture as roselle. However, *H. acetosella* plants are short-lived perennials, but otherwise are similar in appearance and performance.

COMMON MALVA, *Malva verticillata* var. *crispa* L.
EDIBLE MALVA, *Malva neglecta* Wallr.

Family: Malvaceae

The leaves of these semidomesticated weedy Eurasian natives are occasionally used in salads and as cooked vegetables for their mucilaginous properties and the spicy aromatic flavor of the foliage. Plants are seed propagated. Annuals and biennials types occur. The erect plants can become as tall as 1.5 m and bear white or purple flowers. Leaves are rounded, having from five to seven lobes; some plants have strongly curled, crispy foliage (Fig. 25.17). Leaf carotene and protein contents are relatively high.

FIG. 25.17.　Malva, *Malva neglecta,* a vegetable form.

MARTYNIA, *Proboscidea louisianica* (Mill.) Thell. *(P. jussieui)*

Family: Martyniaceae
Other names: Unicorn plant, probosis plant, devil's claw, ram's horn.

An annual plant and native of the southwestern United States, immature fruit of martynia are used as vegetables. Plants are about 60 cm tall and 90 cm in width when fully grown, and thrive in hot climates (30°C and above). Pubescent leaves are large, round, and slightly pointed, about 10–30 cm in diameter with glandular hairs that release an objectionable odor. Flowers are attractive, orchidlike, of a creamy white to lavender color. Fruit are hairy, curved, hornlike seed pods, eaten when tender, often prepared as pickles; flowers are also edible.

Seed pods (capsules) grow rapidly to a length of 10–15 cm or more and achieve a thickness of 2–3 cm. At this stage, they are woody and inedible (Fig. 25.18). Seed are hard and black and dispersed when pods dehisce. The dry pod halves remain attached at the peduncle and spread out to resemble the horns of a ram. Plants are seed propagated.

a   b

FIG. 25.18.   Martynia, *Proboscidea louisianica:* (a) plant, and (b) mature pods.

CEDRUS, *Cedrela sinensis* Juss.

Family: Meliaceae

The shoot tips and tender leaves of this tall (15 m), deciduous tree are eaten as a potherb in parts of China.

HORSERADISH TREE, *Moringa pterygosperma* C.F. Gaertn *(M. oleifera)*

Family: Moringaceae
Other name: Drumstick tree

Of northern Indian and Pakistan origin, the horseradish tree was introduced into Southeast Asia and is now widely cultivated at lower elevations in many tropical areas. The tree is grown for the edible leaves, flowers, and immature fruit. A deciduous perennial, the tree is about 7–8 m tall with large mostly two to three pinnate, 50-cm-long leaves.

Clusters of creamy white or pale yellow bisexual flowers are formed in leaf axils on a spreading panicle. Elongated, narrow, pointed, pendent fruit capsules are as much as 50 cm long. Seed are nearly round, about 1 cm in diameter, and have three thin, attached wings. Propagation is possible with seed, but cuttings are commonly used. The typical fast growth results in productivity within 1 year after planting. Trees have good drought tolerance and prefer fertile, well-drained soils.

Immature fruit are harvested about 50–75 days after anthesis; an additional 50 days are required before mature seed are harvested. The immature fruit resemble the yardlong bean, *Vigna unguiculata,* and are similarly eaten. Like yardlong bean, they rapidly become fibrous. The mucilaginous, slightly pungent, inner tissues of fruit capsules are also eaten, often in curries. Roots have a horseradish taste, but are not eaten. Flowers, shoot tips, and leaves, in addition to fruit capsules, are used as vegetables. Even the seed can be eaten after an objectionable alkaloid has been detoxified by cooking. Seed are commonly prepared by frying, much like groundnuts, and have a similar taste. Extractable seed oil, although also edible, has other uses. In addition to its many vegetable uses, the trees are used for fences, vine supports, and shade, and all parts of the plant seem to have some medicinal application.

POKEWEED, *Phytolacca americana* L.

Family: Phytolaccaceae
Other name: Pokeberry

Pokeweed is a seldom cultivated perennial native of eastern North America. The cultivated and gathered young shoots are used as greens and potherbs. All plant parts are poisonous and have a disagreeable odor, although newly emerged shoots are less poisonous, and these young asparaguslike shoots are used. Before use as greens, shoot bitterness is removed by boiling in water; the cooking water is discarded, with boiling repeated if bitterness remains.

Plants are vigorous, reaching heights up to 3 m. A large, fleshy taproot ensures survival and annual regrowth. Flowers are borne in clusters, bloom in the summer, and produce dark red fruit similar to nightshade. Propagation is with seed, which should be scarified to overcome dormancy. The procedure for forcing offshoots requires lifting roots after annual growth is completed and storing them at low temperatures. After vernalization, growth-promoting temperatures are provided. Emerging shoots are carefully harvested to avoid including the more poisonous root and stem tissues.

PEREJIL, *Peperomia pereskiifolia* (Jacq.) HBK., *P. viridispica*

Family: Piperaceae

Perejil, a small, succulent native of Venezuela and Colombia, is grown for the edible young leaves and shoots. Plants require warm temperatures but little light, and well-drained soils. Propagation is with stem cuttings and root divisions.

RHUBARB, *Rheum rhabarbarum* L. *(R. rhaponticum)*

Family: Polygonaceae

The thick, fleshy petioles of rhubarb are the edible portions consumed for food and medicinal use since ancient times. Although rhubarb was known as a medicinal plant in China for more than 4500 years, the exact origin is not known; Siberia or southeast Russia may be possible centers. Dried rhizomes of Chinese rhubarb, *R. officinale,* endemic to China and Tibet, provide the drug *rhubarb,* an alkaloid. *R. palmatum,* known as east Indian and Chinese rhubarb, is also used medicinally.

New growth of this herbaceous perennial arises from the thickened semiwoody stem (crown) which consists of the fleshy rhizomes, buds, and storage roots. In addition to the thickened storage roots, an extensive, fibrous, root system develops. Buds at the outer portion of the central stem form new shoots. These establish roots, produce leaves, and initiate buds for the next season's growth. The growth pattern results in a horizontal expansion of the plant from the original site.

Leaf blades are large, some as much as 50 cm long and wide. The leaf blades are poisonous; people have died using them as vegetable greens. The edible fleshy succulent petioles can achieve a length of 60–75 cm and can be 4–7 cm wide (Fig. 25.19). The petiole cross-sectional appearance is nearly semicircular, with surface color ranging from green to dark red. Interior flesh can also vary from green to partly or totally red. Flavor is not affected by color.

Rhubarb tolerates a temperature range from 5°C to 25°C; higher temperatures have the effect of reducing petiole red-color formation. At very high temperatures, plant growth is retarded. Plants show fair drought tolerance and are winter hardy. Being a temperate-climate plant, rhubarb is not productive in subtropical climates. Well-drained loam soils with high organic matter are preferred, and a wide pH range is tolerated. The crop is a heavy user of nutrients, and annual fertilization is usually provided.

Overwintering at low temperatures initiates seed stalk development. Small, greenish white, bisexual flowers form on erect hollow panicles; the developed seed is a winged achene. Seed stalks are removed as soon as they are visible, in order to maintain vegetative growth. High plant densities tend to retard floral development.

Rooted crown divisions from 2- or 3-year-old plants are used for propagation; seed is generally not used because of its heterozygosity and because more time is required before the plants are productive.

FIG. 25.19. Petioles of rhubarb, *Rheum rhabarbarum*. The leaf blades are trimmed because they are poisonous and must not be consumed.

Crown divisions for propagation purposes are stored at low temperatures and high humidity until used. Exposure to temperatures less than 4°C for several weeks breaks bud dormancy. About 12,000 crowns per hectare are planted with spacings normally about 90 × 90 cm.

Production practices are somewhat like those for asparagus culture. Seed propagated plants should not be harvested during the first year of growth, but light harvesting can be made during the first year from crown transplants. With well-established, vigorous plants, harvesting can continue for 2 or 3 months.

Leaves of field-grown plants are usually serially harvested, a few at a time over a number of weeks, with new growth replacing those harvested. They are removed by pulling and twisting the petioles away from the base of the crown. Petioles are not removed by cutting; this is to avoid possible virus spread to those remaining. In order not to deplete the food reserves of the crown, harvesting is discontinued, usually in midspring, and further growth is encouraged to replenish the food reserves of the crown for next year's production.

A limited amount of field-produced rhubarb is machine harvested, the majority being hand pulled. Trimming involves removal of the leaf blade from the petiole. Leaf blades are discarded because they contain high levels of oxalic acid as well as an anthraquinone, both of which are toxic. Petioles, usually trimmed to lengths of 30–50 cm, are bundled together for marketing; the longest and deepest colored petioles have a greater market value.

Procedures for forcing production differ from field production. For forcing, second-year plants not previously harvested are removed from the field and placed in high humidity storage at 4°C or less. In preparation for forcing, about five or six crowns per square meter are placed in soil within a dark cellar or similar enclosure. Crowns are watered and temperatures are raised (10–15°C) to encourage sprouting and growth. Warm soil-temperatures are more important than air temperatures, and soil heating often is provided. High temperature (20°C) results in less color development; low temperatures produce a darker petiole color but less growth; about 15°C is optimum.

In about 30 days, or longer when temperatures are cold, the first petioles can be harvested. These are usually relatively long, dark colored, and tender compared to field-grown production. An outer whorl of two to three petioles are removed at each harvest. Petioles continue to develop sequentially. For up to 6 weeks, petioles are harvested about twice a week. By that time, the crowns will have exhausted their storage reserves and are removed and discarded. In some areas, it is possible to extend the production period by using the forcing structure with a second planting of previously stored crowns. Although the value of the forced rhubarb crop is higher than field-grown product, the volume of forced production is steadily declining because of the high labor requirement.

With postharvest temperatures between 0°C and 2°C and high relative humidity (95%) during handling and storage, quality can be maintain for 3–4 weeks. Rhubarb is used as a cooked vegetable, in pies and sauces, and is also processed by canning and freezing. Its acid taste, due to oxalic acid, is often tempered with sugar. Rhubarb contains fair amounts of pro-vitamin A and vitamin C.

Rhubarb is subject to fungal attack from species of *Phytophthora*, *Pythium*, *Rhizoctonia*, and *Botrytis*. Virus, nematodes, and the potato stem borer and rhubarb cucurlio, *Lixus concavus*, are common pests.

GARDEN SORREL, *Rumex acetosa* L.

Family: Polygonaceae
Other names: Sour grass, dock

Garden sorrel is of Eurasian origin. Garden sorrel and other *Rumex* species are perennials grown for their long, thin, light green, slightly crinkled, arrow-shaped leaves for potherb and salad use.

Sorrel leaves are pointed at the apex and somewhat lobed at the base. Basal leaves have long petioles; leaves become almost sessile toward the top of the stem and tend to develop red coloration with maturity (Fig. 25.20). Leaves are periodically removed as they enlarge or the entire plant is harvested. Plants are monoecious; flowers are small, greenish, and formed in clusters on spikelets. Plants are seed propagated.

Other *Rumex* species similar to garden sorrel are French sorrel *(R. scutatus)*, spinach rhubarb *(R. abyssinicus)*, and spinach dock or herb patience *(R. patientia)*. French sorrel differs in being a short plant with branched stems that exhibit a semireclining growth habit. Leaves are arrow or fiddle shaped, more succulent, and smaller than garden sorrel. French sorrel is used like garden sorrel in salads, soups, and as a potherb; it has a slightly more acidic taste. Fitting its name, leaves of spinach rhubarb are eaten like spinach, and the petioles like rhubarb. Spinach dock or herb patience is somewhat similar to sorrel, although the plant is stouter and taller, with larger leaves and a noticeably stronger taproot than sorrel; its foliage is also used for greens.

FIG. 25.20.   Garden sorrel, *Rumex acetosa.*

PURSLANE, *Portulaca oleracea* L.

Family: Portulacaceae

Believed to be a native of Iran or India, purslane is widely known as a weedy plant, frequently gleaned from noncultivated sites for vegetable use. Domesticated garden purslane cultivars such as kitchen purslane (*P. oleracea* var. *sativa*) are cultivated expressly for the fleshy succulent leaves and stems consumed as potherbs or in salads. The taste is somewhat like watercress and spinach. Purslane is a popular vegetable in France, several other European countries, and also in Egypt and Sudan. Like spinach, an undesirable quality is that purslane foliage contains oxalic acid and tends to accumulate nitrates.

The plant is a frost-sensitive, summer annual exhibiting vigorous prostrate growth of many succulent red to purple stems branching from a fleshy taproot. Stems have alternate, obovate thick leaves. Roots readily form at nodes in contact with soil. Flowers in leaf axils are yellow, bloom early, and are prolific seed producers. A large plant is capable of producing 50,000 small (1-mm), black seed, which are also edible.

A related Portulacaceae species is winter purslane, also known as miner's lettuce, *Montia perfoliata (Claytonia perfoliata).* This North

American native is a small annual glabrous herb; the leaves are used as potherbs and for salads.

TALINUM, *Talinum triangulare* (Jacq.) Gaertn.

Family: Portulacaceae
Other names: Surinam spinach, waterleaf

Talinum is a succulent perennial. The glabrous short and narrow leaves and terminal and lateral shoot tips are eaten as vegetables. Their sour and bitter aftertaste is due to the high oxalic acid content. Plants are harvested frequently, as often as every 2–3 weeks. On short triangular stems, leaves about 7 cm long are generally spatulate-obovate in shape. Flowers are red, pink, or yellow and develop in racemes. Although the probable origin of talinum is tropical America, it is found in great abundance growing in partial shade in the wet tropics of Indonesia, West Indies, and west Africa. Propagation is by seed, or with transplanted cuttings.

CURRY LEAF, *Murraya koenigii* (L.) K. Spreng

Family: Rutaceae

Curry leaf is an important leafy vegetable because its leaves are a standard ingredient for curry preparations. Leaves are lightly pungent, bitter, with a mild acidic taste, and the flavor is maintained whether leaves are fresh or dried.

The small curry leaf tree is a perennial and native to Southeast Asia. The lanceolate leaves are about 5 cm long. White flowers develop in clusters on terminal cymes and produce small, black, berrylike fruit.

NEW ZEALAND SPINACH, *Tetragonia tetragonioides* (Pall.) O. Kuntze *(T. expansa)*

Family: Tetragoniaceae

New Zealand spinach, indigenous to New Zealand, is grown for the tender leaves and stems. The crop became more widely cultivated after being introduced to Europe in the late 1700s. The plant is a vigorous, branching annual with a semiprostrate growth habit spreading to a diameter as much as 1 m, with a height of 35–40 cm (Fig. 25.21). Leaves are alternate, nearly triangular, dark green, thick, succulent, and glabrous, measuring about 7–10 cm long; stems are fleshy and light green. Nearly sessile, yellowish green, inconspicuous flowers form in the leaf axils and are self-pollinated. The fruit consists of the hardened, irregularly shaped, receptacle growth surrounding several seed; when dry, about 12–14 such fruit weigh 1 g.

New Zealand spinach is often mentioned as resembling spinach *(Spinacia oleracea)* in culinary qualities. However, it differs in growth characteristics, being tolerant of high growing temperatures (35°C), and is frost sensitive.

Plants are seed propagated, often self-reseeding. Because of the hard covering, soaking of the seed-containing fruit in water for about 24 h before planting is often suggested. Even then, germination can be slow

FIG. 25.21.   New Zealand spinach, *Tetragonia tetragonoides*.

and erratic. "Seed" are planted 3–4 cm deep, about 30–40 cm apart in rows 75–100 cm apart. Although plants grow well on moderately fertile soils, fertilizers are often applied to enhance foliage growth. Plants are relatively drought tolerant but are more productive when water is not limited.

Harvesting usually begins about 40–50 days after planting. Shoot tips, 15–20 cm long, and leaves are harvested and prepared as a boiled vegetable. Shoot tip removal stimulates lateral branching and harvests can be continued for several months. Tender shoots are also eaten raw, often in salads. Like spinach, tissues contain oxalates that render calcium nutritionally unavailable. New Zealand spinach is relatively pest free, although leafminers and snails are occasional problems.

JEW'S MALLOW, *Corchorus olitorius* L.

Family: Tiliaceae

Other names: West African sorrel, jute mallow, tossa jute, molokhia, long fruited jute, bush okra

Of uncertain origin (possibly Africa, with a secondary center in India or Indo-Burma region), Jew's mallow is an important leafy vegetable in the Middle East, Egypt, Sudan, and other parts of tropical Africa and is grown extensively in India. The plant is a annual or short-lived perennial with tall, upright stems that become very fibrous with maturity. There are many local types varying in height, pubescence, leaf and fruit shape, and leaf production. Vegetable cultivars are usually dwarf.

The tender mucilaginous leaves and young shoots are cooked and eaten much like spinach (Fig. 25.22). Leaves are also dried and stored to be rehydrated for later use. Leaf dry weight protein ranges from 1.5% to 5%, and ascorbic acid content is high. Leaves, light green, lanceloate shaped with serrated leaf margins, are about 15 cm long and 5–7 cm wide.

Jew's mallow is a short-day plant that grows well at high temperatures (25–35°C) and high humidity. Plants are sensitive to excessive moisture, but otherwise tolerate drought and a broad range of soil conditions. High density, as much as 250,000 plants/ha, are established with seed; when transplants are used, densities are much lower. Harvesting usually begins 40–60 days after planting. Shoot removal stimulates branching and new shoot growth, thereby extending the harvest period. Yields range from 5 to 8 t/ha.

FIG. 25.22. Jews mallow, *Corchorus olitorius.*

GARDEN NASTURTIUM, *Tropaeolum majus* L.

Family: Tropaeolaceae
Other name: Indian cress

Garden nasturtium are climbing annuals; dwarf forms are also culti-vated. Before domestication, wild forms were found growing in regions ranging from Mexico to Peru. Plants prefer sunny and moist environ-ments and are frost sensitive. The plant has been used for food flavoring for several thousand years. The tart flavor is due to the presence of mustard oil. Most parts of the plant are edible; yellow to red colored flower petals and buds are used, usually raw, in salads and as garnishes. Seeds are pickled, and the nearly round leaves, which have a high vitamin C content, are eaten raw or cooked.

CORN SALAD, *Valerianella locusta* (L.) Latterrade em. Betcke *(V. olitoria)*

Family: Valerianaceae
Other names: Lamb's lettuce, fetticus

Corn salad is a cool season plant of uncertain origin that is popular in the Mediterranean region for salad and cooked as greens; plants are sometimes blanched.

As an annual, plants produce a large rosette of succulent, simple round to somewhat spatulate leaves up to 15 cm long. Flowers are very small, usually white or pale blue with a funnel-shaped corolla. A related species, *V. eriocarpa,* known as Italian corn salad has similar vegeta-ble uses.

SELECTED REFERENCES

Adaniya, S., Shoda, M., and Fujieda, K.K. 1989. Effects of daylength on flowering and rhizome swelling in ginger (*Zingiber officinale* Roscoe). J. Japan. Soc. Hort. Sci. *58,* 649–656.

Akoroda, M.O., Anyim, O.A., and Emiola, I.O.A. 1986. Edible fruit productivity and harvest duration of okra in Southern Nigeria. Trop. Agric. (Trinidad) *63,* 110–112.

Benson, B.L., and Motes, J.E. 1982. Influence of harvesting asparagus the year following planting on subsequent spear yield and quality. HortScience *17,* 744–745.

Bretting, P.K. 1986. Changes in fruit shape in *Proboscidea parviflora* spp. *parviflora* (Martyniaceae) with domestication. Econ. Bot. *40,* 170–176.

Burrows, R.L., and Waters, L.E., Jr. 1989. Fall establishment of asparagus using seedling transplants. HortScience *24,* 611–613.

Carlson, A.W. 1986. Ginseng: America's botanical drug connection to the Orient. Econ. Bot. *40*, 233–249.

Clement, C.R. 1988. Domestication of the pejibaye palm *(Bractris gasipaes):* past and present. Adv. Econ. Bot. *6*, 155–174.

Daloz, C.R., and Munger, H.M. 1980. Amaranth—an unexploited vegetable crop. HortScience *15*, 383.

Dhua, R.S. 1986. Basella. In Vegetable Crops in India. T.K. Bose and M.G. Som, eds. pp. 687–691. Naya Prokash, Calcutta, pp. 687–691.

Fawusi, M.O.A. 1983. Quality and compositional changes in *Corchorus olitorius* as influenced by nitrogen fertilization and postharvest handling. Sci. Hortic. *21*, 1–7.

Fisher, K.J. 1982. Comparisons of the growth and development of young asparagus plants established from seedling transplants and by direct seeding. N.Z. J. Exp. Agric. *10*, 405–408.

Hartung, A.C., Putnam, A.R., and Stephens, C.T. 1989. Inhibitory activity of asparagus root tissue and extracts on asparagus seedlings. J. ASHS *114*, 144–148.

Keng, P.C. 1982. A review of the genera of bamboo from the world. J. Bamboo Res. *1*, 1–19.

Kozukue, E., and Mizuno, S. 1989. Effects of harvest time, size, cultivation phase and storage on changes of homogentisic acid content in bamboo shoots. J. Japan. Soc. Hort. Sci. *58*, 719–722.

Leverington, R.E. 1983. Ginger. In Handbook of Tropical Foods. H.T. Chan, Jr., ed. Marcel Decker, New York, pp. 297–350.

Li, T.S.C. 1995. Asian and American ginseng—A review. HortTechnology *5*, 27–34.

Martin, F.W., and Ruberte, R.M. 1975. Edible Leaves of the Tropics. Mayaguez Inst. Trop. Agric., Sci. Ed. Admin., USDA, Mayaguez, Puerto Rico.

Martin, F.W., Rhodes, A.M., Ortiz, M., and Diaz, F. 1981. Variation in Okra. Euphytica *30*, 697–705.

McClure, F.A. 1993. The Bamboos. Smithsonian Institution Press, Washington, DC.

Murashige, T., Shabde, M.N., Hasegawa, P.M., Takatori, F.H., and Jones, J.B. 1972. Propagation of asparagus through shoot apex culture. I. Nutrient medium for formation of plantlets. J. ASHS *97*, 158–161.

National Academy of Sciences. 1975. Chaya. In Underexploited Tropical Plants with Promising Economic Value. National Academy of Sciences, Washington, DC.

Oke, O.L. 1983. Amaranth. In Handbook of Tropical Foods, H.T. Chan, Jr., ed. Marcel Decker, New York, pp. 1–28.

Proctor, J.T.A., and Bailey, W.G. 1987. Ginseng: Industry, botany, and culture. Hortic. Rev. *9*, 187–236.

Ramachandran, C., Peter, K.V., and Gopalakrishnan, P.K. 1980. Drumstick *(Moringa oleifera):* A multipurpose Indian vegetable. Econ. Bot. *34*, 276–283.

Ramachandran, C., Peter, K.V., and Gopalakrishnan, P.K. 1980. Chekurmanis, a multivitamin leafy vegetable. Indian Hortic. *25*, 17.

Robb, A.R. 1984. Physiology of asparagus *(Asparagus officinalis)* as related to the production of the crop. N.Z. J. Exp. Agric. *17*, 251–260.

Rodrigo, M., Lazaro, M.J., Alvarruiz, A., and Giner, V. 1992. Composition of capers *(Capparis spinosa):* influence of cultivar, size and harvest date. J. Food Sci. *57*, 1151–1154.

Rodriguez, F.A., and Cantwell, M. 1988. Developmental changes in composition and quality of prickly pear cactus cladodes (nopalitos). Plant Foods Human Nutr. *38*, 83–93.

Russell, C.E., and Felker, P. 1987. The prickly-pears *(Opuntia* spp. Cactaceae): A source of human and animal food in semiarid regions. Econ. Bot. *41*, 433–445.

Salikutty, J., and Peter, K.V. 1985. Curry leaf *(Murraya koenigii),* perennial, nutritious, leafy vegetable. Econ. Bot. *39,* 68–73.

Sealy, R.L., Williams, E.L., Novak, J., Fong, F., and Kenerley, C.M. 1990. Vegetable amaranths: cultivar selection for summer production in the South. In Advances in New Crops. J. Janick and J.E. Simon, eds. Timber Press, Portland, OR, pp. 396–398.

Shelton, D.R., and Lacy, M.C. 1980. Effect of harvest duration on yield and on depletion of storage carbohydrates in asparagus roots. J. ASHS *105,* 332–335.

Wen, T.H. 1983. Some ideas about the origin of bamboo. J. Bamboo Res. *2,* 1–10.

26

Aquatic Vegetables

Considerable areas of the earth are flooded, some, all, or part of the time with either fresh or brackish waters. Many plants thrive and reproduce under these conditions and humans have used some of them for food and medicine. Prior to the development of agriculture, aquatic plants were gathered from their natural environments. During the development of agriculture, these began to be cultivated in locations other than their usual habitat through the control of the necessary flooded or waterlogged conditions. Initially, some of these plants used as vegetables were probably first found growing as weeds in rice paddies, although several species are also adapted to upland conditions. Most aquatic and swamp-adapted vegetable species were developed in tropical and semitropical environments, and many seemed to have originated in southern and southeastern Asia. Some aquatic vegetable species are also grown in warmer regions of temperate zones.

Aquatic vegetables provide edible leaves, shoots, rhizomes, corms, and seed. Several important monocotyledon and dicotyledon aquatic vegetable crops are described in this chapter.

MONOCOTYLEDONS

ARROWHEAD, *Sagittaria sagittifolia* L. (*S. sagittifolia* var. *sinensis*)

Family: Alismataceae
Other names: Swamp or swan potato, kuwai

Arrowhead, named because of its sagittate leaves, has been cultivated for centuries in China. Production is largely centered in central and southern China, but the crop is also grown in Japan and Taiwan, and to a lesser extent in Indonesia, Malaysia, and India.

Arrowhead is a perennial that grows to a height of 70–100 cm. Plants produce small white flowers and are cultivated for the starchy, aromatic, mildly sweet corms. The ovoid corms have an elongated beak and are encircled with scales arising from nodes (Fig. 26.1). They are about 4–5 cm in diameter and are formed at the terminus of a stolon. Two major types are grown: an aquatic form with white- or green-colored corms, and another form better adapted to upland moist soils. Other differences are related to time of planting and duration of growth.

Plantings are started in the spring. Propagation is with sprouted corms planted about 10 cm deep, directly into fields much like paddy rice; typical spacings are 45 × 60 cm. Plants are sometimes started in a nursery and later transplanted. The water level during growth should not exceed 6 cm and must be flowing; if stagnant, disease often occurs. Favorable growth temperatures are between 18°C and 22°C in full sunlight. About 75 days after planting, older leaves begin senescence and are removed. The procedure is repeated as needed, with care taken not to injure developing stolons. Cool autumn temperatures cause leaves to yellow, an indication that corms are ready for harvest. Normally 6 or 7 months are required for corms to mature. For harvest, leaves which do not become senescent are cut off, water is drained, and corms are usually dug by hand.

Individual plants usually produces from 6 to more than 10 corms. Scales attached to the corms are removed with a dilute solution of

FIG. 26.1. Arrowhead corms, *Sagittaria sagittifolia.*

alum. Cut corms characteristically exude a milky juice. Corms contain about 20% starch and 5% protein. They are not eaten raw, but are peeled and cut in small pieces and cooked in many Oriental dishes. Young tender leaves are sometimes harvested in summer for use as greens.

The corms of a similar North American noncultivated species, *S. latifolia,* known as duck potato or wapato were used for food by the American Indians. *S. trifolia* and *S. trifolia* var. *sinensis* are known as Chinese arrowhead.

EDIBLE AROIDS

Family: Araceae

The aquatic aroids are discussed in Chapter 13.

CHINESE WATER CHESTNUT, *Eleocharis dulcis* (Birm. f.) Trin. ex Henschel *(E. tuberosa)*

Family: Cyperaceae
Other name: matai

Chinese water chestnuts are native to the Old World tropics, their natural habitat being tropical eastern Asia. They are a widely cultivated crop in the Orient, and a specialty crop elsewhere. Two cultigens are grown: *hon matai,* a sweet form, and *sui matai,* a starchy type. Chinese water chestnut differs from the dicotyledonous *Trapa bicornis* and *T. natans,* which are called water chestnut.

Eleocharis dulcis is characterized by numerous upright tubular septate stems that function as photosynthetic organs in place of leaves. Although plants flower and produce seed, the crop is asexually reproduced by rhizomes and corms.

The crop is often grown in rotation with paddy rice. Corms and cormels can be planted directly, 5–10 cm deep. However, they require more time to be productive compared to nursery transplants. Where early frosts or a short growing season occur, corms are commonly planted in a nursery and plantlets are transplanted when about 20 cm tall. Average plant spacings are about 75–90 cm × 75–90 cm. After planting, the field is flooded and the water level is maintained at 3–5 cm and gradually increased to 10 or 20 cm during growth, but at no time should it exceed 30 cm. Fields are kept flooded until about 30 days before harvest (Fig. 26.2a).

Optimum growing temperatures are from 30°C to 35°C; 40°C is the upper limit. For high productivity, a frost-free growing season of 200–220 days is required. Rhizomes are produced in about 6–8 weeks, from which a series of new secondary plants develop and soon occupy the open spaces. Corm formation occurs after the daughter plants from the main rhizome have grown to maturity, and it is accelerated by short

a

b

FIG. 26.2. Chinese water chessnuts, *Eleocharis dulcis:* (a) flooded field production in China, and (b) harvested corms.

days. The 3–4 cm wide, slightly flattened, small corms form at the end of elongated stolons. The corms usually are mature about the time the tops have died down or are killed by frost. The corms are brown skinned with white flesh (Fig. 26-2b).

Harvests of hon matai are made after the tops have dried and the field has been drained. Corms are plowed out and picked out from the soil by hand. Sui matai plants can be harvested without draining water by simply pulling out the corms by hand. Yields range from 10 to more than 30 t/ha. Corms can be stored in the flooded soil and harvested as needed. The starch level in mature corms is between 20% and 35%. They can be maintained in good condition for 6 months when stored at 0–2°C. However, immature corms are chilling sensitive, and when stored at low temperatures, the sugar content increases slightly. Corms are consumed raw or cooked and have a slightly sweet taste and crisp texture, which is retained even after cooking.

WATER BAMBOO, *Zizania latifolia* L. *(Z. caduciflora)*

Family: Poaceae (Graminae)
Other names: Manchurian wild rice, gau sun, (C), jiao sun (Mandarin), makomo dake (J).

Water bamboo is a perennial that grows in swampy and flooded areas. This crop has been grown since ancient times in regions of eastern Asia from Manchuria to Indochina as well as Japan and Taiwan.

Water bamboo grows to a height of 1–2.5 m (Fig. 26.3). Fully elongated leaves are up to 60 cm in length. The enlarged stems *(culms)* are harvested and upper leaves are cut away to leave the swollen lower culm and the sheath leaves. The edible portion is the succulent enlarged growth of the stem apex that is consumed after sheath leaf removal. This portion of the culm consists of three or more somewhat hollow internodes, each about 5–6 cm in length; the apex and nodes are solid. The enlarged gall-like growth of the lower culm portion is a response incited by the presence of a fungal parasite.

Three cultivar types are known in China. These are as follows: a green stem small early-maturing plant, a large white stem plant, and a large pink or red stem plant; the latter types mature later. Cultivation practices may permit an autumn harvest and in other situations, a harvest is made in the autumn and during the next summer.

Water bamboo is asexually propagated using cuttings or divisions con-

a   b   c

FIG. 26.3.   Water bamboo, *Zizania latifolia:* (a) plant, (b) opened portion of overmature culm showing fungal sporulation which makes this undesirable for food use, and (c) basal stems (culms) bundled for marketing in Taiwan.

sisting of the prostrate underground rhizomes having three to six internodes. These are selected from existing vigorous plants that are without any indication of floral initiation. The attached erect aerial culm portion is trimmed to a length of about 40 cm and placed into mud about 6 or 7 cm deep, so that three or more nodes are buried. Plantings are made similar to that for paddy rice culture. The initial water depth is 4–5 cm and is then raised to a depth of about 10 cm, sometimes more during hot weather. Cuttings with or without roots are selected; those with roots are preferred. Roots will form at nodes below the surface of the mud. Because each cutting produces several culms, spacings are relatively wide, often at 100×150 cm. With each growing season, new rhizomes develop at the base of the culm. Sometimes cuttings are first grown in a nursery for a year before transplanting.

Plants are most productive with high light and at a mean temperature of 25°C. At lower temperatures, plant growth is slower and culms are less tender. Some growers allow for the growth of an aquatic fern, *Lemma minor* L., to reduce light penetration into the water and thus ensure white shoot development. Plants are responsive to fertilizer, especially nitrogen, which often is applied during growth.

Culm enlargement or swelling is the result of gall formation by the host plant due to the parasitic fungus *Ustilago esculenta* L. This organism is related to corn smut, *Ustilago maydis* DC. Cda., which is used as a specialty vegetable in Mexico. The fungus grows in a symbiotic relationship and inhibits floral initiation of the water bamboo host. Black longitudinal steaks that appear in the swollen culm are an indication that the fungus is reproducing. Fungal reproduction results in spore production within the swollen culm, turning it black and reducing its value (Fig. 26.3b). Harvest of water bamboo culms should be made before the fungus becomes reproductive. There are two strains of the fungus, a *T* and *M* strain, the latter forms spores at a later period of gall development, which is desirable when culms are intended for vegetable use. The fungus is difficult to eradicate because water bamboo is asexually propagated and the fungus is transferred from clone to clone. Hot-water immersion for 30 min at 52°C is a successful eradication procedure without killing the host. However, plants will flower following this treatment, which is undesirable because enlarged culms are not produced.

Stem enlargement occurs after about 4 months' growth. Harvest of the green cultivar types are made about 150 days after planting, and after 170 days for other types. Each plant (clump) yields about 30 swollen culms that are usually 10–15 cm long and about 2–4 cm thick (Fig. 26.3c). Yields range from 10 to more than 20 t/ha. Starch does not accumulate, and the principal carbohydrates in the culm are glucose, fructose, and sucrose. Although water bamboo tissues have a softer texture than regular bamboo, they also remain fairly crisp when cooked. Not having the bitter taste of bamboo shoots, they can be eaten raw. In Taiwan, trimmed harvested culms are placed in a water-filled vat to prevent the stalk from drying and to retard the formation of spores. Dried leaves have alternative uses as fodder or fuel. Grain from flowering plants have also been used for food.

Zizania aquatica L., an annual cool weather plant of North American origin, best known as wild rice, is closely related to water bamboo. Previously harvested from the wild for its seed grain used much like rice, it is now an important cultivated crop in the United States. The thickened basal portions of the stem are much like those of water bamboo and can also be eaten.

COMMON CAT-TAIL, *Typha latifolia* L.

Family: Typhaceae
Other names: Bulrush, Cossack asparagus

Cat-tail, a temperate climate monoecious perennial, is adapted to growth in swampy waterlogged soils. Native American Indians have used the rhizomes for food since pre-Columbian times. The plant also grows and is utilized in Europe and Asia. Cat-tail is propagated with seed or rhizome pieces.

The erect plant with unbranched stems grow as tall as 3 m, with narrow leaf blades that are over 1 m long, but only 15–25 mm wide. In the late summer, floral spikes, 10–30 cm long are produced. After flowering, the rhizomes enlarge and are harvested from late summer through early spring. Rhizomes are high in starch and sugars and are used as cooked vegetables but can also be dried and ground into a flour. Young tender leaves, shoots, and inflorescence are also used as cooked vegetables, and fresh pollen is sometimes collected and used as flour.

DICOTYLEDONS

WATER CONVOLVULUS, *Ipomoea aquatica* Forssk. *(I. reptans)*

Family: Convolvulaceae
Other names: Water spinach, swamp spinach, kang kong, weng cai
  (C), you-sai (J)

The precise origin of water convolvulus is unknown, although tropical Africa, Asia, and India are possible sites. The plant has had a long history of cultivation in southeastern China. Water convolvulus is both an aquatic or semiaquatic perennial found in many tropical and subtropical regions. The easily grown, productive, and nutritious crop is often produced on a year-round basis.

Two types of water convolvulus are grown: a narrow-leaf, white-flowered, green stem form called *ching quat* that is adapted to moist soil or semiaquatic culture, and a broad, arrow-shaped-leaf, pink-flowered, white stem plant known as *pak quat,* that is adapted to flooded culture (Fig. 26.4). Ching quat has better cold weather tolerance, and pak quat generally has a higher culinary preference. Water convolvulus differs from sweet potato *(Ipomoea batatas)* in not producing tuberous roots.

Plant growth habit is spreading and prostrate. Leaves, from 7 to 14 cm long, are heart shaped at the base and usually pointed at the tip. Stems are hollow and float on the surface. Adventitious roots readily form at stem nodes when in contact with soil or moisture. Under short-day conditions, erect flower stalks develop in leaf axils. Usually one or two funnel-shaped, purple-throated flowers are formed. Corolla color, white, light pink, or purple, varies with plant types. Seed are readily set and develop in a podlike fruit.

For flooded pak quat culture, crops are propagated with stem cuttings;

FIG. 26.4.   Water convolvulus, *Ipomoea aquatica,* ching quat form, of bunched stems and leaves at a local market in the Philippines.

seed are infrequently used. Cuttings, averaging about 30 cm in length, should have five or more nodes and are obtained from an existing crop or from a propagation nursery. These are inserted horizontally to about half their length, 3–5 cm deep, into well-puddled soil at spacings of 40 × 50 cm. Following planting, the field is flooded. Water should be flowing and the level is adjusted from an initial 3–5 cm and increased as plant growth proceeds to about 15–20 cm. Temperatures averaging between 25°C and 30°C are ideal; plants are damaged at temperatures of 10°C or less. Crops are often well fertilized to increase vegetative growth.

Semiaquatic culture is also practiced on raised beds spaced about 1 m apart. For this production, cuttings, rooted transplants, or directly sown seed are used for propagation. Following planting, furrows are flooded to produce and maintain a high moisture level in beds.

Harvest of shoot tips occurs as soon as 30–40 days after planting of cuttings; about 60 days are required for crops started from seed. Stem tips of about 30 cm length are harvested, washed, and bundled for marketing. Regrowth of new shoots occurs, and in 4–6 weeks harvests are repeated. Three, sometimes more, harvests are made from each planting, with annual yields ranging from 40 to as much as 90 t/ha fresh weight. The tender shoot tips and leaves are used as a potherb much like spinach. Foliage and stems are an excellent source of pro-vitamin A.

JAPANESE HORSERADISH/WASABI, *Eutrema wasabi* (Siebold) Maxim. *(Wasabi japonica)*

Family: Brassicaceae (Cruciferae)

This highly prized, perennial condiment plant is probably native to Japan. The thickened, curved, underground stems (rhizomes) are the product of interest (Fig. 26.5). The long, petiolate leaves have palmate veins. Terminal and axillary inflorescence bear small white flowers.

FIG. 26.5.   Wasabi, *Eutrema wasabi,* harvested rhizomes as marketed in Japan.

Plants are cultivated in semiaquatic conditions and also grow wild near flowing streams. Water temperature and quality is critical for good production; 10–13°C is optimum. Water quality often limits production locations. An upland form, known as "hata," is also grown. Large stem cuttings (20–30 cm) are used for propagation. Seed is not used for propagation because plant development is slow, it requires about 2 years before seedlings are large enough for transplanting. High temperature promotes foliage growth at the expense of stem growth, pungency, and yield but may increase disease susceptibility.

Rhizomes are usually harvested about 18–20 months after transplanting. The harvested rhizomes are freshly ground for condiment use. Leaves also contain the mustardlike compounds that give wasabi the sharp, pungent flavor similar to horseradish, *Armoracia rusticana*. Rhizomes are also dehydrated and ground into powder for use after rehydration.

WATERCRESS, *Nasturtium officinale* R. Be.
(Rorippa nasturtium-aquaticum)
Family: Brassicaceae (Cruciferae)

The origin of watercress is the eastern Mediterranean and adjoining areas of Asia, and possibly Ethiopia. Use is made of both wild and domesticated aquatic forms. A related species, *R. schlechteri,* is endemic to and cultivated in the highlands of Papua New Guinea. Another, *R. heterophylla,* is not cultivated but indigenous to east Asia. Watercress is a cool season perennial. Cool and clean-moving water, preferably slightly alkaline, is required for production. Most water sources contain adequate mineral nutrients for growth. Plants are also grown in moist and wet soils.

Watercress is a labor-intensive crop grown at close, 15 × 15 cm, spacings with the highly branched vegetative growth, creeping along the wet surface or floating on the water. Plants frequently exhibit rooting at stem nodes. Stems are hollow and leaves are petiolate, oblong, and glabrous. Small bisexual and self-compatible flowers have white or yellow petals. The mature pods, called siliques, contain many seed. Being long-day plants, watercress does not flower in the tropics. In tropical locations, production is generally restricted to cool high elevations.

Propagation with stem cuttings is most common, although seed can be used. The diploid form is most frequently produced; tetraploid and triploid forms are also grown. The triploid (known as brown cress) is a sterile hybrid resulting from crossing of the diploid and tetraploid. An anomaly, the diploid form does not need light to germinate, whereas the tetraploid does.

Commercial production often takes place in specially constructed beds where water flow, depth, and quality can be controlled. Flowing water is an important requirement, and growth is improved if water is slightly alkaline. Weeds are controlled mainly by regulating water level. Crookneck disease, caused by species of *Spongospora*, can be a serious cause of production loss that is controllable with zinc or zinc-containing compounds.

Harvests can be made as early as 4–6 weeks after cuttings are transplanted and can continue at approximately monthly intervals. Slow-growing and old foliage tends to taste strong. Harvested stem tips ranging from 10 to 30 cm and attached succulent foliage should be utilized within a few days because postharvest life is relatively short. Low temperature (0°C) and very high relative humidity are required to retain quality. Watercress is used for the pungent flavor it provides to fresh salads and garnishes; it is also cooked as greens.

LOTUS ROOT, *Nelumbo nucifera* Gaertin. *(Nelumbium nelumbo)*

Family: Nymphaeaceae
Other names: East Indian lotus, lian (C), hasu, renkon (J)

Lotus root is a perennial dicot grown in paddy culture for the edible, horizontal, creeping, tuberlike, storage rhizomes. The plant is indigenous to Southeast Asia and possibly is of east Indian origin. It has been cultivated for many centuries in China and Egypt as well as India and Japan, and recently in Hawaii.

Broad leaves, as large as 60 cm or more in diameter, are peltate, nearly round, and concave; most are found standing above water on tall, stout, tubular petioles (Fig. 26.6a). The petioles of the initial one or two leaves are short; thus, the leaf blades float on the surface. Subsequent petioles are longer and leaf blades are above water level, each being slightly higher than the previous. New leaves continue to emerge from rhizome nodes. Roots and secondary stems also develop from the same nodes. Stems bear solitary, large, showy, fragrant flowers which are rose, pink, or white and contain many pistils in a spongy, flat-topped receptacle (Fig. 26.6a). When mature, seed exit from openings at the top of the podlike receptacle. Flower removal is practiced in order to increase rhizome size and yield.

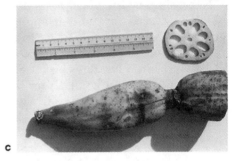

FIG. 26.6.   Lotus, *Nelumbo nucifera:* (a) leaf and seed pods. Note: the developing upper pods and the lower seed pod that has matured and is empty of seed, (b) harvested and washed segmented rhizomes with attached remnants of leaf petioles, and (c) cut section of rhizome showing air passages. Source (b): Courtesy of Han Huang, Taiwan.

A change in the growth of the storage rhizome produces enlarged, thick wall internodes (Fig. 26.6b). Usually three or four, sometimes more, are formed, each successively shorter. These appear to be segmented because the diameter at the node is less than that of the internodes. The proximal segment is long and somewhat tapered, and the distal segment is larger and wider. A prominent characteristic is the large and small tubular longitudinal passages in aerial stems and rhizomes that function as ducts for gaseous exchange with the atmosphere. (Fig. 26.6c)

Rhizome portions are usually used for propagation. The distal end of a rhizome is pushed into muddy soil about 18–20 cm deep at an angle of about 30° above horizontal so that the proximal end remains above water level after flooding. Although less desirable, the crop can also be ratoon propagated from rhizome sections not harvested. Seed can also be used for propagation, either directly or preferably, by first sowing in a nursery for later transplanting.

Rhizomes require about 6 months and sometimes as much as 9 months to mature, and are harvested after leaves have died and the water drained. Rhizomes are carefully dug and washed free of mud. Commonly yields are about 4 t/ha.

The attractive features of lotus root is that the texture of the rhizome tissue remains crisp after cooking. It is a widely used vegetable delicacy in Asian and other dishes. Another characteristic is the threadlike mucilaginous strands that exude from separated or cut surfaces of the tissues. Starch obtained from dried tissues have properties similar to arrowroot. Immature leaves and petioles are also used as greens but contain the alkaloid nelumbine, which is a cardiac poison; detoxification requires washing and cooking. Seed are edible but also must be thoroughly washed to remove bitterness. Dried seed pods are used in medicinal teas and as a floral decoration.

WATER CHESTNUTS, *Trapa bicornis* Osbeck, *T. natans* L., *T. maximowiczii* Korsh.

Family: Trapaceae
Other names: Water caltrop, Jesuit's nut

Water chestnuts are grown for the edible seed of the underwater fruit. Although dissimilar in appearance, they are frequently confused with Chinese water chestnut, *Eleocharis dulcis*. Native to Eurasia and Africa, *Trapa* species are aquatic, floating annuals with roots anchoring the plant in the soil. Petioles have a swollen (inflated) spongy section that buoys the plant on the surface of the water (Fig. 26.7a). Water depths are 30 cm or more, although plants will grow in water 3 or 4 m deep. Leaves float on the surface and are diamond or triangular in shape. Undersides of leaves show prominent veins.

Flowers, which are white, emerge from near the apex of the stem. Fruits set and mature below the water surface. Mature fruit have green and red seed coats or husks. When mature, the seed coat turns black and the seed becomes hard.

Water chestnuts are grown in freshwater ponds free of a high concentration of salts and with soils having a pH between 5.6 and 6.8. Favorable growing temperatures range from 25°C to 33°C. In some situations, urea nitrogen at about 45 kg/ha is broadcasted about 3 weeks after

FIG. 26.7. Water chestnut, *Trapa bicornis:* (a) plant with attached fruit, and (b) crop harvesting in Taiwan.

planting and again several weeks later. Potassium applications have been reported to increase fruit size and quality.

Propagation is with seed, which must be kept moist to remain viable. Seed exhibit a short dormancy of about 1 month. In winter, seed are pregerminated in water and broadcasted into small nursery tanks containing water about 30–60 cm deep. After several months, seedlings are removed from the nursery and planted into a pond at spacings of 1 × 3 m or closer if soils are fertile. Volunteer plants from fruit that previously fell into the soil during harvest often serve as transplants. The transplants gradually spread as their roots grow into the mud and young shoots grow to the surface. Stems terminate in a rosette of triangular-shaped floating leaves.

Fruit harvest begins in the fall about 9 months after planting and continues each 8 or 10 days for about 3 months; the harvest peaks in about the middle of this period (Fig. 26.7b). Yields of *T. bicornis* range from 4 to 14 t/ha; those of *T. natans* from 3 to 6 t/ha.

Trapa natans fruits (nuts) are 5–7 cm wide, have four prominent stout horns, and are most often grown in Eurasian and African regions. *T. bicornis* nuts are about the same size and have two slightly curved horns and are predominantly grown in China, Taiwan, Japan, and India. The fruit of another species, *T. maximowiczii,* are about 2 cm wide and have four thin, slightly curved horns; they are usually grown in Southeast Asia. The fruit of each species are single seeded.

Unlike Chinese water chestnut corms, water chestnuts must be boiled for about 1 h to destroy the toxic substances they contain. Following the heat treatment, the fruit is peeled and the white starchy meat of the nut is eaten, usually after roasting. The nuts contain about 16% starch and 3% protein. In addition to vegetable use, water chestnuts are also used as edible starch.

WATER DROPWORT, *Oenanthe javanica* (Blume) DC.
(O. stolonifera, Sium javanicum)

Family: Apiaceae (Umbelliferae)
Other names: Water celery, seri

Water dropwort is a cool season, semiaquatic and aquatic perennial grown for its foliage. The plant is native to Southeast Asia, where it is often found growing wild in freshwater marshes and along streams. Much of the production is still gathered from the wild. The crop has been used as a potherb or condiment for thousands of years, as early as 2000 B.C. in China; its culture in Japan dates to A.D. 750.

The creeping stolons, when rooted, are used for propagation, although seed can also be used. The hollow green seed stalk grows to a height of about 1 m and is relatively well branched. Leaves are alternate, compound pinnate with petioles that tend to be sheathed about half their length. A compound terminal umbel forms opposite the leaves. The fruit is a schizocarp.

Plants strongly tiller and are easily propagated by division, but stem cuttings also establish quickly. Seed germination is too erratic for reliable cultivation. In Japan, water dropwort is often planted in the early fall, following rice crops. Plants grow in flooded culture into the winter and are harvested in the late winter or early spring about 2–3 months after planting, when plants are about 30 cm in height. Repeated cuttings are possible for several years without the necessity of replanting. Leaves taste like and resemble celery, except the more numerous round petioles are slender and hollow. Leafy petioles and tops are stir fried, eaten raw in salads, and used for garnishes. Water dropwort seed contain the essential oil limonene. Unlike a related water hemlocklike plant, *Oenanthe crocata*, water dropwort does not contain the poisonous oenanthotoxin but does contain myristicine, a hallucinatory drug.

ALGAE, Various species

Commonly known as seaweed, algae are highly appreciated vegetables by many, and especially by those in and near coastal areas of China, Japan, and Korea. Algae are found in freshwater as well as seawater and some species grow on damp soils. Their use for food and medicine is ancient, well over 2000 years, and involves numerous species. For some populations, algae are a daily food, and to meet consumption needs, algae culture as well as gathering from noncultivated populations is practiced.

Algae lack differentiation common to other plants and do not have true roots, stems, or leaves but do carry on photosynthesis. Some species are single cell; others are multicellular bodies. Some algae are free floating, others resemble plants in having rootlike holdfast attachments for anchorage and a thallus consisting of a stemlike stipe, and leaflike blades. Blades range from singular to extensively branched, and may be smooth, heavily waved, chainlike, or filamentous. Propagation is asexual or sexual. Asexual reproduction is by division, zoospores, or immobile spores. Sexual reproduction is by homothallic or heterothallic fusion.

Many algae are used separately or mixed with other foods, both raw and cooked, and also as dried products; preparatory methods vary greatly. Edible vegetable species are found among four major algal groups: Chlorophyta (green), Phaeophyta (brown), Cyanophyta (blue green), and Rhodophyta (red). Although functional, this grouping is not taxonomically accurate.

Some of the important green algae species include the following: *Ulothrix flacca*, which has an appearance like fine, green hair; *Monostroma nitidum*, popular for its flavor and tenderness; *Ulva lactuca*, commonly known as sea lettuce; and the very tender *Enteromorpha linza*. Still another is wakame, the Japanese name for *Undaria pinnatifida*. In the brown group is *Laminaria japonica*, one of the most widely used algae, known as *kobu* or *kombu*, which at one time was a important source of monosodium glutamate, a flavor-enhancing compound. Several others are *Sargassum fusiformis*, *Scytosiphon lomentaria*, and *Ishige okamurai*. The blue-green group is underrepresented with only *Brachytrichia quoyi* having some usage. The red algae group provides *Porphyra tenera*, *P. umbilicalis*, and other *Porphyra* species. Some others which are used most in China, include *Grateloupia filicina* and *Gigartina intermedia*, enjoyed for their gelatinous thickening property that develops during cooking.

Harvested algae are cleaned, washed, drained, and cut or chopped to be eaten raw or cooked in various ways. Those of the *Laminaria* species are used for soup stock. Algae blade sheets of *Porphyra* species, called "nori" in Japan, are frequently used as wrapping for rice balls, sushi, and other foods (Fig. 26.8). A major use of the *Porphyra* species is to flavor bland rice or poi (*Colocasia esculenta*). Algae in salads, as pickled relishes, and flavored agar are other common preparations. Dried and powdered algae products are also used in various food prepa-

FIG. 26.8.   Dried algae, *Porphyra tenera* (nori), in shredded form for use in seasoning. When processed into thin sheets, nori is used to wrap sushi, a popular prepared Japanese food.

rations, including agar production. Algae also have a number of industrial uses, and sometimes are used as fertilizer.

In the consumption of seaweed, the Chinese prefer a variety of cooking methods, whereas the Japanese more often prefer seaweed served cold, following blanching, and often accompanied with some flavoring, such as soy sauce and vinegar, and sugar, or all three combined. In these countries as well as others worldwide about 100 species of algae are utilized some scarce algae species can be a very expensive vegetable, comparable to truffles.

The nutritional value of some species is significant. For example, *Porphyra tenera,* on a dry weight basis, has six times the protein of rice, with a pro-vitamin A content, depending on crop portion tested, ranging from 20,000 to 44,000 IU/100 g. Vitamin C content is also good, as is the high digestibility (75%) of the protein and carbohydrates. Other constituents in appreciable amounts include calcium, phosphorus, iron, and iodine.

SELECTED REFERENCES

Abbott, I.A. 1979. The uses of seaweed as food in Hawaii. Econ. Bot. *32,* 409–412.

Chan, Y.S., and Thrower, L.B. 1980. The host–parasite relationship between *Zizania caduciflora* Turcz. and *Ustilago esculenta* P. Henn. I. Structure and development of the host–parasite combination. New Phytol. *85,* 201–207.

Herklots, G.A.C. 1972. Vegetables in Southeast Asia. George Allen & Unwin, London.

Hodge, W.H. 1956. Chinese water chestnut or Matai—A paddy crop of China. Econ. Bot. *10,* 49–65.

Kanes, C.A., and Vines, H.M. 1977. Storage conditions for Chinese water chestnuts Eleocharis dulcis. Acta Hortic. *62,* 151–160.

Mazumdar, B.C. 1985. Water chestnut—the aquatic fruit. World Crops *37,* 42–44.

Syamal, M.M., and Verma, S.P. 1982. Cultivating the water chestnut. Indian Hortic. *27,* 24–25.

Terrell, E.E., and Batra, L.R. 1982. *Zizania latifolia* and *Ustilago esculenta,* a grass–fungus association. Econ. Bot. *36,* 274–285.

Xia, B., and Abbott, I.A. 1987. Edible seaweeds of China and their place in the Chinese diet. Econ. Bot. *41,* 341–353.

Yamaguchi, M. 1988. Asian Vegetables. In Advances in New Crops. J. Janick and J.E. Simon, eds. Timber Press, Portland, OR, pp. 387–390.

27

Edible Mushrooms

INTRODUCTION

Fungi are broadly identified by their sexual stage characteristics into four main classes: Ascomycetes, Basidiomycetes, Phycomycetes, and Fungi Imperfecti. Mushrooms are filamentous fungi grouped with the Ascomycetes and Basidiomycetes. A major difference in these classes is that the sexual spores of Ascomycetes develop in an asci sack and are released when the asci wall breaks down. Basidiomycetes produce a different structure, the basidium, in which sexual spores are formed and dispersed. Lacking chlorophyll, mushrooms do not carry out photosynthesis and therefore rely on other energy sources. Some are parasitic, others saprophytic, and some live in symbiosis with other plants. The majority develop the fruiting bodies epigeously (aboveground). For others, the fruiting bodies are formed hypogeously (below ground). Saprophytic species obtain their nourishment from nonliving organic matter. In general, these fungi produce fruiting bodies (ascocarps and basidocarps) called mushrooms when specific conditions of temperature, humidity, and nutrition are satisfied.

Not all mushrooms are edible; inedible and/or poisonous forms also occur. Before consumption, the identity and safety of the mushrooms should be determined. Of the edible species, some are cultivated, whereas others are gathered from the wild. Mushrooms found in the wild have been consumed for centuries and are almost everywhere considered a delicacy. Because of the uncertainty of their occurrence and production in the wild, they could not be considered a reliable or staple food source. Mushrooms are generally a complement to meals, enjoyed for their flavor and textual contributions. Depending on consumption frequency and volume, mushrooms can be nutritionally important. In some countries, cultivated mushrooms have an appreciable per capita consumption. Only about 25 of a possible 2000 edible mushroom species are cultivated as human food. The more important cultivated and gathered mushrooms include the following:

Mushroom

Common name(s)	Species
Abalone	*Pleurotus abalonus*
Bear's head	*Hericium erinaceus*
Black truffle	*Tuber melanosporum*
Button/champignon	*Agaricus bisporus/A. bitorquis*
Chanterelle	*Cantharellus cibarius*
Corn smut	*Ustilago maydis*
Enoki/enokitake	*Flammulina velutipes*
Meadow	*Agaricus campestris*
Monkey's head	*Hericium coralloides*
Morels	*Morchella hortensis/M. esculenta,* and other *Morchella* spp.
Nameko	*Pholiota nameko*
Oyster	*Pleurotus ostreatus,* other *Pleurotus* spp.
Pine	*Tricholoma matsutake/Armillaria matsutake*
Shaggy mane	*Coprinus fimetarius*
Shiitake/Black Forest	*Lentinula edodes*
Straw/paddy/Chinese	*Volvariella volvacea*
Stropharia	*Stropharia rugoso-annulata*
Varnish skin	*Ganoderma lucidum*
White jelly/silver ear	*Tremella fuciformis*
Woody ear/Jew's ear	*Auricularia polytricha/A. auricula*

General Botany

The mushroom, or fruiting body of the fungi, has a cap *(pileus)* with thin, bladelike gills on the under surface from which spores are shed, and a stalk *(stipe),* see Fig. 27.1. The below-ground portion is the *mycelium.* During early growth, the gills are covered with a veil, or thin membrane (volva), stretching from cap edge to stalk. As the cap grows, the veil stretches and breaks, exposing the gills. Each fruiting body is capable of producing multitudinous, almost microscopic spores.

CULTIVATION of *Agaricus bisporus* (Lange) Sing.

Most of the world's cultivated mushroom production is commercialized. The majority of that production is by relatively large-scale produc-

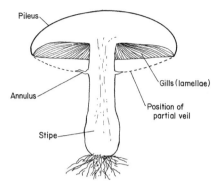

FIG. 27.1. Mushroom fruiting body, *Agaricus bisporus*.

ers; it is estimated that more than 35% of the production of cultivated mushrooms are species of the genus *Agaricus*. The commercial production of other genus and species has greatly increased. Hardly insignificant is the considerable volume of several noncultivated mushrooms species that are also harvested.

The earliest cultivation of *Agaricus* spp. was in caves by French gardeners near Paris about 1700. The cave environment, because of stable temperature and humidity, was ideal for mushroom growing. However, caves were not always available and sanitary conditions were difficult to maintain, so production gradually moved to aboveground structures where cultural manipulations were not restrictive. Nevertheless, some production continues in caves.

About 1935, ground beds began to be replaced with shelf and fixed trays that increased surface areas within the structure. However, beds, fixed shelves, or trays required that all cultural phases be performed in place. Subsequently, conversion to the use of movable trays allowed the various cultural procedures such as pasteurization, spawn inoculation, mycelium establishment, growth, and harvesting to be carried out more efficiently. Often, separate structures are dedicated specifically for conducting different parts of the production procedure. Movable trays, with surface areas of about 3 m², are commonly used, one of several alternative methods uses compost bags rather than tray culture.

The practices involved in the cultivation of the various mushrooms species differs. The discussion that follows is specific for *Agaricus bisporus*. The major phases involved include:

Compost preparation
Filling and pasteurization
Spawn preparation
Spawning
Casing
Fruiting body growth and harvesting
Postharvest

Throughout all phases, maintenance of crop hygiene, proper humidity, temperature, pH, and ventilation is necessary.

Compost Preparation

As a saprophytic fungus, *A. bisporus* requires an external energy base and growing environment that permits mushroom growth preferentially to other fungi. Composting changes the substratum by decomposing the unsuitable complex carbohydrates into forms suitable for mushroom nutrition.

Horse manure remains the most preferred ingredient for producing the compost medium. However, decreased use of work horses as a primary source has diminished the access to this supply. Other manures are used, but, overall, many variables exist in their use. For example, fresh manure ferments better than an older product, and the differences in animal source, moisture content, amount, and composition of bedding materials or other ingredients all contribute to manure variability. Substitution with other organic matter are made when preferred sources are unavailable or too distant from production sites to be economically feasible. Various nonmanure materials such as corn cobs and even waste paper are used to produce compost. The ability to formulate and control the composition of the compost has made important contributions to production consistency.

The typical base components of compost provide about 2% nitrogen on a dry weight basis, preferably with little or no ammonia; a carbon to nitrogen ratio of about 17 : 1 is preferred. The moisture content of the compost should be between 60% and 70%. Good aeration, a fibrous matrix with a nearly neutral (7.0–7.5) pH is desirable. Gypsum is often added (3% fresh weight) as a conditioner in the event of excessive moisture, adverse pH, or ammonia. Gypsum is also added to improve water percolation and reduce soluble-salt accumulation on the compost surface. Other additives include phosphorus, B vitamins, and various other growth substances. Protein-rich materials, in addition to animal manure, are frequently used as an additional source of nitrogen. Protein supplementation, at the time of spawn inoculation, has been shown to increase yields.

In composting, aerobic microorganisms break down the organic matter into products more readily available to the developing mushrooms. The microorganisms are themselves destroyed by the heat of metabolism generated during organic matter decomposition. Composting essentially is the first stage of pasteurization. Although the compost does not get hot enough to eliminate disease and pests, they are greatly reduced. When properly performed, a nearly sterile medium results.

Materials to be composted are placed in a pile, about 2 m high and 2 m wide, moistened, and allowed to ferment. In about 5 days or less, the inner temperature of the pile should reach 70°C or higher. The pile is then turned so that outer portions are turned inward and again allowed to heat up. The pile is turned three or four times at 4–5-day intervals before it is ready for filling. The length of the composting period is dependent on the materials used and temperatures achieved.

The composting procedure should not be interrupted. Excessive moisture or rain exposure should be avoided because it reduces air space. The compost should not become anaerobic, and extended composting will result in loss of nutrients. If composting is not complete, subsequent heating can occur and adversely affect procedures that follow. In the Netherlands, there are firms that specialize in compost production; this gives growers a reliable and consistent medium for growing mushrooms.

Filling and Pasteurization

Filling is the placement of the composted materials into the growing area. The compost should be placed to provide a uniform density and depth. Some growers use filling machines to reduce labor and speed the procedure. One ton of manure compost is usually sufficient to provide a surface of about 11 m^2 to a depth of 15 cm. Depending on the compost, some slight compaction may be required. However, the compost increases in bulk density with time, and if too dense, production is adversely affected. Filling should be completed as rapidly as possible to minimize cooling. Soon after filling, the compost must be heated. Further heating (pasteurization or "sweating") is provided to reduce contamination that may have occurred during filling or preparation of the growing area. Heating destroys microorganisms, pests, and competitive fungi. A high standard of sanitation is a prerequisite for successful production. Heating can be self-generated, but more often, external heating is provided. When the compost does not self-generate a sustained temperature of about 60°C, steam or another heat source is provided to reach and maintain a temperature of about 60°C for several days. The compost should not get too hot; temperatures above 62°C or 63°C decrease nutritional value. However, it is important to have all portions of the compost attain a minimum temperature of 57°C for at least 5 h.

After filling, the compost is allowed to cool slowly. Thermophilic microorganisms survive heating and utilize the ammonia in the compost, converting it into their own protein products, which eventually become available for mushroom growth. In large-scale production, specialized heating houses are used for the "sweating" process.

Spawn Preparation

Spawn is the propagating material used for initiating mushroom production. Many years ago, the spawn used was obtained from the soil and/or other materials in the vicinity of previous mushroom growth. This material contained spores and mycelial remains, as well as many contaminants. Pure culture spawn on a scientific basis was first introduced about 1900 by the Pasteur Institute. Spawn production since then has become a precise laboratory procedure whereby maintaining sanitation and purity of the spawn are critical objectives.

Presently, most spawn consists of spores and mycelial growth supported on sterilized grain, bran, or other materials under aseptic conditions (Fig. 27.2). Spawn growing is performed under sterile conditions in order to eliminate competitive organisms, diseases, and pests. Accordingly, many mushroom growers prefer to obtain spawn from producers that specialize in its preparation.

Spawning

Spawning consists of dispersing the spawn medium uniformly over and into the upper 5 cm of the bed. About a 1 liter volume of the spawn preparation is used per square meter of bed surface. Placement can be by hand but is more effective when done with machines that thoroughly mix the spawn material within the upper surface of the compost.

Spawn is introduced after pasteurization when the compost has cooled. Compost temperatures above 27°C and below 38°C are satisfactory, but temperatures less than 26°C are not and will delay spawn mycelial growth. Because of the metabolic heat caused by the growing mycelium during colonizing of the compost, a slight rise in temperature typically occurs and lasts for several days. It is important to maintain high humidity during this period. Also at this time, CO_2 levels can be

FIG. 27.2. Mushroom spawn developing in bran media. The mycelium in the left bag has fully colonized the medium and is ready for use as spawn.

increased slightly as it improves the growth (running) of the spawn mycelium. At other times, CO_2 levels must be kept low.

Within a few days after spawn introduction, compost temperatures will begin to decline. A temperature between 20°C and 24°C with a compost moisture content 50–70% of water holding capacity is maintained, usually for 8–15 days, and sometimes longer, to allow the mycelium to grow throughout the compost substratum. Temperatures higher than 40°C can kill the mycelium. Ventilation is provided as needed, but ventilation with cold air should be avoided, and when possible, introduced air should be filtered.

Casing

Casing has no nutritional contribution, instead the purpose is to improve and initiate production of the fruiting bodies. This procedure is performed 2–3 weeks after spawning and when the mycelium has fully run through the compost. Casing is accomplished by the uniform placement of a thin, 3–5-cm, layer of disease and pest-free uncontaminated moist peat or material similar to the compost in use; usually, soil is not used. Sometimes the casing material is sterilized before use by heating to 85°C for 30 min or by chemical fumigation. Placement of the casing is mechanized in large-scale operations. Lime is added to the casing material, if needed, to obtain a pH slightly higher than 7. Casing is not a seal over the compost; the layer should be porous and permit aeration and, if necessary, watering.

The physical influence of casing is not fully understood, because fruiting bodies can be formed without casing; however, productivity without casing is greatly reduced. It is theorized that the increase in compost CO_2 concentration (as high as 4%) caused by the casing stimulates the formation of primordia. Casing normally results in a slight temperature increase to about 18°C, and then slowly decreases to range between 10°C and 15°C. It is important to continue monitoring for excessive CO_2 and to ventilate as needed. High CO_2 may have an adverse influence on development of the fruiting bodies. Light watering is performed as required; however, excessive watering may reduce pore space, resulting in poor aeration.

Fruiting Body Initiation, Growth, and Harvesting

Ideally, before fruiting bodies are formed, the mycelium should have fully explored the compost medium, drawing on and utilizing the nutrients contained. Optimum conditions for initiation of primordia (pinheads) are temperatures between 15°C and 17°C and 80–90% RH, with compost moisture content at 70% of water holding capacity, and the CO_2 level at less than 0.1%; excess CO_2 is reduced with ventilation. When compost nutrition is sufficient and the environmental conditions favorable, mycelium is first stimulated to form small knotlike nodules capable of enlargement into fruiting bodies. These dense nodules are primordia and are called "pinheads" (Fig. 27.3). Each mushroom species

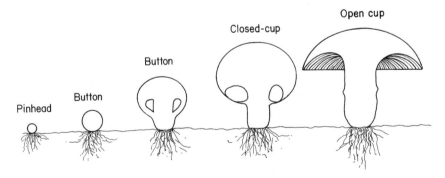

FIG. 27.3. Developmental stages of *Agaricus bisporus* fruiting bodies.

has its specific conditions. *Agaricus* is one species that does not need light for primordia development. Light has been shown to be inhibitory to *A. bisporus* fruiting body primordia and tends to elongated stipe development and limits pileus expansion. The *Agaricus* fruiting bodies are sensitive to CO_2, which is continuously generated from the compost. A CO_2 level in excess of 0.2% (2000 ppm) prevents pinhead formation. Primordia formation is easily interrupted when the environment become unfavorable; therefore, growers exercise precise environment management in order to maintain productivity.

At the stage where pinheads are visible, air temperatures are adjusted to 15–17°C and 85–90% RH to enhance fruiting body production that follows. Growers use temperature and ventilation to regulate relative humidity. Humidity management is important to avoid moisture loss from mushrooms. It is also important to avoid increased heating within the compost and the associated generation of CO_2. Carbon dioxide management together with temperature control is used to initiate fruiting and subsequent growth; high levels favor mycelial growth, and levels less than 0.1% CO_2 favor sporophore initiation and enlargement. A sporophore is the structure upon which spores are borne.

Harvest starts about 17–21 days after casing. Production occurs in flushes called "breaks," the first being the strongest and most productive. After a brief interval of low production, another flush will occur. Flushes occur at 7–10-day intervals, each successive flush being less productive. Air temperatures may be reduced to 16°C to encourage additional primordia formation, and then adjusted to a higher temperature (24°C) to enhance fruiting body growth. Usually, after three or four flushes, harvesting stops.

Once initiated, daily harvest usually continues except when production decreases between flushes. During the harvest period, if moisture is needed, careful light sprinkling is provided; however, it is best applied between flushes.

Mushrooms are harvested before the veil breaks or the stem (stripe) elongates. Harvest terminology for mushrooms includes the following: *buttons* of different sizes, where the membrane (veil) is developing but still closed; *cups,* where the membrane is well developed or just opening; and open or *hats,* where fruiting bodies are beyond the cup stage, fully grown, and generally become flattened. Large-size mushrooms generally command the highest price for the fresh market. No further enlargement occurs after the veil breaks, although a change in the shape of the fruiting body occurs. A delay in harvesting in order to increase the size of the fruiting bodies risks having the veil tear. If the veil is torn and the gills are evident, the mushroom's market value drops. The flavor of the fully open mushroom improves because moisture is lost and dry matter increases. Delaying harvest also tends to reduce subsequent primordia formation. A rapid and complete harvesting of each flush tends to shorten the interval between flushes.

Various strains of *A. bisporus* produce either white-, off-white-, or tan-colored fruiting bodies; white is preferred most often. Color development can also be influenced by airflow, irrigation, light, and crop age.

During harvest, workers simultaneously size grade, cull, and remove trash. The easily damaged mushrooms require careful handling. Mushrooms are twisted from the casing without disturbing smaller buttons or pinheads, but in some operations, use is made of picking aids and machines. When harvest is mechanized, the mushrooms are generally used for processing. Harvesting can occur over as brief a period as 40 days to as long as 200 days. A common goal is to harvest over a 125–130-day period in order to achieve two full production cycles each year. Production procedures and time required are as follows:

Procedure	Range in days
Initial composting	12–20
Final compost pasteurization	5–10
Spawning and running of mycelium	10–15
Casing	10–20
Mushroom appearance	30–60
Harvesting period*	75–125
Total crop cycle	132–250

Short, high-yielding production cycles are advantageous because prolonging any of the procedures or harvest period limits the yield obtained over time. Production yields range from 120 to 240 kg/t of manure compost or 10–20 kg/m^2 of bed surface. Production in the United States during 1994 was reported to average about 27 kg/m^2.

To maintain an effective production schedule, the composing of new material begins about 12–20 days before removal of the spent compost. Some growers renew production beds by flip-flopping the compost (bottom for top), and recasing or skimming off 2–3 cm from the existing top casing and then recasing. The house temperature is raised, the beds watered, and when successful, primordia and fruiting body production resumes after a few days. However, productivity will be less than it was initially. Such renewal practices should only be performed if earlier yields were good and the beds are free of disease and pests.

*Usually 5 flushes occur, each lasting 6–10 days, with a 6–12-day interval between flushes. In the interval between flushes, growers generally will have scheduled another crop that can be harvested, thus allowing for continuous year-round production.

Following the final harvest, the spent compost is removed. Although the nutrients in spent compost are no longer useful for mushroom purposes, the material is a good soil mulch or conditioner and has some fertilizer value.

Postharvest

Following harvest, mushrooms are cleaned but not washed. Some may have the base of the stipe cut off before packing. Mushrooms are easily bruised and rapidly lose weight by drying. Respiration is high; at 20°C, it is about four times that of spinach, known for its high rate of respiration. Postharvest refrigeration at 0°C and 95% RH will maintain mushrooms in good condition for 5 or 6 days; therefore, rapid marketing is important. Dehydration and veil opening are reasons for quality decline.

Use and Nutritional Value

Mushrooms are eaten primarily for their flavor and textural features and are used fresh, dried, or processed by canning, freezing, or pickling. A large proportion of *A. bisporus* is canned. Although an attractive canned product, *Agaricus* does not produce an attractive dried or frozen product because of the initially high moisture content. In addition to fresh use, other mushrooms, such as truffles and shiitake, are often successfully dried because of their lower moisture content.

Market-quality specifications for *Agaricus mushrooms* require that they have a closed veil. Because flavor increases with age, the open or hat stages of mushroom development actually have superior edible qualities. Mushrooms provide about 4–5% of a highly digestible protein on a fresh weight basis, or roughly twice that of most vegetables. However, on a fresh weight basis, they are relatively costly, and some species, such as truffles, are very expensive.

Mushroom nutritional values range from fair to good. They contain ergosterol which can be converted into vitamin D. The B vitamins and many amino acids are present in good quantities. A major attribute is their low caloric levels. The fat content is low, and the content of fiber and minerals is good. Some mushrooms are credited with lowering cholesterol and reducing blood-clotting time.

Diseases and Other Pests

Bacterial blotch, *Pseudomonas tolaasii,* is a frequent disease problem. Other diseases include matt disease (*Myceliophthora* spp.), mildew

(Dactylium dendroids), white mold or bubbles *(Mycogone* spp.), verticillium or dry bubbles *(Verticillium* spp.), and damping off *(Fusarium* spp.) A virus disease known as La France can also affect production. The feeding by rodents and other animals are damaging to mushrooms grown outdoors.

Major insect pests include flies, gnats, and springtails. Other pests include nematodes, mites, slugs, and snails. Strict sanitary practices are very important for disease and pest control. To assure sanitary conditions for disease and pest management, producers fumigate with steam or burn sulfur, and use formaldehyde, or HCN, as well as fungicides and insecticides as required.

STRAW MUSHROOM, *Volvariella volvacea* (Bull ex Fr.) Sing. *(V. esculenta)*

Other names: Paddy mushroom, Chinese mushroom,
    Champignon de Pailla

The straw mushroom is a tropical species, usually represented by *Volvariella volvacea,* although several other species are sometimes identified as straw mushrooms. Southeast Asia is the center of straw mushroom production, most of which is performed outdoors. Structures ranging from simple sheds made of straw to well-constructed buildings are used for indoor production (Fig. 27.4a).

Rice straw is the substratum most frequently used. However, other materials such as water hyacinth, oil palm waste, banana leaves, sawdust, and cotton waste are also utilized. The straw or other materials are soaked in water for 1 or 2 days and then used to build a bed of stacked bundles or layers of material.

Straw mushrooms are produced either with or without compost pasteurization. Outdoor production is generally carried out by placing bundles of straw or layers of other compost material in a pile. Often, a bamboo pole is placed in the center to provide support. The straw bundles or other compost material usually are placed on a mound of soil, bricks, stones, wood planks, or other drainable surfaces. Drainage is very important, especially during tropical rainy periods. The thickness of the compost is about 20 cm and the surface is about 1 m². Another procedure uses wood boxes, about 80 cm square and 10 cm deep, into which the compost is placed.

For indoor production, the compost is also placed into wood boxes or onto shelves at a similar 10 cm depth with a surface area of about 1

FIG. 27.4.   A common low-cost structure (a) in which straw mushrooms, *Volvariella volvacea,* are produced; a wall was removed in order to photograph the interior, and (b) a tray containing straw mushroom fruiting bodies. Note: the mushrooms in the plastic bag at the top of the tray are enoki, *Flammulina velutipes.*

m². Pasteurized compost is more often used for indoor production, and often the compost depth is increased to as much as 30 cm. Indoor production and the use of pasteurized compost is usually more productive.

Production is initiated with spawn, usually obtained from previously "spent" straw or other medium because pure culture spawn is generally not readily available. The straw or other material is cut into 2–5-cm pieces and placed in jars with 1% $CaCO_3$ and 1–2% rice bran for the fungus to colonize and produce the spawn that will be applied between layers of straw bundles during bed building.

The first fruiting bodies appear in 15–25 days after spawn is applied. Relying on natural spawning will take longer, is less reliable, and less productive. Being a "high-temperature mushroom," an ideal temperature at spawning is 36–38°C. During spawn run, temperatures should be maintained at 32–34°C. After spawn running, fruiting temperature is best at 30°C and 80% RH. Proper moisture is critical for successful production, and daily watering as a fine spray is usually required to maintain the compost at about 65–70% of water holding capacity. Light is not required during spawning, but is required at the time of fruiting and must be supplied for indoor production.

Straw mushrooms are not permitted to achieve maximum size but are harvested before or just after the volva ruptures (Fig. 27.4b). In harvesting, the mushrooms are lifted and twisted off while avoiding damage to adjacent developing fruiting bodies. Harvest is carried on for about 20–30 days, sometimes with twice-a-day harvesting, especially for early flushes. During the harvest period, several flushes of about a 4-day duration occur at 5–10-day intervals. Yields range from 2 to 7 kg/m². Straw mushrooms have poor postharvest properties, and much of that which is not consumed fresh is processed as a canned or dried product.

SHIITAKE MUSHROOM *Lentinula edodes* (Berk.) Pegler *(Lentinus edodes)*

Other names: Black Forest, hiratake (Japanese), xiang-gu or shiang-gu (Chinese)

Shiitake, a member of the Polyporaceae family, differs from button (*Agaricus* spp.) and oyster (*Pleurotus* spp.) mushrooms in being wood inhabiting and having a long production cycle. Hardwoods, preferably those of oak (*Quercus* spp.), are used in its production. In Japan, logs from hardwood trees locally known as *kunugi, konara, kuri,* and *shii (sudajii)* are used; these are red oak *(Quercus acutissima)*, white oak *(Quercus serrata)*, Japanese chestnut *(Castanea crenata)*, and live oak *(Castanopsis cuspidata* var. *sieboldii)*, respectively (K. Takayanagi, University of Tsukuba, Japan, personal communication). Trees are cut when dormant in the late fall and when the nutrient content is highest. Logs of 5–15-cm diameter are cut into 1-m lengths; the moisture level of the felled logs must be 40% or more, and the bark in good condition. An intact and healthy bark is important for moisture control. To avoid heating, logs are kept in 70–80% shade.

For inoculation, spawn is obtained from the fungal mycelium; spores obtained from fruiting bodies are not used because they do not produce mushrooms true to type. The mycelium is grown on a grain base and then transferred to a sawdust or grain/bran mixture for inoculation; sawdust of the same species as the bed logs is preferred. For long-term storage, spawn is stored in liquid nitrogen. On a regular basis, fresh spawn is made from inoculum of stored stock cultures.

Logs are inoculated by placing the mycelium sawdust mixture (spawn) into holes bored or cut into the logs or by insertion of mycelium inoculated wood plugs into drilled holes. About one hole is made for each 100 cm^2 of the log surface. Often, inoculation sites are sealed with melted wax to prevent moisture loss and to keep spawn in the hole. In warm climates, the ends of the logs are also waxed to reduce log moisture loss.

Logs are generally crib stacked during the spawn run. In about 5–8 months, while in the laying yard, the mycelium will have penetrated (run through) the logs; sometimes it may require a year or two to thoroughly infest the log. Mycelial growth is optimum at 24°C; light is not required. Following thorough penetration of mycelial growth, the logs are removed to the raising (fruiting) yards or buildings where temperatures of 12–20°C are more suitable for "pinning," which is the initiation of primordia and development of the fruiting bodies (Fig. 27.5a). Light is required for fruiting body development. The logs are placed nearly upright at this time and are regularly turned. Moisture is provided as necessary, usually by immersion in soaking tanks for 12–48 h; sprinkler irrigation is also used but is often less effective. Oversoaking should be avoided. The raising yard areas are generally shaded year-round to minimize moisture losses. In some areas, production is conducted in glass or plastic houses for better environmental control. Under good conditions, yields of 1–2 kg per log are achieved annually.

Logs may produce for 2–4 years, generally providing a major flush during the spring and fall, although small volumes can be produced

a

b

FIG. 27.5.   Shiitake, *Lentinula edodes:* (a) mushrooms developing in inoculated logs in a fruiting yard, and (b) shiitake fruiting bodies developing on a synthetic medium as an alternative to the traditional use of log cultivation.

throughout the year. Such smaller flushes can be forced four to five times per year by regulated soaking. In some commercial production, this practice is carried out so as to have all-year production. Growers generally vary inoculating times in order to have continuous production.

In areas where oak logs are difficult to obtain, producers use bag culture with a medium consisting of oak chips or sawdust; other wood species are also used, and the medium may be in the form of compressed blocks or bricks (Fig. 27.5b). The substrate medium is supplemented nutritionally with various organic amendments such as rice, millet, and wheat brans or other grains. For bag and brick culture, the spore mixture is incorporated into the medium. Measures are taken to avoid possible contamination from competitive fungi; excessive moisture is also avoided. Temperature of 20–24°C are favorable for incubation, and 16°C and 85–95% RH for fruiting. Bag and brick culture generally provides a more rapid development and higher yield than with natural logs, but usually requires more management and production investment.

Harvest is accomplished by twisting the mushroom from the growing medium. The mushroom postharvest life is extended when cooled to 1–2°C and held at high relative humidity. Shiitake fruiting bodies are more resistant to physical damage being less succulent and less fragile than *Agaricus* and other species. Furthermore, the caps are naturally brown and surface injury is less detectable. A large volume of production is directed to fresh marketing. However, most of the world production is dried, especially that proportion that is exported. Shiitake mushrooms are also canned and pickled, but the quality is poor.

ENOKI/ENOKITAKE MUSHROOM, *Flammulina velutipes* (Fr.) Sing.

Enoki mushrooms require low temperatures for fruiting body growth and development. From spawn introduction to harvest, enoki can be produced is a little as 50–60 days. Japan is the major producer and consumer of enoki mushrooms, China and Korea follow, although their popularity has greatly increased in some non-Asian counties. The appearance of cultivated enoki mushroom differs from other mushrooms in having a thin, 7–10-cm-long stripe with a small hemispherical pileus; often the pileus is 1 cm or less in diameter (Fig. 27.4b). The pileus of this wood-inhabiting fungus in a natural setting would measure as much as 10 cm in diameter and assume a flat-plane shape. Another unique feature is the requirement for light in pileus development; light is not required for primordia initiation.

Production commonly involves the use of a medium consisting of hardwood sawdust (usually elm) with rice bran, either placed in bags or wide-neck bottles. The bran serves as the primary source of nutrition. Other brans besides rice can be used, but the mixture is about 80% sawdust and 20% bran.

Mycelial growth can occur over a wide range of temperature, although 25°C and 60–65% RH is optimum. When spawn has thoroughly penetrated the medium, the temperature is lowered to between 8°C and 12°C and relative humidity is raised to 80–85% to initiate primordia formation. Stipe development is favored at 3–8°C and 75–80% RH. Further development of the pileus is optimum at 15°C. The lower temperature at early stipe growth controls stiffness and avoids overelongation. It is also important to keep CO_2 levels at about 5% to encourage the appropriate amount of stipe elongation.

At the latter stage of growth, a cylindrical collar of paper or plastic is placed around the closely grouped stipes in order to produce the characteristic erect, thin, long shape. Stipes elongate upward through the collar or bottle neck. When the desired amount of growth is achieved, the mushrooms are harvested by cutting below the base of the grouped stipes.

OYSTER MUSHROOMS, *Pleurotus* spp.

Among the rapidly growing, wood-inhabiting mushrooms are several species of *Pleurotus*; *P. ostreatus* (Jacq. ex Fr.) Kummer is the best known and most widely grown species. The gray or tree oyster *(P. saforsajou)*, abalone mushroom *(P. abalonus)*, white oyster *(P. citrinopileatus)*, and several other species are also produced.

Oyster mushrooms are generally recognized because of the shell-like appearance of the pileus and the eccentric placement of the stipe. These are rapidly growing fungi that are naturally tree inhabiting but are easily grown on substrate mediums consisting of various kinds of plant waste materials. The mushrooms are fragile and prolific spore producers; strains have been developed that produce very few spores.

Mycelial growth is optimum at about 27°C. However, species and strains vary in temperature response for fruiting body development. The high-temperature group prefer 25–30°C, and for the low-temperature group, 12–15°C is optimum.

Mycelium has a strong CO_2 tolerance and will grow in an environment containing as high as 15–20% CO_2, but fruiting bodies lack this tolerance. At CO_2 levels greater than 0.06%, the stipe will elongate and pileus (cap) growth decreases or is prevented.

Light is not necessary for mycelium growth; in fact, growth is better in the dark. However, light, even if brief, is necessary for primordia development. Absence of light decreases pileus size and low light results in pale-colored pileus.

Cultivation using cylindrical bags, jars, or compressed blocks are common. Natural logs are still used, but infrequently. Many kinds of plant waste products are used, some of which include sawdust, straw, cotton waste, straw, bagasse, and even banana leaves. These are often used without additional nutritional enrichment.

Spawn is introduced into or onto the medium and requires 3–4 weeks for complete penetration. With plastic bag culture, the wrapping is removed and light is provided. In about 5 days, some mushrooms are ready for harvest. They should be harvested before the pileus edge begins to curl. The mushrooms are easily pulled away from the medium, but care is necessary because the fragile pileus is easily torn. Proper watering is very important during the cropping period because many flushes occur.

Oyster mushrooms are subject to rapid desiccation, and postharvest life is short without low temperature and high humidity. Those not consumed in the fresh state are processed by air drying and brine canning.

TRUFFLES, several *Tuber* spp.

Truffles, *Tuber melanosporum, T. magnatum,* and other *Tuber* species, are Ascomycetes, and fruiting bodies are produced hypogeously. They have an intimate and probably obligatory symbiotic relationship with the roots of many different living tree species; those usually preferred are oak (*Quercus* spp.) trees. Truffle production, whether cultivated or gathered from the wild, probably involves more forestry management than agronomy or horticulture. Truffle propagation begins through natural inoculation or with the introduction of spawn placed near tree roots. The major distribution of *Tuber* species is from Portugal to the Baltic countries, but these and other truffle species are found in North America, Africa, and China.

Propagation is via spore inoculation, contact with mycorrhized roots, or rarely by mycelial inoculation. These procedures are used to produce truffle-supporting plants in nurseries, which are then transplanted for truffle fruiting body production. During cultivation, it is necessary to protect seedlings from animals, diseases, and weeds. Mulching is highly desirable as is a reliable source of moisture, preferably during the winter period. Soil organic matter is also desirable, but agriculturally marginal calcareous soils appear to favor production more than highly fertile soils.

Truffles are known to be harvested with the aid of dogs or hogs whose keen sense of smell helps to locate the underground fruiting bodies. A major component of the many aromatic compounds that provide the noticeable aroma is dimethyl sulphide. Actually, experienced truffle gatherers do most of the harvesting, and the use of aromatic detection equipment is increasing. Nevertheless, although animals are infrequently used, this impression persists.

Truffles are less perishable than most other mushrooms and also are very suitable for drying and for processing by canning. Production statistics are difficult to obtain because producers prefer not to disclose supply information about this highly prized fungus in order to keep prices high. World production is estimated at about 400,000 kg/year, but year-to-year variability is high. The possible expansion of production areas, already in process in North America and China, and the possible use of synthetic mediums for cultivation could further increase truffle supplies. However, production using synthetic mediums relies on being able to learn the conditions necessary for primordia formation and fruiting body development.

MOREL MUSHROOMS, *Morchella* spp.

Morels are another group of cultivated mushrooms that are Ascomycetes; they are similar to truffles except the fruiting bodies are produced aboveground (epigeous). Species of *Morchella,* mostly *M. deliciosa* (white morel), *M. elata* (black morel), and *M. esculenta* (yellow morel) are the most popular. Their cultivation is also dependent on an association with trees. Morels were not cultivated until recently but were gathered from tree-associated noncultivated sites; species of elms *(Ulmus)* are preferred. As further cultural information is acquired, cultivation of this highly appreciated mushroom will expand.

Morel mushrooms are characterized by the pitted surface of the upper portion of the pileus which is often globose shaped. Morels have an advantage of being less perishable and considerable more flavorful than the common mushroom, *A. bisporus.*

CHANTERELLE MUSHROOM, *Cantharellus cibarius* Fr.

Family: Cantharellaceae

Chanterelle mushrooms can be found growing wild in the autumn, winter, and spring in temperate regions. Their habitat is among conifers, live oak, and in pastures of North America, especially in the high-rainfall regions of the Pacific Northwest and California. They are also found in similar habitats during the summer in eastern North America. Commercial cultivation has not been successful.

This edible mushroom has a thick yellow-orange cap, 3–15 cm in diameter, which is broadly convex when young, turning plane to concave with age. The margin is lobed or wavy, the pileus underside with deeply decurrent gills and the tapered stipe ranges from a length of 2–10 cm and thickness of 0.5–3 cm. They tend to be fibrous and are better when cooked.

CORN SMUT, *Ustilago maydis* (DC.) Cda.

Other names: Cuitlacoche, huitlacoche, Mexican truffle

Corn smut is a common fungal pest that invades the ear and kernels of corn, *Zea mays* (Fig. 15.8). Fruiting bodies are used as mushrooms and some cultivation is performed by intentionally infecting corn ears. Supersweet cultivars of sweet corn appear to be more susceptible to infection. The enlarged grayish fruiting bodies are eaten, fresh or lightly cooked, as a delicacy vegetable; some are also processed by canning. Quality is best when fruiting bodies are harvested about 12–15 days after corn silking but before the development of the black dustlike spores which occurs 2 or 3 days later.

A related fungus having mushroomlike qualities is *Ustilago esculenta,* which invades and is an important factor in water bamboo production; see Chapter 26.

PRODUCTION

It is estimated that the 1990 world production of cultivated mushrooms was more than 3,764,000 tons and was comprised of the species listed in Table 27.1.

Although difficult to ascertain, the amount of gathered, noncultivated mushroom is considered to be substantial but cannot be accurately

TABLE 27.1.   ESTIMATED WORLD PRODUCTION OF MAJOR EDIBLE MUSHROOM SPECIES, 1990

Species	Production (×1000 t)
Button (*Agaricus* spp.)	1424
Shiitake *(Lentinula edodes)*	383
Straw (*Volvariella* spp.)	207
Oyster (*Pleurotus* spp.)	909
Wood ear (*Auricularia* spp.)	400
Enoki *(Flammulina velutipes)*	143
Tremella *(Tremella fuciformis)*	106
Other species	192

determined. Major countries producing cultivated *A. bisporus* mushrooms are the United States, France, the Netherlands, United Kingdom, Belgium, Germany, Ireland, Italy, Poland, Hungry, and several Balkan countries. Production in Korea, Japan, Taiwan, China, Canada, Argentina, Australia, and New Zealand is also significant. Production of shiitake, straw, enoki, and oyster mushrooms is highest in the Orient. Surprisingly, mushroom production in South America and Africa is small.

SELECTED REFERENCES

Arora, D. 1986. Mushrooms Demystified, 2nd ed. Ten Speed Press, Berkeley, CA.

Chang, S.T., and Miles, P.G. 1989. Edible Mushrooms and their Cultivation. CRC Press, Boca Raton, FL.

Chang, S.T., and Quimio, T.H., eds. 1982. Tropical Mushrooms: Biological Nature and Cultivation Methods. Chinese University Press, Hong Kong.

Chen, M. 1994. Personal communication. University of California, Berkeley.

Flegg, P.B., Spencer, D.M., and Wood, D.A., eds. 1985. The Biology and Technology of the Cultivated Mushroom. John Wiley & Sons, New York.

Giovannetti, G., Roth-Bejerano, N., Zanini E., and Kagan-Zur, V. 1994. Truffles and their cultivation. Hort. Rev. *16,* 71–107.

Maher, M.J., ed. 1991. Science and Cultivation of Edible Fungi. Mushroom Science XIII, Vol. 2. A.A. Balema, Rotterdam.

Pataky, J.K. 1991. Production of Cuitlacoche [*Ustilago maydis* (DS) Corda] on sweet corn. HortScience *26,* 1374–1377.

Sabota, C., and Nall, H. 1994. Shiitake Mushroom Production on Logs. Program Pub. No. AGR-H-A109R93. Alabama A&M Cooperative Extension, Normal, AL.

Singer, R., and Harris, B. 1987. Mushrooms and Truffles: Botany, Cultivation, and Utilization, 2nd ed. Koeltz Scientific Books, Koenigstein, Germany.

van Griensven, L.J.L.D., ed. 1988. The Cultivation of Mushrooms. Darlington Mushroom Laboratories Ltd., England/Somycel SA, France.

Wuest, P.J., Royse, D.J., and Beelman, R.B., eds. 1986. Cultivating Edible Fungi. Elsevier, Amsterdam.

28

Condiment Herbs and Spices

INTRODUCTION

As commonly used, the term "herb" is not restricted to the botanical definition. Herbs are widely recognized as a broad class of plants, generally associated with growth in the temperate zone, that are useful for flavor, aromatic, dye, or medicinal purposes. The Chinese and other cultures have used herbs medicinally for thousands of years, and the practice continues. Savory herbs have aromatic properties and are utilized principally as flavoring agents.

Spices are frequently identified as products of dried vegetative substances derived from the roots, bark, fruit, or berries of perennial plants, mostly of tropical or subtropical origin. They have aromatic qualities and are used for seasoning foods.

Herbs and spices represent a broad category of plants with an equally broad array of uses. They make foods and beverages more pleasurable, rather than providing major caloric or other nutritional benefits. Relatively small amounts give zest to foods, usually without fear of adverse effects. However, not to be overlooked is the role of herbs and spices in food preservation, reducing spoilage, and disguising or overpowering the odor and taste of spoiling foods.

Because there is no clear determination, a particular plant could be considered a vegetable, herb, or spice. How the plant or its products are used usually determines the category. Plant tissues involved in herb and spice use may include foliage, stem, bark, root, bulb, rhizome, bud, flower, fruit, and seed.

A large number of commonly used culinary herbs are found in temperate growing regions and are members of the following families: Boraginaceae, Asteraceae (Compositae), Brassicaceae (Cruciferae), Lamiaceae (Labiatae), and Apiaceae (Umbelliferae). Many commonly used spices are in the families: Lauraceae, Myristicaceae, Piperaceae, and Zingiberaceae. By no means is this family listing complete; medicinal herbs are especially diverse with regard to their botanical taxonomy.

Flavor and Medicinal Characteristics

The value of herbs, spices, and medicinal plants results from the variety of unusual chemical compounds contained in the various plant tissues. These compounds are responsible for the specific flavors, odors, or medicinal qualities associated with each species. A distinguishing feature common to most are the essential oils. Essential oils are highly aromatic compounds, usually of a benzene or terpene chemical structure or of moderate-length, straight-chain hydrocarbons. Other active compounds found can include alkaloids, tannins, terpenoids, steroids, and glycosides. Because of the complex assemblage of chemicals, complete synthetic duplication of the active ingredient, whether flavor or medicinal, often remains elusive or economically impractical. However, advances in analytical instrumentation have identified and quantified many of the components, and some can be produced synthetically or closely approximated.

Origin and Production Areas

A major portion of culinary herbs such as anise, basil, bay leaf, coriander, dill, fennel, marjoram, mint, oregano, rosemary, sage, terragon, and thyme originated in Asiatic Europe, with various species ranging from southern Europe eastward to Iran. On the other hand, many commonly used spices, such as cinnamon, cloves, ginger, mace, nutmeg, and pepper *(Piper nigrum)*, were primarily Asiatic in origin, ranging across a vast area from India, its immediate surrounding areas, and eastward to Malaysia and Indonesia. These locations were not exclusive origins of herbs and spices because Central and South America and the East Indies contributed chiles *(Capsicum* spp.), vanilla, allspice, and cacao.

Many herbs and spices are still either harvested in the wild or produced in the region of origin. However, the present production of many herbs has less relevancy to their centers of origin because they have been transferred to other parts of the world where comparable growth can be achieved, and/or where resources such as labor make production feasible. For example, Central and South America and the West Indies have become important producers of spice plants that were exclusively grown in India, Southeast Asia, and South China.

Historical Background

Human interest in herbs and spices is ancient. The long association of herbs and human activity has resulted in an abundance of folk lore passed on through the ages. Although some information about herbal effects may be questionable, many attributes are valid and require serious consideration (Chapter 5). Excessive consumption of some herbs or spices can introduce hazardous ingredients. Medicinal properties can range from highly beneficial to extreme toxicity, as well as inconclusive effects.

Considerable mention about spices and herbs, often with spiritual and social connotations, exists in old writings of more than 5000 years ago. Thousands of years of usage are traceable with regard to their contributions in food preparation. Additional nonfood uses are for medicine and fragrance and as ornamentals.

Spices have had a profound indirect effect beyond their relevancy to food. The search for spices was one of several priorities of Columbus' voyages and played a role in widening further world explorations. The spice trade was an important activity in stimulating trade between Europe and South China, Malaysia, India, and the East Indies. Although these areas were not necessarily centers of origin, they were major supply sources. Early dominance in this trade was held by Arabia and subsequently acquired by merchants in Venice and Constantinople (Istanbul), followed by Portuguese, Dutch, and English traders. Intentional deception was used by merchants in these regions so that sources would not be revealed to competitors. The search and possession of spices also lead to conflicts, large and small, and the making and dissolution of empires. The monopolylike control lessened when improved communication and travel increased opportunities for broadened and less restrictive trade. Herb and spice trade continues as an important and unique worldwide activity.

Regional and international trade developed because people throughout the world crave flavor and utilize herbs and spices to enhance the enjoyment of their foods. Trade exists because, in many countries, domestic production is not possible and producing regions frequently do not coincide with areas of consumption. Worldwide immigration and travel also provide opportunities for many people to sample and acquire tastes for the many existing herb and spice products.

Production

Considerable amounts of herbs and spices for commerce and individual use are and likely will continue to be gathered from native or natural noncultivated sources. Much is also produced in home gardens. Accordingly, accurate production statistics are difficult to obtain. These herb and spice sources are often inadequate to meet total world or even regional needs. In some situations, intensive production is carried out to meet market demands. The great diversity of species and relatively little available information presents serious challenges to growers. However, some general agronomic principles for growing these crops are often similar to those of many other crop plants. Access to accurate modern technical information is often limited or guarded as proprietary information.

Propagation

Generally, annual and biennial herbs are grown from seed, whereas perennials more often are vegetatively propagated. However, vegetative propagation is not exclusive to perennials, and in many situations seed are used. Growers should use high-quality seed or vegetative materials that are true to type and free of disease or pest contamination.

Cultural and Pest Management

Information on the appropriate cultural needs of many herb and spices evolved largely from the experience and experimentation of growers; for some species, information remains inadequate. Large-scale commercial producers, because of the value of their production, can in some cases justify developing information specific for their production.

Herbs and spice plants are affected by many fungal, bacterial, and viral diseases as well as nematodes, which reduce quality and yield and, when serious, can prevent production. Diseases associated with herbs are often difficult to control because the informational base to properly identify and control the pathogens is limited. Because most herbs and spices are considered minor crops, approved pesticides are often unavailable. When pesticide usage is restricted, pest management relies on an integration of cultural and/or biological control practices. Some herb and spice species, because of their chemical constituents, may repel and/or exhibit resistance to some pests.

Harvesting, Postharvest, and Curing

Successful harvesting is dependent on timing, stage of plant development, and intended usage. Each species has an optimum harvest period that provides the highest amount and/or quality of its active principle.

Harvested plants should be healthy and clean. Physical damage must be avoided, as well as contact with free moisture, exposure to temperature extremes, and rapid desiccation. Accordingly, harvests are scheduled to coincide with the stage of plant development that will maximize yield and/or product quality. In some situations, plants may be intentionally stressed, as lush growth may dilute the concentration of active compounds.

Generally, fresh herb and spice products will have the strongest flavor and aroma. Quality is judged by the quantity and quality of essential oils or other constituents which are usually highest in the fresh product. Some commodities have unique harvest or postharvest handling peculiarities. Proper harvest and postharvest procedures in handling and holding these products can significantly slow the rate of quality decline.

In many situations, harvests are made as early in the day as possible after allowing for drying of dew. Large piles of foliage or other tissues, which might cause respiratory heating within the pile, should be avoided. Also avoided are contamination and exposure to free moisture that may encourage decay, and exposure to direct sunlight that will bleach, heat, and desiccate the product.

Some products are intentionally dried as a necessary procedure in their handling and processing. Slow drying at low temperatures and, if possible, in the shade is preferred to rapid, high-temperature drying to minimize losses of volatile essential oils.

Postharvest handling of dried herbs and spices is less complicated than that for the fresh product, but the general principles of careful handling, proper temperature, and relative humidity maintenance, and avoiding contamination still apply.

For many dried products, storage conditions should provide a cool, dark, low-moisture, and airtight environment. The use of moisture barriers and absorbers, and modified atmosphere packaging has increased because their use significantly increases quality retention. Preservation of herbs and spices are not limited to drying; some are frozen, salted, or pickled. Freeze drying is an excellent but expensive method, frequently used for products of high value.

CULINARY HERBS AND SPICES

The following listing is not inclusive, but includes species most representative of North American and western European usage. Some are also vegetable commodities for which additional information is available in appropriate chapters.

	Part used	Propagation	Growth habit
Allspice *Pimenta officinalis* MYRTACEAE	Fruit	Seed	Tree
Angelica *Angelica archangelica* APIACEAE	Stem, shoots, seed	Seed, veg.	Perennial
Anise *Pimpinella anisum* APIACEAE	Foliage, seed	Seed	Annual
Balm, Lemon *Melissa officinalis* LAMIACEAE	Foliage	Seed, veg.	Perennial
Basil, Sweet *Ocimum basilicum* LAMIACEAE	Foliage	Seed	Annual
Bay, Sweet *Laurus nobilis* LAURACEAE	Foliage	Veg., seed	Tree
Black Caraway (Burnet Saxifrage) *Pimpinella saxifraga* APIACEAE	Foliage	Seed	Perennial
Borage *Borago officinalis* BORAGINACEAE	Foliage, flower	Seed, veg.	Biennial
Burnet, Salad *Sanguisorba minor* ROSACEAE	Foliage	Seed, veg.	Perennial
Cacao *Theobroma cacao* BYTTNERIACEAE	Bean	Seed	Tree
Caper *Capparis spinosa* CAPPARACEAE	Flower buds	Veg., seed	Perennial
Caraway *Carum carvi* APIACEAE	Seed	Seed	Biennial
Cardamom *Elettaria cardamomum* ZINGIBERACEAE	Fruit	Veg., seed	Perennial
Celery *Apium graveolens* APIACEAE	Foliage, seed	Seed	Biennial
Chamomile *Anthemis nobilis* ASTERACEAE	Flower	Seed, veg.	Perennial
Chervil, Sweet *Anthriscus cerefolium* APIACEAE	Foliage, seed	Seed	Annual
Chili, Pepper *Capsicum annuum, Capsicum frutescens* SOLANACEAE	Fruit, seed	Seed	Annual

(continued)

	Part used	Propagation	Growth habit
Chives *Allium schoenoprasum* ALLIACEAE	Foliage	Seed, bulbs	Perennial
Cicely, Sweet *Myrrhis odorata* APIACEAE	Foliage	Seed, veg.	Perennial
Cinnamon *Cinnamomum zeylanicum* LAURACEAE	Bark, fruit	Veg., seed	Tree
Citron *Citrus medica* RUTACEAE	Fruit	Veg., seed	Tree
Cloves *Syzygium aromaticum* MYRTACEAE	Flower buds	Veg., seed	Tree
Coriander, Cilantro *Coriandum sativum* APIACEAE	Foliage, shoot, seed	Seed	Annual
Costmary *Chrysanthemum balsamita* ASTERACEAE	Leaves	Seed	Annual
Cress, Garden *Lepidium sativum* BRASSICACEAE	Foliage	Seed	Annual
Cumin *Cuminum cyminum* APIACEAE	Seed	Seed	Annual
Dill *Anethum graveolens* APIACEAE	Foliage, seed	Seed	Annual
Fennel *Foeniculum vulgare* APIACEAE	Stem, leaves, seed	Seed	Perennial
Fenugreek *Trigonella foenum-graecum* FABACEAE	Seed, foliage	Seed	Annual
Garlic *Allium sativum* ALLIACEAE	Cloves, leaves	Cloves	Annual
Ginger *Zingiber officinale* ZINGIBERACEAE	Rhizome	Veg.	Perennial
Hops *Humulus lupulus* CANNABACEAE	Stems. flowers	Veg., seed	Perennial
Horehound *Marrubium vulgare* LAMIACEAE	Leaf oil	Seed, veg.	Perennial
Horseradish *Armoracia rusticana* BRASSICACEAE	Rhizome	Veg.	Perennial
Horseradish, Japanese			

(continued)

	Part used	*Propagation*	*Growth habit*
Eutrema wasabi (Wasabi japonica) BRASSICACEAE	Rhizome	Rhizome	Perennial
Hyssop *Hyssopus officinalis* LAMIACEAE	Flower, foliage	Seed, veg.	Perennial
Juniper *Juniperus communis* CUPRESSACEAE	Fruit	Veg.	Tree
Lavender *Lavandula angustifolia* LAMIACEAE	Flower	Veg., seed	Perennial
Lemon grass *Cymbopogon citratus* POACEAE	Foliage	Seed	Annual
Lemon Verbena *Aloysia triphylla* VERBENACEAE	Foliage	Veg.	Perennial
Licorice *Glycyrrhiza glabra* FABACEAE	Root	Veg., seed	Perennial
Lovage *Levisticum officinale* APIACEAE	Foliage, seed	Seed, veg.	Perennial
Marigold *Calendula officinalis* ASTERACEAE	Flower	Seed, veg.	Annual
Marjoram, Sweet *Majorana hortensis* LAMIACEAE	Foliage, stems	Seed, veg.	Perennial
Mioga *Zingiber mioga* ZINGIBERACEAE	Flower bud stems	Rhizome	Perennial
Mitsuba *Cryptotaenia canadensis* APIACEAE	Leaves	Seed	Annual
Mugwort *Artemisia vulgaris* ASTERACEAE	Foliage	Veg., seed	Perennial
Mustard, Black *Brassica nigra* BRASSICACEAE	Foliage, seed	Seed	Annual
Mustard, White *Sinapis alba* BRASSICACEAE	Foliage, seed	Seed	Annual
Nasturtium *Tropaeolum majus* TROPAEOLACEAE	Foliage, seed	Seed	Annual
Nutmeg, Mace *Myristica fragrans* MYRISTICEAE	Nut, aril	Seed, veg.	Tree
Oregano *Origanum vulgare*	Foliage	Seed, veg.	Perennial

(continued)

	Part used	Propagation	Growth habit
LAMIACEAE			
Paprika *Capsicum annuum* SOLANACEAE	Fruit	Seed	Annual
Parsley *Petroselinum crispum* APIACEAE	Foliage, seed	Seed	Biennial
Pepper, Black, White *Piper nigrum* PIPERACEAE	Fruit	Veg., seed	Perennial
Peppermint *Mentha piperita* LAMIACEAE	Foliage	Veg.	Perennial
Poppy *Papaver rhoeas* PAPAVERACEAE	Seed	Seed	Annual
Rocket *Eruca vesicaria* sub sp. *sativa* BRASSICACEAE	Foliage	Seed	Annual
Rosemary *Rosmarinus officinalis* LABIATAE	Foliage, flower, seed	Seed, veg.	Perennial
Rue *Ruta graveolens* RUTACEAE	Foliage	Veg., seed	Perennial
Saffron *Crocus sativus* IRIDACEAE	Flower	Veg.	Perennial
Sage *Salvia officinalis* LAMIACEAE	Foliage, seed	Seed, veg.	Perennial
Savory, Summer *Satureja hortensis* LAMIACEAE	Foliage	Seed	Annual
Savory, Winter *Satureja montana* LAMIACEAE	Foliage	Veg.	Perennial
Sesame *Sesamum indicum* PEDALIACEAE	Seed	Seed	Annual
Shiso, Perilla *Perilla frutescens* LAMIACEAE	Leaves, seed	Seed	Annual
Sorrel, French *Rumex scutatus* POLYGONACEAE	Foliage	Seed	Perennial
Sorrel *Rumex acetosa* POLYGONACEAE	Foliage	Seed	Perennial
Spearmint *Mentha spicata* LAMIACEAE	Foliage, oil	Veg.	Perennial

(continued)

	Part used	Propagation	Growth habit
Star Anise *Illicium verum* MAGNOLIACEAE	Seed	Veg., seed	Tree
Tamarind *Tamarindus indica* FABACEAE	Seed	Seed	Tree
Tansy *Tanacetum vulgare* ASTERACEAE	Foliage	Seed, veg.	Perennial
Terragon *Artemisia dracunculus* ASTERACEAE	Foliage	Veg., seed	Perennial
Thyme, Garden or Lemon *Thymus vulgaris* LAMIACEAE	Foliage	Veg.	Perennial
Turmeric *Curcuma longa* ZINGIBERACEAE	Rhizome	Veg.	Perennial
Vanilla *Vanilla planifolia* ORCHIDACEAE	Bean pods	Veg.	Perennial
Zedoary *Curcuma zedoaria* ZINGIBERACEAE	Rhizome	Veg.	Perennial

POISONOUS HERBS

It is necessary to indicate that caution be taken with regard to the consumption of some herbs and spices. Depending on the amount consumed, some herb and spice species pose a health risk because of toxic effects. Whether for culinary or medical use, uncontrolled consumption must be avoided. *Herbs that are extremely toxic and should not be consumed include the following:*

Species	*Common name(s)*
Acorus calamus	Sweet root/Calamus
Aesculus hippocastanum	Buckeye/Horse Chestnut
Aristolochia serpentaria	Snakeweed
Arnica montana	Wolf's-bane/Mountain tobacco/ Arnica
Artemisia absinthium	Wormwood/Absinthe
Atropa belladonna	Belladonna
Chrysanthemum vulgare	Tansy
Cinicifuga racemosa	Black cohosh
Colchicum autumnale	Autumn crocus

Conium maculatum	Hemlock
Convallaria majalis	Lily of the Valley
Corynanthe yohimbi	Yohimbine
Cytisus scoparius	Scotch Broom
Datura stramonium	Jimson weed/Datura
Digitalis purpurea	Foxglove
Euonymus autropurpurea	Wahoo Bark/Euonymus
Euonymus euroaeus	Spindle-tree
Eupatorium rugosum	White Snakeroot
Galium odoratum	Sweet woodruff
Gaultheria procumbens	Wintergreen
Heliotropium europaeum	Heliotrope
Hyoscyamus niger	Henbane
Hypericum perforatum	St. Johnswort
Ipomoea purpurea	Morning Glory
Kalmia latifolia	Mountain Laurel
Lobelia inflata	Indian Tobacco/Lobelia
Mandragora officinarum	Mandrake
Mentha pulegium	Pennyroyal
Phoradendron flavescens	Mistletoe
Podophyllum pelatum	May apple
Prunus laurocerasus	Cherry laurel
Ricinus communis	Castor bean
Ruta graveolens	Rue
Sanguinaria canadensis	Bloodroot
Solanum dulcamara	Bittersweet/Dulcamara
Symphytum peregrinum	Comfrey
Vinca major, V. minor	Periwinkle/Vinca

SELECTED REFERENCES

Ashurst, P.R., ed. 1995, Food Flavoring, 2nd ed. Blackie Academic and Profession/Chapman & Hall, Glasgow.

Duke, J.A. 1985. CRC Handbook of Medicinal Herbs. CRC Press, Boca Raton, FL.

Craker, L.E., and Simon, J.E., eds. 1988. Herbs, Spices, and Medicinal Plants—Recent Advances in Botany, Horticulture and Pharmacology. Oryx Press, Phoenix, AZ.

Hayes, E.S. 1961. Spices and Herbs Around the World. Doubleday Co., Inc., Garden City, NY.

Purseglove, J.W., Brown, E.G., Green, C.L., and Robbins, S.R.J., eds. 1981. Spices, Vols. 1 and 2. Longman Group Ltd., New York.

Appendix A

Conversion Equilivents From Metric To U.S. Units

Length
1 cm = 0.394 inch
1 m = 3.28 feet
1 km = 0.621 mile

Area
1 m^2 = 10.76 feet^2
1 ha = 2.47 acres

Weight
1 g = 0.0353 ounce
1 kg = 2.205 pounds
1 t = 1.102 tons

Volume (fluid)
1 ml = 0.033 ounce
1 liter = 0.264 gallon

Yield
1 kg/ha = 0.893 pound/acre
1 t/ha = 0.446 tons/acre

Temperature

°C	°F	°C	°F	°C	°F	°C	°F	°C	°F	°C	°F
−20	−4.0	1	33.8	11	51.8	21	69.8	31	87.8	41	105.8
−15	5.0	2	35.6	12	53.6	22	71.6	32	89.6	42	107.6
−10	14.0	3	37.4	13	55.4	23	71.6	33	91.4	43	109.4
−5	23.0	4	39.2	14	57.2	24	75.2	34	93.2	44	111.2
0	32.0	5	41.0	15	59.0	25	77.0	35	95.0	45	113.0
		6	42.8	16	60.8	26	78.8	36	96.8	46	114.8
		7	44.6	17	62.6	27	80.6	37	98.6	47	116.6
		8	46.4	18	64.4	28	82.4	38	100.4	48	118.4
		9	48.2	19	66.2	29	84.2	39	102.2	49	120.2
		10	50.0	20	68.0	30	86.0	40	104.0	50	122.0
										100	212.0

For other conversions the following formulas can be used:

$$°F = 9 \div 5°C + 32 \qquad \text{and} \qquad °C = 5 \div 9 \ (°F\text{–}32)$$

Appendix B

VEGETABLE CLASSIFICATION

I-A Fungi

Classification and scientific name	Common name(s)[a]	Edible part(s)	Regions/countries grown
AGARICACEAE *Agaricus bisporus*	Button mushroom F: champignon S: seta C: yang gu J: mashirumu, hara take	Fruiting body	Most temperate regions
Agaricus campestris	Meadow mushroom	Fruiting body	Most temperate regions
AURICULARIACEAE *Auricularia polytricha*	Jew's ear mushroom C: mu er, mo er J: kikurage	Fruiting body	China, Taiwan
CANTHARELLACEAE *Cantharellus cibarius*	Chanterelle mushroom	Fruiting body	N. California, Pacific Northwest

COPRINACEAE *Coprinus comatus*	Shaggy mane mushroom	Fruiting body	Temperate Northern Hemi-sphere
HERICIACEAE *Hericium coralloides* *Hericium erinaceus*	Monkey's head mushroom Bear's head mushroom, Medusa fungi	Fruiting body Fruting body	Eastern N. America Eastern N. America
MORCHELLACEAE *Morchella hortensis* *M. esculenta*, and other *Morchella* spp.	Morel mushrooms	Fruiting body	Many temperate regions
PLUTEACEAE *Volvariella volvacea*	Straw/paddy/Chinese mushroom C: cao gu J: fukuro take	Fruiting body	S.E. Asia
STROPHARIACEAE *Pholiota nameko*	Nameko mushroom J: namekotake	Fruiting body	Japan
Stropharia rugoso annulata	Stropharia mushroom	Fruiting body	Eastern Europe
TREMELLACEAE *Tremella fuciformis*	White jelly/silver ear mushroom	Fruiting body	Warmer parts of the world, Orient
TRICHOLOMATACEAE *Flammulina velutipes*	Winter/enoki mushroom C: jin tsen gu J: enoki take	Fruiting body	Japan, Taiwan
Lentinula edodes	Shiitake/Black Forest mushroom C: xiang gu, shan ku J: shiitake	Fruiting body	Japan, China, Taiwan, many temperate regions
Pleurotus atus	Oyster mushroom C: how gu J: hiratake	Fruiting body	China, Taiwan, many areas
Pleurotus citrinopiletus *Pleurotus cystidiosus*	White oyster mushroom Abalone mushroom C: bao yu gu	Fruiting body Fruiting body	China, Taiwan, many areas China, Taiwan, many areas

(continued)

APPENDIX B *(cont.)*

Classification and scientific name	Common name(s)[a]	Edible part(s)	Regions/countries grown
Pleurotus safor-cajou *Pleurotus sapidus* *Tricholoma matsutake*	Gray oyster mushroom Black oyster mushroom Pine mushroom: C: song er J: matsutake	Fruiting body Fruiting body Fruiting body	China, Taiwan, many areas China, Taiwan, many areas Temperate pine forest
TUBERACEAE *Tuber melanosporum* and other *Tuber* spp.	Black truffle	Fruiting body	Spain, Portugal, and Baltic region
USTILAGINACEAE *Ustilago maydis*	Corn smut, coche, cuitla, huitla-coche, Mexican truffle	Fruiting body	Mexico
I-B ALGAE			
ALARIACEAE *Undaria pinnatifida*	Wakame	Blade lamina	Japan
BANGIACEAE *Porphyra ambilicalis* *Porphyra tenera*	Laver J: chishima kuro nori Nori J: asakusa nori	Blade lamina Blade lamina	Britain, Japan, Iceland Japan
LAMINARIACEAE *Laminaria japonica*	Makombu	Blade lamina	Japan
MONOSTROMATACEAE *Monostroma latissimum*	Awo nori	Blade lamina	Japan
SARGASSACEAE *Hizikia fusiformis*	Hijiki	Segmented blades	Japan
ULVACEAE *Ulva lactuca*	Sea lettuce, sea grass J: aosa	Blade lamina	Britain, Japan, Iceland

II FERNS

BLENCHNACEAE *Blechnum capense*	Loplop	Immature frond	Southern Hemisphere, New Zealand
OSMUNDACEAE *Osmunda japonica*	Japanese flowering fern J: zen mai	Immature frond	Japan
POLYPODIACEAE *Pteridium aquilinum*	Braken/brake fern C: chu en J: warabi	Immature frond	China, Japan, France, Canada

III GYMNOSPERM

GINKGOACEAE *Ginkgo biloba*	Ginkgo J: ichiyo, gin nan	Mature fruit (nut)	East Asia, China, Japan

IV ANGIOSPERM
I-A *MONOCOTYLEDON*

AGAVACEAE *Cordyline australis* *Cordyline indivisa* *Yucca elephantipes*	Cabbage tree, palm lily Blue dracaena Spanish bayonet, ozote	Young leaves Immature leaves Flowers, tender shoots	New Zealand New Zealand Mexico, Guatemala, Central America
ALISMATACEAE *Sagittaria sagittifolia*	Arrowhead, swamp potato F: flechiere S: saeta de agua C: ci gu J: kuwai	Corm	Warm temperate, subtropical, and tropical Asia
ALLIACEAE (AMARYLLIDACEAE) *Allium ampeloprasum* var. *aegypticum*	Kurrat F: kurrat J: kurato	Leaves	Egypt, Eastern Mediterranean

(continued)

APPENDIX B (cont.)

Classification and scientific name	Common name(s)[a]	Edible part(s)	Regions/countries grown
Allium ampeloprasum var. *holmense*	Great headed garlic F: ail d' orient S: puerro agreste C: da tou suan J: gureto hedo gariku	Bulb, leaves	Temperate regions
Allium ampeloprasum var. *porrum*	Leek F: poireau S: puerro, ajo porro C: jiu cong J: riki	Pseudostem, leaves	Temperate regions
Allium ampeloprasum var. *sectivum*	Pearl onion	Small daughter bulbs	Temperate regions
Allium cepa	Onion F: oignon S: cebolla C: yang cong J: tama negi	Bulb, leaves	Temperate and subtropical regions
Allium cepa var. *aggregatum*	Multiplier/potato/ever-ready onion	Bulb, leaves	Temperate regions
Allium cepa var. *ascalonicum*	Shallot F: echalote S: chalote C: feng cong, fen nie yang cong J: shiyaroto	Bulb, leaves	Temperate regions
Allium cepa var. *perutile*	Ever-ready onion	Bulb, leaves	Temperate regions
Allium cepa var. *solanina*	Potato onion	Bulb, leaves	Temperate regions
Allium cepa var. *viviparum*	Topset onion	Bulbils, leaves	Orient
Allium cepa var. *bulbiforum*	Tree onion	Bulbils, leaves	
Allium cepa var. *proliferum*	Egyptian/topset onion F: oignon d' Egypte S: cebolla del Egipto C: ding sheng yang cong J: kitsune negi	Bulbils, leaves	
Allium cepa var. *wakegi*	Wakegi onion F: ciboule, oignon d'hiver S: cebolleta C: da gaun cong J: wakegi	Leaves	S. Korea, Taiwan, Japan

Allium chinense	Rakkyo, Baker's garlic F: echalotte chinesa S: chalota chinesa C: jiao tou, chiao tou J: rakkyo	Bulb	Temperate China and Japan
Allium fistulosum	Japanese bunching onion, Welsh onion F: ciboulette S: cebolleta C: da cong J: negi	Pseudostem, leaves	Temperate China and Japan
Allium sativum	Garlic F: ail ordinaire S: ajo vulgar C: da suan, ta suan J: ninniku	Cloves, leaves	Temperate and subtropical regions
Allium schoenoprasum	Chive F: ciboulette, civette S: cebollino C: xi xiang cong J: asatsuki	Leaves	Temperate regions
Allium tuberosum	Chinese chive, or leek F: ail chinoise S: cive chino C: jiu, kau tsoi J: nira	Leaves, young scapes	China, S.E. Asia
AMARYLLIDACEAE (see **ALLIACEAE**)			
ARACEAE			
Acorus calamus	Sweet flag	Rhizome	India
Alocasia macrorrhiza	Giant taro, alocasia C: hai yu J: doku imo, kuwazu imo	Corm, immature leaves, and petioles	S.E. Asia
Amorphophallus campanulatus	Elephant foot yam	Corm	S.E. Asia, India
Amorphophallus rivieri var. *konjak*	Konjak F: kouniak d'annam S: patata de Telinga C: mo yu J: konniyaku	Corm	Japan, China

(continued)

Classification and scientific name	Common name(s)[a]	Edible part(s)	Regions/countries grown
Colocasia esculenta	Taro, dasheen, cocoyam F: taro, aronille S: colocasca C: yu, yu tao J: sato imo	Corm, immature leaves, and petioles	Warm temperate and sub-tropical regions
Cyrtosperma chamissonia	Giant swamp taro F: faux taro S: falso taro J: te babai	Corm	S.E. Asia, Pacific Islands
Xanthosoma brasiliense	Tanier spinach, belembe F: calalou S: blembe	Immature leaves	Subtropical and tropical America, Brazil
Xanthosoma sagittifolium	Tannia, yautia, new cocoyam F: yautia des anglo S: malanga C: ya yu J: yochiya	Corm	Subtropical and tropical regions
ARECACEAE (PALMAE) *Bractris gasipaes*	Pejibaye/peach palm	Apical shoot tip, leaves, ripe fruit	Ecuador, Colombia
Euterpe edulis	Assai palm, palm cabbage, palmito Brazil: palmito	Apical shoot tip	Subtropical Brazil
Metroxylon rumphii (M. bougainvillii)	Sago plam	Starchy inner stem	New Guinea, South Pacific Islands
Sabal palmetto	Swamp cabbage, heart of palm	Apical shoot tip	Caribbean
CANNACEAE *Canna edulis*	Edible canna, Queensland arrowroot F: achira, canne d'Inde S: achera C: shi yong mei ren jiao J: shokuyou kana	Rhizome	Subtropical and tropical regions
CYPERACEAE *Cyperus esculentus*	Yellow nut grass	Tuber	W. Asia and Africa

Species	Common names	Type	Distribution
Cyperus esculentus var. *sativus*	Chufa/tiger/rush nut F: amande de terre S: chufa C: yang du li	Tuber	Tropics, N. Africa, Spain
Eleocharis dulcis	Chinese water chestnut, matai F: chataigne d'eau S: cabezas de negrito C: pi chi, bi qi J: shina kuro kuwai, okuro kuwai	Corm	Subtropical and tropical Asia
DIOSCOREACEAE *Dioscorea alata*	Greater/water/white yam F: igname de chine S: name de agua C: da shu J: daisho, oyama imo S.E. Asia: ubi	Tuber	Subtropical and tropical Asia, Africa, S. America
Dioscorea bulbifera	Aerial/potato yam F: igname pousse en l'air S: name del aire C: huang du	Tuber	Africa, Asia
Dioscorea cayenensis	Yellow yam F: igname de la Guinee S: affoo, mane amarollo C: fei zhou shan yao	Tuber	Africa
Dioscorea dumetorum	Cluster/bitter/trifoliate yam F: igname sauvage S: name amargo	Tuber	Africa
Dioscorea esculenta	Lesser/Chinese/potato yam F: igname des blancs S: name papa C: tian shu J: ama yama imo	Tuber	Asia
Dioscorea hispida	Intoxicating/starch yam	Tuber	India, S.E. Asia
Dioscorea japonica	Japanese yam J: jinenjo	Tuber	Japan

(continued)

APPENDIX B *(cont.)*

Classification and scientific name	Common name(s)[a]	Edible part(s)	Regions/countries grown
Dioscorea opposita	Chinese/cinnamon yam F: igname de Chine S: name de le China C: shan yao J: naga imo, yama no imo	Tuber	Temperate and subtropical China, Japan
Dioscorea rotundata	White/white guinea yam F: igname de la Guinee S: name Guieno blanco C: yuan xing shu yu	Tuber	Africa
Dioscorea trifida	Cush cush/Indian yam F: igname couche-couche S: name morado	Tuber	Tropical S. America
GRAMINEAE (see POACEAE)			
LILIACEAE *Asparagus officinalis*	Asparagus F: asperge S: esparrago C: shi diao bai J: asuparagasu, matsuba undo	Emerging shoot	Temperate and subtropical regions
Lillium spp.	Lily F: lis S: azucena, lirio C: bai he J: yuri ne, yama oni yuri	Bulb	China, Japan
MARANTACEAE *Calathea allouia*	Leren, calathea, bamboo tuber, sweet corn root F: topi, alleluia S: tupinambur, alcluia	Tuberous roots	Caribbean, tropical America, India, S.E. Asia
Maranta arundinacea	Arrowroot, West Indian arrowroot F: marante, enforme de bambou S: maranta, cara maco C: zhu yu J: kuzuukon	Rhizome	West Indies, northern S. America

MUSACEAE			
Ensete venticosum	Abyssinian banana	Fruit, starchy corm, inner pseudostem	Ethiopia
Musa acuminata	Plantain/starchy banana F: plantain S: iianten J: mibashiyo banana	Fruit, male flower blossom	Subtropical and tropical regions
Musa balbisiana	Plantain/starchy banana F: plantain S: platano de cocinor J: ryori banana	Fruit	Subtropical and tropical regions
PALMAE (see ARECACEAE)			
POACEAE (GRAMINEAE)			
Bambusa beecheyana	Bamboo shoot J: daisan chiku, takenoko	Immerging shoot	Subtropical regions, Japan
Dendrocalmus latiflorus	Bamboo shoot J: machiku, takenoko	Immerging shoot	Warm temperate and subtropical regions
Phyllostachys edulis also *P. bambusoides* *P. dulcis* *P. nuda* *P. pubescens* *P. viridis*	Bamboo shoot C: mao zu shun J: takenoko, mouso uchiku	Immerging shoot	Warm temperate and subtropical Asia
Sasa spp.	Bamboo shoot J: takenoko	Immerging shoot	Warm temperate regions and Japan
Zea mays var. sacchorata	Sweet corn F: mais sucre S: maiz dulce C: tian yu mi J: toumorokoshi, suito kon	Immature seed, immature ear (cob)	Temperate, subtropical and tropical regions
Zizania latifolia	Water bamboo, Manchurian wild rice F: viz sauvage S: arroz silvestre C: gau sun, jiao bai J: makomo take	Swollen, emerging "flower" stalk	China, Japan, Taiwan, other temperate and subtropical regions

(continued)

APPENDIX B (cont.)

Classification and scientific name	Common name(s)[a]	Edible part(s)	Regions/countries grown
TACCACEAE *Tacca leontopetaloides*	East Indian arrowroot, tacca F: hydropire S: yabia C: ju ruo shu	Starchy tuber (rhizome)	Tropical Asia, Oceania
Tacca pinnatifida	Polynesian arrowroot	Starchy tuber (rhizome)	S.E. Asia
TYPHACEAE *Typha latifolia*	Cat-tail, bull rush, Cossack asparagus C: pu tsai J: gama	Rhizome	Temperate regions
ZINGERBERACEAE *Curcuma zedoaria*	Shoti, zedoary	Starchy rhizomes, tender shoots	N.E. India, Sri Lanka, S.E. Asia
Zingiber mioga	Japanese giner, mioga C: hsiang ho J: mioga, miyoga dake	Immature flower bud, tender stems	Temperate Japan
Zingiber officinale	Ginger F: amome S: jengibre C: jiang J: shioga	Rhizome	Subtropical and tropical re- gions
IV-B DICOTYLEDON			
AIZOACEAE *Mesembranthemun crystallinum*	Ice plant F: ficoide glaciale S: escarhosa C: bing hua J: tsurana	Leaves, tender shoots	Temperate regions, United States, S. Africa, Greece

AMARANTHACEAE
Amaranthus tricolor other sp. include:
  A. caudatus
  A. blitum
  A. dubius
  A. gangeticus
  A. hybridus
	Edible amaranth, tampala, Chi-
	nese spinach
	F: amaranthe, amarante
	S: amarantos
	C: xian cai, lao quiang gu
	J: yasai hiyu, hage ito
		Leaves, shoots	Most temperate regions

Amaranthus lividus	Bondue	Leaves, shoots	Tropical Africa
Amaranthus spinosus	Uray	Leaves, shoots	Philippines
Celosia argentea	Cock's comb F: crete de cog, celosie S: cresta de gallo C: ji guan hua J: keito	Leaves, tender shoots	Tropical regions and S.E. Asia
APIACEAE (UMBELLIFEREAE)			
Anethum graveolens	Dill	Foliage, seed	Temperate Europe, N. America
Angelica archangelica	Angelica F: angelique S: angelica C: bai zhi	Root, stem, seed oil	Southern Europe
Anthriscus cerefolium	Chervil, salad chervil F: cerfeuil S: perifoleo, cerefoleo G: kerbel C: xi ye xinaga cai	Leaves	S.E. Russia, W. Asia and temperate regions
Apium graveolens var. *dulce*	Celery F: celeri a cotes S: apio C: qin cai J: seruri, oranda mitsuba	Petioles, leaves	Temperate regions
Apium graveolens var. *rapaceum*	Celeriac, celery root knob/turnip celery F: celeri-rave S: apio nabo C: gen qin cai J: seruri atsuku	Swollen hypocotyl and root, leaves	Temperate Europe, N. America
Apium graveolens var. *secalinum*	Smallage, leaf celery	Leaves	Asia, Mediterranean region
Arracacia xanthorrhiza	Arracacha, Peruvian carrot F: arracacha, pane me, pomme de terre celeri S: arracacha, zanahoria J: imo zeri P: mandioquina-salsa	Root, leaves	Subtropical S. America, Brazil
Carum carvi	Caraway	Seed, leaves, stem, root	Europe, W. Asia

(continued)

Classification and scientific name	Common name(s)[a]	Edible part(s)	Regions/countries grown
Centella asiatica	Asiatic/Indian pennywort F: ecuelle d'eau S: hierba de clavo C: ji xue cao J: gotukora, tsubo kusa	Leaves	Sri Lanka, S.E. Asia
Chaeropyllum bulbosum	Turnip-rooted chervil F: cerfeuil tubereux S: perifollo tuberoso G: knolliger C: pu ching J: yama ningin	Tuberous root, leaves	Temperate Europe
Coriandrum sativum	Coriander, Chinese parsley, cilantro F: coriandre S: cilantro C: yuan sui J: koendoro	Leaves, seed	Temperate regions
Cryptotaenia japonica	Japanese honewort F: mitsuba C: ya er chin J: mitsuba	Leaves	Temperate China, Japan, Korea
Cumin cyminum	Cuminum	Seed	Mediterranean
Daucus carota var. *sativa*	Carrot F: carrotte S: zanahoria C: hu luo bo J: ninjin	Root, leaves	Most temperate regions
Foeniculum vulgare var. *dulce*	Florence fennel F: fenouil S: hinojo I: finocchio C: hui xiang J: uikyo	Petiole base, leaves	Temperate Europe, N. America
Levisticum officinale	Lovage	Leaves, root, seed	Mediterranean region
Myrrhis odorata	Sweet cicely, myrrh F: cerfeuil musque S: perifollo oloroso C: ou zhou mo yao	Root, seed	Europe

Oenanthe javanica	Water dropwort, oriental celery F: oenanthe S: cicuta C: shui chin J: seri, kawa na	Leaves	Temperate and subtropical regions, Japan, S.E. Asia
Pastinaca sativa	Parsnip F: panais S: pastinaca. chirivia G: pastinak C: mei guo fang feng J: pasunipu, shiro ninjin	Root	Temperate Europe, N. America
Perideridia gairdneri	Squaw root, yampah, epos root	Storage root	United States; Oregon, N. California
Petroselinum crispum	Parsley F: persil S: perejil G: petersilie C: xiang qing cai J: paseri, oranda seri	Leaves	Temperate regions
Petroselinum crispum var. *tuberosum*	Hamburg, turnip rooted parsley, petrouska F: persil tubereux S: perejil tuberoso G: knollen petersillie C: gen xiang qin J: ne paseri	Tuberous root, leaves	Europe
Pimpinella anisum *Sium sisarum*	Anise Skirret F: berle á sucre S: escaravia C: shen qin	Seed, leaves Tuberous root	Mediterranean region E. Asia
ARALIACEAE *Aralia cordata*	Udo, spikenard F: aralia cordiforme S: aralia, zazaparilha C: tu dang qui J: udo	Tender shoot	Japan
Panax ginseng	Chinese ginseng	Root	Temperate Asia, Korea

(continued)

APPENDIX B (cont.)

Classification and scientific name	Common name(s)[a]	Edible part(s)	Regions/countries grown
Panax quiquefolius	American ginseng	Root	Central and N.E. United States, Canada
ASTERACEAE (COMPOSITAE)			
Arctium lappa var. *edule*	Edible burdock F: grande burdane S: bardana mayor C: ci cai ji, niou pang J: gobo	Root	Temperate regions
Artemisia lactiflora	White mugwort F: armoise S: ajenjo C: ai J: yomogina	Leaves	China
Chrysanthemum coronarium	Garland chrysanthemum F: chrysantheme S: crisantemo, margariata C: tong hao, ou tong hao J: shungiku, kikuna	Leaves, tender shoots, flower buds	Temperate China, Japan
Chrysanthemum morifolium	Edible chrysanthemum J: shokuyo giku	Flower	Japan
Cichorium endivia	Endive, escarole F: endiva, chicoree S: endivia, escarola I: indivia-cicoria, escarola C: ku ju, ku qu J: endaibu, kikujisha	Leaves	Temperate regions
Cichorium intybus	Chicory, leaf chicory F: Chicoree sauvage S: achicoria I: cicoria C: ju ju, ye ku qu J: kiku ngana, chikori	Leaves, root	Temperate regions
Cichorium intybus	Belgium/French endive, witloof chicory F: chicoree a feuilles vertes S: achicoria silvestre D: witloof	Forced apical bud	Temperate Europe

Cosmos caudatus	Cosmos	Leaves	United States, Mexico
Crassocephalum crepidodes	Sierra Leone bologi	Leaves, shoots	W. Africa
Cynara cardunculus	Cardoon F: cardon S: cardo comestible I: cardone C: ci cai ji, shi yong ji J: karudon	Petiole	Italy, temperate regions
Cynara scolymus	Globe artichoke F: artichaut I: carciofo S: alachofa C: chao xian ji J: achichiyoku	Immature flower bud	Italy, Mediterranean region, United States and some temperate countries
Gynura bicolor	Gynura F: gynura a' deux couleurs C: zi tien kui, bae tian J: suizen jina	Shoot	China
Helianthus tuberosus	Jerusalem artichoke, girasole, sunchoke F: topinambour S: topinambo, aguaturma C: ju yu J: kiku imo	Tuber	N. America, other temperate regions
Lactuca indica	Indian lettuce F: laitue d'Inde S: lechuga dela India C: shan wo ju J: aki no nogeshi	Leaves	Asia, China, Japan, S.E. Asia
Lactuca sativa var. asparagina	Celtuce, asparagus lettuce F: laitue asperge S: lechuga esparrago C: wo ju sun J: kuki chisha	Stem, young leaves	China, temperate regions
Lactuca sativa	Lettuce F: laitue S: lechuga I: lattuga G: kopf salat	Leaves	Temperate regions

(continued)

Classification and scientific name	Common name(s)[a]	Edible part(s)	Regions/countries grown
Lactuca sativa (Con't)			
Petasites japonicus	C: wo ju J: chisha, retasu Japanese butterbur F: petasites japonais S: farfara japonesa C: kuan dong, feng dou cai J: fuki	Petiole	Japan, Orient
Polymnia sonchifolia	Yacon, earth apple F: poir de terre Cochet S: yacon, jacon, arboloco J: yakon	Tuberous root	S. American Andes
Scorzonera hispanica	Scorzonera, black salsify F: scorzonere S: escorzonera, salsifi negro C: hei pi ho luo men shen J: kiku gobo	Root, blanched leaves	Southern Europe
Sonchus oleraceus *Spilanthes oleracea* *Taraxacum offinale*	Sowthistle Paragrass Indonesian: jotang Dandelion greens F: dent de lion S: diente de leon C: pu cong ying J: shokuyo tanpopo	Leaves Leaves Leaves	Temperate regions Southeast Asia Northern temperate regions
Tragopogon porrifolius	Salsify, vegetable oyster F: salsifi blanc S: salsifi blanco C: po luo meng shen J: sarushifuai, baramonjin	Root	Temperate Europe
BASELLACEAE *Basella alba*	Malabar spinach F: espinard de Malabar S: espinaca basela C: bai luo kui	Leaves, tender shoots	Subtropical and tropical regions
Basella rubra	J: tsuru murasaki Ceylon/Indian spinach (Other names as above)	Leaves, tender shoots	Subtropical and tropical regions
Ulluacus tuberosus	Ulluco	Tuber	Andean region

BORAGINACEAE			
Borago officinalis	Borage F: bourrache S: borraja C: bo li qu J: ruri jisa	Young leaves, petioles	Temperate Europe
Symphytum officinale	Comfrey F: consoule officinale S: consuelda mayor	Leaves, tender shoots	Europe
BRASSICACEAE (CRUCIFERAE)			
Armoracia rusticana	Horseradish F: raifort sauvage S: termayo G: meerrettig, kran C: hsi yang, shan yu tsai J: wasabi daikon	Root, leaves, sprouted seed	Temperate Europe, N. America
Barbarea verna	Upland/spring cress F: cresson de terre S: hierba de Santa Barbara C: mei kuo shan chieh J: kibana kuresu	Leaves	Temperate Europe, N. America
Barbarea vulgaris	Winter cress	Leaves, tender shoots	Temperate regions
Brassica carinata	Ethiopian/Abyssinian mustard F: moutarde de l'Ethiopie S: mostaza de Abisinia	Leaves	Temperate and subtropical regions
Brassica juncea	Indian mustard F: moutarde S: mostaza C: jie cai, jia tsai lai J: karashina, takana	Leaves	Temperate and subtropical regions
Brassica juncea var. *japonica*	Mizuna, kyona	Leaves	Temperate regions, Japan
Brassica juncea var. *rugosa*	Large/leaf brown/Indian mustard F: chou faux jone S: mostaza de la China C: gai choi, kai choi J: taniku takana	Leaves	Temperate and subtropical regions

(continued)

Classification and scientific name	Common name(s)[a]	Edible part(s)	Regions/countries grown
Brassica napus var. *napobrassica*	Rutabaga, Swede turnip F: chou-navet, navet de Suede G: kohlrube C: fu qin gan lan S: colinabo J: sueden kabu, rutabaga	Enlarged root	Temperate regions
Brassica napus var. *pabularia*	Siberian kale, Hanover salad F: chou de Siberie S: col de la Siberia C: xi bai li ya yu yi gan lan	Leaves	Temperate regions
Brassica nigra	Black mustard F: moutarde noir S: mostaza negra C: hei jie	Leaves, seed	Temperate regions
Brassica oleracea gp. Acephala	Kale, collard F: chou vert S: col crepa G: blatterkohl C: yu yi gan lan J: kuro garashi, nikera	Leaves	Temperate regions
Brassica oleracea gp. Alboglabra	Chinese kale, Chinese broccoli, kailan F: kailan C: kai lan, jie lan J: kanran chiyo J: kyokuyo kanran	Flowering stalk	China, S.E. Asia
Brassica oleracea gp. Botrytis	Cauliflower F: chou fleur S: coliflor G: blumenkohl C: hua ye cai J: karifurawa, hana yasai	Immature infloresence	Temperate and some sub-tropical regions, India
Brassica oleracea gp. Capitata	Cabbage F: chou S: repollo G: kraut C: kan lan, jie qiou gan lan J: kiyabetsu, tamana	Leaves (head)	Temperate and some sub-tropical regions

Species	Common names	Part used	Distribution
Brassica oleracea gp. Gemmifera	Brussels sprouts F: choux de Bruxelles S: col de Bruselas G: rosenkohl C: bao zi gan lan J: me kiyabetsu	Axillary buds	Temperate Europe and some temperate regions
Brassica oleracea gp. Gongylodes	Kohlrabi F: chou-rave S: col rabano G: oberkohlrabi C: qiu jing gan lan J: korurabi, kyukei kanran	Swollen stem	Temperate regions
Brassica oleracea gp. Italica	Broccoli, sprouting broccoli F: chou brocoli S: brocoli C: jing ye cai, chin hua tsai J: burokori, midori hana yasai	Immature floral parts, tender stems and leaves	Temperate regions, some subtropical regions
Brassica oleracea gp. Tronchuda	Portuguese cabbage	Leaves, stems	Portugal, Spain
Brassica rapa var. *chinensis*	Chinese mustard, pak choi F: pak choi S: pak choi C: xiao bai choi J: taisai, taina	Leaves	Temperate and subtropical regions
Brassica rapa var. *oleifera*	Turnip rape	Leaves, sprouted seeds	Temperate regions
Brassica rapa var. *parachinesis*	Mock/flowering pak choi, choi sum F: choy sum S: choy sum C: cai tai, cai xin J: saishin, shiroguki taisai	Leaves, stems	S.E. Asia
Brassica rapa var. *pekinensis*	Chinese cabbage, Chinese cabbage, pe-tsai F: chou de Chine S: col de Chino C: da bai cai J: hakusai	Leaves (head)	Temperate and subtropical regions
Brassica rapa var. *perviridis*	Tendergreen mustard F: moutarde epinard S: mostaza espinaca J: komatsuna	Leaves	Temperate regions

(continued)

Classification and scientific name	Common name(s)[a]	Edible part(s)	Regions/countries grown
Brassica rapa var. *rapifera*	Turnip F: navet S: nabo C: wu qin, wu ching J: kabu	Swollen root, leaves	Temperate regions
Brassica rapa var. *ruvo*	Brocoli raab, rapini F: brocolis de raves S: nabo de brotes I: cima di rapa	Leaves	Temperate Europe, Italy
Brassica rapa var. *septiceps*	Seven top turnip F: raves de feuille S: nabizas	Leaves, stems	Temperate regions
Crambe maritima	Sea kale, crambe F: chou marin S: col marina I: cavolo marino C: bin cai J: hamana	Leaves, stems	Temperate Europe
Eruca sativa	Rocket salad F: roquette S: roqueta I: eruca C: huo chien seng tsai, zhi ma cai J: roketto, kibana suzushiro	Leaves	Temperate Europe
Eutrema wasabi	Japanese horseradish, wasabi F: raifort japonais S: rabano japones C: shanyu tsai, ahan kui J: wasabi	Rhizome	Temperate Japan
Lepidium meyenii	Maca S: maca	Enlarged root	Andean regions
Lepidium sativum	Garden/land cress F: cresson alenois S: mastuerzo C: du hang cai J: koshoso	Leaves	Temperate Europe
Nasturtium officinale	Watercress F: cresson d'eau	Leaves, tender shoots	Temperate regions

Raphanus sativus	S: berro de agua G: garten kresse C: dou ban cai J: mizu garashi Radish F: radis S: rabanito C: luo bu J: daikon	Root, leaves, sprouted seed	Temperate and subtropical regions
Sinapis alba	White mustard F: moutarde blanche S: mostaza blanca C: bai jie J: shiro garashi	Leaves, seed	Temperate regions
Sinapis arvensis	Charlock	Leaves, seed	Mediterranean region
CACTACEAE *Opuntia ficus-indica*	Cactus pad, prickly pear, Indian fig, tuna F: chardon d'Indie S: higuera de las Indias C: xian ren zhang J: hirochiwa	Young cladode (stem), mature fruit (tuna)	Subtropical, some temperate regions, Mexico, N. and S. America
CAMPANULACEAE *Campanula rapunculus*	Rampion F: campanule raiponce S: raponchigo C: jie geng cai J: kabura kikyo	Leaves, young roots	Britain, Northern Europe
CAPPARACEAE *Capparis spinosa*	Caper F: caprier S: alcaparra C: je geng cai, ci shan gan	Immature flower buds	Mediterranean region

(continued)

APPENDIX B *(cont.)*

Classification and scientific name	Common name(s)[a]	Edible part(s)	Regions/countries grown
CHENOPODIACEAE			
Atriplex hortensis	Orach, mountain spinach F: arroche S: amuella C: si yong bin li J: yama horenso	Leaves	Temperate regions, North- ern India
Beta vulgaris var. *cicla*	Swiss chard F: poiree ou bette S: bleda, acelga C: tian cai J: fudanso, tojishiya	Leaves, petioles	Temperate regions
Beta vulgaris var. *orientalis*	Palax, spinach beet	Leaves	Northern India, Indochina
Beta vulgaris var. *crassa*	Table beet, beetroot F: betterave rouge S: remolacha hortelana C: zi cai tou J: kaensai, teburu bito	Root, leaves	Temperate regions
Chenopodium album	Lambs quarter	Leaves	Temperate regions
Chenopodium ambrosioides	Epazote, Mexican tea	Young leaves	Warm temperate, subtropi- cal regions
Chenopodium bonus-henricus	Good King Henry, mercury	Leaves, young stems	Temperate regions
Chenopodium capitatum	Strawberry blite	Leaves	Temperate regions
Chenopodium nuttalliae	Huauzontle	Young plants, immature fruit clusters	Warm temperate, subtropi- cal regions
Chenopodium quinoa	Quinoa	Mature seed, leaves	Warm temperate, subtropi- cal regions
Spinacia oleracea	Spinach F: epinard S: espinaca G: spinat C: bo cai J: horenso	Leaves	Temperate regions
COMPOSITAE (see ASTERACEAE)			
CONVOLVULACEAE			
Ipomoea aquatica	Water convolvulus, water spinach F: liseron d'eau	Leaves, tender shoots	Warm temperate, S.E. Asia and subtropical China

Ipomoea batatas	S: batatilla aquatica C: weng cai, yong tsai J: yon sai, asagao na S.E. Asia: kang kong Sweet potato F: patate douce S: batata C: ken shu, gan zu J: satsuma imo	Root, tender shoots	Warm temperate, subtropical, and tropical regions

CRUCIFERAE (see BRASSICACEAE)

CUCURBITACEAE

Benincasa hispida	Wax/white gourd, Chinese winter melon F: courge a la cire S: calabaza branca C: mo kwa (immature) don kwa (mature) J: tougan	Immature and mature fruit	Warm temperate and sub-tropical Asia
Citrullus colocynthis	Colocynth, bitter apple	Seed	Africa
Citrullus lanatus	Watermelon F: melon d'eau S: sandia C: shi kwa, xi gua J: suika	Ripe fruit, seed	Temperate and subtropical regions
Citrullus lanatus var. *citroides*	Citron/preserving melon F: citre C: xiao xi gua	Mature fruit	Temperate and subtropical regions
Coccinia grandis	Ivy gourd, tindora S: pepino cimarron C: hong gua J: yasai karasu uri	Fruit, leaves, tender shoots	India, S.E. Asia
Cucumeropsis manni	White seeded melon F: ononde S: calabaza pamue C: gang guo gua J: egusi ito	Seed, fruit, leaves	W. Africa

(continued)

781

APPENDIX B *(cont.)*

Classification and scientific name	Common name(s)[a]	Edible part(s)	Regions/countries grown
Cucumis anguria	West Indian gherkin, bur cucumber F: concombre des antilles S: pepino des sabana C: xi yin du huang gua J: gakin	Immature fruit	Warm temperate, subtropical regions
Cucumis melo	Melon F: melon S: melon C: tian gua, tien kwa J: makuwa uri, meron	Immature and mature fruit	Temperate and subtropical regions
Cucumis melo gp. Cantalupensis	Netted melon, cantaloupe F: cantaloup S: melon cantaloupe C: ying pi tian gua J: kantaropu	Mature fruit	Temperate and subtropical regions
Cucumis melo gp. Chito	Mango/lemon melon F: melon mango S: melon mango C: mi gan tian gua	Immature fruit	Warm temperate, subtropical regions
Cucumis melo gp. Conomon	Oriental pickling melon S: melon manzana C: yu gua J: uri, shiro uri, ao uri	Immature fruit	Warm temperate, subtropical regions, China, Japan, Taiwan
Cucumis melo gp. Dudaim	Pomegranate/Queen's pocket melon	Mature fruit	Warm temperate, subtropical regions
Cucumis melo gp. Flexuosus	Armenian cucumber, snake/serpent melon F: concombre serpent S: melon serpie, tsai gua C: she tian gua	Immature fruit	Middle Eastern countries
Cucumis melo gp. Inodorus	Winter melon F: melon d'hiver S: melon C: song tian gua J: uinta meron	Mature fruit	Temperate and subtropical regions
Cucumis melo gp. Momordica	Snap/phut melon	Mature fruit	India

Cucumis sativus	Cucumber, pickling cucumber, gherkin F: concombre, cornichon S: pepino, pepinillo C: hu gua, hu kwa J: kiuri, kyu uri	Immature fruit	Warm temperate, subtropical regions
Cucurbita argyrosperma	Winter squash, pumpkin, cushaw F: potiron S: zapallo C: mo xi ku J: mikusuta kabocha	Mature fruit	Warm temperate, subtropical regions
Cucurbita ficifolia	Malabar/fig-leaf gourd F: gourge de Malabar S: chilacayote C: hei zi nan gua J: kuro tane kabocha	Mature fruit, seed	Warm temperate and subtropical Americas
Cucurbita foetidissma	Buffalo gourd	Storage root, seed	S.W. United States, Northern Mexico
Cucurbita maxima	Winter squash, pumpkin F: potiron, grosse courge S: zapallo, calabaza grande C: sun gua, yin du nan gua J: seiyo kuri kabocha	Mature fruit	Warm temperate, subtropical regions
Cucurbita moschata	Winter squash, pumpkin F: courge muscarde S: calabaza moscada C: wo gua, chung ku nan gua J: nihon kabocha	Mature fruit	Warm temperate, subtropical regions
Cucurbita pepo	Summer and winter squash, pumpkin, vegetable marrow F: courge pepon, courgette S: spizapallo, calabaza C: xi hu lu J: pepo kabocha	Immature and mature fruit	Warm temperate, subtropical regions
Cyclanthera pedata	Caihua, wild cucumber F: concombre grimpant S: pepino de comer C: sai huang gua J: yasai kiuri	Immature fruit	Subtropical and tropical S. America

(continued)

APPENDIX B (cont.)

Classification and scientific name	Common name(s)[a]	Edible part(s)	Regions/countries grown
Hodgsonia macrocarpa	Chinese lard fruit	Seed	N. India, S. China
Lagenaria siceraria	White flower/bottle/calabash gourd, zucca melon F: courge bouteille S: calabaza de peregrinos C: hu lu J: yugao, kanpiyo, hiyotan	Immature fruit	Warm temperate, subtropical regions
Luffa acutangula	Angled loofa, vegetable gourd, Chinese okra F: concombre papengaie S: calabaza de aristas, dringi C: ling jiao si gua J: tokado hechima	Immature fruit	Warm temperate, subtropical, and S.E. Asian regions
Luffa aegyptiaca	Smooth loofa, sponge/dish cloth gourd F: loofah, eponge vegetable S: calabaza de aristas, esponja C: si gua J: hechima	Immature fruit	Warm temperate, subtropical, and S.E. Asian regions
Momordica charantia	Bitter melon, bitter gourd, balsam pear F: assorossie S: balsamina C: ku gua, fu kwa J: tsuru reishi, nega uri	Immature fruit leaves, shoot tips	Warm temperate, subtropical, tropical regions, and S.E. Asia
Momordica cochinchinensis	Sweet gourd	Immature fruit	India, S.E. Asia
Sechium edule	Chayote, vegetable pear, mirliton F: chayotte, chou-chou S: cayote C: fe shou gua J: hayato uri	Fruit, tender shoots, roots	Subtropical and tropical regions
Telfairia occidentalis	Fluted pumpkin, fluted gourd C: xi fei li	Seed, leaves, tender shoots	Subtropical regions, and Nigeria, Ghana, W. Africa
Telfairia pedata	Oyster nut F: bane S: kueme C: wen li	Seed, leaves, shoots	E. Africa

Trichosanthes cucumerina var. *anguinea*	Snake/serpent gourd F: courge serpent S: anguina calabaza C: she gua, sui gua J: hebi uri, kekarasu uri	Immature fruit	Subtropical and tropical Asia
Trichosanthes dioica	Pointed gourd, parval F: patole S: patole C: ye she gua	Immature fruit	India
EUPHORBIACEAE *Cnidoscolus chayamansa*	Chaya, tree spinach	Leaves	Mexico, Central and S. America
Manihot esculenta	Cassava, manioc, tapioca, yuca F: manioc S: yuca C: shu shu J: kyatsusaba, tapioka noki	Root, leaves	Subtropical and tropical regions
Ricinus commuis	Castor bean	Detoxified shoots and leaves	Africa, many tropical regions
Sauropus androgynus	Katuk, chekurmanis, souropus S: katuk C: sou gong mu J: ruridama no ki	Shoot tops	Indonesia, India, S.E. Asia
FABACEAE (LEGUMINOSAE) *Apois americana*	Apois	Rhizome	Eastern N. America
Arachis hypogaea	Ground nut, peanut F: arachide, pois de terre S: araquida, cacahuete C: lo hua sheng J: nankin mame, rakasei	Immature and mature seed, shoot tips, and leaves	Warm temperate, subtropical and tropical regions
Cajanus cajan	Pigeon/Congo pea, red gram F: pois d'Angola S: frijol del monte C: shu dou J: kimame, juto	Immature pods and seeds, mature seed	India, subtropical, and warm temperate regions

(continued)

APPENDIX B *(cont.)*

Classification and scientific name	Common name(s)[a]	Edible part(s)	Regions/countries grown
Canavalia ensiformis	Jack bean F: haricot de Madagascar S: judia sable C: ai xing dao dou J: tachi nata mame	Immature pod, seed	Warm temperate, subtropical regions, Central America, West Indies
Canavalia gladiata	Sword bean F: pois sabre rouge S: haba de burro C: dao dou J: nata mame	Immature pod, seed	India, subtropical and tropical Africa and Asia
Cicer arietinum	Chick pea, garbanzo, Bengal gram F: pois-chiche S: garbanzo C: ji er dou J: hiyoko mame	Immature pod, seed, tender shoots	Warm temperate, subtropical regions, India, Middle Eastern countries
Cyamposis tetragonolobus	Cluster bean, guar F: guar S: guar C: sui dou	Immature pod	Warm temperate, subtropical and dry tropical regions
Dolichos lignosus *Dolichos uniform* *Glycine max*	Australian pea Horse gram Soya, soy bean F: soya S: soja C: da tou, mao dou J: daizu, eda mame	Mature seed Mature seed Mature, immature, and sprouted seed	Australia Mediterranean region Temperate and subtropical regions
Lablab purpureus	Hyacinth/Indian/lablab bean F: pois indien S: dolico lablab C: pien tou, que dou J: fuji mame	Immature pod, seed	Subtropical regions, India
Lathyris sativus	Chickling/grass pea F: gesse blanche S: arvejo J: garasu mame	Mature seed, leaves	Warm temperate, subtropical regions, India, W. Asia, Africa
Lens culinaris	Lentil F: lentile S: lenteja	Seed, immature pod	Warm temperate, subtropical regions, India, Egypt, Southern Europe

Scientific name	Common names	Part used	Region
	C: jin mai wan J: hira mame		
Lupinus spp. *L. albus* (white) *L. luteus* (yellow) *L. augustifolius* (blue) *L. mutabilis* (sweet)	Lupine F: lupin S: altramuz, lupino C: bai hua yu shan dou	Mature seed	Mediterranean, Egypt, Sudan, Eastern Europe
Pachyrhizus ahipa	Yam bean, ahipa F: dolique des Andes S: ahipa	Swollen root	S. and Central America
Pachyrhizus erosus	Mexican yam bean, jicama F: dolique bulbeux S: jicama, achipa C: dou shu J: kuzu imo, mame imo	Swollen root, immature pod	Warm temperate, subtropical and tropical regions, S.E. Asia, Mexico, Central America
Pachyrhizus tuberosus	Potato bean F: dolique bulbeux S: cohen, sincama C: da di gua	Swollen root	Subtropical and tropical America
Phaseolus acutifolius	Tepary bean F: haricot tepary S: judia tepari C: jian ye cai dou	Immature pod, seed	Subtropical and tropical regions
Phaseolus coccineus	Scarlet runner bean F: haricot d'Espagne S: judia escarlata C: hong hua cai dou J: beni bana ingen	Seed, immature pod, tuber	Temperate, subtropical regions, Central America, England
Phaseolus lunatus	Lima/sieva bean F: haricot de lima S: judia de lima C: cai ma dou J: raima mame	Immature and mature seed	Warm temperate and subtropical regions
Phaseolus vulgaris	Snap/pole/French bean F: haricot vert S: judia comun C: cai dou J: ingen mame	Immature pod and seed	Temperate and subtropical regions

(continued)

Classification and scientific name	Common name(s)[a]	Edible part(s)	Regions/countries grown
Pisum sativum	Garden/English pea F: pois S: guisante C: wan dou, hua tou J: endo	Immature and mature seed, tender shoots	Moderate to cool temperate regions
Pisum sativum var. *macrocarpon*	Edible podded/sugar/China pea F: pois sans parchemin S: chicharo C: shi jia wan dou J: saya endo	Immature pod	Temperate and some sub-tropical regions
Psophocarpus tetragonolobus	Winged/goa/four-angled/Manila bean F: pois aile S: frijol alado C: si leng dou, yeh tou J: shikaku mame, tau sai	Immature pod, seed, tuberous root, leaves	Subtropical, tropical regions, S.E. Asia, Papua, New Guinea, Philippines
Pueraria lobata	Kudzu F: kudzu S: kudzi comun C: ge teng J: kuzu	Swollen root	Orient
Sphenostylis stenocarpa	African yam bean F: haricot igname C: suan pan zi	Storage root, mature seed	Tropical Africa, Ethiopia
Trigonella foenum-graecum	Fenugreek, metha F: trigonelle S: alholva C: hu lu ba J: koroha	Leaves	Temperate and subtropical regions, India
Tylosema esculentum	Marama bean	Tuberous roots, seed	S. Africa
Vicia faba	Broad/horse/Windsor/fava bean F: feve, gourgane S: haba C: can dou, ye hu lu ba J: sora mame	Immature and mature seed	Mediterranean and temperate regions
Vigna acontifolia	Moth/mat bean F: haricot papillon C: e dou	Seed, immature pod	India, Pakistan, Indochina, China

Vigna angularis	Adzuki bean F: haricot á feuilles angulairis, pois azuki S: judia adzuki, poroto arroz C: hong dou, hong xiao dou J: azuki	Seed, immature pod	Japan, China, Korea, warm temperate and subtropical regions
Vigna mungo	Urd, black gram F: haricot velu S: frijol mungo C: ji dou	Immature pod	India, warm temperate and subtropical regions
Vigna radiata	Mung bean, green/golden gram F: amberique, haricot dore S: judia de mungo C: liu dou J: riyokutou, yaenari	Seed, sprouted seed, immature pod	Warm temperate and subtropical Asia
Vigna subterranea	Bambara/Madagascar ground nut, African peanut F: pois bambara S: bambarra C: ban ba la hua sheng	Immature, mature seed	Tropical Africa
Vigna umbellata	Rice/red bean F: haricot sauvage S: frijoi arroz C: mi dou J: shima tsuru azuki	Seed, immature pod, leaves	Subtropical, tropical regions
Vigna unguiculata cultigroup *cylindrica*	Catjung/Bombay cowpea F: dolique de Chine S: judia catjang C: jiang dou J: hata sasage	Seed, immature pod	Tropical, subtropical regions, India, Africa
Vigna unguiculata cultigroup *sesquipedalis*	Yardlong/asparagus bean F: dolique, haricot asperge S: judia esparrago C: chang jiang dou J: juroku sasage	Immature pod	Warm temperate, subtropical regions, China, S.E. Asia
Vigna unguiculata cultigroup *unguiculata*	Cowpea/black-eyed pea F: haricot a oeil noir S: frijol de vaca	Seed, immature pod	Warm temperate, subtropical regions, Africa

(continued)

APPENDIX B *(cont.)*

Classification and scientific name	Common name(s)[a]	Edible part(s)	Regions/countries grown
Vigna unguiculata cultigroup *unguiculata* (con't)	C: pu tong jiang dou J: sasage		
ICACINACEAE *Icacina senegalensis*	False yam F: bankanas	Enlarged hypocotyl	W. and Central Africa
LABIATAE (see LAMIACEAE)			
LAMIACEAE (LABIATAE) *Coleus amboinicus* *Coleus parviflorus*[b] (*C. rotundifolius*, *C. tuberosus*)	Spinash thyme, Indian borage Coleus/hausa/Madagascar/Sudan potato Malaysia: ubi kembili India: kourka	Shoots Tuber	Tropical America Subtropical, tropical regions, S.E. Asia, India, E. Africa
Perilla frutescens	Perilla, beefsteak plant F: melisse, perille S: perilla C: zi su, bai su J: shiso	Leaves, seed	Warm temperate and sub-tropical regions, Japan
Plectranthus esculentus[b] (*Coleus esculenta, C. parviflorus, C. rotundifolius, C. tuberosus*) *Soleostemon rodundifolius*[b]	Livingstone/Kaffir potato F: geunaine C: ka fei shu Hausa potato F: pomme de terre des Hausa S: patata de los Hausas	Tuber Tuber	Subtropical, tropical regions, S.E. Asia, India S.E. Asia, India, Shi Lanka, Africa
Stachys affinis	Chinese artichoke, knot root F: stachys du japon S: ortiga japonesa C: cao shi can, chao shu zei J: chorogi	Tuber	Subtropical regions, China, Japan
MALVACEAE *Abelmoschus esculentus*	Okra, gumbo, lady's finger, bhindi F: gombo S: gombo, ocra C: huang qui kui J: okura	Immature pods	Temperate, subtropical and tropical regions

Abelmoschus manihot	Sunset hibiscus F: ketmic á feuilles de manioc J: tororo aoi	Young leaves, shoot tips	Tropical Asia, China, Indonesia, Pacific islands
Hibiscus sabdariffa	Roselle, Jamaican, Indian sorrel F: oseille de Guinee S: canamo de Guinea C: lou shen kiu J: rozeru	Calyx	Subtropical regions, West Indies
Malva neglecta	Edible/Chinese mallow C: dong han cai J: oka nori, hatakena	Leaves, young shoots	Temperate regions, China
Malva parviflora	Egyptian mallow	Leaves, young shoots	Subtropic regions, Egypt, Sudan
Malva rotundifolia	Mallow F: mauve S: malva C: yuan ye jin kui J: fuyu aoi	Leaves, young shoots	
MARTYNIACEAE *Proboscidea louisianica*	Martynia, Unicorn/probosis plant, devil's claw, ram's horn	Immature pod	S.W. United States
MELIACEAE *Cedrela sinensis*	Cedrus	Tender leaves and shoot tips	China
MORACEAE *Artocarpus altilis*	Breadfruit F: arbe a pain S: fruta de pan Polynesia: ulu	Fruit, seed	Subtropical and tropical regions, S.E. Asia, Pacific Is.
Artocarpus heterophyllus	Jackfruit, jak, jaca	Fruit, seed	Subtropical and tropical regions, S.E. Asia, S. Pacific islands, India
MORINGACEAE *Moringa pterygosperma*	Horseradish/drumstick tree	Leaves, shoot tips, immature fruit, flowers, seed	Northern India, Pakistan

(continued)

APPENDIX B *(cont.)*

Classification and scientific name	Common name(s)[a]	Edible part(s)	Regions/countries grown
NYCTAGINACEAE *Mirabilis expansa*	Mauka, yuca inca F: bellede nuit S: arracacha de toro	Tuberous root, leaves	Andean region
NYMPHAEACEAE *Nelumbo nucifera*	Lotus root F: lotus sacre S: semillas de loto C: lian J: hasu, renkon	Rhizome, seed	Warm temperate, subtropical and tropical regions, S.E. Asia, China, Japan
OXALIDACEAE *Oxalis tuberosa*	Oca F: oxalide tubereuse S: oca C: nam mei cu jiang cao	Tuber	Subtropical, temperate regions, Peru, New Zealand
PHYTOLACCACEAE *Phytolacca americana*	Pokeberry, pokeweed, poke F: raisin d'Amerique, morelle a grappe S: fitolaca C: yang shang lu J: yoshu yama gobo	Emerging young	Eastern United States
PIPERACEA *Peperomia pereskiifolia*	Perejil	Young leaves, shoots	Venezuela, Colombia
POLYGONACEAE *Rheum rhabarbarum*	Rhubarb, pie plant F: rhuarbe S: ruibarbo C: shi yong da huang J: rubabu, shokuyo daio	Petiole	Temperate Europe, N. America
Rumex abyssinicus *Rumex acetosa*	Spinach rhubarb Sorrel, dock, sour grass F: oseille S: acedera	Leaves Leaves	Ethiopia Temperate Europe, N. America

Rumex patientia	C: suan me J: suiba, sukan po Spinach dock, herb patience F: patience S: hierba de la paciencia C: ba tian suan mo J: wase suiba	Leaves	Temperate Europe, N. America
Rumex scutatus	French sorrel F: rumex á ecusson S: acedera romana C: yuan ye juan mo J: maruba, suiba	Leaves	Temperate Europe, Asia
PORTULACACEAE *Montra perfoliata* *Portulaca oleracea* var. *sativa*	Miner's lettuce Purslane F: pourpier potager S: verdolaga C: ma chi xian J: tachi suberi hiyu	Leaves Leaves, tender shoots	Temperate N. America Temperate and subtropical regions
Talinum triangulare	Waterleaf, Lagos bologi, talinum spinach F: grand pourpier S: espinaca de Filipinas C: jia ren shen	Leaves, shoot tip	Tropical America, Asia, West Indies, W. Africa
RUTACEAE *Murraya koenigii*	Curry leaf	Leaves	S.E. Asia
SAURUACEAE *Houttuynia cordata*	Saururis F: hottuynie C: ji cai J: dokudami	Leaves	S.E. Asia, Indonesia
SOLANACEAE *Capsicum annuum*	Pepper, chile F: poivre d'Espagne, paprica S: pimiento, chile C: la jiao, tian jiao	Fruit, leaves	Warm temperate, subtropical and tropical regions

(continued)

APPENDIX B *(cont.)*

Classification and scientific name	Common name(s)[a]	Edible part(s)	Regions/countries grown
Capsicum annuum (con't)	J: togarashi, piiman Philippines: (leaves) dahon ng sili		
Capsicum frutescens	Tabasco pepper F: piment S: tabasco, chile C: xiao mi jiao J: tougarishi	Fruit	Warm temperate, subtropical and tropical regions
Cyphomandra betacea	Tree tomato, tamarillo F: arbe a tomates S: arbol tomate C: shu fan qie J: shinoman dora, kodachi tomato	Mature fruit	Subtropical and tropical America, New Zealand
Lycium chinense	Chinese boxthorn, matrimony vine, wolf berry F: lyciet de la Chine S: cambronera de la China C: chu chi, gao chi J: kuko	Leaves	Warm temperate, subtropical regions, S.E. Asia, China, Taiwan
Lycopersicon lycopersicum	Tomato F: tomate S: tomate I: pomodoro C: fan qie J: tomato	Fruit	Temperate, subtropical and tropical regions
Lycopersicon pimpinellifolium	Currant/cherry tomato F: tomate cerise S: tomate creza C: cu li fan qie J: hozuki tomato	Mature fruit	Temperate, subtropical and tropical regions
Physalis alkekengi	Chinese lantern, strawberry tomato F: coqueret alkekenga S: tomate ingles C: suan jiang J: sennari hozuki	Fruit	Warm temperate, subtropical Asia
Physalis ixocarpa	Husk tomato, tomatillo F: tomatillo	Fruit	Warm temperate, subtropical and tropical America

Species	Common names	Part	Distribution
Physalis peruviana	S: tomate de cascara C: da suan jiang J: obudo hozuki Cape gooseberry, uchuba, Peruvian ground cherry F: coqueret du Perou S: capuli C: xiao suan jiang J: shima, hozuki	Fruit	Warm temperate, subtropical and tropical America
Physalis pruinosa	Dwarf cape gooseberry C: suan jiang, sui jian J: shokuyo hozuki	Fruit	Subtropical and tropical America
Physalis pubescens	Ground cherry	Fruit	Subtropical and tropical America
Solanum aethiopicum	Mock tomato, golden apple F: tomate amere, pomme d'amour	Leaves, immature fruit	Tropical Africa
Solanum americanum	Glossy nightshade	Leaves, green fruit	Central eastern United States
Solanum gilo	Jilo, gilo, garden egg F: grande morelle S: gilo	Fruit, shoots	Subtropical regions, Western Africa, Brazil
Solanum hirsutissimum *Solanum hygrothermicum*	Lulita Peruvian potato S. Amer.: urahji, cachariqui, moshaki	Mature fruit Tuber	Central and S. America Peru
Solanum incanum	Sodom apple S: berenjena silvestre	Immature fruit	W. Africa, India
Solanum indicum *Solanum integrifolium*	Indian nightshade, tiberato Tomato/scarlet eggplant F: aubergine mouchete S: berenjena escarlata C: hong qie J: hira nasu, aka nasu	Fruit Fruit	S.E. Asia Warm temperate, subtropical Asia
Solanum macrocarpon	African eggplant F: aubergine indigene S: berenjena africana C: fei zhou qie	Fruit, leaves	Subtropical and tropical regions, W. and central Africa

(continued)

APPENDIX B *(cont.)*

Classification and scientific name	Common name(s)[a]	Edible part(s)	Regions/countries grown
Solanum melanocerasum	Garden huckleberry, wonder berry, black nightshade F: morelle noire S: hierba mora C: long kui J: inu hozuki	Mature fruit	Warm temperate regions, and subtropical America, W. Africa
Solanum melongena	Eggplant, brinjal, aubergine F: aubergine S: berejena C: qie zi, chieh tzu J: nasu, nasubi	Immature fruit	Warm temperate, subtropical and tropical regions
Solanum muricatum	Pepino, melon pear F: melon poire S: pepino C: xiang qua qie J: pepino	Mature fruit	Subtropical and tropical America
Solanum quitoense	Naranjillo, lulo F: narangille S: naranjilla	Mature fruit	Subtropical and tropical America
Solanum sessiliflorum	Cocona Venezuela: topiro Colombia: lulo	Mature fruit	Northern S. America
Solanum torvum	Turkeyberry F: morelle diable S: berengena cimarrona C: shui qie J: suzume nasubi	Immature fruit	S. America tropics
Solanum tuberosum	White/Irish potato F: pomme de terre S: patata, papa C: ma lin shu J: jaga imo, bareisho	Tuber	Most temperate regions
TETRAGONIACEAE *Tetragonia tetragonioides*	New Zealand spinach F: tetragone cornue S: espinaca de Nueva Zealandia C: fan xing J: tsurana, hamajisha	Leaves, tender shoots	Temperate and subtropical regions

TILIACEAE			
Corchorus olitorius	Jew's/jute mallow, molokhia, bush okra F: corette potagere S: yute C: chang shuo huang ma J: Taiwan tsunaso, moroheyia	Leaves, tender shoots	Subtropical and tropical regions, Egypt, Sudan, India
TRAPACEAE			
Trapa bicornis	Water chestnut, trapa nut F: chataigne d'eau S: castana de agua C: lin ku, er jiao wan ling J: tobishi, tsuno bishi	Seed	Subtropical and tropical regions
Trapa maximowiczii	Maximowicz water chestnut J: hime bishi	Seed	S.E. Asia
Trapa natans	Water caltrop, Jesuit's nut F: noix aquatique S: castana de agua singara C: szu chiao ling J: hishi, oni bishi	Seed	Eurasia, Africa
TROPAEOLACEAE			
Tropaeolum majus	Graden nasturium, Indian cress F: cresson d'Inde, capucine S: nasturcio C: jin lian hua J: nasutashimu, nozen haren	Leaves, seed	Central and S. America, Mexico, Peru
Tropaeolum tuberosum	Mashua, anu F: capucine tubereuse S: mashua C: kuai jin zan hua J: kyukonkin renka	Tuber	Andean region
UMBELLIFERAE (see APIACEAE)			
VALERIANACEAE			
Valerianella eriocarpa	Italian corn salad F: valerianella á fruits velus S: valerianela de Italia C: yi da li ye ju	Leaves	Europe

(continued)

797

APPENDIX B *(cont.)*

Classification and scientific name	Common name(s)[a]	Edible part(s)	Regions/countries grown
Valerianella locusta	marche, corn salad, lamb's lettuce, fetticus F: mache, doucette S: valeriana C: ye ju J: kon sarada, nojisha	Leaves	Mediterranean region

[a]Common name languages: F = French, S = Spanish, C = Chinese, J = Japanese, P = Portugese, G = German, I = Italian, D = Dutch.
[b]See Coleus potatoes, Chapter 14.
Reference sources: Primary references used for this table were:
L.H. Bailey Hortorium Staff. 1976. Hortus Third, revised, initially compiled by L.H. Bailey and E.Z. Bailey. Macmillan Publishers, New York.

Hung, L., Huang, H., and Yen, H-E., eds. 1992. The Nomenclature of Vegetable Crops. National Taiwan University, Taiwan.
Kays, S.J., and Silva Dias, J.C., 1995. Common names of commercially cultivated vegetables of the world in 15 languages. Econ. Bot. 49(2);115–152.

Appendix C

APPROXIMATE NUTRIENT COMPOSITION OF VARIOUS VEGETABLES*

	% Water	Calories	CHO	Pro	Fat	Fib	Ash	A	C	B1	B2	Niacin	Ca	P	K	Na	Mg	Fe
FUNGI																		
Agaricus bisporus Button mushroom	91.5	23	4.5	2.4	0.20	0.80	1.00		4	0.10	0.50	4.40	8	112	392	10	11	1.1
Auricularia polytricha Jew's ear mushroom	92.6	25	6.8	0.5	0.40	2.10	0.20		1	0.80	0.20	0.07	16	14	43	9	25	0.6
Cantharellus cibarius Chanterelle mushroom	91.5	25	3.0	1.5	0.50		0.80		6	0.02	0.23	6.50	8	44	507	3	14	6.5
Lentinula edodes Shiitake mushroom (dried)	9.5	296	75.4	9.6	1	11.5	4.6		4	0.30	1.30	14.10	11	294	1534	13	132	1.7
Morchella spp. Morel mushroom	90.0	31	5.4	1.7	0.30	0.80	1.00		5	0.13	0.06		11	162	390	2	11	1.2
Tuber melanosporum Truffle	75.5	56	7.4	5.5	0.50	6.40	1.90						24	62	526	77	24	3.5
ALGAE																		
Laminaria spp. Brown algae	81.6	43	9.6	1.7	0.60	1.30	6.60	116		0.05	0.15	0.47	168	42	89	233	121	2.9
Porphyra spp. Red algae "nori"	85.0	35	5.1	5.8	0.30	0.30	3.80	5200	39	0.10	0.50	1.50	70	58	356	48	2	1.8

(continued)

	% Water	Calories	CHO	Pro	Fat	Fib	Ash	A	C	B1	B2	Niacin	Ca	P	K	Na	Mg	Fe
FERNS																		
POLYPODIACEAE																		
Pteridium aquilinum Bracken fern	88.6	36	6.5	2.3	0.40	1.00	1.20	66	30		0.30	3.50	11	19				1.1
OSMUNDACEAE																		
Osmunda japonica Japanese flowering fern	88.3	38	4.3	3.1	0.20	3.80	0.30	16	15		0.40	0.70	9	18				0.8
GYMNOSPERM																		
GINKGOACEAE																		
Ginkgo biloba Ginkgo (mature seeds)	12.4	349	69.9	12.20	2.70	0.70	2.80						20	269				1.6
MONOCOTS																		
ALISMATACEAE																		
Sagittaria sagittifolia Arrowhead	72.3	99	20.2	5.3	0.29	0.82	1.67		1	0.17	0.07	1.65	10	174	922	22	51	2.6
ALLIACEAE																		
Allium ampeloprasum Great headed garlic (bulb)	86.3	45	10.3	2.2	0.30		0.90						52	50				1.1
Allium ampeloprasum Leek (pseudostem and leaves)	83.0	56	12.6	1.9	0.30	1.40	1.10	75	15	0.08	0.05	0.45	55	43	230	16	25	1.9
Allium cepa Shallot (bulb)	79.8	72	16.8	2.4	0.10	0.70	0.80		8	0.06	0.02	0.20	37	60	334	12		1.2
Allium cepa Shallot (leaves)	91.0	30	5.0	1.8	0.90	1.60			19	0.07	0.12	0.40	86	25				3.7
Allium cepa Onion (bulb)	90.8	36	8.0	1.3	0.18	0.52	0.40	40	9	0.05	0.03	0.15	26	33	156	6	11	0.4
Allium cepa Onion (leaves "scallions")	91.9	25	5.6	1.7	0.14	0.90	0.70	5000	45	0.07	0.14	0.20	60	33	257	14	20	1.1

Allium chinense Rakkyo (bulb)	86.0	50	12.0	2.2	0.30	0.50	0.50		10	0.05	0.03	1.0	22	66			0.5
Allium fistulosum Japanese bunching onion	90.5	34	6.5	1.9	0.30	1.00	0.70	380	27	0.05	0.09	0.40	55	49	200	10	1
Allium sativum Garlic (bulb)	64.0	131	27.9	6.20	0.35	1.00	1.30	0	11	0.20	0.08	0.50	32	187	465	17	1.3
Allium sativum Garlic (leaves)	86.4	12	9.5	2.60	0.50	1.50	1.00	680	38	0.08	0.16	0.70	82	66	326	4	0.5
Allium schoenoprasum Chives (leaves)	92.0	26	3.9	2.80	0.60	0.90	0.80	6400	70	0.10	0.12	0.60	82	46	250	6	1.2
Allium tuberosum Chinese chives	93.1	23	2.8	2.10	0.10	0.90	1.00	1800	25	0.06	0.19	0.6	50	32			0.6
Allium wakegi Wakegi onion	91.2	29	5	1.90	0.30	1.00	0.60	500	30	0.05	0.15	0.4	38	35		20	1.2
ARACEAE																	
Alocasia macrorrhiza Giant taro, "Alocasia" (corm)	84	71	17	0.60	0.20	1.50	1.10		5	0.08	0.02	0.4	50	50			1
Amorphophallus campanulatus Elephant foot yam (corm)	78	81	18	1.60	0.20	0.70	1.00	434	6	0.07	0.06	1	50	36			1.1
Colocasia esculenta Taro (corm)	72.9	99	23.4	2.00	0.20	0.80	1.20	20	5	0.11	0.04	0.9	30	63	512	10	0.9
Colocasia esculenta Taro (leaves)	86	42	7.4	4.40	0.90	1.60	1.50	8100	45	0.18	0.4	1.4	156	80	648	3	1.7
Colocasia esculenta Taro (shoots)	94	18	15	0.60	0.40	0.50		107	10	0.04	0.05	0.5	50	25	332	1	1.1
Cyrtosperma chamissonis Giant swamp taro (corm)	65	116	32	0.80	0.30	1.30	1.00		1	0.04	0.1	0.9	370	55			1.2
Cyrtosperma chamissonis Giant swamp taro (leaves)	86	45	4	5.00	0.70	1.50	2.00										
Xanthosoma sagittifolium Tannia (cormels)	65	134	32	2.10	0.30	0.90	1.20	10	10	0.10	0.90	0.5	14	48			0.8

(continued)

APPENDIX C (cont.)

	% Water	Calories	CHO	Pro	Fat	Fib	Ash	A	C	B1	B2	Niacin	Ca	P	K	Na	Mg	Fe
Xanthosoma sagittifolium Tannia (leaves)	89.9	34	5.3	2.50	1.00	2.10	1.30	3300	37				95	388				2
Xanthosoma sagittifolium Tannia (shoots)	89.5	33	5.7	3.10	0.60	3.20	1.10		82				49	80				0.3
ARECACEAE (PALMAE)																		
Bactris gasipaes & Euterpe spp. Heart of palm, pejibaye	89.8	26	5.2	2.20	2.10	0.80	1.30		17	0.04	0.09	0.7	86	79	38	1	9	0.7
Sabal palmetto Swamp cabbage	85	21	3.9	2.20	0.20	0.09		5200	16	0.05	0.08	0.5	55	32	88			1.5
CYPERACEAE																		
Cyperus esculentus Chufa, "tiger nut" (tuber)	34.5	306	45	4.00	15.00	7.30	1.50		3	0.28	0.09	1.3	40	175				2.4
Eleocharis dulcis Chinese water chestnut	76.4	86	19.6	1.45	0.18	0.85	1.20		10	0.17	0.2	1	11	65	542	17	17	2.1
DIOSCOREACEAE																		
Dioscorea spp. Tubers/bulbils, various spp.	70	108	25.1	2.05	0.20	0.95	1.00		12	0.11	0.04	0.5	20	62	600	9	26	0.7
LILIACEAE																		
Asparagus officinalis Asparagus, green spear	92.2	22	3.8	2.60	0.21	0.77	0.79	950	33	0.20	0.14	2	22	67	271	2	18	0.8
Asparagus officinalis Asparagus, white spear	93.6	20	2.9	1.90	0.14	0.84	0.62	50	21	0.11	0.12	1	21	46	207	4	20	1
Lilium spp. Lily bulbs	66	128	26.1	4.80	0.60	1.10	1.40		20	0.05	0.04	0.2	5	240		10		0.5
MARANTACEAE																		
Maranta arundinacae West Indian arrowroot (rhizome)	64.9	142	29.7	1.90	0.10	1.70	1.30		9	0.08	0.03	0.7	20	24				3.2

MUSACEAE																		
Musa spp. Plantain and cooking banana	64.7	127	31.4	1.10	0.40	0.40	0.90	990	18	0.07	0.04	0.6	7	35	420	1	33	0.7
POACEAE																		
Bambusa and other spp. Bamboo shoots	91	27	5.2	2.60	0.30	0.70	0.90	20	4	0.15	0.07	0.6	13	59	532	4	3	0.5
Zea mays Sweet corn	75.3	95	20.3	3.60	1.20	0.75	0.71	480	12	0.17	0.12	1.7	35	109	273	8	45	1.1
Zizania latifolia Water bamboo	92.6	26	5.5	1.20	0.20	1.00	0.50		2	0.09	0.04	0.02	5	36				0.6
TACCACEAE																		
Tacca spp. East Indian arrowroot	68.8	121	28.8	1.50	0.10	0.40	0.80						77	64				
ZINGIBERACEAE																		
Zingiber officinale Ginger, fresh	86.2	50	9.8	1.45	1.00	1.15	1.10	10	4	0.02	0.04	0.75	22	37	264	6		1.1
DICOTS																		
AIZOACEAE																		
Mesembryanthemum crystallinum Ice plant	94	5	0.3	0.70	0.20			2000	23	0.04	0.06	0.3	90	26				0.6
AMARANTHACEAE																		
Amaranthus spp. "Tampala" (leaves)	87	36	5.6	3.80	2.90	1.40	2.20	4320	56	0.05	0.3	1	305	77	527	20	55	5.5
Celosia spp. Legos or Silver spinach	86.5	39	6.9	3.70	0.50	1.50	3.20	9000	10		0.1		207	37				6.4
APIACEAE [UMBELLIFERAE]																		
Anthriscus cerefolium Chervil (leaves)		57	11.5	3.40					9						102			0.5
Apium graveolens var. *dulce* Celery, green (leaves and petioles)	94.7	17	3.8	0.80	0.11	0.65	0.89	184	8	0.03	0.03	0.6	52	32	313	107	17	0.4

(continued)

APPENDIX C   (cont.)

	% Water	Calories	CHO	Pro	Fat	Fib	Ash	A	C	B1	B2	Niacin	Ca	P	K	Na	Mg	Fe
Apium graveolens var. *dulce* / Celery, blanched (leaves and petioles)	93	18	3.8	0.85	0.15	0.70	1.10	90	8	0.04	0.05	0.6	35	34	308	116	14	0.5
Apium graveolens var. *rapaceum* / Celeriac, celery root (root and leaves)	88.3	40	8.4	1.70	0.32	1.28	1.00	16	8	0.04	0.07	0.8	50	107	305	94	15	0.6
Apium graveolens var. *secalinum* / Smallage (leaves)	90.9	27	4.6	2.20	0.60	1.40	1.70	2685	49	0.08	0.12	0.6	326	51	318	151		1.5
Arracacia xanthorrhiza / Arracacha, yellow root	73	104	24.9	0.80	0.20	0.60	1.10	yes	28	0.06	0.04	3.4	29	58				1.2
Arracacia xanthorrhiza / Arracacia, white root	73.4	102	24.4	0.80	0.20	1.00	1.20	no	23	0.07	0.06	2.8	26	52				0.9
Centella asiatica / Asiatic pennywort	87.7		6.7	2.00	0.20	1.60	1.60	730	7	0.09			170	32				5.5
Coriandrum sativum / Cilantro, Chinese parsley	90	28	5	2.40	0.50	1.00	1.60	2800		0.1	0.1	1.1	134	48	542	28	37	5.5
Cryptotaenia japonica / Japanese hornwort, mitsuba	93.5	18	2.1	2.00	0.10	1.30	1.00	1300	60	0.15	0.2	0.5	81	45				1.8
Daucus carota / Carrot	87.8	42	9.9	1.10	0.20	1.00	0.90	1100	9	0.35	0.06	0.8	32	40	332	41	19	0.6
Foeniculum vulgare var. *dulce* / Florence fennel	88	33	6.1	1.90	0.23	0.50	1.70	100	20	0.3	0.1	0.2	73	51	445	86	36	1.5
Oenanthe javanica / Water dropwort, Oriental celery	91.6	110	4.4	1.10	0.04	1.00	1.50		61		0.31		138	43				2.3
Pastinaca sativa / Parsnip (root)	80	74	17	1.47	0.44	2.00	1.10	30	17	0.09	0.09	0.5	47	75	355	10	29	0.7
Perideridia gairdneri / Squaw root	60	154	30.8	4.60	1.80				13	0.11	0.12	3	440	165	340	12	32	6.5
Petroselinum crispum / Parsley (leaves)	85	46	8.5	2.80	0.46	1.40	2.00	6050	123	0.12	0.23	1.1	160	57	508	90	61	4.3

Petroselinum crispum Parsley (root)	88	26	2.3	2.88	0.60		1.62	trace	41	0.1	0.09	2	13	57				0.3
ARALIACEAE																		
Aralia cordata Udo, spikenard	95.3	14	2.4	1.00	0.20	0.50	0.60		5	0.06	0.02	0.8	50	25	308	3	38	1
ASTERACEAE [COMPOSITAE]																		
Arctium lappa Edible burdock, gobo (root)	76.9	103	6.7	1.60	0.11	2.10	1.00	0	4	0.02	0.04	0.3	65	50	571	5	17	2.8
Chrysanthemum coronarium Garland chrysanthemum (leaves)	93	17	4	1.70	0.19	0.80	1.30	14000	31	0.1	0.25	0.9	66	32		52	13	1.3
Cichorium endiva Endive, escarole (leaves)	93.8	19	3.8	1.50	0.15	0.90	1.40	2140	8	0.08	0.11	0.5	93	41	304	50	13	0.9
Cichorium intybus Chicory (leaves)	92	22	4.3	1.80	0.30	0.80	1.30	4000	24	0.06	0.1	0.5	16	43	420	45	22	0.5
Cichorium intybus Chicory (witloof, chicon)	95.1	16	3.2	0.90	0.10		0.60							20	177	8	13	0.8
Cichorium intybus Chicory (root)	80	73	17.5	1.40	0.20	1.95	0.89	6	8	0.05	0.03	0.4	41	61	290	50	22	
Cosmos caudatus Cosmos (leaves and shoots)	93		0.4	3.00	0.40	1.60	1.60						270					
Crassocephalum crepidodes Sierra Leone bologi (leaf, shoots)	79	64	14	3.20	0.70	1.90							260	52				
Cynara cardunculus Cardoon (petioles)	94	20	4.9	0.70	0.10		0.31	120	2	0.02	0.03	0.3	70	23	400	170	42	0.7
Cynara scolymus Globe artichoke	85.7	45	10.2	2.80	0.19	2.00	1.03	250	10	0.08	0.05	0.9	50	88	388	61	41	1.3
Helianthus tuberosus Jerusalem artichoke	79	72	16.7	2.20	1.30	0.85	1.60	20	7	0.19	0.08	1.2	18	80	478	15	15	3

(continued)

APPENDIX C   *(cont.)*

	% Water	Calories	CHO	Pro	Fat	Fib	Ash	A	C	B1	B2	Niacin	Ca	P	K	Na	Mg	Fe
Lactuca sativa Lettuce, head or iceberg	95	14.5	2.4	1.05	0.16	0.51	0.70	470	7	0.06	0.07	0.3	22	26	166	7	11	1.5
Lactuca sativa Lettuce, butterhead	95.5	13.8	2.4	1.25	0.21	0.50	0.90	1065	8	0.06	0.06	0.3	35	26	260	7	11	1.8
Lactuca sativa Lettuce, cos	94.9	17.4	2.9	1.45	0.25	0.70	0.90	1925	22	0.08	0.09	0.5	44	35	277	9	9	1.3
Lactuca sativa Lettuce, leaf	94	18.3	3.5	1.30	0.30	0.70	0.90	1900	18	0.05	0.08	0.4	68	25	264	9	11	1.4
Lactuca sativa (stems) Celtuce, asparagus lettuce	94.5	22	3.7	0.85	0.30	0.40	0.70	70	6	0.06	0.07	0.6	39	39	330	11	28	0.6
Petasites japonicus Butterbur	94.5	14	3	0.50	0.04	1.30	1.40	50	32	0.02	0.02	0.2	103	12	655	7	14	0.1
Polymnia sonchifolia Yacon (root)	76			1.30	3.50	3.80									high			
Scorzonera hispanica Scorzonera, black salsify	77	81	17.9	2.65	0.40	1.40	0.96	10	10	0.06	0.07	0.4	51	65	365	13	23	1.55
Taraxacum officinale Dandelion	85.7	45	8.9	2.70	0.70	1.60	1.90	13900	34	0.19	0.22	0.8	179	67	409	76	36	3.1
Tragopogon porrifolius Salsify, vegetable oyster	77	34	5.2	3.30	0.20				8	0.08	0.22	0.5	60	75	380	30	23	0.7
BASELLACEAE																		
Basella alba, B. rubra Ceylon, Malabar spinach	93.1	20	3.6	1.80	0.30	0.70	1.40	7000	95	0.05	0.12	0.5	130	52	505	21		1.5
Ullucus tuberosus Ulluco	85.5	51	13	1.00		0.60	0.60		23	0.04	0.02	0.3	3	35				0.8
BORAGINACEAE																		
Borago officinalis	93	17	1.1	1.80	0.70			4200	35	0.06	0.15	0.9	93	53	470	80	52	3.3
BRASSICACEAE [CRUCIFERAE]																		
Armoracia rusticana Horseradish	71	100	21	3.30	0.23	2.50	2.20	trace	107	0.17	0.1	0.6	111	53	442	11	57	2.2

Species / Common name																		
Barbarea verna Cress, upland cress	90	30	6	3.00	1.00	1.00		9300	70	0.1	0.3	1	80	80	610	10		1
Brassica juncea Indian mustard	91.6	27	4.9	2.69	0.42	0.99	1.26	6000	93	0.1	0.19	0.8	181	46	374	33	29	2
Brassica napus Rutabaga	89.3	41	9.6	1.10	0.12	1.20	0.79	455	38	0.07	0.07	1	59	41	257	8	16	0.4
Brassica napus Siberian kale	87	42	8.3	2.80	0.60	1.23	1.28	3100	130	0.07	0.06	1.3	205	62	450	70	88	3
Brassica oleracea Collard	89.7	36	6.6	3.40	0.60	0.95	1.20	5350	86	0.15	0.24	1.2	202	55	420	43	45	1.2
Brassica oleracea Kale	85	41	6.8	4.00	0.80	1.40		7150	119	0.11	0.21	1.7	177	71	383	55	35	2
Brassica oleracea Chinese kale, broccoli, kailan	88							7540	115				62					2.2
Brassica oleracea Cauliflower	91.1	27	5.2	2.60	0.22	0.88	0.80	75	75	0.1	0.1	0.7	25	57	328	16	20	1
Brassica oleracea Cabbage, white	92.5	24	5.1	1.40	0.22	0.09	0.66	150	50	0.06	0.05	0.3	49	29	272	17	16	0.5
Brassica oleracea Cabbage, red	91.7	29	6.1	1.60	0.22	1.00	0.68	40	58	0.08	0.05	1.4	43	36	257	17	17	0.7
Brassica oleracea Cabbage, savoy	90.5	27	4.9	2.50	0.24	0.60	0.90	600	47	0.06	0.06	0.3	54	51	262	20	20	0.8
Brassica oleracea Brussels sprouts	84.8	46	8.6	4.20	0.40	1.60	1.36	520	98	0.11	0.15	0.9	61	80	404	22	26	1.5
Brassica oleracea Kohlrabi	91	29	6.2	2.00	0.10	1.10		25	64	0.56	0.38	0.5	42	50	369	17	27	0.5
Brassica oleracea Broccoli	89.2	32	5.7	3.60	0.35	1.40	1.10	3150	109	0.1	0.2	1	101	77	389	23	23	1.2
Brassica perviridis Tendergreen mustard	92.2	22	3.9	2.20	0.30	1.00	1.40	9900	130				210	28				1.5
Brassica rapa Pak choi, Chinese or mustard celery	93.8	17	3.1	1.70	0.23	0.67	0.83	3300	57	0.06	0.1	0.8	118	39	234	71	19	0.8
Brassica rapa Choy sum, mock pak choy	95		1.2	1.20	0.20			5800	53	0.04	0.07	0.5	102	37	181	100	27	2
Brassica rapa Chinese cabbage, Pe tsai	95	14	2.7	1.20	0.15	0.57	0.69	625	25	0.05	0.04	0.6	71	35	235	44	13	0.6

(continued)

807

APPENDIX C *(cont.)*

	% Water	Calories	CHO	Pro	Fat	Fib	Ash	A	C	B1	B2	Niacin	Ca	P	K	Na	Mg	Fe
Brassica rapa Turnip (root)	91.5	30	6.6	1.00	0.19	0.90	0.74	trace	30	0.05	0.06	6	39	32	207	50	15	0.5
Brassica rapa Turnip greens	90.5	28	5.2	2.10	0.35	0.80	1.60	5740	93	0.14	0.33	0.8	224	66	309	32	45	1.5
Brassica rapa Rappini, Broccoli raab	92	18	2	1.80	0.20	1.00	1.20	2700	70	0.05	0.07	0.5	125	45	250	40	45	1.5
Crambe maritima Sea kale, blanched	94	13	1	2.00	0.30			100	26	0.04	0.04	0.3	35	34				0.5
Crambe maritima Sea kale, green leaves	90	22	1.3	3.50	0.30			4600	87	0.16	0.1	0.5	110	63	360	30	64	0.9
Eruca sativa Roquette, rocker salad	91.8	21	3.7	2.70	0.20	0.90	1.60						352	46				0.8
Eutrema wasabi Japanese horseradish, wasabi	76.7	80	15.3	5.10	0.20	1.40	1.30	50	80	0.15	0.1	0.5	93	72				0.8
Lepidium sativum Garden cress	88.2	40	5	3.75	1.10	1.10	1.83	9300	73	0.11	0.16	1.2	179	48	592	12		2.5
Nasturtium officinale Watercress	93.5	19	3	2.15	0.30	0.85	1.10	4800	67	0.07	0.15	0.6	143	57	293	39	24	1.9
Raphanus sativus Radish (root)	94.4	18	3.7	0.90	0.18	0.66	0.72	9	26	0.03	0.03	0.3	29	28	237	25	11	1
Raphanus sativus Radish (pods)	90.5		4.7	2.30	0.30	1.40		50	69	0.07	0.05	0.2	80	100				2.8
Raphanus sativus Radish (sprouted seed)	90	41	3.1	3.80	2.50		0.53	391	29	0.1	0.1	2.9	51	113	86	6	44	0.9
Raphanus sativus Black radish	93.5	21	3.9	1.05	0.15	0.70	0.75	trace	30	0.30	0.03	0.4	33	29	322	18	15	0.8
Raphanus sativus cv. *longipinnatus* Daikon	94.6	19	4.2	0.75	0.10	0.67	0.58	10	27	0.03	0.02	0.3	31	24	203	21	16	0.5
CACTACEAE																		
Opuntia ficus-indica Cactus (cladode)	91.9	37	6.9	1.30	0.40	2.10	0.90		21	0.27	0.5	0.3	110	20				1.4
Opuntia ficus-indica Cactus (tuna)	85			0.50	0.10	1.80	1.60	50	30				60					

CHENOPODIACEAE																		
Beta vulgaris — Table beet	87.7	43	9.6	1.67	0.11	0.90	0.97	20	12	0.03	0.05	0.36	20	40	310	105	23	0.8
Beta vulgaris — Table beet greens	92.3	23	4.5	2.00	0.35	1.70	1.90	6250	31	0.1	0.21	0.4	121	40	565	148	89	3.3
Beta vulgaris, var. *orientalis* — Palak, spinach beet	86.4		6.5	3.40	0.80	0.70		9770	70	0.26	0.56	3.3	380	30				16.2
Beta vulgaris var. *cicla* — Swiss chard	92.3	24	4.2	2.20	0.28	0.77	1.64	5000	35	0.07	0.17	0.5	93	39	380	205	73	2.3
Chenopodium ambrosioides — Epaspote, Mexican tea	85.5	42	7.6	3.80	0.70	1.30	2.40		11	0.60	0.28	0.6	304	52				5.2
Chenopodium album, C.quinoa — Lambsquarter, quinoa	84	44	7.3	4.30	0.80	2.10	3.30	11300	90	0.15	0.4	1.3	280	81				2.1
Spinacia oleracea — Spinach	91.5	25	4	3.20	0.33	0.65	1.70	7045	50	0.4	0.2	0.6	107	57	605	110	92	2.7
CONVOLVULACEAE																		
Ipomoea aquatica — Water spinach, Kang kong	90.6	26	4.6	3.10	0.38	1.20	1.60	4600	50	0.07	0.17	1.1	84	49	385	43	60	2.7
Ipomoea batatas — Sweet potato, orange root	69.4	118	27	1.63	0.40	0.83	0.85	14000	23	0.90	0.06	0.6	35	60	343	60	30	0.8
Ipomoea batatas — Sweet potato (leaves and shoot tips)	87	48	8	3.50	0.60	1.70	1.40	3300	23	0.12	0.28	1.2	147	78	540	9	35	2.6
CUCURBITACEAE																		
Benincasa hispida — Mo kwa, immature fruit	96	13	2.9	0.71	1.17	0.70	0.30	210	39	0.05	0.07	0.4	22	20	111	6		0.5
Benincasa hispida — Chinese wax/winter gourd, mature fruit	96	9	2.3	0.20	0.10			trace	14	0.02	0.03	0.5	14	7	200	2	16	0.4
Citrullus colocynthis — Colocynth (dried seed)	6.7	556	19.5	23.60	47.20	1.50	3.00			0.11		0.2	46	580				
Citrullus lanatus — Watermelon (fruit)	92.8	24	6.4	0.50	0.16	0.50	0.30	235	6	0.04	0.05	0.2	8	11	106	2	10	0.2

(continued)

APPENDIX C (cont.)

	% Water	Calories	CHO	Pro	Fat	Fib	Ash	A	C	B1	B2	Niacin	Ca	P	K	Na	Mg	Fe
Citrullus lanatus Watermelon (dried seed)	5.7	567	15.1	25.80	49.70	4.00	3.70			0.1	0.12	1.4	53	30			7.4	1.4
Coccinia grandis Ivy gourd (fruit)	93.5	18	3.1	1.20	0.10	1.60			15	0.07	0.08	0.7	40	38				
Cucumis anguria West India gherkin	93	17	2	1.40	0.30	0.60	0.50	270	51	0.1	0.04	0.4	26	14	290	6	32	0.6
Cucumis melo Melon, Casaba	90	26	7.3	1.00	0.10	0.50	0.50	trace	19	0.06	0.02	0.4	15	14	210	10	10	0.5
Cucumis melo Melon, Honeydew	87	41	10.3	0.90	0.10	0.50	0.50	500	32	1.06	1.02	0.6	14	206	330	17	10	0.3
Cucumis melo Melon, cantaloupe (orange flesh)	93.7	20	4.5	0.50	0.18	0.43	0.55	2245	31	0.06	0.04	0.6	20		6	12	0.7	
Cucumis sativus Cucumber (fruit)	96.3	14	2.4	0.75	0.17	0.60	0.50	250	11	0.30	0.40	0.2	19	25	166	6	11	0.9
Cucumis sativus Cucumber, pickling fruit	96	12	2.4	0.70	0.10	0.60	0.50	270	19	0.04	0.20	0.4	13	24	190	6	14	0.6
Cucumis sativus Cucumber (leaves)	90.4	26	3.4	4.20	0.40	1.50	1.60		58	0.17	0.17	1.8	127	96				5.8
Cucumis sativus Cucumber (flowers)	94.8	16	2.7	1.40	0.30	0.60	0.80		18	0.02	0.11	0.6	47	86				1
Cucurbita spp. Winter squash (ave. of fruit)	88.8	43	10.5	1.40	0.17	1.32	0.80	1980	10	0.06	0.1	0.6	25	36	300	2	22	0.7
Cucurbita spp. Winter squash (ave. of leaves)	88	34	4.9	3.90	0.50	1.80	1.70	1942	45	0.07	0.22	0.9	303	117	436	11	38	1.5
Cucurbita spp. Winter squash (ave. of flowers)	92.2	27	4.5	1.60	0.08	0.65	0.48	1947	28	0.04	0.08	1.7	80	55	173	5	24	0.7
Cucurbita spp. Winter squash (ave. of dried seeds)	4.3	548	19.7	27.10	44.70	2.00	4.20		2	0.2	0.13	1.7	45	1095				2.8
Cucurbita ficifolia Malabar gourd	93.6	21	5.1	0.80	0.10	0.40	0.40	4000	11	0.40	0.03	0.3	15	19				0.4
Cucurbita pepo Pumpkin (fruit)	91.4	27	6	1.00	0.10	1.20	0.80	1600	9	0.16	0.25	0.6	22	44	360	1	10	0.8

Species (part)																	
Cucurbita pepo — Summer squash	93.5	20	4.3	1.10	0.15	0.65	0.65	319	19	0.05	0.09	0.8	22	27	188	2	19	0.5
Cyclanthera pedata — Cyclanthera, caihua (fruit)	94.1	17	4	0.60	0.10	0.70	0.70		14	0.04	0.04	0.3	14	26		3	11	0.8
Lagenaria siceraria — Bottle gourd (immature fruit)	94.7	15.5	3.6	0.46	0.10	0.50	0.40	10	12	0.04	0.02	0.4	23	18	143	3	11	0.5
Lagenaria siceraria — Bottle gourd (dried seed)	3.2	574	15	28.20	49.80	2.00	4.20			0.4	0.26	4.6	75	1100				5.3
Lagenaria siceraria — Kanpyo (dried fruit wall strips)	19.97	258	65	8.58	0.56	9.13	5.86		1		0.04	2.9	280	188	1582	15	125	5.1
Luffa acutangula — Angled loofah (immature fruit)	94.6	17	3.7	0.90	0.15	0.50		315	12	0.04	0.04	0.4	18	26				0.5
Luffa aegyptiaca — Smooth loofah, sponge gourd (fruit)	93.7	20	4.1	1.00	0.20	1.00	0.40	430	10	0.03	0.05	0.4	24	26	118	3	19	0.7
Momordica charantia — Balsma pear, bitter melon (fruit)	93.3	21	4.7	1.00	0.16	1.20	0.80	380	137	0.05	0.06	0.4	29	32	296	5	17	0.7
Momordica charantia — Balsam pear (leaves)	80.1	60	12.3	5.10	0.40	0.50	2.10		247	0.05	0.47		264	666				7.1
Momordica charantia — Balsam pear (shoot tips)	89.2	30	3.29	5.30	0.69	2.28	1.47	1735	88	0.18	0.36	1.1	84	99	608	11	85	2
Momordica cochinchinensis — Sweet gourd (fruit)	90.2	29	6.4	0.60	0.10	1.60							27	38				
Momordica dioica — Balsam apple (fruit)	84.1	52	7.7	3.10	1.00	3.00				0.05	0.18	0.6	33	42				4.6
Sechium edule — Chayote (fruit)	91.7	26	6.3	0.71	0.15	0.70	0.40	43	20	0.60	0.1	0.5	19	20	107	4	14	0.5
Sechium edule — Chayote (leaves)	89.1	60	4.7	3.70	0.40	3.00	1.50	4560	16	0.08	0.18	1.1	58	108				2.5
Sechium edule — Chayote (roots)	79	79	17.8	2.00	0.20	0.40	1.00	trace	19	0.05	0.03	0.9	7	34				0.8
Telfairia occidentalis — Fluted gourd (leaves)	86.4	47	7	2.90	1.80	1.70	1.90											

(continued)

APPENDIX C (cont.)

	% Water	Calories	CHO	Pro	Fat	Fib	Ash	A	C	B1	B2	Niacin	Ca	P	K	Na	Mg	Fe
Telfairia pedata Oyster nut (dried seed)	6.2	543	23.5	20.50	45.00	2.20	4.80						84	572				
Trichosanthes cucumerina Snake gourd (immature fruit)	94.3	18	3.6	0.70	0.30	0.80		235	12	0.04	0.05	0.5	26	34				0.8
Trichosanthes dioica Pointed gourd (immature fruit)	92	20	2.2	2.00	0.30	3.00			29	0.05	0.06	0.05	30	40				1.7
EUPHORBIACEAE																		
Cnidoscolus chayamansa Chaya (leaves)	79.8	64	10.7	6.20	0.60	2.00	2.00		194	0.2	0.2	1.6	234	76	270	58	88	2.8
Manihot esculenta Cassava, manioc, yuca (root)	60	148	35	1.20	0.30	1.40	1.56	trace	36	0.05	0.03	0.7	35	75				0.7
Manihot esculenta Cassava (leaves)	84.2	91	18.3	7.00	1.00	3.20	2.30	12450	316	0.26	0.5	3	297	116				7.8
Sauropus androgynus Katuk, common sauropus (leaves)	79.8	310	6.9	7.60	1.80	1.90	2.00	10000	136	0.23	0.15		234	64				3.1
FABACEAE [LEGUMINOSEAE]																		
Arachis hypogaea Peanut groundnut (dry seed)	6.5	549	23	23.20	44.80	2.90	2.50	1	1	0.79	0.14	15.5	49	409				3.8
Arachis hypogaea Peanut (leaves)	78.5	69	14.9	4.40	0.60	4.60	1.60	7735	98	0.23	0.58	1.6	262	82				4.2
Cajanus cajan Pigeon pea (mature seeds)	67.5	122	23.1	7.40	0.90	3.30	41.60	140	39	0.39	0.17	2.1	44	136	552	5		1.7
Cajanus cajan Pigeon pea (fresh seed)	9.9	343	64.7	19.95	1.35	7.15	3.80	115	20	0.72	0.14	2.9	161	285	26			15
Canavalia ensiformis Jack bean (dry seeds)	10	347	59	24.50	2.60	7.40				0.77	0.15	1.8	158	298				7
Canavalia ensiformis Jack bean (fresh pods)	8	37	8	2.40	0.30	1.80												
Canavalia gladiata Sword bean (dry seeds)	10.7	347	59	24.50	2.60	7.40	3.20	1	1	0.15	1.8		158	298				

Canavalia gladiata Sword bean (fresh pods)	83.6	59	10.7	4.60	0.40	2.60	0.70	40	32	0.22	0.1	2	33	66				1.2
Cicer arietinum Chick pea, garbanzo (dry seeds)	10.8	360	61	20.40	4.50	4.80	2.90	50	3	0.4	0.16	1.8	125	365	688	26	108	7
Cyamopsis tetragonolobus Cluster bean (fresh pods)	81		10.8	3.20	0.40			316	47	0.09	0.09							
Glycine max Soybeans (fresh seeds)	67.5	141	12.5	13.70	5.10	1.80	1.70	410	25	0.44	0.18	1.7	107	205				2.8
Glycine max Soybean sprouts	90	46	5.3	6.20	1.40	0.80		80	13	0.23	0.2	0.8	48	67				1
Lablab purpureus Hyacinth bean (fresh pods)	87	46	8	2.90	0.45	1.50	0.64	210	11	0.9	0.08	0.6	130	59	163	2	37	1.2
Lablab purpureus Hyacinth bean (dry seeds)	10.9	340	62.1	22.80	1.00	8.60	3.20			0.54	0.14	2.3	90	328				9
Lathyrus sativus Chickling, grass pea (dry seeds)	9.2	348	59	27.80	0.80	7.30	3.20						127	410				10
Lens culinaris Lentil (dry seeds)	11.8	342	59.5	24.50	1.15	3.85	3.20	60	5	0.44	0.23	2.1	73	397	895	17	79	6.9
Pachyrhizus erosus Jicama, yam bean (tuberous stem)	87.4	40	8.8	1.20	0.20	0.60	0.30	trace	20	0.05	0.03	0.3	17	18	144	4	16	0.5
Pachyrhizus erosus Jicama, (immature pods)	86.4	45	10	2.60	0.30	2.90	0.70	575	1056	0.11	0.09	0.8	121	39				1.3
Phaseolus acutifolius Tepary bean (dry seeds)	9	353	63.6	20.70	1.30	4.10	3.60			0.33	0.12	2.8	112	310				
Phaseolus coccineus Scarlet runner bean (fresh seeds)	34.2	250	46.3	16.40	0.30	12.20	2.80			0.54	0.14	2.3	61	227				4.1
Phaseolus lunatus Lima beans (fresh seeds)	68.3	105	22.2	7.90	1.20	1.60	1.75	250	29	0.22	0.11	1.4	102	165	460	3	51	2.8

(continued)

APPENDIX C (cont.)

	% Water	Calories	CHO	Pro	Fat	Fib	Ash	A	C	B1	B2	Niacin	Ca	P	K	Na	Mg	Fe
Phaseolus lunatus Lima beans (dry seeds)	12.7	334	61.3	21.00	1.40	4.30			2	0.42	0.17	2	68	381				7.5
Phaseolus vulgaris Snap beans (fresh green pods)	90.9	33	6.6	2.00	0.19	1.20	0.76	550	19	0.11	0.11	0.7	50	41	220	8	34	0.9
Phaseolus vulgaris Snap beans (ave. of dried seeds)	12.4	340	57	21.00	1.50	4.00	4.20		24	0.5	0.27	2.8	150	425	1540	4	132	6.5
Phaseolus vulgaris Field beans (ave. of dried seeds)	11.5	343	62.7	22.20	1.40	4.00	3.90		2	0.07	0.02	2.1	163	437				6.9
Pisum sativum Peas (immature seeds)	74.6	87	15.4	6.50	0.55	2.20	0.86	780	33	0.33	0.15	2.6	30	118	285	4	35	1.8
Pisum sativum Peas (dried seeds)	11.3	343	60.1	23.80	1.35	4.75	2.80	120	2	0.75	0.29	3	60	362	798	25	128	3.1
Pisum sativum Edible podded peas	87.6	38	7.9	2.60	0.13	1.80	0.58	405	52	0.13	0.09	0.7	50	50	185	4	24	1.3
Psophocarpus tetragonolobus Winged bean (fresh pods)	90	27	4.9	2.60	0.50	1.90	0.67	332	15	0.21	0.1	0.8	64	37	214	4	34	0.8
Psophocarpus tetragonolobus Winged bean (dry seeds)	11	1697	32	33.00	16.00	5.00	3.00											
Psophocarpus tetragonolobus Winged bean (fresh leaves)	76.9	74	14.1	5.85	1.10	2.50	2.10						224	63	176		8	4
Psophocarpus tetragonolobus Winged bean (root)	66.4	125	24.2	7.20	0.75	2.40	1.50						30	45				0.2
Pueraria lobata Kudzu (tuberous root)	68.6	113	27.8	2.10	0.10	0.70	1.40											
Sphenostylis stenocarpa African yambean (tuberous root)	64.8	30	12.9	3.80	0.20	0.40	0.80						10	80				
Trigonella foenum-graecum Fenugreek (leaves)	86.9	35	6.1	4.50	0.55	1.20	1.40		54	0.50		1.1	255	50	51	8	67	17.2

Species																		
Vicia faba Broad bean (fresh seeds)	75.4	88	15.8	7.30	0.47	3.10	1.00	350	33	0.18	0.17	1.6	30	121	200	50	38	2.1
Vigna angularis Adzuki bean (mature seeds)	15.5	339	54.4	20.30	2.20	4.30	3.30			0.45	0.16	2.2	75	350	1500	1		5.4
Vigna mungo Urd, black gram (mature seeds)	10	334	59.4	24.00	1.10	4.90	5.50			0.48	0.21	2.3	110	382				8.9
Vigna radiata Mung bean (mature seeds)	10	340	60.3	24.20	1.30	4.40	4.00	80		0.4	0.2	2.6	118	340	1028	6		7.7
Vigna radiata Mung bean sprouts	91.4	31	6.2	3.10	0.14	0.70	0.42	21	17	0.1	0.08	0.7	18	48	158	4	21	0.8
Vigna subterranea Bambara groundnut (mature seeds)	10	367	60	20.00	6.90	6.80	3.70	30	1	0.25	0.13	2.1	75	415	1290	12	260	10
Vigna umbellata Rice bean (mature seeds)	11.9	327	60.7	20.90	0.90	4.50	4.30			0.56	0.16	2.2	300	331				8.7
Vigna unguiculata cultgp. *sesquipedalis* Yardlong bean (immature pods)	88.6	40	6.8	3.00	0.33	1.30	0.63	750	26	0.09	0.11	0.8	56	60	240	4	44	1.1
Vigna unguiculata cultgp. *unguiculata* Cowpea (immature pods)	85.3	39	7.9	3.00	0.28	1.80	1.25	1250	20	0.1	0.1	0.9	64	60	205	3	54	1.7
Vigna unguiculata cultgp. *unguiculata* Cowpea (fresh seeds)	66.8	127	22	9.10	0.75	1.80	1.60	365	29	0.34	0.13	1.1	29	145	518	3	51	2.1
Vigna unguiculata cultgp. *unguiculata* Cowpea (mature seeds)	10.8	343	61.6	22.90	1.45	4.30	3.30	30	2	0.99	0.19	2.2	75	438	987	37	230	6.2
LAMIACEAE (LABIATAE) *Coleus, Plectranthus, Solenstemon* spp. Coleus, kaffir, hausa, Sudan potato	77	19	22	1.30	0.30	1.10	1.00	540	1	0.05	0.02	1	17					6

(continued)

APPENDIX C *(cont.)*

	% Water	Calories	CHO	Pro	Fat	Fib	Ash	A	C	B1	B2	Niacin	Ca	P	K	Na	Mg	Fe
Perilla frutescens Perilla, beefstake plant, shiso	88.6	34	4.7	2.80	0.60	1.50	1.80	6600	85	0.1	0.4	0.5	197	76				10.1
MALVACEAE																		
Abelmoschus esculentus Okra (immature pods)	88.9	36	7.8	2.20	0.24	1.00	0.80	610	30	0.18	0.15	0.9	89	57	234	7	46	0.9
Abelmoschus manihot Sunset hibiscus (leaves)	90	150	4	4.10	0.40	1.00		900	118				580					3
Hibiscus sabdariffa Roselle (fresh calyces)	86.2	44	11.1	1.60	0.10	2.50	1.00		14	0.04	0.06		160	60				3.8
Hibiscus sabdariffa Roselle (leaves)	85.6	43	9.2	3.30	0.30	1.60	1.60		54	0.17	0.45	1.2	213	93				4.8
Malva spp. Edible malva (leaves)	83.9	47	8	4.90	0.70	4.00	2.40		35	0.13	0.2	1	287	68				12.7
MORACEAE																		
Artocarpus altilis Breadfruit (fruit)	71.9	109	25.4	1.50	0.30	1.80	0.90		31	0.80	0.05	0.7	28	34				2
Artocarpus altilis Breadfruit (dry seeds)	4.2	391	76.3	11.30	4.90	2.80	3.30						110	350				
Artocarpus heterophyllus Jackfruit (fruit)	72.5	98	24.4	1.00	0.50	0.90	0.80		8	0.30		0.4	22	38				
Artocarpus heterophyllus Jackfruit (fresh seeds)	51.6		38.4	6.60	0.40	1.50	1.50											
MORINGACEAE																		
Moringa pterygosperma Horseradish tree (leaves)	75	75	11.7	6.70	1.10	0.90	2.38	10615	177	0.06	0.08	3.2	472	120	337	20	147	6.7
Moringa pterygosperma Horseradish tree (fruit)	87.6	37	6.1	2.30	0.15	3.00	1.50	129	120	0.05	0.07	0.4	30	80	461	30	45	4.5
NYMPHAEACEAE																		
Nelumbo nucifera Lotus root	79.1	63	16.4	2.70	0.10	0.80	0.97	0	60	0.15	0.16	0.4	41	102	643	40	23	1.3
Nelumbo nucifera Lotus root (dry seeds)	8.3	365	74.1	12.80	2.30	4.40	2.40						136	294				2.3
OXALIDACEAE																		
Oxalis tuberosa Oca (tubers)	83.8	63	13.8	1.00	0.60	0.80	0.80		37	0.05	0.07	0.4	4	34				0.8

PHYTOLACCACEAE																	
Phytolacca americana Poke (leaves and shoots)	91.6	23	3.7	2.60	0.40	1.70	1.70	8700	136	0.08	0.33	1.2	53	44				1.7
POLYGONACEAE																	
Rheum rhabarbarum Rhubarb (petioles)	93.3	18	3.8	0.74	0.13	0.75	1.10	100	10	0.03	0.04	0.3	130	21	360	6	21	0.9
Rumex acetosa, R. patientia Sorrel and dock (leaves)	92.8	25	4.63	1.93	0.40	0.82		8000	62	0.07	0.16	0.5	63	50	364	5	103	2.8
PORTULACACEAE																	
Portulaca oleracea Purslane (leaves and young shoots)	93.9	19	3.6	1.50	0.25	0.85	1.25	1320	25	0.04	0.1	0.5	84	41	494	45	68	2.7
Talinum triangulare Water leaf (leaves)	90.5	26	4.2	2.50	0.50	0.90	2.20		39	0.09	0.17	0.4	120	40				5.3
SOLANACEAE																	
Capsicum annuum Pepper (mature pungent red fruit)	86	65	15.8	2.30	0.40	2.30		11000	260	0.1	0.2	2.9	9	49	420	17	27	1.1
Capsicum annuum Pepper (mature green fruit)	92.8	24	5.06	1.05	0.33	1.30	0.62	500	144	0.08	0.07	0.5	8	22	204	5	16	0.8
Capsicum frutescens Tabasco pepper	82.6	61	12.9	2.60	1.00	1.70	0.90	770	163	0.14	0.13	1.5	33	60	340	7	25	2
Cyphomandra betacea Tree tomato, tamarillo (fruit)	85.5	53	12.9	1.50	0.30	2.60	1.00	150	22	0.04	0.04	1	33	22				0.7
Lycium chinense Chinese boxthorn (leaves)	88.9	33	4.6	4.10	0.50	1.40	2.20		8	0.08	0.03	0.8	187	50	498	184		4.3
Lycopersicon lycopersicum Tomato	93.5	22	4.75	1.05	0.20	0.55	0.50	900	25	0.06	0.04	0.7	12	26	244	3	14	0.5
Lycopersicon pimpinellifolium Cherry tomato	93.2	22	4.9	1.00	0.20	0.40	0.70	2000	50	0.05	0.04		29	62				1.7

(continued)

APPENDIX C (cont.)

	% Water	Calories	CHO	Pro	Fat	Fib	Ash	A	C	B1	B2	Niacin	Ca	P	K	Na	Mg	Fe
Physalis ixocarpa Husk tomato, tomatillo, fruit	89	38	7.8	2.10	0.40	1.90	1.00	380	10	0.49	0.3	3.3	20	40	420	16	81	1.8
Physalis peruviana Cape gooseberry	87.8	40	8.2	2.20	0.60	2.00	1.20		9	0.09	0.04	2.6	15	45				1
Solanum americanum Glossy nightshade (green fruit)	90	140	7.4	1.90	0.10				17	0.1			274					4
Solanum americanum Glossy nightshade (leaves)	85	190	8.1	4.70	0.50				40	0.14			210					6.1
Solanum macrocarpon African eggplant "brunei" (fruit)	89	40	8	1.40	1.00	1.50	0.60						13					
Solanum macrocarpon African eggplant (leaves)	86	42	6.2	4.60	1.00	1.60	2.40						391	49				
Solanum melongena Eggplant, aubergine (fruit)	91.8	26	6.1	1.10	0.25	1.05	0.06	30	6	0.05	0.05	0.7	18	28	214	2	16	0.8
Solanum muricatum Pepino (fruit)	92		7			1.00	0.30		47	0.06	0.04	0.4	4	166				0.6
Solanum melanocerasum Black nightshade, wonder berry	89	24	1.1	2.00	0.10			570	12	0.1	0.06	0.7	24	42	510	2	40	0.6
Solanum torvum Turkeyberry	89		7.9	2.00	0.10			750	80	0.08			50	30				2
Solanum tuberosum White or Irish potato	78.4	82	17.9	2.10	0.10	0.50	0.90	20	18	0.1	0.04	1.4	12	51	420	10	27	0.7
TETRAGONIACEAE *Tetragonia tetragonioides* New Zealand spinach (leaves)	93	28	3	2.00	0.30	0.70	1.60	4350	31	0.04	0.15	0.6	60	36	290	145	40	1.5
TILIACEAE *Corchorus olitorius* Jew's mallow	84.1	46	9.1	4.60	0.28	1.60	2.00	5985		0.14	0.54	1.2	284	103	559	8	64	6

	% Water	Cal	CHO	Pro	Fat	Fib	Ash	A	C	B1	B2	Niacin	Ca	P	K	Na	Mg	Fe
TRAPACEAE																		
Trapa bicornis, T.natans Water chestnut (fresh seeds)	79.5	73	17	2.50	0.20	0.50	0.90	16		0.05		0.6	10	85				0.7
TROPAEOLACEAE																		
Tropaeolum tuberosum (dry wt.) Mashua, tuber		371	78.6	11.40	4.30	5.70	5.70	476		0.43	0.57	4.2	50	300				8.6
VALERIANACEAE																		
Valerianella locusta, V.eriocarpa Corn salads, leaves	93.1	22	3.4	2.00	0.40	0.80	1.00	35		0.07	0.08	0.4	35	49	421	4	13	1.7

*All column values are averages based on 100 grams of fresh edible portion. Columns, % Water and Calories represent grams of water and kilocalories of energy, respectively. Columns, CHO, Pro, Fat, Fib, and Ash represent grams of carbohydrate, protein, fat, fiber and ash, respectively. Column A represents Vitamin A as International Units. The other vitamins and minerals in columns, C, B1, B2, Niacin, Ca, P, K, Na, Mg, and Fe each are respectively, milligrams of vitamin C, vitamin B1 (thiamine), vitamin B2 (riboflavin), niacin, calcium, phosphorus, potassium, sodium, magnesium, and iron.

Most of the proximate values presented in this table were obtained from the following references:

Haytowitz, D.B., and Matthews, R.H., 1984. Composition of Foods: Vegetables and Vegetable Products Raw, Processed, revised. USDA Agricultural Handbook 8-11. USDA, Washington, D.C.

Howard, F.D., MacGillivray, J., and Yamaguchi, M., 1962. Nutrient Composition of Fresh California Grown Vegetables. California Ag. Expt. Sta. Bull. 788.

Pennington, J.A.T., and Church, H.N., 1980. Biwes and Church's Food Values of Portions Commonly Used. J.B. Lippincott Co., Philadelphia.

Souci, S.W., Fachmann, W., and Kraut, H., 1981. Food Composition and Nutrition Tables 1981/82. Verlagsgesellschaft mBH., Stuttgart, Germany.

Many other sources were examined to arrive at what are believed to be representative values for the various vegetable edible portions. Because of variations in the analytical procedures used and the condition of the analyzed products, these values cannot be exact and, therefore, are offered as a guide.

Appendix D

RECOMMENDED POSTHARVEST TEMPERATURE, RELATIVE HUMIDITY AND
APPROXIMATE STORAGE LIFE OF VARIOUS VEGETABLES

Commodity	Storage temperature (°C)	% Relative humidity	Approximate storage life
Adzuki bean sprouts	0–5	95–100	10–11 days
Alfalfa sprouts	0	>90	3–7 days
Amaranth	0–2	95–100	10–14 days
Artichoke, Chinese	0–2	90–95	1–2 weeks
Artichoke, globe	0	95–100	2–3 weeks
Artichoke, Jerusalem	0	90–95	3–5 months
Asparagus	2–3	90–95	2–3 weeks
Bamboo shoot	5	80	5–7 days
Beans, dry	4–10	40–50	6–10 months
Beans, snap	6–8	90–95	10–12 days
Beans, Lima	3–5	95	5–7 days
Bean, sprouts	0	95–100	7–9 days
Beets, bunched with tops	0	98–100	10–14 days
Beets, topped	0	98–100	4–6 months
Bitter melon	12–15	85–90	2–3 weeks
Bok choy (pak choy)	0	95–100	14–21 days
Breadfruit	13–15	85–90	2–6 weeks
Broad bean, fresh beans	0–5	95–98	7–10 days
Broccoli	0	90–95	10–14 days
Broccoli raab	0	95–100	1–2 weeks
Brussels sprouts	0	90–95	3–5 weeks
Burdock	0	95–100	4–7 weeks
Cabbage, early-fresh	0	98–100	3–6 weeks
Cabbage, late-stored	0	98–100	5–6 months
Cabbage, Chinese	0	95–100	1–3 months
Cactus, cladodes	5–10	95–98	2–3 weeks
Cactus, tuna	5–7	90–95	2–3 weeks
Cape gooseberry	12–15	80	1–2 months
Carrots, bunched	0	98–100	10–14 days
Carrots, mature (topped)	0	98–100	1–6 months
Carrots, immature (topped)	0	98–100	4–6 weeks
Cassava	14–15	85–90	2–3 weeks
Cardoon	2–5	95–100	2–3 weeks
Cauliflower	0	95–98	3–4 weeks
Celeriac	0	97–99	6–8 months
Celery	0	98–100	2–4 weeks
Chard, Swiss	0	95–100	10–14 days

(continued)

820

APPENDIX D   *(cont.)*

Commodity	Storage temperature (°C)	% Relative humidity	Approximate storage life
Chayote	8–10	85–90	2–3 weeks
Chicory, leaf	0	95–100	2–3 weeks
Chicory, raddichio	0	95–100	2–3 weeks
Chicory, witloof	2–3	95–98	2–4 weeks
Chinese water chestnuts	0–2	98–100	1–2 months
Chinese winter melon	10	85–90	2–3 months
Chives	0	98–100	5–7 days
Chrysanthemum greens	0–2	95–98	1–2 weeks
Collards	0	95–100	10–14 days
Corn, sweet	0	95–98	5–8 days
Corn, sweet (super sweet)	0	95–98	10–15 days
Cowpea, pods	2–5	95–98	5–7 days
Cowpea, fresh beans	0–2	98–100	7–10 days
Cucumbers	10–13	90–95	10–14 days
Daikon	0	95–100	2–3 months
Dandelion	0	95–100	7–10 days
Eggplant	10–12	90–95	10–14 days
Eggplant, Japanese	8–10	90–95	6–8 days
Endive	0	95–100	2–3 weeks
Endive, (Belgium/witloof)	2–3	95–98	2–4 weeks
Escarole	0	95–100	2–3 weeks
Fennel, Florence	0	90–95	2–4 weeks
Garlic	0	65–70	4–7 months
Gherkin, West Indian	5–10	85–90	1–2 weeks
Ginger	12–14	80–90	3–6 months
Greens, leafy, cool season	0	95–100	10–14 days
Horseradish	−1–0	95–98	8–10 months
Jackfruit	13–15	85–90	2–6 weeks
Jicama	13–18	70–80	1–2 months
Kale	0	95–100	10–14 days
Kohlrabi	0	98–100	3–4 weeks
Lagenaria, young fruit	10–12	85–90	7–10 days
Leek	0	95–100	2–3 months
Lettuce, crisphead	0	98–100	2–3 weeks
Lettuce, butterhead	0	95–98	8–12 days
Lettuce, romaine	0	98–100	2–3 weeks
Lettuce, looseleaf	0	95–98	2 weeks
Lettuce, stem	0	98–100	1–2 months
Luffa, cylindrical	10–12	95–98	1–2 weeks
Luffa, angular	10–12	95–98	1–2 weeks
Melons			
Cantaloupe ¾ slip	4–5	85–90	12–15 days
Cantaloupe full slip	2–4	85–90	5–12 days
Casaba	10	85–90	3–4 weeks
Crenshaw	7–10	85–90	14–21 days
Honey dew	7–10	85–90	3–4 weeks
Persian	7–10	85–90	2–3 weeks
Mushrooms *(Agaricus)*	0	90–95	7–10 days
Mustard greens	0	95–100	2–3 weeks
New Zealand spinach	0	95–100	2–3 weeks
Okra	8–10	90–95	7–10 days
Onion, green	0	95–100	7–14 days
Onion, dry bulb	0	65–70	1–6 months
Onion, sets	0	65–70	4–6 months
Parsley	0	95–100	1–2 months
Parsnip	0	98–100	3–5 months

(continued)

APPENDIX D   *(cont.)*

Commodity	Storage temperature (°C)	% Relative humidity	Approximate storage life
Peas, edible podded	0	95–98	1–2 weeks
Peas, green in pods	0	95–98	1–2 weeks
Peas, southern (cowpeas)	4–5	95	6–8 days
Pepino	5	85–90	3–4 weeks
Pepper, dry chiles	0–10	60–70	6 months
Peppers, fresh chiles	5–10	90–95	2–3 weeks
Peppers, sweet	7–10	90–95	2–3 weeks
Plantain, starchy banana	13–15	85–90	1–5 weeks
Potatoes, early crop	7–10	90–95	2–3 weeks
Potatoes, late crop	4–10	90–95	2–5 months
Pumpkins	10–13	70–75	2–5 months
Radish, Chinese	0	95	2–3 months
Radish, garden	0	95–100	3–4 weeks
Radish, winter (black)	0	95–100	2–3 months
Rape, foliage	0	95–98	1–2 weeks
Rhubarb	0	95–100	2–4 weeks
Rocket salad	0–2	95–98	7–10 days
Rutabaga (Swede)	0	98–100	2–4 months
Salsify	0	95–98	2–3 months
Scorzonera	0	95–98	3–6 months
Spinach	0	95–100	10–14 days
Squash, summer	5–10	95	1–2 weeks
Squash, winter	10–13	70–75	2–5 months
Sweet potato	13–16	85–90	4–7 months
Sweet potato, leaf tips	10	95–98	1 week
Tamarillo	3–4	85–95	3–10 weeks
Tannia (Malanga) corms	8–10	70–80	2–4 months
Taro leaves, stems	5–10	85–90	2–3 weeks
Taro corms	12–13	85–90	2–3 months
Tomato, mature green	13–21	90–95	3–4 weeks
Tomato, firm-ripe	8–10	90–95	7–14 days
Turnip	0	95	2–4 months
Turnip greens	0	95–100	10–14 days
Watercress	0	95–100	7–10 days
Water convolvulus	10	95–98	1 week
Watermelon	10–12	85–90	3–4 weeks
Winged bean, pods	7–10	85–90	10–15 days
Yams (*Dioscorea* sp.)	15	80–85	2–3 months
Yardlong bean pods	10–12	85–95	7–10 days

Sources: Hardenberg, R.E., A.E. Watada and C.Y. Wang. 1986. The Commercial Storage of Fruits, Vegetables, and Florist and Nursery Stocks. Agricultural Handbook No. 66. Washington, D.C.: United States Department of Agriculture, and Marita Cantwell-De-Trejo. 1996. University of California, Davis, California.

Glossary

Abaxial: Away from the axis (bottom or lower).

Abscise: Separation of leaves, flowers, fruits, or other plant parts, usually following the formation of a separation (abscission zone) layer.

Achene: A small, usually single-seeded, dry, indehiscent fruit.

Acclimatization: Adaptation of an plant to a changed climate, or adjustment of a species or population to a changed environment, often over many generations.

Adaxial: Side of or toward the axis (top or upper).

Adventitious: Plant organs, such as shoots or roots, produced in an unusual position or at an unusual time of development.

Afteripening: Metabolic changes occurring in some dormant seed before germination can occur.

Aleurone: Outer layer of cells surrounding the endosperm of a cereal grain (caryopsis).

Alkaloid: Nitrogen-containing organic compounds produced by plants, bitter in taste, often poisonous.

Allelopathy: Excretion of chemicals by plants of some species that are toxic to plants of other species.

Allotetraploid: A polyploid having four genomes or basic sets of chromosomes, with one or two sets (usually two) from a species different from that of the other sets.

Amino acids: Organic acids with nitrogen derived from ammonia. Serves as building units for proteins.

Amphidiploid: Tetraploid with two genomes or two sets of chromosomes from each parent species.

Anaerobic: An environment or condition in which molecular oxygen is deficient for chemical, physical, or biological processes.

Androeceum: Male- or stamen-bearing part of a flower.

Angiosperm: Plants having seed in an enclosed ovary.

Annual: Of one season's duration, completes life cycle from germination to maturity and death.

Anthocyanin: Water-soluble nitrogenous plant pigment, usually red, blue, or purple.

Anoxia: Damage due to lack of oxygen.

Anthesis: When flowers first open or anthers dehisce.

Apetalous: Without petals.

Apical dominance: Suppression of axillary growth in favor of apical growth; removal of apex encourages axillary growth (branching).

Apical meristem: Meristematic cells of the apex of a shoot or root.

Arable: Land capable of producing crops requiring tillage.

Arid: Dry, low rainfall, usually less than 250 mm of annual precipitation.

Asexual reproduction: Production of a new plant by any vegetative means not involving meiosis and the union of gametes.

Autotetraploid: Organism with four sets of a similar genome.

Auxin: Growth substance naturally present in plants, indoleacetic acid (IAA).

Available water: Portion of water in soil that can be readily absorbed by roots. That soil moisture held in the soil between field capacity and permanent wilting percentage.

Berry: Fleshy, many-seeded fruit.

Biennial: A plant that normally requires two growing season for its life cycle; vegetative the first season, flowers the second season, and dies.

Blanching: (Horticultural) Prevention of stems and leaves from turning green; to etiolate. (Food Science): To inactivate enzymes by heating using hot water or steam.

Bolt: Emergence of a seed stalk.

Bud: Embryonic or undeveloped shoot.

Bulb: Compressed, often underground stem to which numerous storage scales (modified leaves) are attached. Often surrounded by protective scale leaves; can be used for vegetative reproduction.

Bulbil: Small bulb, usually formed at a leaf axil; can function as a propagule.

Bulbing: Formation of a bulb.

Callus: Mass of large, thin-walled parenchyma cells, often protective tissue formed in the area of a wound.

Calorie (gram calorie): Unit for measuring energy, defined as the heat necessary to raise the temperature of 1 g of water from 14.5°C to 15.5°C at standard pressure.

Calyx: Collective term for the sepals.

Cambium: A zone of meristematic cells which give rise to vascular tissues.

Capitula: Cluster of sessile flowers on a flattened axis.

Capsule: A dry dehiscent fruit.

Carotene: A yellow or orange pigment produced by a plant; depending on structure may have pro-vitamin A value.

Caryopsis: Small, one-seeded, dry fruit of grasses, e.g., corn.

Center of origin: Geographical area in which a species is thought to have evolved through natural selection from its ancestors.

Chilling injury: Physiological injury caused by low or nonfreezing temperature. For sensitive crops, temperatures of 0–10°C, upon sufficient exposure, cause tissue injury.

Chlorophyll: Green pigment in chloroplasts necessary for photosynthesis.

Chlorosis: Condition whereby a plant or plant part is light green or greenish yellow because of poor chlorophyll development or destruction of chlorophyll.

Climacteric: Period in development of some plant parts involving a series of biochemical changes associated with the natural respiratory rise and autocatalytic ethylene production.

Climate: Summation of all weather conditions for a particular region of the earth.

Clone: A group of individuals of common origin, produced by asexual or vegetative means.

Clove: A small bulb developed in the axil of a large bulb; there may be several within a bulb, e.g., garlic.

Compound leaf: A leaf whose blade is divided into a number of distinct leaflets.

Control atmosphere (CA): Storage in which the atmospheric content is regulated.

Corm: A short, thickened, underground stem, usually upright, in which food reserves are stored; it contains undeveloped buds.

Cormel: A daughter corm arising from a parent corm.

Corymbose: Describes an inflorescence in which the outer pedicels are longer than the inner, resulting in a flat or bowl shape.

Culm: Stem of a monocotyledonous plant, often hollow except at swollen nodes, e.g., grass or sedge.

Cultigen: A plant or group with no known place of origin except having originated under cultivation. Not synonymous with cultivar.

Cultivar: A horticultural variety or race that originated under cultivation and not essentially referable to a botanical species; abbreviated as cv.

Curing: Treatment of plants or plant parts to protect from infection, desiccation, or to modify internal composition.

Cutting: A plant part, usually stem, used for propagation.

Cyanogenic glycoside: Toxic and bitter compound; when hydrolyzed releases hydrogen cyanide (HCN).

Day-neutral plants: Plants that are not affected by length of day or dark period with regard to floral initiation.

Deciduous: Falling of leaves or fruit at end of season or functional period.

Dehiscent: Opening or splitting of organ usually at maturity.

Determinate: Plant growth in which the shoot terminates in an inflorescence and further growth is arrested.

Dew point: Temperature at which water vapor in the atmosphere condenses to liquid.

Dioecious: Plants that have male and female flowers on separate plants.

Dormancy: A physiological state of some plant parts, e.g., bud and seed, in which sprouting or rooting do not occur due to unfavorable conditions.

Edaphic: Pertaining to soil conditions.

Endemic: Native to region or a place; not introduced.

Endogenous: Developed or produced internally.

Endosperm: Triploid tissue of seed containing food for the embryo.

Epigeal: Cotyledons appear above soil when seedlings emerge.

Epicotyl: Part of embryo above the attachment of the cotyledon consisting of stem tips and several embryonic leaves.

Ethylene: Gas (C_2H_4) having growth regulating capabilities. Induces physiological responses. Produced by plant tissues, especially by many ripening fruits. Hastens fruit ripening and abscission.

Evapotranspiration: The total loss of water by evaporation from a given area of the soil surface and plant transpiration.

Exogenous: Produced from without.

Family: Category of classification above genus and below order. Suffix of family name is usually "aceae."

Field capacity (FC):   Amount of water retained by soil after gravitational drainage of excess water.

Forcing: A cultural procedure used to hasten flowering or growth of plants outside their natural season.

Genome: Chromosomes corresponding to the haploid set (n set).

Genotype: Genetic constitution, latent or expressed of an organism.

Genus: A group of closely related species clearly different from other groups.

Geotropism: Directional growth of organ in response to gravity.

Glycoalkaloids: Usually a bitter-tasting, nitrogen-containing compound found in tissues of some plant species.

Green manure: Crop grown and plowed under to increase soil organic matter.

Group: Horticultural designation of a botanical variety that includes many cultivars.

Growth regulator: Natural or synthetic hormone affecting plant growth.

Hard pan: A hard layer of soil beneath the tilled zone through which water and root penetrations are difficult.

Hardening: Procedure to acclimate plants to adverse environmental conditions, e.g., low temperature or low moisture.

Hermaphrodite: Flower having both male and female parts (perfect).

Hilum: Scar where seed was attached to the ovary.

Hybrid: A cross between parents that are genetically unlike.

Hydrophyte: Plant that is normally grown in water or wet conditions.

Hypocotyl: Stem of embryo of young seedling below attachment of cotyledon and above the radicle.

Hypogeal: Cotyledons remain below soil when seedlings emerge.

Indeterminate: Shoot axis remains vegetative; does not terminate with an inflorescence.

Indehiscent: Not opening at maturity.

Indigenous: Native or natural to an area or region.

Inflorescence: An axis bearing flowers, or a flower cluster.

Internode: Region between nodes.

Juvenile phase: Stage of plant in which the vegetative growth proceeds and the plant is insensitive to conditions which induce flowering.

Landrace: A sexually reproducing crop variety developed under local natural and human selection.

Leaf area index (LAI): Leaf foliage density expressed as leaf area subtended per unit area or land.

Line: Uniform sexually reproduced population, usually self-pollinated.

Long-day plants: Plants that are inhibited from flowering when the dark period exceeds some critical length.

Lux: Unit of light intensity.

Lycopene: A carotenoid pigment; has no pro-vitamin A value.

Macronutrient: Essential element required by plants in relatively large amounts.

Meristem: A group of undifferentiated cells capable of division for plant growth.

Mesophyte: Plant which is grown in an environment that is neither extremely wet nor extremely dry.

Microclimate: Atmospheric environmental conditions in the immediate vicinity of the plant.

Micronutrient: Chemical element necessary in extremely small amounts for the growth of plants.

Monoecious: Both male and female flowers are produced on the same plant.

Mulch: Materials such as straw, sawdust, leaves, plastic film, or loose soil that is spread on the surface of the soil to protect the soil and plant roots from the effects of rain, soil crusting, freezing, or evaporation and to improve growth.

Mycoplasm: Viruslike infectious agent (now known as phytoplasm in plants).

Necrosis: Plant tissue turning dark due to disintegration or death of cells, usually caused by disease.

Node: Enlarged region of a stem that is generally solid where a leaf is attached and buds are located.

Olericulture: Science and culture of vegetables.

Ovary: Enlarged basal portion of pistil containing egg cell(s).

Ovules: A rudimentary seed containing unfertilized seed gametophyte(s).

Parthenocarpy: Development of fruit without fertilization, therefore seedless.

Pedicel: Stalk of an individual flower.

Peduncle: A stalk or stem of a flower borne singularly or on a main stem of an inflorescence.

Peltate: Shaped like a shield.

Pepo: Many-seeded berry with a hard rind.

Pericarp: Fruit wall, often with three distinct layers; endocarp, mesocarp, and outer exocarp.

Periderm: Outer corky tissues (skin).

Permanent wilting point (PWP): The soil moisture content at which a plant wilts and cannot recover when placed in an environment of 100% RH. This is usually about 15 atmosphere diffusion pressure deficit (DPD).

Petiole: Stalk or stemlike structure of leaf.

pH: Negative log of hydrogen ion concentration. pH 7.0 denotes neutrality, < 7.0 denotes acidity, > 7.0 denotes alkalinity.

Phenotype: Physical or external appearance of an organism.

Phloem: Conductive tissue that transports synthesized substances to other plant parts.

Photoperiod: Relative length of period of light and darkness.

Photoperiodism: Physiological response of an organism to periodic intervals of light or darkness, e.g., flowering.

Physiological disorder: Disorder not caused by pathogens, but rather due to physiological dysfunction of the organism.

Phytoplasm: Viruslike infectious agent affecting plants.

Photosynthesis: Process in which CO_2 and H_2O with light are combined in chlorophyllous tissues to form carbohydrates and O_2.

Phototropism: Response of plants to light stimulus.

Pith: Region in center of stems consisting of loosely packed, thin-walled parenchyma cells.

Pistillate: Female flower.

Protandrous: Maturation of male floral organs before female floral organs.

Protogynous: Maturation of female floral organs before male floral organs.

Pseudostem: Stemlike structure, usually composed of tightly enfolding leaf sheaths; not a true stem, e.g., plantain, some alliums.

Pulse: Dry, edible seed of pod-bearing plants.

Receptacle: Enlarged end of pedicle or peduncle to which flower parts are attached.

Reconditioning: Treatment given to change the physiological or chemical condition of a plant organ.

Relative humidity: Ratio of actual amount of water in air to the maximum amount (saturation) that air can hold at the same temperature expressed in percent.

Respiration: Biological oxidation of organic matter by enzymes to obtain energy.

Rest: A physiological condition of some plant parts (seed or bud) in which sprouting and/or rooting is inhibited even when the environment is conducive to growth.

Rhizome: Usually elongated, underground stem capable of producing new roots and shoots at the nodes.

Ripening: Chemical and physical changes in a fruit that follow maturation.

Rogue: Removal of inferior, diseased, or off-type plants.

Saggittate: Shape like a arrowhead.

Saponin: Bitter-tasting glycoside producing foam in water; causes lysing of red blood cells.

Scape: Flower stalk.

Schizocarp: Dry dehiscent fruit derived from two or more carpels, each with a single seed.

Seed: The mature ovule of a flowering plant containing an embryo, a food supply, and a seed coat.

Seedbed: Soil that has been prepared for planting seed or transplants.

Seed priming: A specialized treatment to improve seed germination and performance under adverse conditions.

Sepal: Part of a calyx.

Sets: Small bulbs used for propagation of onions, and some other alliums. Also other vegetative portions for propagation.

Setts: Plantlets used in vegetative propagation, e.g., aroids.

Senescence: A physiological aging process in which tissues in an organism deteriorate and finally die.

Sessile: Refers to flowers, florets, leaves, leaflets, or fruits that are attached directly to a shoot and not borne on any type of a stalk.

Short-day plants: Plants that flower when the dark period exceeds some critical length.

Silique: Podlike fruit, e.g., Brassicaceae.

Spadix: A spike of flowers on a swollen axis, e.g., aroid inflorescence.

Species: A group of similar organisms capable of interbreeding and more or less distinctly different in morphological characteristics from other species in the same genus.

Spine (spicule):   a hard, sharp, pointed outgrowth of an organ such as a leaf, stem, or fruit.

Stamen: A structure of the flower consisting of anther and filament (stalk).

Staminate: Male flower.

Starch: A complex polysaccharide of glucose; the form of food commonly stored by plants.

Stigma: Part of female reproductive organ on which pollen grains germinate.

Strain: Plants of a cultivar possessing similar characteristics but differing in some minor features or qualities. Also, may form storage organs, eg. tuber.

Stolon: A stem growing horizontally along or below the surface capable of producing a shoot or root at a node.

Style: Extended portion of carpel supporting the stigma.

Sucker: Adventitious shoot from lower part of the plant.

Symbiosis: Association of lving organisms involving benefit to both.

Tannin: Chemical substances produced by some plants that have an astringent and bitter taste.

Tendril: A slender coiling modified leaf or stem arising from stems, often aiding in their support.

Testa: Seed coat.

Thermoquiescence: No activity (germination) due to high temperature.

Thinning: Removal of young plants to provide remaining plants more space to develop.

Tiller: A shoot from the axis of lower leaves in monocots.

Tissue: A group of cells of similar structure that perform a specific function.

Transpiration: Water loss from plant tissues, usually leaves.

Trichome: Hair or bristle from the epidermis.

Tuber: Enlarged fleshy portion of stem or rhizome usually having buds (eyes).

Umbel: Inflorescence with many individual pedicels arising from the apex of the peduncle, umbrellalike.

Utricle: A thin, baglike pericarp which usually contains one seed.

Variety (botanical):   A subdivision of a species with distinct morphological characters and give an Latin binomial name according to rules of the International Code of Botanical Nomenclature.

Vegetable: Food plant, most often herbaceous annuals, cultivated or gathered. Edible portions can be roots, stems, leaves, floral parts, fruits, and seed, usually high in water content, eaten raw, or cooked.

Vegetative propagation: Asexual reproduction by use of plant parts other than seed.

Vernalization: Low-temperature induction of floral initiation.

Vivpary: Seed germination while in an intact fruit.

Windrow: To set aside in rows, usually to dry or cure and to facilitate harvest.

Xerophyte: Plants resistant to arid conditions, grown in soils with low available moisture.

Xylem: Plant conductive tissue that transports water and absorbed nutrients from roots to other tissues.

Zygote: Fertilized cell resulting from the fusion of gametes in sexual reproduction.

Index